REASONING

REASONING

THE NEUROSCIENCE OF HOW WE THINK

DANIEL C. KRAWCZYK

The University of Texas at Dallas, Richardson, TX, United States; University of Texas Southwestern Medical Center, Dallas, TX, United States

Academic Press is an imprint of Elsevier
125 London Wall, London EC2Y 5AS, United Kingdom
525 B Street, Suite 1800, San Diego, CA 92101-4495, United States
50 Hampshire Street, 5th Floor, Cambridge, MA 02139, United States
The Boulevard, Langford Lane, Kidlington, Oxford OX5 1GB, United Kingdom

Library of Congress Cataloging-in-Publication Data
A catalog record for this book is available from the Library of Congress

British Library Cataloguing-in-Publication Data
A catalogue record for this book is available from the British Library

ISBN: 978-0-12-809285-9

Printed in the United States of America
Last digit is the print number: 10 9 8 7 6 5 4 3 2

For information on all Academic Press publications visit our website at
https://www.elsevier.com/books-and-journals

Publisher: Nikki Levy
Acquisition Editor: Nikki Levy
Editorial Project Manager: Timothy Bennett
Production Project Manager: Stalin Viswanathan
Cover Designer: Miles Hitchen

Typeset by TNQ Books and Journals

Cover image courtesy of Dominic Krawczyk

Contents

Foreword

Across the millennia, many great thinkers have addressed the topic of reasoning and its importance to the human existence (see Chapter 2). But human reasoning likely began long before any written record. For example, early evidence of reasoning at work is found in the toolmaking of *pre-Homo sapiens* transitional humans *Lomekwi 3 hominins'* knapping dated to 3.3 Ma in West Turkana, Kenya (Lewis & Harmand, 2016). And if these prehuman beings possessed reasoning capacities, are there other species with similar abilities (see Chapter 4)? Understanding the origins of reasoning tells us much about the depth and breadth of reasoning.

Elucidating the likely constituent necessary and sufficient cognitive factors that contribute to reasoning (see Chapters 1, 3, and 5) such as memory (especially working and long-term memory), attention, sensory and speeded processing, executive functions, symbolic processing, or fluid intelligence, has taught us that reasoning abilities are not entirely a fixed trait (Plomin & Spinath, 2002), but rather a cognitive state that enables decision making (see Chapters 6 and 11) and successfully and flexibly adapts to a complex and changing environment (Bolger, Mackey, Wang, & Grigorenko, 2014; Mackey, Hill, Stone, & Bunge, 2011). It is fair to state that there is hardly an area of the brain that doesn't contribute to some type/form of reasoning, further reinforcing the notion that reasoning is a convergence of genetic factors melded to learning and experience to form as unique a brain as the individual reasoning with it.

By studying the life-course development of reasoning (Chapters 5 and 6), as well as the biological bases of reasoning (Chapters 3 and 7), we begin to grasp opportunities to improve reasoning skills (Bergman Nutley et al., 2011; Mackey et al., 2011). Understanding the behavioral characteristics and cognitive foundations of the various types of reasoning (Chapters 8–12) provides new insight into the potential shared and unique inflection points for the engagement of scaffolding, intervention, or educational strategies (Jaeggi, Buschkuehl, Jonides, & Shah, 2011; Mackey et al., 2011).

The advancement of new technologies, especially in computational domains such as widely available robust computing and information processing, mobile computing and social media, context-aware systems, artificial intelligence, computational modeling, and virtual and augmented reality formats, have changed both the nature and the quantity of information reasoned about while also often taxing human reasoning with the quantity and complexity of these data (see Chapter 13). At the same time, computational technologies increasingly assist human reasoning in everyday activities such as the fastest path to drive to work, which store offers the best bargain on the product that most suits your needs and preferences, or using crowd-sourced reviews to inform which is the best Thai restaurant in the area.

Although decades of research have produced great strides in the understanding of the cognitive and neurobiological bases of reasoning in healthy persons, many gaps remain. Far less is known about the typical and atypical human development of this ability and the biological (e.g., genetic and neurobiological) underpinnings of this developmental process. With the emergence of an "information society," strong STEM (science/technology/engineering/math) reasoning abilities have become essential for a healthy population and the economic success of the society. Understanding how these reasoning skills develop, identifying developmental challenges and sensitive periods, recognizing risk factors to normal and optimal development, and identifying key prevention and remedial interventions have emerged as critical priorities in the study of developmental cognition and learning.

To move this goal forward, a multidisciplinary working group of leading scientists were brought together as the NICHD Reasoning Work Group during the Winter 2014–2015 to identify current knowledge and advancement gaps. Speakers focused on issues that could be addressed in both the short term, as well as those requiring significantly more time and/or preliminary data before the field could fully address the issues. The overall goal was to move the field toward a better understanding of the development of reasoning ability with the hope to improve the health and health decision making, the welfare, and the STEM success of Americans, when possible, by improving teaching/learning of domains, such as science and mathematics, that are highly dependent on reasoning abilities, and provide effective remediation for individuals with suboptimal reasoning abilities.

Following these sessions, I had the great pleasure to organize the 2015 Cognitive Neuroscience Annual

Conference Mini-Symposium "Reasoning: Origins and Development" (March 29, 2015), where selected members of the NICHD Reasoning Work Group were invited to present their work and disseminate the Reasoning Work Group's conclusions, as well as to stimulate research interests in identified gap areas.

The following book is the culmination of that groundwork laid by the NICHD Reasoning Work Group and the extraordinary efforts of one of its members, Dr. Daniel Krawczyk, to reach far beyond the Reasoning Work Group's discussions and provide a vehicle to more broadly disseminate our current understanding of reasoning in its many instantiations and to provide the next generation of reasoning researchers a firm base and rich understanding of their science, it's impacts, and gap opportunities. This work is the first of its kind to present an extremely broad swath of the science of reasoning in a comprehensive and easily comprehensible fashion that peaks the interests of the reader with every turn of a page. Whether student or scholar, this volume is certain to become an indispensable tome. It is of course also among the few extant tools to engage junior scientists and open before them an enticing field ready and awaiting their exploration.

It is an honor to be asked to write this Foreword, to introduce you, the reader, to the reasoning puzzles I have come to love and explore, and to welcome all readers to the science than enables our reasoning to solve these puzzles. I hope you will join me in this increasingly critical research.

This is also my opportunity to thank Daniel for taking up the challenge I posed to the NICHD Reasoning Work Group back in 2014 to write a significant volume to cohere what is currently known about reasoning and where are the gaps, and to use a writing style appreciated by the student, the investigator, as well as the interested casual learner. The writing of this book has been a monumental effort; I will never be able to thank him enough. I believe he has succeeded splendidly.

Kathleen Mann Koepke, PhD
Director, Math & Science Cognition, Reasoning, & Learning - Development & Disorders Program
Eunice Kennedy Shriver National Institute of Child Health & Human Development (NICHD)
National Institutes of Health (NIH)
Bethesda, Maryland USA
June 2, 2017

References

Bergman Nutley, S., Söderqvist, S., Bryde, S., Thorell, L. B., Humphreys, K., & Klingberg, T. (2011). Gains in fluid intelligence after training non-verbal reasoning in 4-year-old children: A controlled, randomized study. *Developmental Science, 14*(3), 591–601. http://dx.doi.org/10.1111/j.1467-7687.2010.01022.x.

Bolger, D. J., Mackey, A. P., Wang, M., & Grigorenko, E. L. (2014). The role and sources of individual differences in critical-analytic thinking: A capsule overview. *Educational Psychology Review, 26*(4), 495–518. http://dx.doi.org/10.1007/s10648-014-9279-x.

Jaeggi, S. M., Buschkuehl, M., Jonides, J., & Shah, P. (2011). Short- and long-term benefits of cognitive training. *Proceedings of the National Academy of Sciences of the United States of America, 108*(25), 10081–10086. http://dx.doi.org/10.1073/pnas.1103228108.

Lewis, J. E., & Harmand, S. (2016). An earlier origin for stone tool making: Implications for cognitive evolution and the transition to homo. *Philosophical Transactions of the Royal Society B: Biological Sciences, 371*(1698). http://dx.doi.org/10.1098/rstb.2015.0233.

Mackey, A. P., Hill, S. S., Stone, S. I., & Bunge, S. A. (2011). Differential effects of reasoning and speed training in children. *Developmental Science, 14*(3), 582–590. http://dx.doi.org/10.1111/j.1467-7687.2010.01005.x.

Plomin, R., & Spinath, F. M. (2002). Genetics and general cognitive ability (g). *Trends in Cognitive Sciences, 6*(4), 169–176. http://dx.doi.org/10.1016/S1364-6613(00)01853-2.

Acknowledgments

This project was initiated in part by my participation in a Eunice Kennedy Shriver National Institute of Child Health & Human Development Reasoning Work Group that was convened by Kathy Mann Koepke in 2014. The group consisted of scholars representing a diversity of perspectives on reasoning (Renee Baillargeon, Aaron Blaisdell, Silvia Bunge, Kevin Dunbar, Lisa Freund, Susan Jaeggi, Frank Keil, Ben Rottman, and Fei Xu). Their presentations were most enlightening and helped to set the stage for this book. Special thanks go to Kathy Mann Koepke who was most helpful and supportive throughout the writing process and to Aaron Blaisdell for his contributions to Chapter 4 on reasoning in other species. I thank Jonathan Fugelsang, Michael Vendetti, and two anonymous reviewers for their insights on the organization of the book. My colleagues John Hart, Francesca Filbey, Sandi Chapman, and Bart Rypma helped to motivate this project and inspire the writing.

Many of my colleagues with whom I have studied reasoning and discussed the topic have helped to shape the book content. For this I thank Keith Holyoak, Mark D'Esposito, Barbara Knowlton, John Hummel, Dan Simon, Nancy Gee, Bob Morrison, Lindsey Richland, Dan Levine, Michelle McClelland, Gloria Yang, Ehsan Shokri-Kojori, Kevin Murch, Don Kretz, Amy Boggan, Leanne Young, Adam Teed, Tiffani Jantz Fox, Barry Rodgers, Jameson Miller, Jelena Rakic, Guido Schauer, Kihwan Han, Zhengsi Chang, Mandy Maguire, and Jim Stallings.

Teaching a course on reasoning has afforded me excellent feedback on the topics covered in this book. I especially thank Daniel Mark, David Martinez, Linda Nguyen, Matt Kmiecik, and Michael Lundie for their valuable suggestions. I thank Jim Bartlett and Bob Stillman for being supportive of the project. I also thank Ed Krawczyk, Liz Krawczyk, Lee Drew, George Baxter, Galen Westmoore, John Sterling, Craig Caravaglio, Adam Green, and Bret Grasse for the encouragement, support, and the interesting discussions we have had about the many facets of reasoning.

I thank the staff at Elsevier who have helped to move this project forward and see it to completion. They include April Farr, Joslyn Paguio, Timothy Bennett, and Stalin Viswanathan. I thank Kristine Miranda at the Center for BrainHealth for her critical contributions to the figures, captions, and editing, along with her organization skills and persistence. I thank Dominic Krawczyk for his inspiring cover artwork.

This book is dedicated to my wife Linda Drew who made exceptional contributions to this project from start to finish and to Joshua and Dominic who frequently inspire me to think about a wide range of topics in human reasoning.

CHAPTER

1

Introduction to Reasoning

KEY THEMES

- The study of reasoning is a multidisciplinary area involving researchers in neuroscience, psychology, economics, computer science, philosophy, and business.
- Reasoning involves moving from multiple inputs to a single output, which can be a conclusion or an action.
- Reasoning involves multiple steps through a state space to achieve a final outcome. There are numerous ways through the process to achieve different conclusions.
- Reasoning involves a mixture of previous knowledge and novel information.
- Reasoning capability is determined by our cognitive operations. It depends upon attention, working memory, and long-term memory.
- Reasoning is determined by the capabilities of an organism's nervous system. Species with complex brains tend to show greater levels of complexity in reasoning and behavior.
- There are several different categories of reasoning, and these are defined by the complexity of the inputs, the particular operations that occur, and the types of conclusions that can be reached.
- Technological advances have opened up a variety of exciting new avenues in reasoning research including augmenting human reasoning.

Reasoning
http://dx.doi.org/10.1016/B978-0-12-809285-9.00001-6

OUTLINE

INTRODUCTION TO REASONING

About This Book

The field of reasoning has a long and diverse history. There was a time when scholars believed that the mind and body were largely separate and that the functions of the mind were somehow unique, possibly defying the laws of nature. This led thinkers to use their own imaginations in order to create theories about how we reason. They did this using those powers of reasoning to construct theories and myths about the operations of the world around them. Over time people began to notice links between the natural world and human behavior. In our not so distant past, a wealth of new measures emerged and it became clear that the body and mind were intimately linked. Brain research flourished, and people began to construct biological theories about the mind and test them using scientific methods. At this time the scholars applied a new set of interesting tools to help try to make sense of our remarkable brains. Today we have a wide array of incredible tools with which to investigate our thinking and reasoning processes. We do so in the context of both biology and psychology. This book describes many of the breakthroughs that have led to our current thinking on the topic of reasoning in relation to the mind and brain.

Writing a book is a vast undertaking, and this may be particularly true about a book on reasoning. There are still many unknowns in this field. We do not fully understand how different cognitive processes are defined at the level of the brain. We do not yet have methods that allow us to record from large enough populations of neurons across the brain. We cannot yet read neural codes to understand what areas of the brain are actually signaling to one another when information is relayed across the cortex. We do not yet have a strong enough grasp on which cognitive processes are involved in which types of reasoning. Added to these challenges is the fact that reasoning occurs in a dynamic way. Thoughts occur to us seemingly out of the blue. We connect ideas together because we notice events occurring in a particular order. We make errors in our inference and quickly correct them. When we gain expertise in an area, we have a difficult time describing to a novice how we are accomplishing our plans in what appears to be a remarkably fast and efficient way to the beginner. Due to these challenges and limitations the field remains quite diverse. There is always more research to be done.

The challenges in studying reasoning may stem from the fact that we are ultimately limited by our own reasoning processes in how we study the topic. Reasoning emerges out of a remarkably complex series of events that we can only glimpse at with the current technology.

The brain is a classic example of a complex system, and reasoning must be examined from a certain point of view or level of analysis. We cannot hope to understand it from a purely top-down perspective, using our own introspection to look inward upon our thought processes. Likewise, we cannot view it from a purely bottom-up perspective by examining the biochemistry of the brain and attempting to grasp our thinking at the level of neurochemical activity. Understanding reasoning requires a balanced viewpoint. We must take a top-down approach to define a limited aspect of cognitive processing constrained to certain circumstances and combine this with a bottom-up view that is informed by our growing knowledge of neuroscience. Because our top-down definitions can be imprecise and there remain many gaps in our understanding of the brain, there are no easy answers in this field.

This book will walk you through some of the major steps on the journey toward understanding thinking, reasoning, and decision making. We will examine these topics from a series of different perspectives. Together, the chapters add up to inform the reader about the current state of the field, how we got to where we are now, and possibly where we are going as new methods develop and technological capabilities begin to approach human levels of capacity on certain tasks.

Features of the Book

Each chapter features research that is organized around a core topic. The topic serves as an anchor point or general domain, but remember that this is all about thinking and reasoning. Due to the sheer diversity of research that makes up this fascinating field, there is a rather wide distribution of research topics and themes in each chapter. Rather than focus on every topic and provide a very detailed and complete description of that specific area, the chapters emphasize providing an overview, or survey, of several aspects of each topic. I will freely admit that no book on this topic can hope to offer a complete guide to reasoning. There is simply too much research in too many disciplines in this field to achieve such a goal. Therefore, I offer you, the reader, a flavor of some of the important research in each area and invite you to dig deeper and search the relevant literature in the areas that you find captivating.

The book is organized in a bottom-up manner. We will begin with basic definitions of reasoning presented in this chapter. From there we will proceed to examine reasoning from a historical perspective in Chapter 2. I felt this was important to set the stage and offer context to the topics that come later. To appreciate the research in this field, one has to have a sense of its context, what was going on during different time periods, and why certain conclusions were drawn. We then move to an introduction to the study of reasoning from the perspective of neuroscience. This chapter offers the reader a sense of the growing body of work connecting the mind to the brain. This chapter may be easier to follow if you have had a course in an area of neuroscience, but I have not assumed an extensive neuroscience background and have constructed it in a way that should make sense to a diverse audience. Our reasoning and thought processes are shaped by our nervous system and the experiences we have had in life. Having a grasp of the types of neuroscience research currently available will help the reader to follow much of the content in the later chapters, all of which include some elements of neuroscience. Chapter 4 examines the fascinating field of animal reasoning research. From there, we move through the lifespan in Chapters 5 and 6 beginning with reasoning in young children, moving through adolescence, and lastly examining reasoning in older adults. Chapter 7 complements the previous three chapters by discussing the impact of neurological and psychiatric disorders on reasoning. Chapter 7 includes extensive information about the brain and many disorders that impact people throughout the lifespan. Chapters 8–11 discuss different categories of reasoning, such as deduction, induction, reasoning about contingencies and causes, analogical reasoning, and how we make decisions. The final chapters take a broader view examining how we reason within the context of society and the impact of technology on reasoning.

Throughout the book I invite the reader to identify consistencies and areas of overlap among the topics presented. The discipline of reasoning has blurry edges. Research does not always fit neatly under one particular theme or area. There are times where the same cognitive processes that constrain reasoning in young children also impact older adults. There are situations in which historical concepts from ancient philosophy are studied using modern brain imaging techniques. There are also domains of research that fall well outside of psychology, neuroscience, or biology that strongly impact the topic of reasoning. The development of computers is one such area. Because of these overlapping conceptual frameworks, I encourage the reader to take an active role in noticing the connections that occur in the research from chapter to chapter. I have tried whenever possible to refer the reader to other chapters in the book that have strong conceptual overlap with a given topic. This overlap at times makes content within multiple chapters merge, but I suppose that is just a part of making sense of a diverse and evolving field of study.

The chapters all contain a mix of research from different disciplines and historical periods. This differentiates the text from other neuroscience books on offer. Some chapters focus heavily on the constraints on reasoning imposed by the brain. Other chapters focus on research at the level of conceptual thought, memory, and human behavior. This diversity reflects a core feature of reasoning research. Few other disciplines within research on the mind and brain are so defined by the context and a confluence of multiple cognitive operations. Throughout the reference sections of each chapter you will notice research spanning a wide period of time. This reflects the evolution of our field. For example some of the core research on deductive reasoning took place over 50 years ago. There have been improvements in our understanding and expanded methodology, but one has to view the research with an understanding of what had been accomplished during the peak periods of discovery. There are times when the chapters will introduce concepts from diverse disciplines. For instance we benefit greatly from the perspective of evolutionary biology when we consider some aspects of complex behavior in animals. We need to understand that perspective to decide if those behaviors should be defined as reasoning. We also rely upon work from behavioral economics in order to understand social aspects of decision making. Throughout the book I have attempted to offer the reader some of these diverse perspectives that connect to further inform our understanding of reasoning and decision making.

Each chapter begins with a section describing *key themes*. These are helpful hints previewing some of the major areas associated with each chapter. Readers may find it helpful to read the key themes over before reading the chapter. Likewise, it may be helpful to reread these in association with the summaries at the end of each chapter, as they will recap major ideas that were covered.

All of the chapters contain *boxes*, which are independent sections that present a different perspective on a topic relevant to reasoning. Each box is intended to complement the material covered in the sections surrounding it. These can be read in sequence with the text, or may be better read after completing a surrounding section of the chapter. Reasoning is very much a daily life activity that is most interesting and engaging when placed into a specific context. The boxes are all self-contained and many of them present unique, engaging, or fun examples of the lessons of reasoning as they are carried out in our daily lives, or at some interesting point in the history of the field.

Each chapter concludes with a series of *end-of-chapter thought questions*. These questions are intended to stimulate further thought in the reader. There may be clear answers to some of these questions after reading the chapter, while others may force you to ponder over them for some time. For some thought questions, we may not yet have answers, but further research into the area may prove fruitful. All references are provided at the end of each chapter, and I encourage the reader to look these papers up and read them if you feel inspired. There is no substitute for reading the original sources as they appeared in the literature and framed in the context in which the authors originally presented their work.

DEFINING REASONING

What Counts as "Reasoning"?

Everyone has some general sense of what reasoning entails. We have all heard this term in a wide variety of contexts across various aspects of our lives. Other terms such as "thinking," "pondering," "problem-solving," and "decision making" are frequently interchanged for "reasoning." When I tell people that I study reasoning, I often feel that is just not descriptive enough as a stand-alone statement. I usually feel I have to follow-up immediately by stating what that really means in the constrained and limited context of the research lab. The topic of reasoning is taught in schools and at the college level. Again, defining what we mean by "reasoning" appears warranted in these contexts as well. A psychology of reasoning class is likely to be very different than a philosophy of logic class or a scientific reasoning class taught through a chemistry or biology department. Lastly, a literature search for reasoning articles may yield artificial intelligence papers, engineering papers, developmental psychology papers, and clinical assessment articles. Again, a clear definition is called for.

For the purposes of this book and more broadly for the actual study of reasoning in human participants within a laboratory environment, I will define reasoning according to three characteristics. There are certainly other definitions possible, but I believe these three features capture the essence of what most of us mean when we use the term "reasoning." You can think of these as the "A, B, and C" of reasoning:

A. *Reasoning uses multiple inputs to produce one output.* That output can emerge in the form of a physical action or simply a new thought that emerges from mental processing.
B. *Reasoning involves multiple steps.* It is helpful to think of these steps as occurring within a space. There may be multiple routes through the space toward an output. Reasoning has individual elements that often form sequences.
C. *Reasoning is a hybrid.* It relies upon a combination of prior knowledge and new information. Some outputs follow from the novel combination of multiple elements of prior knowledge. Some outputs follow mostly from assembling new elements of information.

Let's consider each of these characteristics in a bit more detail beginning with A, *reasoning uses multiple inputs to produce one output*. It is critical that there be multiple inputs into a reasoning process. Reasoning takes in a wider set of premises, conditions, or possibilities and transforms or distills these into one output. There are many behaviors that are evoked by a single input, but these do not qualify as reasoning. The sensation of pain makes us withdraw our hand. The visual input of an oncoming snowball makes us duck. The sight of a green traffic signal leads us to drive forward. These are somewhat complex behaviors, especially when one considers how they are carried out at the level of the nervous system. Most of us would call withdrawing our hand due to pain, or ducking from an oncoming projectile, to be reflexes. These behaviors do not involve conscious deliberation. We simply act before we have taken the time to assess the situation. Deciding to drive forward in response to a green light is probably a better fit to the category of learned behavior. We rarely find ourselves pondering whether to go or remain stationary in response to the traffic signal. It is not a good example of a reasoning task (Fig. 1.1).

Meanwhile, imagine that you are a new driver and are not familiar with the meaning of the lights. Many of us have had some situation like this when visiting a foreign country and looking at a road sign. In this situation, we take in several inputs (the visual elements of the sign, the placement of the sign, the physical surroundings, etc.), and we attempt to find a single output to the best of our abilities (stop, go, pause, etc.). The output may be spot on. We drive our rental car forward after a short pause and receive positive feedback in the form of not causing an accident or making anyone angry. At this point we have learned something new and with enough repetition of encountering this novel sign and repeating the successful "drive forward after pausing" behavior, we can move this scenario on into the "learned behaviors" section of our mental lives. Imagine instead that you drive forward in response to this sign and are met with other oncoming vehicles blocking your path and pedestrians looking at you with irritable expressions. This feedback suggests that your output was incorrect. You'll have to try reasoning again, this time integrating the newly acquired negative feedback into your approach and narrowing down your search for a solution by excluding the faulty "drive forward" possibility.

FIGURE 1.1 Deciding how to respond to a stimulus in the environment depends on our familiarity with it. When we are highly familiar with a traffic signal, then our response is a learned behavior. If we are unfamiliar, such as when visiting a foreign country, we may have to reason in order to discover the appropriate action that is signaled. *From Wikimedia Commons.*

You could take issue with the idea that reasoning involves only one output. Why not several outputs? This becomes somewhat tricky. There can be situations in complex reasoning, in which the output may be to carry out a particular sequence of multiple steps and in some specified order. I would argue that this happens in daily life through the process of iterative reasoning in which we have carried out several acts of reasoning and then assembled the outputs of those independent acts in order to accomplish something greater. I maintain that reasoning moves from multiple elements of information toward a single action or conclusion. That action, or conclusion, can be nuanced and have its own conditional qualities, but reasoning rarely involves multiplying information into many conclusions. Such an act of thought would probably fall into the topic of idea generation or brainstorming, which we will cover in greater detail in Chapter 12. The process of idea generation almost always follows with a second act, in which we narrow down the numerous candidate ideas according to a rank order and often end up selecting the best single output or solution.

Let's now further consider B, *reasoning involves multiple steps*. Reasoning is just not simple. Like my examples of reflexes and learned behaviors, when we can reach an intuitive conclusion, or act without deliberating, then we are not reasoning. Reasoning nearly always involves moving from one consideration to the next, to eventually reach an output. This may be due in part to the characteristics of our brains. We can only process a limited set of information at any one time. In complex environments, such as navigating city streets, we are faced with numerous sensory inputs, moving cars, buses, bikes, and people, along with a complex set of interconnecting streets and visual features. When we add our emotions, time pressure, and goal states to this context, we are simply forced to do things in a sequence. We have limited powers of perception, attention, and memory. The mismatch between an environment loaded with inputs and our limited capacity of information processing results in multiple steps that need to take place for reasoning to occur.

If behaviors occur without multiple steps, then they probably would not count as reasoning. This is again debatable and one would have to consider the specifics of a situation, but I maintain that reasoning involves making transformations to the information that we take in. Taking in inputs and producing them verbatim should be categorized as memory or learning. There must be multiple steps and these often occur flexibly. There is not always one solution to a problem. Rather, as the premises or initial state becomes more complex, we can usually find multiple ways forward toward an acceptable end state.

Researchers in several disciplines have considered *state spaces* to be helpful ways to analyze problems and to conceptualize the possibilities for different situations. A state space is a helpful spatial analogy with which to describe our mental processing (Fig. 1.2). State spaces are defined as being the set of all possible configurations that a given problem can achieve. This term is used in computer science to analyze software capabilities and in economics to analyze the possible utility, or value, of different possibilities. The analogy between a search space and problem solving highlights the fact that there are many ways to progress toward a solution. A thinker would like to move in a direction that enables the most optimal solution among the possibilities existing in the state space (Zhang, 1999). This captures an essential feature of reasoning: there is not one way to do it. There may not even be a logically correct answer. Through processing and evaluating numerous inputs, we can transform information using a series of steps that move us toward a successful solution. You can think of many examples in which reasoning fails. We try different approaches, but find that we are not successful. Fortunately, reasoning can be iterative, and we can often try again armed with new knowledge and a narrowed search space (with fewer possible routes achieved through the process of elimination).

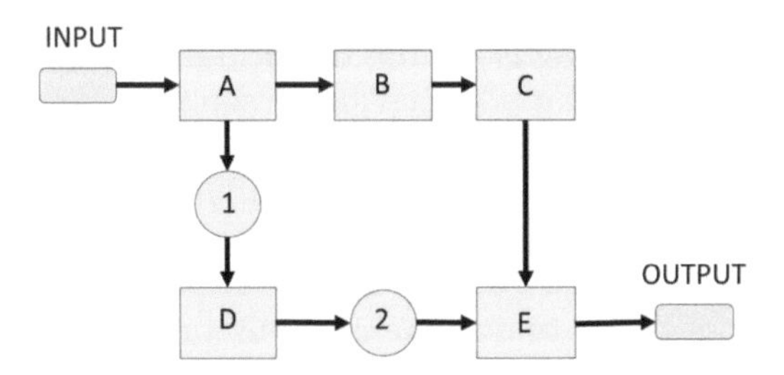

FIGURE 1.2 In engineering, state spaces can be represented by mathematical models of a physical system. These include inputs, output, and paths that can be taken. Reasoning can be thought of in terms of a state space that moves us from inputs to an output down a variety of different paths.

Now let's turn our attention to C, *reasoning is a hybrid*. There are rare instances in which we reason about unfamiliar information to reach a completely novel conclusion. We may call this approach creative. Reasoning can be heavily weighted toward taking a very new approach and "thinking on our feet" and "making it up as we go along." There are also times when we reason using well-known facts or prior knowledge. Either situation can be defined as reasoning. I would argue that most of the time we reason by combining some degree of previous knowledge with other elements of new information. At minimum, we must assemble previous knowledge elements in a novel way for a mental process to qualify as reasoning. Almost any time we are operating with new information, we will consider that to be reasoning as well. Complexity is inherent in feature C of the definition. Reasoning cannot involve the simple output of previous knowledge, or it reduces to a learned behavior, or possibly even a reflexive behavior. Implicit memories (Knowlton, Mangels, & Squire, 1996) would not count as reasoning for most of us, as they evoke a behavior without deliberation or the necessary complexity. Rather, they are classic examples of learning.

Factors That Influence Reasoning

We can also consider some other important mediating factors when we define a behavior as being an act of reasoning. One of these factors is the biology of an organism. We are not purely agnostic information processors. While we can attempt to work toward an optimal solution, we have drive states that govern our thinking. Simple states such as pain, satiety, hunger, and thirst can dominate our thinking and dictate our next behavior. The need to interact with others influences us. We can gain by making other people happy. Seeking social rewards such as recognition, approval, and the acceptance of others limits people from acting in a purely selfish manner (Fig. 1.3). We will discuss social reasoning situations in detail in Chapter 12. Furthermore, our needs can evolve as we age and develop our knowledge through experience. Hormone levels are another biological factor that can influence reasoning and behavior. We may reason purely based on information, but many times our desired outputs or solutions are intertwined with our biological needs and drive states.

Our cognitive capacities also strongly influence our reasoning abilities. Our limited mental resources prevent us from being purely rational thinkers. Consider attention as an example. Classic research demonstrating the concept of change blindness indicates that people have a very difficult time noticing a change in visual features present in two images presented directly one after another (Simons & Levin, 1997). For example, if you look at the images in Fig. 1.4 you may have to alternate back and forth several times before you notice the change. Go ahead and try this short demonstration for yourself. The fact that we cannot quickly isolate the change in visual features indicates that our attention has limitations (part of the letter on the right lower half of the field has changed from blue to gray). This means that we can only take in some of our visual environment at any given time. Similarly, we can only focus on a few memories from our lives in a detailed way before we become overwhelmed. Once we attend to information we can move it into a short-term store of information often referred to as *working memory* (Baddeley, 1986). Our working memory capacity is limited, and we simply cannot add more into our buffers. This limitation can be overcome to some degree by performing an operation known as *chunking*. When we chunk information we can use our long-term memory to group sets of objects, events, or locations into a set, effectively expanding our active working memory capacity. There is no getting around the limits of our cognition. If we cannot fully perceive or attend to all possible information, then our reasoning will suffer as a result.

FIGURE 1.3 We seek social rewards including recognition, approval, and the acceptance of others. These incentives influence our reasoning when other people are involved. *From Shutterstock.com*

The topic of limited mental resources leads us to another relevant capacity for reasoning and that is consciousness. Many people would consider the act of deliberating about a topic to be the essence of reasoning. The extended consciousness of humans enables us to enrich our mental representations and examine their contents. We can consciously imagine future states of the world, reflect on previous situations, cue our own memories of the past, and in doing so discover novel connections relevant to a problem. This is often thought to be a dividing line among different species. Is consciousness necessary for reasoning? I would argue that it is not necessary, although it does increasingly accompany acts of reasoning as it becomes complex and probably occurs to a greater extent in species with large and complex nervous systems. Consciousness may help or inhibit our efforts in reasoning. The fact remains that all species have limitations of consciousness and many aspects of our mental processing occur without our conscious

FIGURE 1.4 Can you spot the change between these two images? Change blindness occurs when we fail to notice a change in a visual feature between two images presented directly one after another.

supervision. Similarly, it is difficult to find a clear role for consciousness in acts of reasoning carried out by computers (although they have a strong helping hand from conscious human programmers!).

The brain may be the final arbiter of reasoning ability. As we will see in the coming chapters, the characteristics of our brains heavily shape our reasoning. This becomes evident when we examine reasoning across the childhood years. The reasoning of very young children lacks the benefit of experience and brain maturation. The very young tend to be self-focused in their thinking. Most situations are novel to them, they are highly emotional, and they are driven by impulses that differ from older children and adults. There is great biological change and brain change over the course of childhood, and we will discuss these aspects of reasoning in Chapter 5. Our reasoning skills can also be directly compromised by disruption to our brains. Strokes and brain injuries to particular locations will reliably impair our power of reason. Some types of brain damage affect our emotional processing and change the drive states and impulsivity levels of a thinker. Reliably different behaviors and outputs occur in such individuals. Reasoning impairments can also follow damage to the white matter of the brain, which enables regions to communicate effectively. This can occur because of aging or injury and may lead to slower processing, which compromises attention and working memory resources and thereby limits the number of inputs that can be perceived and considered in solving a problem.

HOW WE STUDY REASONING

Historical Considerations

One of the first perspectives presented in Chapter 2 is an overview of the history of reasoning research. It is necessary in any field to have a sense of how we got to the current state of knowledge. We can learn much about the evolving definition of reasoning by examining the way people conceptualized this topic in the past. We can also gain a valuable perspective by noticing how the development of new methods in science has changed our thinking. Throughout the history of reasoning the state-of-the-art has depended upon the methods that were available at the time. Early philosophers were equipped only with their conscious awareness of the inputs and outputs of their thinking. They could categorize reasoning based almost entirely upon the subjective qualities and conceptual basis of the tasks. A new set of measures became possible and theories of human reasoning changed dramatically as the field adopted the methods of experimental psychology. Fascinating research comparing humans to other species has occurred for centuries. The work of comparative psychologists and ethologists further changed the way the field thinks about reasoning. Eventually we entered into an age in which neuroscience and information processing dominate our thinking on the topic of reasoning. Technological developments have made it possible to image the brains of individuals as they think and have enabled researchers to develop machines that reason. These machines may not be conscious, but computers can now process information with remarkable speed and accuracy far exceeding that of human processing in many cases. We must remain aware that humans still program machines and must set up the conditions for machine-assisted reasoning. As we will discuss in Chapter 13, machines are evolving to the point at which they can outperform even expert humans at many rule-based reasoning tasks.

A Multidisciplinary Approach

The text is largely written from my own perspective. I am a cognitive neuroscientist. In my undergraduate years I studied psychology and biology. My PhD is from an experimental psychology area and I did postdoctoral training with a cognitive neurologist studying brain imaging methods as they apply to human cognition. My current academic position enables me to regularly interact with a diverse set of scholars including developmental psychologists, social psychologists, behavioral economists, neurologists, psychiatrists, neurobiologists, and rehabilitation specialists. I believe this lack of strong disciplinary boundaries is helpful for the study of reasoning. I have gained from the wealth of different methods and perspectives that I am exposed to, and I believe this shows in the chapters of this book.

It takes a village to study reasoning. You will find that we cover each topic from a variety of perspectives. I find biology to be particularly useful in describing the reasoning behavior of other species. In some of the chapters, economic perspectives are particularly valuable for describing social aspects of reasoning and how we decide under uncertain conditions. Developmental psychology and gerontology offer incredibly rich perspectives on the changes in reasoning observed through the lifespan. Cognitive psychology helps to delineate the important mental operations that underlie reasoning abilities. At the core of all of these perspectives is neuroscience. The brain characteristics of humans and other species are central to all of the reasoning capacities that we observe in the natural world. Even computerized reasoning operations are heavily informed by the information processing approach that is used by the biological neural networks of the brain.

CATEGORIZING REASONING

Defining Diverse Modes of Thinking

Before we proceed on to the content areas of the book, it is worth considering some of the different categories of reasoning. The first several chapters (Chapters 2–7) cover topics including the history of reasoning research, the neuroscience of reasoning, the reasoning of other species, and lifespan approaches to reasoning. Since these chapters focus on specific characteristics of particular groups of people and organisms that reason, it may be helpful to foreground some of the major categories of reasoning that are agreed upon within the field. Many of these categories will be discussed in advance of the later chapters that more exclusively focus on each category of reasoning. These categories are summarized in Box 1.1.

Determining Cause and Effect

Causal reasoning is the process by which we establish cause-and-effect relationships between two or more entities. This can be as simple as inferring that a marble moving toward another marble caused the second marble to move after colliding. Although the causal attribution feels almost perceptual, it is not a trivial inference. Likewise, people draw a causal association between thunder and lightning due to their co-occurrence in time; however many people remain fuzzy on the actual mechanistic relationship between these features of a storm. Causal reasoning enables us to isolate causes and effects in more complicated situations as well. Detectives regularly have to establish the most plausible explanation for who caused a crime to happen. Our entire legal system is based upon causal relationships between actions and outcomes. We will discuss causal reasoning in animals in Chapter 4 and cover the topic in greater breadth and depth in Chapter 8.

Reasoning About Rules

Very early in the history of human thinking philosophers considered the validity or soundness of conclusions. The method of *deduction* allows a thinker to move from a set of inputs called premises toward a specific output or conclusion. The premises set the rules for the deductive process. Deductive reasoning yields a valid conclusion provided that it has been carried out effectively. As we will discuss in Chapter 3, there are

BOX 1.1

CATEGORIES OF REASONING

This list of areas of reasoning will occur throughout the book. Below are some basic definitions you may find useful to refer to.

- *Abductive reasoning*: Abduction refers to a situation in which we do not have all of the possible information available as premises or inputs. Abduction involves reasoning to the best possible output or explanation without having all of the information necessary to yield an objectively correct output.
- *Analogical reasoning*: Analogical reasoning involves using relational information about a known situation and applying that set of information toward a new situation that shares relational similarity. Drawing an inference on the basis of shared relations is a type of inductive inference.
- *Causal reasoning*: The process by which we establish cause-and-effect relationships between two or more entities.
- *Decision making*: An area of study that focuses on how we arrive at a single choice among other options. Decision making and reasoning are highly interrelated.
- *Deductive reasoning*: Reasoning from a set of inputs called premises toward a specific output or conclusion. The premises set the rules for the deductive process. Deductive reasoning yields a valid conclusion provided that it has been carried out effectively.
- *Insight*: Insight describes a feeling of confidence that we have discovered a novel and creative answer to a problem.
- *Inductive reasoning*: Reasoning about a situation in which we draw a general inference about a set of items on the basis of a limited set of premises. Unlike deduction, inductive inferences are not guaranteed to be valid.
- *Relational reasoning*: The term relational refers to the way that multiple items are interrelated. Relational reasoning is concerned with the relational connections among objects, or items.
- *Social reasoning*: This term describes any situation in which we reason about a social situation involving other people. It can include cooperating with others or competing against them in order to generate an output, or solution.

neuroscience studies of deduction that help to inform us about the brain basis for this process. In Chapter 9 we will discuss how people reason deductively and compare this to inductive reasoning.

Inductive Inferences

Much of our reasoning involves using our knowledge about specific examples to draw conclusions about a wider set of items that we see as occupying that same category. This is *inductive reasoning*, a situation in which we draw a general inference about a set of items on the basis of a limited set of premises. Induction is discussed with regard to the brain in Chapter 3 and again in more detail in Chapter 9. *Analogical reasoning* involves using relational information about a known situation and applying that set of information toward a new situation that shares relational similarity. The term relational refers to aspects of multiple items and how they are interrelated. For example, terms such as above, below, causes, kisses, and helps are all relational, as they describe a type of action or connection that occurs between two or more people, objects, or items. Drawing an inference on the basis of shared relations is a type of inductive inference. We discuss *relational reasoning* throughout this book. We emphasize analogical reasoning in Chapter 3 by discussing relevant neuroscience research. We discuss the topic of analogical reasoning again in Chapter 5; in Chapter 7 we discuss disorders of reasoning and in Chapter 10 we discuss in depth this form of reasoning.

Reasoning and Decision Making

Reasoning and decision making are often linked under the broader heading "higher cognition." Decision making also fits parts A, B, and C of the definition for reasoning that I provided in this chapter. Decisions almost always involve multiple inputs in the form of choice options or multiple attributes of the options and result in one output, the decision in favor of one or more options. Decision making is also a multiple step process with a state space that is not predefined. Rather we work toward an optimized decision from the possible choices. Decision making involves some hybrid characteristics combining new information with previous knowledge. Chapters 2, 3, 5–7 all cover aspects of decision making. While Chapter 11 is fully devoted to this topic, we will revisit decision making again in Chapter 12 covering social reasoning.

PUTTING IT ALL TOGETHER

Reasoning in Society

Reasoning can become even more complex and interesting in the context of other people. As we will discuss in Chapter 3, animals can reason in groups both benefiting and competing with one another. People also influence one another's reasoning. In Chapter 12 we will discuss *social reasoning*, which pertains to the way that people reason with one another, sometimes deciding to compete and other times cooperating. We will also focus on *group reasoning* in this chapter. Sometimes we benefit from a group effort, while other times we create conflicts when we think in groups.

What Does the Future Hold?

We will conclude the book with a discussion of the role of technology in reasoning. A particularly innovative and exciting part of the field is focused on using new methods in computing to augment, or expand, human reasoning capabilities. In the past several decades we have witnessed unparalleled advances in computer processing power and software sophistication. Now computers programmed to reason against human experts can dominate even the most astute humans in some of the most complex games of advanced reasoning. Interestingly, the act of building intelligent machines continues to tell us a great deal about the nature of reasoning and even about how the brain may operate as well. Chapter 13 covers new directions in technologically enhanced reasoning.

SUMMARY

Reasoning is a complex and multidisciplinary area of study. This book presents a series of overviews of an array of topics in the field. The book proceeds in a bottom-up manner by first introducing research on the brain and reasoning and working up toward social aspects of reasoning and the impacts of technology. Along the way we will cover many categories of reasoning and consider the implications of the nervous system, development, aging, and brain injury on reasoning.

Reasoning involves some core features. Three important aspects of reasoning were discussed in this chapter. First, reasoning involves moving from multiple inputs to a single output, which can be a conclusion or an action. Second, reasoning involves multiple steps through a state space to achieve a final outcome. There are numerous ways through the process to achieve different outputs. Third, reasoning involves a mixture of previous knowledge combined with novel information. The precise combination of these types of information will vary based on the type of problem.

Reasoning is a fascinating topic of study. The research community must take a broad view and incorporate many research methods to understand the process. It is also important that researchers propose clear operational definitions of the reasoning type that they are planning to study. The future of reasoning research looks bright with exciting new methods advances occurring regularly, along with innovative developments in technology.

END-OF-CHAPTER THOUGHT QUESTIONS

1. The study of reasoning is a multidisciplinary area. Which academic disciplines do you most associate with the study of reasoning? Are there cases in which two or more disciplines essentially study the same topic?
2. Reasoning involves moving from multiple inputs to a single output, which can be a conclusion or an action. Think of how this applies to a simple reasoning task and a complex one. Does this aspect of the definition apply to both?
3. Reasoning involves multiple steps through a state space to achieve a final outcome. Does the state space idea work better for simpler acts of reasoning, more complex ones, or both?
4. Reasoning involves a mixture of previous knowledge combined with novel information. At what point does reasoning reduce to being learned behavior?
5. Reasoning capability is determined by our cognitive operations. It depends upon attention, working memory, and long-term memory. Which of these appears to be the most important?
6. Species with complex brains tend to show greater levels of complexity in reasoning and behavior. Which other species engage in behavior that fits the definition of reasoning?
7. Technological advances are beginning to change human reasoning capabilities. Think of some examples of this in your daily life.

References

Baddeley, A. (1986). *Oxford psychology series, no. 11. Working memory.*

Knowlton, B. J., Mangels, J. A., & Squire, L. R. (1996). A neostriatal habit learning system in humans. *Science, 273*, 1399.

Simons, D. J., & Levin, D. T. (1997). Change blindness. *Trends in Cognitive Sciences, 1*(7), 261–267. http://dx.doi.org/10.1016/s1364-6613(97)01080-2.

Zhang, W. (1999). *State-space search: Algorithms, complexity, extensions, and applications.* Springer. ISBN: 978-0-387-98832-0.

CHAPTER

2

The History of Reasoning Research

KEY THEMES

- The study of reasoning made remarkable progress in the 20th century.
- This field moved from being primarily a philosophical area of inquiry to a psychological one.
- The methods of experimental psychology enabled researchers to examine the role of context in reasoning accuracy and speed. Additionally, these methods enabled researchers to find consistencies across people forming principles of reasoning.
- Early studies of reasoning were carried out in Europe by the Gestalt school of psychology. These researchers focused on the role of insight in problem solving.
- In the first half of the 20th century behaviorism dominated psychological inquiry in the United States. With a primary focus on observable states of learning and conditioning, behaviorism limited the progress on understanding reasoning.
- Early studies of deductive reasoning were carried out in the early 20th century. These studies clarified the importance of familiarity with materials on the process people used for deduction.
- Modeling cognitive phenomena requires a sensitivity to both the architecture of the nervous system and the interactivity of information within the mind.
- Technological advances in brain imaging opened up a wide array of new avenues for linking neuroscience principles with reasoning research.

Reasoning
http://dx.doi.org/10.1016/B978-0-12-809285-9.00002-8

OUTLINE

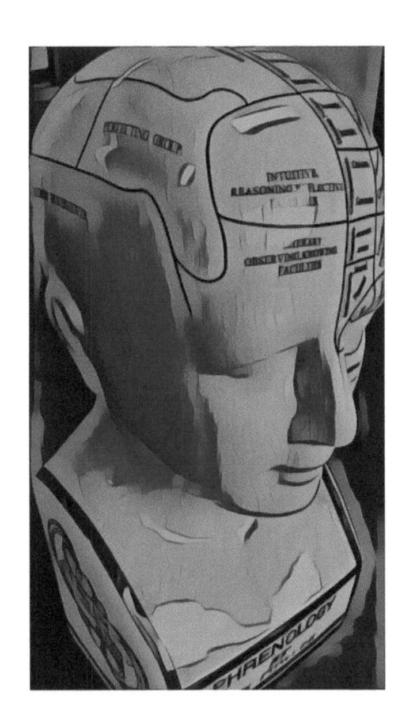

HISTORY OF REASONING

Introduction

There is a rich and interesting history that preceded the current research approaches that we use to study reasoning. This chapter explores views on reasoning that date back to the early periods of human history. There were times of intense inquiry where philosophers pursued the meaning of existence. There were also long periods of relative dormancy in which the study of the mind did not advance significantly. Over time, knowledge grew and was distributed much more widely. The methods of psychology gradually replaced introspection and purely thought-based experiments. It is remarkable how much of this development has occurred within the past century. The rapid advancement of reasoning research has co-occurred with the advancement of education and increased ease of sharing of ideas made possible by technological growth.

We will begin with an examination of some of the early approaches to the study of reasoning. There are some early clues as to when reasoning abilities began to advance and take hold for humans. The developments of technology and writing enabled people to preserve a record of their thinking about reasoning and cognition. These early times were mostly limited to thought experiments and philosophical writing. Despite the limitations in the methodology available early on, the roots of many important ideas can be traced quite far back to earlier periods in human history.

The late 1800s saw radical increases in intellectual development. This was the time during which Charles Darwin published his landmark book *On the Origin of Species* spelling out the theory of evolution by natural selection (1859). The book and the ensuing debate radically changed the perspective of the biological world, or natural philosophy, as it was referred to at the time. Neuroanatomical knowledge was growing at a rapid pace and advances in microscope technology led to new understanding about the characteristics of the nervous system and its relationship to thinking. Alongside the new perspectives developing in biology, the first psychology labs began enabling principled empirical studies of the mind. These approaches would provide the means to address many of the challenges that had previously hindered the field. Such challenges included subjectivity and a lack of systematic research.

We will next consider the growth in reasoning research that took place over the 20th century. This century saw unprecedented advances in thinking, methodology, knowledge, and technology. Many critical ideas about reasoning were grappled with in the early 1900s, as the first psychologists gathered their thoughts about thinking, reasoning, and deciding and began to carry out studies. Some of these studies would fit within the Gestalt school of psychology (Box 2.1), which was centered largely in Europe. The Gestalt psychologists were keenly interested in the nature of insight, and problem solving was one of their well-researched areas of emphasis. In the United States, the behaviorist approach came to dominate the field in the first half of the 1900s, and this had a profound impact on reasoning research, as certain topics were relatively underexplored due to the restrictive nature of the behaviorist methods.

The mid-20th century saw key developments take place that would shape the rest of the century. Among these was the development of cognitive psychology, an approach that led to the development of new methods. These methods could be used to explore the contents of the mind and led to further studies of reasoning and decision making. The development of the computer was also occurring at this time. With that technology as a reference point, an intellectual position evolved maintaining that we are information processors and that our brains may operate similar to computers. Radical increases in knowledge about physiology and neuroscience also led to new ways of thinking about how we reason.

The developmental psychology perspective thrived in the 20th century as well. Cognitive development became a major interest within the discipline and remarkable progress occurred in both theoretical views of development and experimental approaches to the study of children's cognition. Reasoning ability emerged as a major component of the theories and methods that were applied in developmental studies. Studies of cognitive development also led to models of infant and child reasoning abilities. These developments have practical implications for parenting practices, early child care, and education.

As the 20th century came to an end, remarkable new technological developments were changing the face of the fields of cognition, neuroscience, and biology. The emergence of brain imaging techniques had a major impact on the types of questions that investigators were able to ask. Advancing knowledge about the brain and nervous system also led to new theories and perspectives on the way that reasoning is accomplished within the nervous system. The ability to compare cognition across species was also advancing.

As we look toward the future, there is reason to believe that our capabilities in thinking and reasoning will advance through the knowledge gained and the technologies that are currently in development. Over the past 50 years people's Intelligence Quotient (IQ) has been rising steadily. This is referred to as the Flynn Effect after James Flynn (1987). One theory is that we now reason about such abstract content that we actually have enhanced our capabilities for reasoning to meet these demands. Also relevant to this topic is the rapidly expanding information search abilities that have become possible in the past 20 years. Never before have people had so much knowledge available at such a rapid pace. Our capabilities for reasoning will likely change as we move into the future, and we conclude this chapter with a discussion about the future of reasoning, which is a major topic of Chapter 13 of this text.

EARLY APPROACHES TO STUDYING REASONING

Reasoning Through Human History

At what point did people begin to reason? This is an intriguing question that follows from some of the discussion of the introductory Chapter 1. As modern humans evolved, we developed the capacity to represent the world in ways that were not apparent in previous ancestors. Humans have an advanced capability to represent information abstractly and elaborately. Abstract reasoning involves storing and manipulating representations that are removed from concrete perceptual information and actions. Abstract information can be replayed in sequences through mental simulation. Such information can be manipulated and changed. We can also selectively recall abstract information in a manner that helps to inform us about future possibilities.

Some brief discussion of human evolution is warranted before we examine some of the features of modern humans that are linked to our advanced reasoning capacity. There are several factors that impact our reasoning ability. Among these are brain size and the ability to manipulate factors in the environment. Little is known about either of these for much of our history.

BOX 2.1

GESTALT PSYCHOLOGY

The Gestalt school of psychology was a movement that took place in the field centered primarily in Europe in the early 1900s. Gestalt refers to a *whole* being different than the sum of its parts. Gestalt psychologists were interested in how people perceived incoming information, how they interpreted that information, and how they solved problems. The Gestalt psychologists started in Germany and were important historically, as they conducted research at the same time as behaviorism had taken hold in the United States. The behaviorist model of psychology placed strong limitations on the study of topics such as reasoning and decision making.

The Gestalt psychologists were interested in how situations or structures could be viewed as being unified. This idea applies throughout modern psychology. We perceive a face to be a whole entity, rather than merely a collection of features. We can group information into meaningful units, such as when a chess expert perceives a line of figures as being a "pawn chain," because he has seen this formation so many times before. In many situations in life, numerous independent items can take on a singular overall meaning of its own. Gestalt psychologists were interested in how people group information perceptually. Fig. 2.1 shows a series of grouping principles that had been observed and noted by Gestalt psychologists. The arrangement and perceptual features of dots or lines can lead us to group those features into predictable arrangements. Gestalt psychology noted the regularities in grouping that people will tend to carry out and these came to represent Gestalt principles.

The Gestalt school emphasized rigorous research methods. Earlier experimental psychologists interested in mental representations included Wilhelm Wundt and Hermann Ebbinghaus, who had used the introspection method and subjective interpretations of thought as the basis for developing views on mental representations. The Gestalt psychologists, by contrast, emphasized conducting experiments with naïve human subjects who had not previously encountered the experimental materials before. This emphasis enabled the experimenter to take on a more detached and objective viewpoint. These methods allowed researchers to draw conclusions that could be more widely generalized across populations.

Gestalt psychologists were interested in how people solved problems. There was a particular interest in the process of *insight* in problem solving (Sharps & Wertheimer, 2000). Gestalt psychologists, such as Karl Duncker, noted that there appeared to be several distinct steps that occurred in problem solving. These included four separable stages in problem solving:

1. *Preparation*: recognize the problem and make a first-pass attempt
2. *Incubation*: if at first you don't succeed ... try again later, but possibly continue to work on the problem at an unconscious level
3. *Illumination*: insight is achieved and this is accompanied by a level of confidence that the answer has been found
4. *Verification*: confirm that the solution works

The Gestalt psychology school made enduring contributions that helped to shape our understanding of both perception and problem solving. Their methods and research interests foreshadowed much of what would follow, as cognitive psychology emerged in the 1950s.

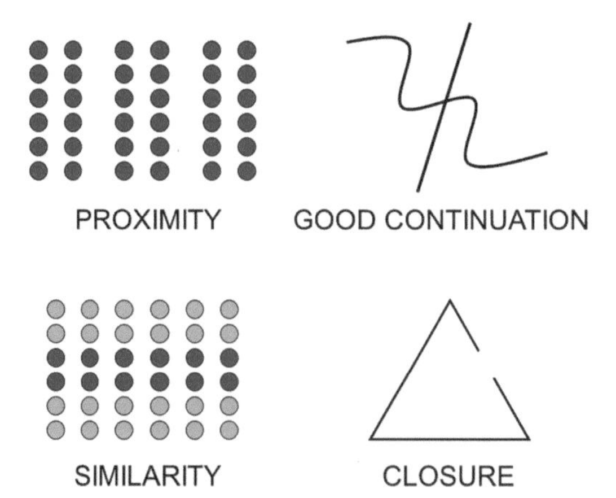

FIGURE 2.1 The principles of Gestalt grouping. Proximity: objects close to one another are grouped. Similarity: objects that are similar are grouped. Good Continuation: people tend to see two intersecting lines as continuing rather than being four separate pieces. Closure: people see a gap in a shape and tend to close that gap in their minds.

Notable Gestalt Psychologists (Fig. 2.2):

Max Wertheimer (1880–1943)—Developed an interest in apparent movement in perception. He also focused on the distinction between *reproductive thinking*, concerning learning by repetition, conditioning, and habit formation, and *productive thinking*, which is characterized by insights and the development of new ideas.
Kurt Koffka (1886–1941)—Articulated the distinctions between levels of learning. For example, Koffka claimed that *sensorimotor learning* occurs when our actions are linked to outcomes. This is different from his concept of *ideational learning*, which deals with language concepts. Koffka also contributed to developmental psychology.
Wolfgang Kohler (1887–1967)—Kohler worked with Wertheimer and Koffka on perception and apparent motion. He also studied problem solving and was keenly interested in the similarities and differences between chimpanzees and humans.

FIGURE 2.2 Considered by many to be the founders of Gestalt psychology, Max Wertheimer, Kurt Koffka, and Wolfgang Kohler (pictured left to right) were studying mental representations and the process of insight problem solving at a time when much of the psychological community was restricting itself to the study of stimulus–response based learning. *Max Wertheimer (1880–1943) By Anonymous [Public domain], via Wikimedia Commons. Kurt Koffka (1886–1941) Courtesy of Smith College Archives; photograph by Katherine E. McClellan C. 1928. Wolfgang Köhler, from (Reval, January 21, 1887–New Hampshire, June 11, 1967).*

We can roughly estimate the likely size of the brains of our primate ancestors, and brain size is observed to have increased over time. The difficulty is that fossils can only give limited clues as to the reasoning abilities of our ancestors. What we can say is speculative and can only be viewed through the lens of comparison to modern humans.

Increases in brain size point toward the capacity for more complex behavior and the tendency to engage in social groups. Modern humans appear to have developed larger brains relative to earlier hominids. We have a relatively long postnatal growth period, and we require considerable parental care. Coupled with these features, people have relatively few offspring and have to invest considerably in those progeny. Fig. 2.3 shows a tree indicating the relatedness of different extant species of primates.

One of the closest ancestors of the modern human is *Australopithecus*, a name meaning southern ape. There are several different fossil discoveries of *Australopithecus* that vary to some degree in their size and characteristics. These early hominids lived between 2.5 to 3 million years ago and were smaller on average than modern humans, weighing approximately 40–140 pounds depending on the specific variant of *Australopithecus*. Perhaps the most famous fossil is named "Lucy," a female *Australopithecus afarensis* who measured three-and-one-half to four feet in height and appears to have engaged in bipedal locomotion. Other fossils of *Australopithecus* have been found over the past century. The skulls of these fossils indicate that *Australopithecus robustus* and *Australopithecus africanus* would have had estimated brain volumes of approximately 550 cc. The precise lifestyle of *Australopithecus* is a matter of some speculation, but based on the traits of these fossils it appears that considerable evolutionary change was occurring (Strickberger, 1996).

Our closest ancestors are of the genus *Homo*. Fossils of these hominids originate from Africa dating between 2.2 and 1.8 million years ago. *Homo habilis* was a relatively short species estimated to be roughly the size of *Australopithecus*. The brain volume of *H. habilis* is estimated to have been between 600 and 700 cc, an enlargement over the earlier hominids (Fig. 2.4). *H. habilis* also appears to have used stone tools for hunting and butchering large

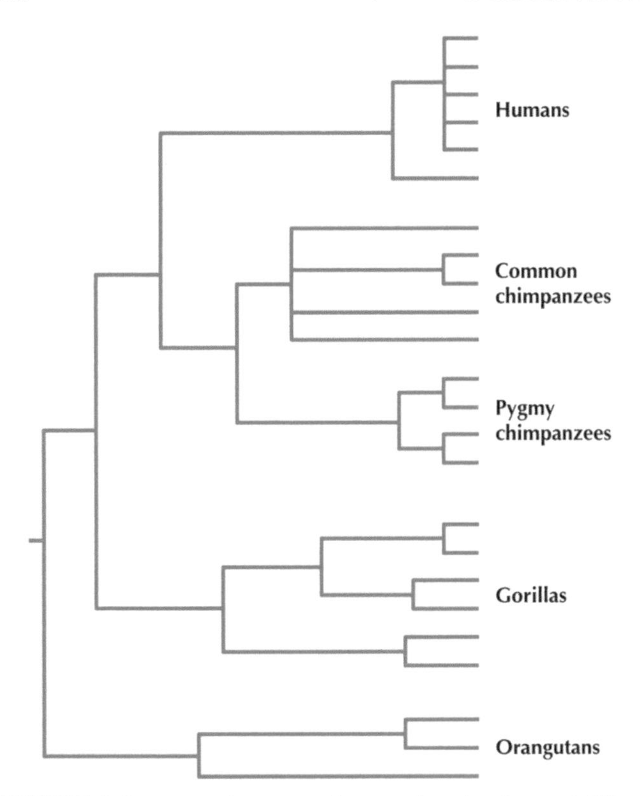

FIGURE 2.3 A tree diagram indicating the relatedness of different species of primates. *Ruvolo 1994 figure on human ancestry. Copyright (1994) National Academy of Sciences, U.S.A.*

carcasses. These skills imply a level of social organization that would have been needed in order to take down large prey and to invent tools. Both of these lifestyle properties are characteristic of a complex species. Other members of the genus *Homo* include the larger *Homo erectus*, whose brain volume ranged from 750 to 1000cc and dating to approximately 250,000years ago. Meanwhile *Homo sapiens* who appeared around 50,000years ago are our direct ancestors having the largest hominid brain volume of approximately 1300–1500cc (Strickberger, 1996). There is a further distinction between the *Homo sapiens sapiens* (modern humans) and the *Homo sapiens neanderthalesis*, who are commonly known as Neanderthals, hominids possessing a somewhat smaller skull than *H. sapiens sapiens*, a shorter and more squat body plan, but otherwise bearing high similarity to modern humans.

While we cannot solidly establish the reasoning capacities of these earlier hominids, the scientific community has established several important features that they possessed. Among these is bipedalism, the ability to walk on two legs. Bipedalism likely allowed for improvements in the ability to detect predators due to a height advantage, as well as the ability to carry food over long distances (Lovejoy, 1981). The ability to use fire as a tool likely led to a reduction in digestive load, as cooked meat could be more easily digested. Such a capacity could have led to less time spent sleeping and enhanced productivity. Lastly, hunting appears to have been cooperative at this point and likely involved the use of tools such as flaked stone hatchets and spears. These advances in hunting are characteristic of advancements in reasoning ability. As we will discuss in Chapter 4 on animal reasoning, we can learn about the complexity and capabilities of a species based on its capacity to obtain food in novel and flexible ways.

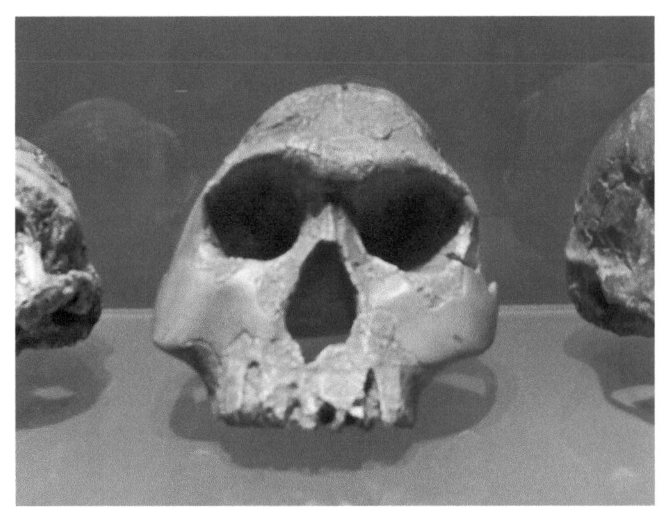

FIGURE 2.4 A skull of the primate *Homo habilis*. *From Wikimedia Commons.*

There are some other clues to the development of the human ability to think abstractly. One clue is cave art. A species that is capable of producing detailed drawings has a well-developed level of dexterity that enables this activity. Perhaps more importantly producing cave art demonstrates an elaborated capacity for memory. The act of creating art requires a brain that has great flexibility, as situations must be learned in a single trial. Emotionally salient situations are good examples of such one-trial learning. If the artist wishes to convey the dynamics of a hunt, then the ability to store a sequence of events surrounding the hunt that evolved over time is needed. Modern humans have an impressive ability for storing and recalling episodic memories. This refers to the ability to store situations that occur over an extended period of time. The act of representing a scene, such as a cooperative hunt, is made possible by the ability to store episodes in time. Memory psychologist Alan Baddeley (2000) has referred to our ability to recall and examine episodes in time as an episodic buffer within our working memory system. We do not see cave art in any other existing species, which may suggest that they do not possess the type of memory for episodes that humans do.

Philosophy of Reasoning

Since the dawning of consciousness, there has likely been a human desire to understand the nature of thinking

FIGURE 2.5 Plato (427–326 BCE) wrote about different types of knowledge and how certain we should be about the conclusions yielded by different forms of reasoning. *From Shutterstock.com.*

and reasoning. It would take many years to develop the experimental methods and tools to measure reasoning abilities. There are few clues about the earliest periods of inquiry into knowledge and reasoning, as few cultures wrote down their thoughts and few writings survived to the present day.

One of the early impressive contributions to human reasoning came from the Greek philosopher Plato (427–326 BCE) (Fig. 2.5). Plato wrote about different types of knowledge, how it was obtained, and how certain we should be of different forms of reasoning. The writings took the form of a dialogue between Plato and a student. In one such writing, Plato discusses knowledge of the subject of geometry. He draws a square in the sand and asks a student how many sides it has and whether they are equal. The boy responds appropriately. Plato then asks his students to imagine that the sides had been doubled in size. He asks how much larger this new imagined square would be than the one that had been drawn in the sand. The student responds that the new larger square would be equal in size to four squares of the original size. Plato points out that ideas can emerge innately and without the person consciously planning out the rules, or using any sophisticated formal geometry. While this demonstration is a far cry from later psychological experiments, it indicates that scholars such as Plato had a keen interest in the nature of our mental representations and how they may be applied toward reasoning.

Other notable philosophers of the mind would emerge later in history. Around the time of the Italian Renaissance, thinkers such as Rene Descartes (1596–1650) were active (Fig. 2.6). The Renaissance was a time of great curiosity and creativity. Alongside remarkable

FIGURE 2.6 Rene Descartes (1596–1650) studied physiology, physics, geometry, and language. Some of his most notable observations were on the relationships between the mind and body. *From Wikimedia Commons.*

artworks, philosophers were once again striving to understand the nature of the mind and how it interacted with the body. The contributions of Descartes were myriad, as he studied physiology, physics, geometry, and language. Some of his most notable observations were on the relationships between the mind and body. Descartes believed that geometry represented the purity of thought and that when each person considered a particular shape, such as a square or triangle they would be certain to agree upon the properties of the representation. These thoughts were considered to be the essence of reasoning for Descartes, a subject matter that was not open to the challenges of interpretation that plagued other aspects of human judgment. For Descartes, the complexities of judgment extended to how people perceive information from the senses. It was clear to philosophers for some time that illusions could occur. People could experience distortions under different lighting conditions, or when facing comparisons between objects of different sizes. Even the interpretation of marks on paper varied depending upon one's vantage point and the lighting of the room. Indeed, Descartes concluded that the substance of the mind was different and in some sense more pure than the translation of physical stimuli onto the senses and subsequent perception by the mind.

Descartes is famous for claiming that the mind and body were separable. He believed that the human mind was unique, and that it transcended the physiology of the rest of the body. This would include the brain, which he regarded to be separable from the mind at a core level. In some cases, people still hold such views, as the subjective nature of introspection appears to be in some way separate from the body and the rest of the physical world. This view also implied to Descartes that there was a strong difference between humans and the rest of the natural order. This implied that humans were special and differed fundamentally from other species due to their ability to think in pure terms. Descartes was aware of the ventricles of the brain. He believed that they were important for enabling movements of the body and this led him to form an early, though inaccurate, theory about the relationship between the brain and body. Since humans bleed when cut, Descartes claimed that the mind was able to force fluid from the brain via the pineal body to the muscles of the body. He claimed that movement was achieved through hydraulic pressure. This theory was in part informed by the actions of mechanical statues that existed in Paris at the time, which could be controlled via hydraulic pressure when visitors stepped upon pressure plates.

One of the other major contributions of Descartes to the science of the mind was his interest in mental representations. Descartes referred to these as *models*, and he was also aware that we can think in images and symbols. The models could be manipulated and transformed by the mind. For instance, a model of a chair could be rotated in different ways to imagine the view from the top or underside. Additionally, Descartes contributed to the methods of introspection, or self-referential thinking, which would be a cornerstone method for early psychologists in the centuries to come. Not all of Descartes ideas have stood the test of time, and indeed many are considered to have been refuted many years ago. We cannot, however, overlook his contributions to inquiry on the nature of mental representations, and how they relate to the physical world through our physiology. His writings continue to influence the directions of both philosophy of the mind and experimental psychology.

Other important philosophical contributions to the study of reasoning came from David Hume (1711–1776) (Fig. 2.7). Hume was a pioneer of many ideas that would form the basis for cognitive psychology. Hume claimed that there existed both an external world and an internal one. He claimed that humans were capable of reasoning by combining simple ideas by comparing them. For example, Hume believed we could develop new knowledge by making comparisons between objects. The comparison between two shades of red, for example, led to the concept of identity, a new and abstract form of knowledge (Haberlandt, 1994). This relational comparison ability forms the basis of many later studies of reasoning that took place in the 20th century. Many psychologists still consider relational thinking to be at the top of a hierarchy of our reasoning abilities (Hummel & Holyoak, 1997).

FIGURE 2.7 David Hume (1711–76) claimed that there existed both an external world and an internal one. He claimed that humans were capable of reasoning by combining simple ideas by comparing them. *From Shutterstock.com.*

Hume was also keenly aware of the associations that could be made between sensory events or objects. As an example, he noted that lightning preceded thunder, and thus people would be likely to draw the inference that lighting actually caused thunder. Causal inferences are particularly powerful relational comparisons. Whether we understand the mechanisms of cause ultimately determines whether we are correct in how we attribute cause, but Hume's thinking foreshadows much of the interesting psychology of causal reasoning that we will describe later in Chapter 8. The concept of association would also form the basis of many early programs of psychological research in the late 19th and the 20th century.

Immanuel Kant (1724–1804) was another important philosopher who made enduring contributions to the study of reasoning (Fig. 2.8). Kant was interested in describing the types of knowledge that could exist. He was an early proponent of distinguishing between innate mental abilities and those that were acquired by experience. For Kant, experience was a major driving force that led people to possess knowledge. Kant proposed multiple forms of knowledge. Among these were *dimensions, categories*, and *schemas*. Dimensions were the concepts of space and time which people's minds understood innately. Categories meanwhile were the tools of reasoning. Kant described 12 innate categories of reasoning

FIGURE 2.8 Immanuel Kant (1724–1804) made enduring contributions to the study of reasoning. Kant was interested in describing the types of knowledge that could exist. He was an early proponent of distinguishing between innate mental abilities and those acquired through experience. *From Shutterstock.com.*

including aspects such as quality and quantity of information. Lastly, schemas referred to units of knowledge. A schema was a general form of knowledge such as animals or trees and did not necessarily refer to particular specific examples. While these are not the modern terms that we would apply to mental representations, the ideas are quite similar to ones that psychologists would later propose. Kant did not believe that one could understand the nature of knowledge without examination of the mind, a concept very similar to the proposals made later by physiologists and psychologists.

What We Can Learn From the Philosophical Approach

We can benefit greatly by examining some of these historical precursors to the study of reasoning that would flourish later in the 20th century. Many of the ideas contributed by early philosophers of the mind would form the basis for experimental procedures that would develop later. We can see from writings as early as those produced by Plato the concept that the mind was somehow different from the physical aspects of matter and of the world. Plato, Descartes, and others believed that the contents of the mind were in some ways more pure and reliable than aspects of perception or sensation. The idea that mental representations were important and critical for reasoning can be traced throughout the works of those interested in the philosophy of the mind. We see in the later ideas of Hume and Kant a strong desire to understand and categorize knowledge structures and a realization that reasoning is directly linked to the acquisition and manipulation of knowledge structures. Also foreshadowed by the philosophers of the mind was a realization that events that occurred closely in space and time would come to be associated. These events may even lead to causal inferences, which help us to make sense of the world by enabling predictions to be made between events and their possible outcomes. This idea would continue to influence the psychology of the mind long after these philosophers were gone.

Limitations of Philosophical Inquiry

While philosophers of the mind were busy describing how things may work in the world and more intimately, within our minds, the approach was still missing something critical. There were no strong methods to test the ideas of Hume, Kant, and others. Introspection or thinking about one's own thinking is an important aspect of psychological theorizing. The challenge is that we cannot distinguish objectively between different theories unless we are able to test these under controlled conditions. This is certainly a challenge as subjectivity cannot help but enter into our deductions, as we are limited to using the mind and brain in order to study these same subjects. A variety of experimental approaches would later enter into popular usage enabling people to be more objective about their conclusions and to be able to test their theories. In order for this to happen, the field had to transition to using a psychological approach to inquiry about the nature of the mind and human reasoning.

EARLY PSYCHOLOGY OF REASONING

Reasoning in Early Psychological Research

The late 1800s saw the emergence of psychology as an independent discipline. The earliest psychologists were trained as medical doctors. They became interested in aspects of the mind and were keenly interested in mental representations, memory, and how these processes linked to emerging biological knowledge about the brain.

Wilhelm Wundt (1832–1920) is often considered to be the first experimental psychologist (Fig. 2.9). Wundt founded the first psychology lab in Leipzig, Germany, in 1879. Wundt was trained originally in medicine and physiology, and he was interested in turning the study of the mind into a rigorous discipline that would have the

FIGURE 2.9 Wilhelm Wundt (1832–1920) is often considered to be the first experimental psychologist. He founded the first psychology lab in Leipzig, Germany, in 1879. *[Public domain], via Wikimedia Commons.*

FIGURE 2.10 William James (1842–1910) is credited with writing the first text of psychology entitled *Principles of Psychology* in 1890. *By Notman Studios (photographer)—[1]MS Am 1092 (1185), Series II, 23, Houghton Library, Harvard University, Public Domain, Wikimedia Commons.*

same inferential power as physiological study. Wundt practiced a style of research that is known as structuralism. He was interested in the structure of how mental events occurred and the regularity of such events.

Introspection was the experimental approach employed by Wundt in order to study the mind. He would ask his research subjects to think inwardly about situations that were described by the experimenter. Ideas that came to the minds of the subjects were reported and scrutinized by the experimenter.

Wundt had a clear impact on the methods used in order to understand the mind, but one of his other decisions would prove challenging for the study of reasoning. Wundt claimed that there were two classes of psychological phenomena, one of them being higher cognition, which depended on language and culture, while the other had more to do with stimulus and response properties. Wundt believed that the higher psychological phenomena that depended upon variables such as culture and language could not be adequately studied using the limited methods that were possible at that time. Rather, he claimed that such thinking could only be adequately characterized by viewing an entire culture. This points out the challenges that early psychologists would face when trying to study abstract processes such as those involved in reasoning. Indeed the evolution of measurement and methodology has had a profound impact upon the success and the nature of theorizing that has accompanied the study of reasoning.

William James (1842–1910) is often considered to be the founding father of psychological research in the United States (Fig. 2.10). James is credited with writing the first text of psychology entitled *Principles of Psychology* in 1890. This influential book encapsulated many of the ideas that would make up the fabric of psychological research in the 20th century. James wrote with a visionary quality outlining his views on topics ranging from development to emotion. James even anticipated the Hebb rule of learning and memory, even speculating that pathways within the brain that were regularly activated would become strengthened. Clearly this was almost pure speculation at the neuroscience level, as there were no data or methods by which to determine what was actually occurring at a cellular level in physiology. The ideas could be traced to the nature of human habit formation, which would have been evident to a keen observer. James, like Wundt, was also a prolific mentor of trainees. Many modern researchers can trace their academic training lineage directly back to either Wundt or James. Some of his famous students include physiology pioneer Edward Thorndike, Walter Cannon, who famously described the "fight or flight" response, memory researcher Mary Calkins, and writer Gertrude Stein.

William James described an array of viewpoints on many topics in psychology. Among those was reasoning. James was not exhaustive in his treatment of the topic and ignored many critical distinctions among classes of reasoning, its development in childhood, and deduction versus induction. James was highly concerned with some critical aspects of reasoning however. Among these was whether humans possessed reasoning abilities that were not present in other species. James was an associationist, like others before him, such as David Hume. James did not view the retrieval of memories in response to sensory stimuli as being unique to humans. When we smell a certain odor and call to mind a particular experience, or concept linked directly to the smell, we are not likely exhibiting anything other than mere association. This type of association, James believed, was akin to what other species may do. James indicated that inferences were perhaps unique to human reasoning in instances when people draw together a new association, particularly one between concepts, or ideas, that had not previously been considered to be related in the past. He referred to reasoning as being *productive* in this regard, rather than merely reproductive, which would describe the simpler associations that had been experienced together repeatedly in past instances (Nickerson, 1990). James also claimed that reasoning enabled people to deal with novel data. This was one of the properties that marked reasoning as being different from other mental processes and perhaps constitutes what is unique about human thinking.

Another feature of reasoning that William James discussed at length was the ability of people to abstract or symbolize concepts or ideas. He claimed that "key to abstraction is the ability to take away the essence of an object, rather than storing the object as a literal copy." "Every possible case of reasoning involves the extraction of a particular partial aspect of the phenomena thought about," "whilst empirical thought simply associates phenomena in their entirety, reasoned thought couples them by the conscious use of this extract" (p. 967). James was aware of the differences between perception and conceptualization. He also emphasized that those people who are effective in reasoning are able to focus on the relevant properties of objects and ignore the irrelevant ones. Deciding on relevance is another matter, but assuming one has a clear goal in mind, this would be a capability that is reminiscent of modern researchers' focus on the role of the frontal lobes in reasoning, as being important for filtering the associations needed for accomplishing a task. An appreciation of similarity and the ability to communicate ideas were other areas of emphasis for James when describing the processes of human reasoning. His influence can still be felt over 100 years later in the work of modern cognitive psychologists, cognitive scientists, and neuroscientists of reasoning.

The Emergence of Knowledge About the Brain

The 1800s saw the dawning of many important insights about the brain and its physiology. These ideas would lay the groundwork for much of the remarkable growth and progress in neuroscience observed in the 20th century. Among the key figures in physiology were those studying early animal models, examining the microstructure of the nervous system, and those involved in treating patients who had sustained brain damage.

With regard to reasoning, one particular case endures as one of the most famous instances of a change in biology that brought about a psychological change linked to thinking. Railroad worker Phineas Gage sustained a remarkable injury to his brain during an accident involving a tamping iron. Gage was struck by a 6-ft-long iron rod that had a pointed end. The rod was fired forcibly through his forebrain passing just under his left eye and emerging out of the middle of his frontal lobes. Remarkably Gage survived this catastrophic injury. While details are somewhat incomplete, Gage evidently regained consciousness shortly after the explosion had fired this rod through his brain. Onlookers were shocked that Gage was able to move around and speak shortly after the injury. While the case of Phineas Gage was not studied with the degree of precision that a modern brain injury would be, John Harlow, a local medical doctor did examine Gage and took a strong interest in his case. Phineas Gage apparently escaped with his life, along with many of his faculties remaining intact. Gage's speech, motor action, and sensory processing were spared, but Harlow did note some rather surprising and ultimately devastating effects of the injury. There is little clear evidence of Phineas Gage's behavior prior to the injury, but the limited reports of the day indicate that he was moderate, respectful, polite, and a family man prior to the injury. His life after the accident was strikingly different.

Gage underwent an apparent personality change after the tamping iron damaged his frontal lobes. The man who had once been a model citizen fell into difficult times. After the injury, he made poor financial decisions, acted impulsively, used vulgarity, and generally became difficult for others to be around. These same types of traits are also reported in modern day cases of brain injury, in which once highly successful and disciplined individuals fall into making decisions that are unwise, rash, and destructive (Bechara, Tranel, Damasio, & Damasio, 1996). While Phineas Gage did survive, he was apparently never quite the same and suffered from a form of insult to his intellectual faculties. He appears to have had a compromised ability to wisely and effectively plan for the future. He also exhibited difficulties in carrying out his plans, and became irritable and ill-tempered. Gage's life ended some 14 years later, and he never appeared to have regained his former abilities (See Box 2.2).

BOX 2.2

THE REMARKABLE AND TRAGIC LIFE OF PHINEAS GAGE

Phineas Gage (1823–1860) was a railroad worker who suffered a remarkable accident on September 13, 1848. Gage was involved in clearing rock for the railway by preparing for dynamite blasts. He worked as a foreman for the Rutland & Burlington Railroad south of the town of Cavendish, Vermont. One particular day he set to work placing a dynamite charge within a hole in the rock. Standard practice of the day was for the charge to have sand poured over it and then the charge could be tamped down further into the rock prior to detonation. The tamping was accomplished using an iron rod. In this case, Phineas Gage set to work using a tamping rod measuring approximately 6ft in length with a diameter of roughly 3 in. across. One end tapered to a point, while the other was blunt. Gage was apparently ramming down the charge into the rock using the blunt end when a spark presumably detonated the charge prematurely, while Phineas Gage was still working over the hole.

What happened next was stunning to medical personnel at the time. The tamping iron shot upward piercing Phineas Gage just below his left eye. The rod shot through the eye socket passing directly through Gage's frontal lobes. The 3 inch diameter of the bar destroyed a large portion of his medial prefrontal cortex (see Fig. 2.11). The bar then passed up through the middle and dorsal portions of his frontal lobes before launching out of the top of his head and into the air some 50ft. Remarkably, Phineas Gage regained consciousness shortly afterward and was up and moving about. John Harlow, a Vermont doctor saw Phineas Gage and made a general examination of the wound. It is remarkable that Gage survived the initial effects of this traumatic injury to his brain and secondly that he did not succumb to an infection afterward. Harlow was fascinated by the injury and that Gage was able to survive.

FIGURE 2.11 Phineas Gage experienced a devastating brain injury that changed his personality. *Author of underlying work unknown. (File:PhineasPGage.jpg) [Public domain], via Wikimedia Commons.*

Little is known for sure about the earlier life of Phineas Gage, but in the aftermath of the injury it became clear that he had acquired some difficulties with his ability to plan, carry out plans, reason, and decide. In modern terms, we would call this an executive function deficit. Executive functions are the mental abilities that allow us to both appreciate and devise plans and to carry them out without becoming distracted or losing focus. In the years following his frontal injury, Phineas Gage had great difficulty in life. He could no longer hold down a job due to problems controlling his temper. He engaged in profanity and was childlike and irritable. According to Harlow, Gage had also become capricious, devising many plans only to scrap them quickly afterward. Gage also had great difficulties with financial decision making and was unable to maintain a stable family life.

Over the course of the next 12 years Phineas Gage led an erratic and challenging life. He ended up finding work as a carnival performer in P. T. Barnum's Circus, in which he carried around and displayed the tamping iron that had so nearly ended his life. Eventually Gage found his way west and died in San Francisco in 1860 from complications related to epilepsy (O'Driscoll & Leach, 1998). He was buried there in a rather infamous cemetery, which would be dug up and relocated in the early 1900s. Realizing that Phineas Gage's body was to be exhumed in the cemetery relocation, researchers at Harvard Medical School asked permission from Gage's relatives to exhume the body. The folks at Harvard were able to acquire Gage's skull, which bears the marks of that incredible injury, as well as the tamping iron itself, which had been buried with Phineas Gage in a rather macabre twist to the story. Both of these items are on display at the Harvard Medical Library in Boston. Having visited this collection, I was amazed at the devastation that the tamping iron had caused to Gage's skull (Fig. 2.12). It is quite apparent that the bones had to fuse back together after massive cracking to enable the enormous width of the tamping iron to pass cleanly through the front of Gage's head.

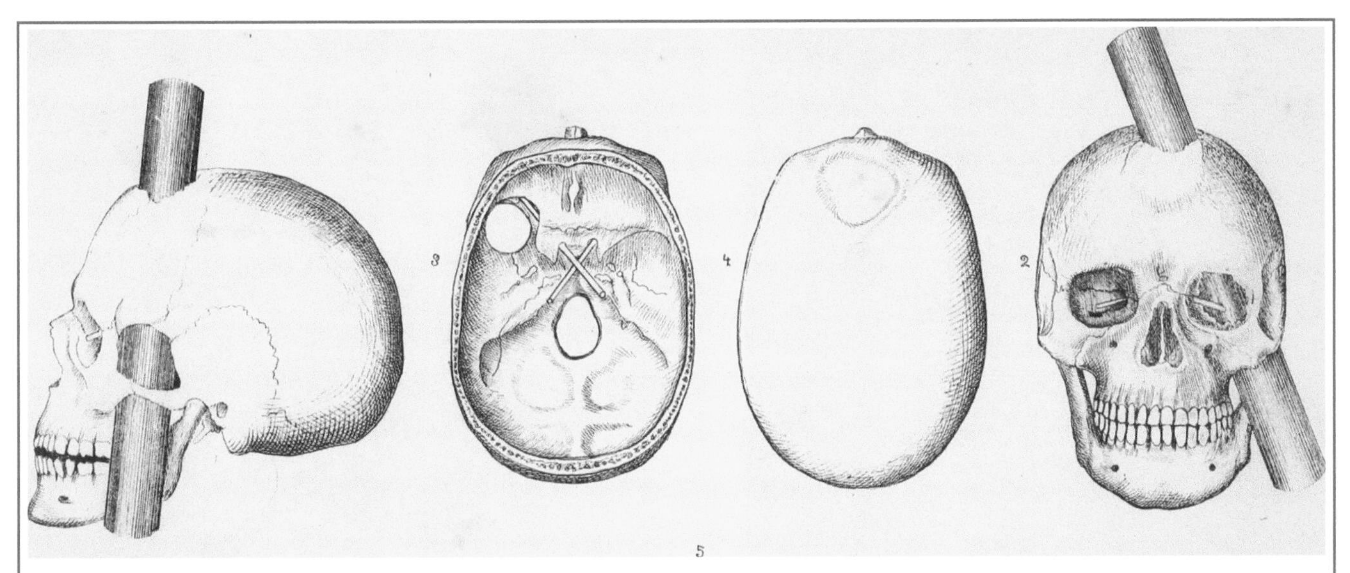

FIGURE 2.12 Researchers have tried to estimate the damage sustained by Phineas Gage using his skull and the tamping iron, which were recovered from his grave. *By Piper, Richard Upton, 1816–97; Bigelow, Henry Jacob, 1818–90. Anaesthetic agents, their mode of exhibition, and physiological effects [No restrictions], via Wikimedia Commons.*

As a side note, the medical community did not quite know what to make of Gage's injury for some time. Some felt the frontal lobes were not especially important in our daily functioning, as Phineas Gage was still able to speak and move after the injury. Others pointed to Harlow's writings as an indication that damage to the frontal lobes can disrupt cognitive abilities including reasoning and decision making.

Other famous neurological case studies occurred in the 1800s. Two patients contributed significantly to the understanding of how language functioned in the brain. Like reasoning, language was another elusive property of the human mind that appeared to vary extensively in the comparison between humans and other species. Paul Broca was one of the intellectuals of the time. An anthropologist, surgeon, and anatomist in Paris, Broca had been pondering the claim that language functions were localized to the frontal lobes (Dronkers, Plaisant, Iba-Zizen, & Cabanis, 2007). This possibility had been hypothesized by the phrenology community in the middle of the 1800s.

Phrenology was a discipline that had proposed that the functions of the mind and brain could be determined by feeling the protrusions of the head and mapping the largest areas to traits that had been laid out in a map of the head (Fig. 2.13). The claims of phrenology were made prior to any formal analysis of the brain, thus the mapping of function to brain area was very crude and generally inaccurate. An opportunity to test this hypothesis presented itself for Broca in 1860.

Broca was asked to consult about the condition of a patient who had a striking and peculiar speech and language deficit. The patient in question was named Leborgne. He was 51 years old and had a variety of neurological problems. When Leborgne tried to speak or answer a question, he was limited to the production of just one single repetitive syllable "tan." Leborgne was able to vary his monosyllabic statements by changing the intonation, but was not able to produce anything resembling normal speech and language. The patient did not survive long and upon autopsy, it was discovered that Leborgne had sustained damage to his left frontal lobe. Broca presented this case as evidence for the linkage between language and the frontal lobes (Broca, 1861). Broca confirmed this relationship between anatomy and speech function by examining other similar individuals who had sustained strokes and presented with similar symptoms (Broca, 1865). Broca's modern approach to neurological investigation with these patients advanced the state of the science dramatically from the speculative localization suggested by Franz Joseph Gall's phrenology approach of the early to mid-1800s. The left frontal region would come to be known as Broca's area, and people who sustain this type of injury are described as having Broca's aphasia.

From these early neuropsychology studies, it became clear that damage to different areas of the brain disrupted different functions. Though brain imaging was over 100 years away, the process of documenting the deficits acquired by the victim of an accident or stroke and then viewing the brain or skull after death enabled people to begin to understand how functions

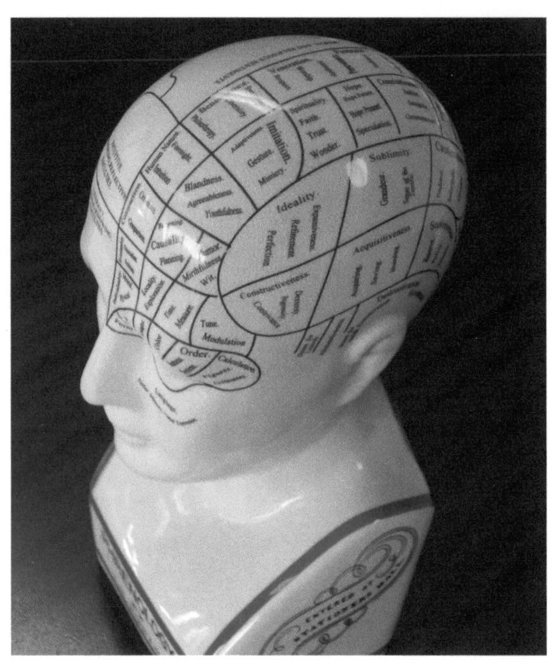

FIGURE 2.13 A phrenology statue. The head is labeled with a variety of mental faculties and propensities. Phrenology predicted that mental abilities could be located within different areas of the head.

FIGURE 2.14 Ivan Pavlov (1849–1946) was a Russian physiologist who conducted experiments on learning with dogs. Pavlov trained his dogs using a set of methods known as *classical conditioning. By Deschiens ([1] [2]) [Public domain or CC BY 4.0], via Wikimedia Commons.*

are locally organized in the brain. Phrenology had suggested a strong form of localization, in which different cognitive abilities and personality traits were able to be completely mapped to single regions of the brain. We now appreciate that there are localization differences in the brain and lesion mapping can yield important insights into brain function, but in addition to localized collections of neurons that are heavily involved in a particular function, there is also a wide distribution of function that is evident in the neural networks of the brain. These networks indicate that there is also a strong degree of distributed function. As we explore the neurological basis of reasoning in Chapters 3 and 4, we will find that these are both important principles of brain function that help us to understand how reasoning is organized within the brain.

Behaviorism in the United States and Its Effects on Reasoning Research

As the 20th century began, the fledgling field of psychology was still attempting to agree upon the methods that would become the standard for investigating the properties of the mind and brain. One of the strong debates centered upon the nature of experiments that would be most informative in these endeavors. The early psychologists, such as Wilhelm Wundt and Hermann Ebbinghaus, had been largely conducting experiments using *introspection*, a method that allowed a window into the complex process of mental representations. The early memory research of Hermann Ebbinghaus is rather famous, as he served as his own experimental subject. Ebbinghaus had assigned himself a variety of nonsense trigrams, which were meaningless collections of letters, such as "dax" and "bap". Ebbinghaus could not easily connect these letter strings to anything in his mind, and thereby avoided the challenges of existing knowledge influencing his studies of memory. Such subjective approaches led to criticism from other psychologists. These were interesting speculative results, but it was unclear whether they would generalize to other individuals. Furthermore, the measures themselves were not tangible. It was difficult to operationally define these concepts and many people were concerned about the data and interpretations becoming contaminated by the intimate closeness of the person, the data, and the fact that the whole endeavor was occurring within the mind of the experimenter.

Another approach that gained momentum was stimulus response learning. One of the most famous researchers utilizing this approach was Ivan Pavlov (1849–1946) (Fig. 2.14). Pavlov was a Russian physiologist who famously conducted experiments on learning with dogs. Pavlov trained his dogs using a set of methods known as *classical conditioning*. The dogs were initially presented with meat powder. This was termed the unconditioned stimulus, a substance that the dogs found naturally interesting and rewarding. The dogs salivated in response to the meat powder, an unconditioned response, as they found the powder appetizing, without training. The presentation of the meat powder was paired with a tone, which initially lacked meaning for the dogs. This tone represented the conditioned stimulus, as conditioning was needed in order for the tone to acquire meaning. Over time, after repeated presentations of the tone, the dogs learned that the tone was predictive of the meat powder and they became conditioned to salivate in response to the tone alone (Fig. 2.15).

Pavlov's famous study is representative of a very different approach to psychological inquiry than that used by Wundt and Ebbinghaus. Pavlov's work had some strong advantages. First, it relied on careful experimental control. The dogs did not have prior experience with the tone. Second, the stimulus was concrete and could be delivered in the same manner across subjects. Third, the presence of a response such as salivation was concrete and uncontroversial. This experimental approach focusing on observable behaviors that could be linked to new stimuli became known as *behaviorism*. Behaviorism had its roots in associationism, which was advocated as a key psychological mechanism during the philosophical era by David Hume. The type of associationism demonstrated by behaviorists was powerful, as it enabled previously unprecedented levels of experimental control and the ability to manipulate single independent variables allowed for strong causal inferences about the outcomes of the experiments.

The widespread adoption of behaviorism across experimental psychologists in the United States came with a price. The focus on purely observable behavior limited the subject matter that could be plausibly studied under the restrictions of the discipline. Behaviorist studies prevented the study of mental representations of any kind outside of the mechanisms of association. The approach also limited the ability of researchers to study the psychological processes that are involved in everyday life, which had been a key focus of individuals such as William James. The dominance of behaviorism from the early to mid-1900s largely limited the study of reasoning, an area of inquiry that is highly dependent upon understanding the nature of mental representations, semantic memories, and the use of explicit and conscious processing. In time, these limitations prevented American experimental psychology from making progress in critical areas within the field. Change would come in the World War II era, when psychologists began to focus on the practical problems of the war effort. This shifted the focus of experimental psychology in the United States squarely back toward the study of mental representations.

Behavioral Views on Reasoning and Gestalt Psychology (1920s–1940s)

While behaviorism was dominant in the United States, there was a distinctly different approach to psychology being carried out in parts of Europe. As described earlier in this chapter (Box 2.1), Gestalt psychology was highly active in the study of mental representations. The Gestalt psychologists included Hermann Ebbinghaus, Kurt Koffka, and Wolfgang Kohler. These psychologists were interested in cognitive phenomena that did not involve explicit associationist learning. In fact, one of the most notable contributions of the Gestalt movement was the demonstration that certain types of perceptual phenomena operated almost universally across individuals and that these phenomena did not

FIGURE 2.15 Pavlov's dogs were presented with a tone followed by meat powder. They would salivate in response to the meat powder. Over time, the dogs learned that the tone predicted the meat powder and therefore salivated simply at the presentation of the tone. Depicted is one of the many dogs Pavlov used in his experiments. Note the saliva catch container and tube surgically implanted in the dog's muzzle. *By Rklawton—English Wikipedia, see below, CC BY-SA 3.0, https://commons.wikimedia.org/w/index.php?curid=601191.*

require instances of learning or association to occur. Phenomena such as grouping stimuli based on similarity or proximity were so commonly observed among experimental participants that they were termed principles. A key difference between Gestalt psychology and behaviorism is that the principles implied that the mind was not a blank slate that was merely populated by associations learned by stimulus-response conditioning. Rather, the fact that people so automatically and consistently experienced incoming information in the same manner implied that something structurally significant about our perceptual systems and cognitive abilities was enabling this phenomena to occur. The perceptual principles of Gestalt psychology did not require reward-based association to occur.

For the psychology of reasoning, some of the most noteworthy contributions of Gestalt psychology came about in the domain of insight problem solving. Gestalt psychologists such as Karl Duncker and Wolfgang Kohler were fascinated by the process by which people experienced a leap of insight as they solved a challenging problem. Further, there appeared to be distinct steps that occurred leading to insight. Duncker noted that we appear to undergo a phase of problem exploration. This is followed by a period of attempting a solution and possibly failing. After an incubation period in which the person takes a break from the problem, they may then experience a feeling of insight, an "aha" moment as they obtain a solution. Lastly, the problem solver verifies that his or her solution does indeed solve the problem (Wallas, 1926).

The Gestalt psychologist Karl Duncker (1903–1940) (Fig. 2.16) contributed several important ideas to the problem-solving literature. Behaviorists and earlier psychologists had maintained that problem solving arises from trial-and-error learning and from the experience that the problem solver had. By contrast, Duncker sought to investigate problem solving from an alternative viewpoint, that the solutions come about from the context or situation in which the problem occurs. This different point of view was important in driving Duncker's research forward (Newell, 1980).

Duncker viewed problem solving in the following manner (summarized from Duncker, 1945):

1. A problem arises when one has a goal, but does not know how to reach it. When one cannot go from the given situation to the desired situation simply by action, then there has to be recourse to thinking.
2. The solution to a new problem typically takes place in successive phases, which takes on the character of a solution over several iterations.
3. The functional value of a solution is critical to the understanding of it being a solution. It can be called the sense, or the principle, of the solution.

FIGURE 2.16 The Gestalt psychologist Karl Duncker (1903–1940) contributed several important ideas to the problem-solving literature. *Created December 31, 1866 By Unbekannter Grafiker der Epoche. (Illustrirte Zeitung, Bd. 49 (1867), S. 108.) [Public domain], via Wikimedia Commons.*

4. The final form of an individual solution is, in general, not reached by a single step from the original setting of the problem; on the contrary, the principle, the functional value of the solution, typically arises first and the final form of the solution in question develops only as this principle becomes successively more and more concrete.
5. In the transition to phases in another line of thinking, the thought process may range more or less widely. When every transition involves a return to an earlier phase of the problem, a new branching off from an earlier point in the family tree occurs. Sometimes one may return to the original setting of the problem, or just to the immediately preceding phase.
6. The essential nature of generating a solution is through thinking about sub-elements of the problem and how they meet the goal. Such relatively general procedures, are 'heuristic methods of thinking'.
7. Every solution takes place, based on the substrates of its particular problem situation.

From this summary, we can get a sense of the thinking of the Gestalt psychologists and how their views differed from the behaviorist viewpoint. Note that the context of the problem was seen to be critical to its solution. The idea of heuristics, or general methods, that can be successfully applied toward solving problems appears. Problem solving is observed to occur in multiple phases and the problem solver may refer back to the original problem, or potentially other points in their solution, in order to move forward. Many of these phenomena would appear in later forms in the second half of the 20th century when cognitive psychology began to take on these same challenges.

Karl Duncker's methodology was also noteworthy in distinguishing the Gestalt psychologists from the methods commonly used by behaviorists. Duncker frequently used "think aloud techniques,", in which a subject was asked to state aloud all of the ideas that came to mind when solving a problem. This approach resembles the introspection analyses that had been popular in the late 1800s as psychological research was just getting started. Among the notable work of Karl Ducker was the introduction of his X-ray problem, in which someone is asked how radiation can be used to destroy a tumor, which is below the surface of the body. A key condition is that there could be no contact between rays and healthy tissue. We will revisit this radiation problem when we discuss reasoning by analogy in Chapter 10.

Reasoning Research (1930s–1950s)

There were several interesting approaches that did not squarely belong to either behaviorist or Gestalt psychology during the time periods in which those approaches were dominant. Steps were being taken toward understanding the mental processing that makes up reasoning.

One key area of interest centered upon the psychological distinction between inductive and deductive reasoning. These concepts will be the focus of Chapter 9, discussing the differences between these two forms of reasoning and also tracing the full history of research into induction and deduction up to the present. For our purposes in the current chapter of tracing the history of reasoning, we need only discuss some of the developments in this field that took place in the early to mid-1900s.

The key differences between induction and deduction can be summarized on the basis of premises and conclusions. If we were to reason about the possibility of rain, we could do so deductively by evaluating a set of rules. For instance we could have a rule as follows:

If there are dark clouds, then it will rain.

When coupled with the premise *there are dark clouds*, we can conclude that it will rain. This is deduction in action. Notice that we can confidently reach a valid conclusion based on the rule. This rarely happens in our daily life reasoning, due to the probabilistic nature of events in the world and the fallibility of our perception. More commonly, we will make inductive arguments about a domain such as the weather that is determined by numerous complex and interacting factors. An inductive argument about rain could go as follows:

Most of the time that dark clouds are present, rain will follow

The thinker in this case will probably infer that dark clouds are linked to rain. The key difference is that this inference is not guaranteed to be valid. There is not a formal rule present in inductive cases, so they serve as a likely guideline that may prove true much of the time, but they are not necessarily valid. You will probably notice that the deductive case is different only if we can always rely on the rule linking dark clouds to rain. If we cannot (which is the state of affairs in daily life), then we are essentially in a similar place as we would have been with the inductive inference having to make do with some guesswork.

Philosophers long noted that there were fundamental differences between inductive and deductive reasoning, but these differences had not been held up to the scrutiny of behavioral testing. Experimental psychologists began formal investigations into the nature of deduction and induction in the 1920s and 1930s. Minna Cheves Wilkins (1928) reported on a deductive reasoning task, in which participants were asked to evaluate deductive arguments about concrete situations (locations of geographical places) and abstract situations about arbitrary rules involving letters. She found that real-world problems enabled the best inferences. This finding demonstrated the value of human experiments, as the context of the problem contributed greatly toward whether someone could answer soundly. Robert S. Woodworth and Saul B. Sells (1935) followed up Wilkins research, finding further evidence that people struggled to assess the validity of deductive reasoning problems. These researchers noted that people had difficulty processing the specific direction that particular rules specify. They termed this directionality error *an atmosphere effect*, as the atmosphere of the premises appeared to influence how the problem was interpreted. Irving Janis and Frederick Frick (1943) found additional evidence of logical errors in deductive problems and emphasized that people's prior beliefs are one of the major reasons that they fail to reason accurately about validity. This phenomenon became known as the *belief-bias effect*. These early studies helped to inform later work on deductive reasoning carried out by Peter Wason and Philip N. Johnson-Laird in the 1960s and beyond. Less formal work would be carried out on the study of inductive inference in psychology labs until later in the 20th century. It should be noted however, that the scientific method of carrying out hypothesis testing to falsify a hypothesis relies considerably on inductive inference. The philosopher Karl Popper (1959) pointed out that scientists should carry out their research in order to falsify, rather than to confirm, their hypotheses.

Another important early contributor to the study of reasoning in the early 20th century is Sir Frederic Bartlett (1886–1969) of Cambridge University (Fig. 2.17). Bartlett was a famous British psychologist who is perhaps best known for his 1932 book entitled *Remembering: A Study in Experimental and Social Psychology*. Bartlett was widely interested in perception, reasoning, memory, and even topics such as fatigue and skill learning as relevant to the World War II efforts of enhancing human performance (Roediger, 2000). Bartlett was a remarkable scholar and an accomplished mentor, giving him a wide influence in the field. Bartlett set up his lab in the 1920s placing him as a contemporary of the behaviorist approach in the United States and the Gestalt school of psychology in Europe. He was well acquainted with numerous researchers of the day, but his work had its own character that was ahead of its time. Like other early experimental psychologists, Bartlett did not employ a formal experimental method, in which one limits themselves to the manipulation of one or more independent variables and carefully measures the results. Rather, Bartlett's group conducted less constrained experiments, in which people were observed performing a task. It should be noted that Bartlett's significant achievements have been replicated under more formal conditions (Bergman & Roediger, 1999; Roediger, 2000). Perhaps Bartlett's most famous contributions about memory can be summarized by his description of a recall exercise entitled "the War of the Ghosts" (we will also discuss this study in more detail in Chapter 3). Participants in this experiment were presented with a relatively complex story about a story about Native American warfare (1932). They were asked to recall the story at a variety of later points in time. The attempts by participants indicated that memory is a reconstructive process, rather than a literal one. People changed details within the story in their attempts to recall it, but fundamentally did get the gist of the story consistently correct. Bartlett referred to the gist of a memory as a schema. The term schema would later become standard in cognitive psychology in the late 20th century, when researchers began to categorize semantic memory. Bartlett was a rare early psychologist who was able to take a broad perspective on human behavior, while also using experimental methods (Roediger, 2000). The enduring contributions of his work on schemas and the reconstructive nature of memory are likely due to his appreciation that numerous factors enable us to solve problems and view the world and that the complexity of human cognition cannot be easily reduced to simple stimulus–response pairings. Indeed, few studies conducted today approach the everyday realism of Bartlett's work from that time period in the 1930s.

Edward Chace Tolman (Fig. 2.18) (1886–1959) is another important transitional figure between the behaviorist and cognitive eras. While Tolman conducted numerous learning experiments on rodents, he was of the opinion that classical conditioning and purely stimulus–response-based learning could not explain all of behavior. Tolman indicated that he believed that the Gestalt psychologists such as Kurt Koffka, with whom he studied, were making important contributions to

FIGURE 2.17 Sir Frederic Bartlett (1886–1969) was a famous British psychologist who is perhaps best known for his 1932 book entitled *Remembering: A Study in Experimental and Social Psychology*. *Courtesy of the Department of Psychology Archive, University of Cambridge.*

FIGURE 2.18 Edward Chace Tolman (1886–1959) was another important transitional figure between the behaviorist and cognitive eras. Tolman conducted learning studies and postulated that animals used mental representations of their environment. *By http://faculty.frostburg.edu/mbradley/psyography/edwardtolman.html [FAL], via Wikimedia Commons from Wikimedia Commons.*

psychology. Tolman wrote of "mental maps" and demonstrated through his experiments that rats can take shortcuts in maze running. Such behavior indicated to Tolman that the rat is capable of more than simply running through a learned pattern from memory, but rather acted as if it understood the location of a food reward, based on a map-like system of knowledge. He also indicated that rats could alter their behavior within a maze when their starting point had been changed. All of these findings contributed to his proposal that these animals had something analogous to a mental representation of the world in the form of a map. Later research would confirm that the hippocampus has spatially sensitive neurons, a finding that would later result in the awarding of a Nobel Prize for neurophysiologists John O'Keefe, Edvard Moser, and May-Britt Moser in 2014.

COGNITIVE PSYCHOLOGY OF REASONING

Cognitive Psychology and Reasoning Research (1950s to Present)

Trends in psychological research in the United States began to change with the experience of World War II. Amid great pressure to develop scientific technology for the war effort, American researchers were active contributors bringing about a stronger focus on information processing. This took the form of two dominant directions: the advancement of computer science and a renewed interest in studying the unobservable aspects of the mind that had been largely ignored in the United States during the era in which behaviorism was dominant. Ultimately, these changes led to the formation of cognitive psychology as a discipline that would change the type of research being conducted on human reasoning.

Computer science was emerging as an important discipline. This development was in part due to the need for computers in code breaking. The 1940s and 1950s saw an expansion of the capabilities for computing on a large scale due to both hardware development and the ability to program the large computing devices of the era. Over a relatively short period of time from the mid-1940s through the 1950s computers shrank in size and expanded in computing capability. The reduction in size was largely due to the changes in hardware. Early research computers such as the Colossus and the ENIAC relied upon vacuum tube technology to implement binary code. The vacuum tubes are similar to light bulbs and consist of a heated filament inside a glass surround. The vacuum tube was both unreliable and demanding of resources. The invention of transistors and then integrated circuits enabled computers to become much smaller, more computationally powerful, and cheaper to run (we discuss computer development further in Chapter 13). Meanwhile, computer scientists were taking advantage of the new developing technology to write and execute ever more complex and useful software code. Fundamentally, the computer science movement was based on advancing new ways to store, process, and utilize information. People could not help but take notice that these same operations appeared to apply similarly to the operations of the brain as we carry out cognitive processes.

Experimental psychologists were active in helping with the war effort as well. The work of Donald Broadbent is illustrative of the way that psychologists were able to help ensure that new technological developments operated at maximal efficiency. Broadbent (1958) was interested in the basic perceptual and processing mechanisms of cognition (Newell, 1985). Broadbent noted that humans were guided by feedback, or information that was provided to them by machines (Best, 1989). An example of this occurred within aircraft cockpits in which numerous gauges and controls were available to the pilot. By analyzing pilot errors in working with the devices, Broadbent was able to determine that some arrangements of the cockpit were superior to others. The work indicated that there was not a single type of design that could be learned equally well by all pilots, as would be implied by behaviorist learning. According to behaviorist principles, success in flight should have been equally likely regardless of the arrangement of the cockpit, as all learning could be reduced to stimulus–response acquisition. Instead it appeared to Broadbent that different arrangements of the controls would yield different outcomes with the human pilot. Broadbent noted that limitations of attention were likely responsible for the awkward ways that pilots engaged with certain control devices.

Soon a variety of experimental psychologists were beginning to study mental representations. Some of the earliest cognitive psychologists included George Sperling (1960) and Ulrich Neisser (1967) who examined a form of memory known as *sensory memory*. Sensory memory is thought to be separable depending upon sensory modality. Iconic memory is the visual form of sensory memory which has a relatively large capacity of as many as nine items, but which lasted for a very brief duration of under 1 s and is almost constantly being overwritten by new information being taken in by the visual system (Neisser, 1967). These investigators were also interested in the auditory equivalent of sensory memory, which was a fleeting aftereffect of an incoming sound that lasts approximately 4 s (Neisser, 1967).

George Miller was another early cognitive psychologist who studied the capacity of short-term memory. This is the type of memory which lasts for a relatively short time, but can be rehearsed in the mind. A classic example of short-term memory occurs when someone reads out a phone number to you, and you say it to yourself several times until you can type it into your phone.

Miller estimated the capacity of short-term memory to be roughly seven items give or take two. His paper on this subject bears the now famous title "The Magic Number Seven Plus or Minus Two" (Miller, 1956). While short-term storage also had a clear upper limit on capacity, it also appeared to be relatively fleeting provided that an individual did not rehearse the items regularly. Other psychologists were interested in the conditions under which the contents of short-term memory would be lost. Waugh and Norman (1965) using a digit-span paradigm demonstrated that memory loss from short-term storage occurred primarily due to interference from other items, rather than purely disuse or decay over time.

Other researchers were interested in higher cognitive functions including reasoning and language. One of the most famous examples of early cognitive psychology was the debate between eminent behaviorist B. F. Skinner and the young linguist Noam Chomsky. In this case, Skinner had published a book entitled "Verbal Behavior" (1957). Skinner's position was that language was able to be acquired using a classical conditioning method in which a small child looks to the parent or caregiver for reinforcement of each of his or her utterances. The adult would provide reinforcement for the correct behaviors by a head nod or look of approval. Skinner claimed full human language was possible through shaping by reinforcement over time. This position was widely refuted after Noam Chomsky provided a critical review of the book (Fig. 2.19). Chomsky would become well-known for advocating a nativist position on language that it would have to be learned in part by a biological propensity, which he referred to as a language acquisition device. Things had shifted solidly toward a cognitive approach to the study of psychology. Rather than emphasizing learning by context and reward, the field became much more interested in the nature of mental representations and the processes that we engage in to reason and decide.

The Information Processing Approach to Studying the Mind

The study of psychology would move toward understanding the features of mental representation and mental processing. These features have led many to describe the research following the cognitive revolution as focusing on an information processing approach. From this point forward theories were formed around the types of operations that people carry out and the characteristics of our mental processing.

The Atkinson-Shiffrin modal model of memory is an excellent example of the information processing approach in psychology. Richard Atkinson and Richard Shiffrin organized a viewpoint on memory types into a "boxes and arrows" model. The model specified that we have a *sensory store*, as had been studied by researchers such as Sperling (1960) and Neisser (1967). Information within the sensory store was commonly lost, as depicted in Fig. 2.20. Alternatively, the information could be passed on to another box in the model referred to as *short-term storage*. This box differed in quality, compared to the sensory store, as by this point the capacity and duration of storage was believed to vary in comparison to sensory storage. Lastly, a separate box for *long-term memory* represented all of the information that a person maintains indefinitely. This store could both accept new information from short-term storage and place older information back into short-term memory. This model captures the essential qualities of the information processing approach. Rather than describing memory as one unitary process, this model represents the position that multiple types of memory exist, that these forms of memory vary in their duration and capacity, and that information can move interactively from one to another within the set of modules. Other cognitive psychological theories also included these properties. Soon, other models began to

B. F. SKINNER NOAM CHOMSKY

FIGURE 2.19 B. F. Skinner and Noam Chomsky had a famous dispute in 1956 over how language was acquired. Skinner was a behaviorist and advocated for a reinforcement learning approach, while Chomsky claimed that children acquire language through biological guidance. *Professor B.F. Skinner, (1904–90). Photo: https://psychbehaviorism.wikispaces.com/CC By-SA. Noam Chomsky from wikispaces.com/CC By-SA/3.0.*

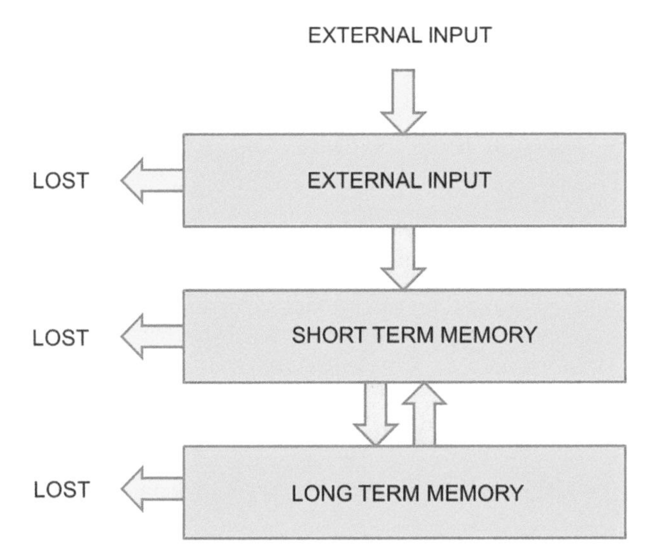

FIGURE 2.20 Atkinson-Shiffrin modal model of memory demonstrated many of the principles of the information processing approach to cognition.

be programmed formally into computerized simulations called *computational models*, which took the information processing approach to a degree to which actual information processing within a computer could be used to examine and simulate psychological processes.

Cognition and Reasoning

Alan Newell, J. C. Shaw, and Herbert Simon described some of the early influential cognitive psychology processes believed to be important in reasoning. In a comprehensive paper published in 1958 they emphasized several key components that are necessary for a theory of problem solving. This paper emerged from the authors' observation that no strong theory adequately explained how people solve problems. Their work centered upon the idea that the human mind is an information processing system and that it can be likened to the operations of a computer. Below is a quote from that original paper describing a set of properties that should be able to be specified in order to understand problem solving.

> What questions should a theory of problem solving answer? First, it should predict the performance of a problem solver handling specified tasks. It should explain how human problem solving takes place: what processes are used, and what mechanisms perform these processes. It should predict the incidental phenomena that accompany problem solving, and the relation of these to the problem-solving process… It should show how changes in the attendant conditions - both changes "inside" the problem solver and changes in the task confronting him - alter problem-solving behavior. It should explain how specific and general problem-solving skills are learned, and what it is that the problem solver "has" when he has learned them. [p. 151].

Notice the set of observations contained within this passage. The specification of "how" a problem is to be solved should occur. To get at the larger "how" question, the authors went on to specify that the processes and mechanisms need to be understood. Let's imagine this occurring as you grapple with the problem of entering your first floor apartment after having misplaced your key. The mechanisms and process imply an understanding of the physiology of the brain and what types of information processing will be accomplished along the way toward solving a problem. In our example, relevant internal factors may include your perception of the situation, being stuck outside of a locked apartment, memory in the form of what you know about the building, and perhaps what you know about the security of the building.

In addition to this neurobiological and cognitive specification, the authors indicate that "incidental phenomena" must be specified. In modern terms, we might call this the context under which the problem must be solved. For example, if it is the middle of the night and the neighbor is a light sleeper, then this may indicate that stealth is needed to avoid alerting the neighbor. Furthermore, imagine that it is raining in our example. Perhaps some greater urgency is needed to accomplish the goal and enter the apartment quickly.

The next specification indicates that a dynamic set of circumstances needs to be accounted for. This includes not only the changes that occur within the brain (learning or possibly changes in perception), but also the changes occurring within the situation itself. For example, if you recall that you may have left your bathroom window unlocked, but find that it appears to be jammed, then you have achieved two things: learning that your window is stuck and realizing that this is probably a dead end for the problem. Lastly, assuming that you realize you have a loose window pane that can be pried slightly in order to open the latch and enter the apartment, then you have learned something about yourself and how to approach this problem. You are likely to have more confidence that you can enter your apartment through some means other than the front door. Further, you may now realize that you will be able to accomplish this under other similar circumstances in the future. These features of information processing became the keys to a program known as the general problem solver (GPS).

The GPS idea became influential for two reasons. First, it presented a clear approach offering concrete guidelines and objectives. Second, Newell and Simon programmed an early computer to carry out the GPS in an attempt to duplicate human thinking and in doing so learn about human reasoning. The programmers could give the GPS program a description of the problem to be solved. This is somewhat different than what a human problem solver would do, as the computer did not have to perceive the problem or specify the goal state and context. This simplified the task of the GPS dramatically. Next, the program would carry out information processing by applying some heuristics and a decision tree could be formed with points for evaluation. The program could implement a *means-ends analysis*, a method that the program could employ to reduce the difference between the current state and end state. This program added depth to the theory and allowed these psychologists to demonstrate several important features of problem solving. Among these features was a correct representation of the problem space, the limitations of our search capabilities, and that the limitations of human problem solving can directly stem from our limitations in formulating the correct representation of the problem (Best, 1989).

Another influential early cognitive psychology approach to reasoning was occurring in the domain of deduction. Philip Johnson-Laird and Peter Wason were interested in how we go about applying rules to achieve outcomes. Wason and Johnson-Laird were studying how deductive arguments are processed. This work followed from the earlier investigations of researchers including Wilkins, Sells, and Woodworth. Johnson-Laird proposed that we use *mental models* to solve such problems. For Johnson-Laird, mental models are representations of the world that can include both visuo-spatial information stored in an analog code, as well as propositional knowledge that is stored in the form of words or symbols. Johnson-Laird et al. claimed that mental models underlie reasoning performance and rely on general knowledge and other relevant representations that move beyond what is explicitly asserted in a set of premises (Johnson-Laird, 1983). In an example representing this approach, Ehrlich, Mani, and Johnson-Laird (1979) provided participants with a set of information involving spatial relations among objects as follows:

The knife is in front of the spoon.
The spoon is on the left of the glass.
The glass is behind the dish.

This set of propositions can be linked (in the following order: knife, spoon, glass, dish). A mental model can be constructed additively in this case, unlike an alternative version in which the order was scrambled such as:

The glass is behind the dish.
The knife is in front of the spoon.
The spoon is on the left of the glass.

Participants were asked to draw the appropriate spatial layout of these propositional descriptions. Participants with this later set of scrambled propositions presented out of order would have to construct two mental models and then combine them after receiving the third proposition. Sixty-nine percent of participants' drawings based on continuous premises were correct compared to 42% of the drawings based on the scrambled order of descriptions. This result indicates that people build mental representations of space within their minds in order to solve problems and that the formation of these models can be facilitated or hindered based on the manner in which information is presented.

A Cognitive Approach to Human Judgment and Decision Making

The psychology of judgment and decision making is a multidisciplinary field that seeks to understand how and why people decide the way that they do. Philosophers, economists, legal scholars, and psychologists have actively studied the nature of our decisions. The study of decision making flourished within the framework of cognitive psychology. Two of the most notable psychologists active in the study of how people make decisions in their daily lives were Daniel Kahneman (born 1934) and Amos Tversky (1937–1996) (Fig. 2.21). Kahneman and Tversky approached the study of judgment and decision making from the perspective that people were not always rational decision makers. This was a challenge for the field of behavioral

FIGURE 2.21 Daniel Kahneman and Amos Tversky are highly influential researchers in the area of judgment and decision making. Amos Tversky and Daniel Kahneman toast to their partnership in the 1970s. *From Vanity Fair, Courtesy of Barbara Tversky.*

economics, as actual behavior of people out in the world is nonrational. People make errors, decide inconsistently from one context to another, and often respond differently when information is varied or deviates from their prior beliefs. These two researchers revolutionized the study of decision making by demonstrating that people are both nonrational and tend to act differently under different circumstances.

One of Tversky and Kahneman's most notable contributions was the development of Prospect Theory (1981). This served as a general model of how people approach decisions with regard to their perceived gain or loss. Notably, Kahneman and Tversky had established that people tend to behave in a more risky fashion when they are faced with the possibility of a loss compared to when they are faced with the possibility of a gain through extensive experimentation. The gain can be equivalent to the loss, but for most people the loss will be felt with a greater psychological magnitude than the gain. In accordance with this theory, people will take greater risks in order to avoid a loss than they will in order to obtain a gain. This theory also accounts for the finding that people demonstrate an endowment effect, in which they will insist upon a higher selling price for an item that they own than an equivalent item that they are attempting to attain (Kahneman, Knetsch, & Thaler, 1990).

Kahneman and Tversky's work on decision making centered upon mental representations. These researchers demonstrated that contexts matter, as they will influence the way in which a person sees his or her options. This emphasis on mental representations was key to establishing a variety of rules of thumb, referred to as heuristics, by which people see the world and operate on information. The fields of economics, law, and ultimately neuroscience have been dramatically influenced by this approach. This work led to the Nobel Prize for Daniel Kahneman in 2002. Amos Tversky died at the relatively young age of 59 and therefore could not share in this prize, which Kahneman shared with economist Vernon Smith.

THE EVOLUTION OF COGNITIVE MODELING

Modeling the Cognitive Processes Involved in Reasoning (1980s–Present)

Computational modeling is an important movement in the evolution of cognitive psychology. Modeling represents the ultimate convergence between the information processing approach, as applied to the mind and that same approach instantiated within machines. Early computational models of reasoning include the general problem solver (GPS) of Alan Newell and Herbert Simon. Experimental psychologists have employed cognitive modeling as a method for furthering our understanding of human cognition for over five decades. Some of the best instances of the approach combine computational models with human experiments that can test the predictions made by the model. This approach bears strong resemblance to artificial intelligence (AI), an endeavor in which people attempt to build a computer that captures and carries out aspects of intelligent behavior. The major difference between computational models of cognition and AI models is that computational cognitive models are usually intended to simulate aspects of the human nervous system and are therefore constrained by biological characteristics of the human brain. AI models, by contrast, can at times exceed the capabilities of biological cognition, or carry out operations in a completely different manner.

Parallel Distributed Processing

One of the most influential approaches employed within cognitive modeling is known as parallel distributed processing (PDP). PDP modeling originated with simulations by David Rumelhart and James McClelland (1986). These researchers wanted to build computational models of cognition that aligned with the biology of the mind. They set out a series of principles by which PDP models should operate. One of the most important of these principles was the dense interconnectivity of processing nodes within the models. This interconnectivity is intended to simulate the axons, which relay neural impulses within biological neural systems. These critical connections have also led PDP models to be referred to as *connectionist models*.

The principles of connectionist models are listed below. These models can be built to simulate a variety of processes within cognition. The principles align with the idea that simple binary on/off processing units can be built up into organized networks that can then carry out sophisticated processing such as that observed in human thinking and reasoning. These principles align with those described years earlier by Oliver Selfridge's (1959) *Pandemonium* architecture. An advance by McClelland and Rumelhart was developing the principles into a programming method that could be used to simulate numerous areas of cognition. PDP models include the following elements:

1. Primitive processors with an on/off binary code
2. Processors are stacked in layers
3. The layers connect via inhibitory and excitatory connections
4. Energy is transmitted through the connections
5. Iterations can simulate information processing

Note that each of these aspects simulates a property of the nervous system. The on/off primitive processors correspond to the action potentials, or spike trains of neurons that activate in an "all-or-none" fashion. The stacked layers correspond to topographic functional units of cortical neurons that carry out a similar aspect of processing. The connections simulate the axons of a nervous system, while energy simulates the movement of information transmitted through the connections of neurons. Lastly, the modification of the network strength by iterative tuning simulates the principle that neural connections can be modified by experience.

Fig. 2.22 presents a diagram of a PDP connectionist architecture. In this case, the layers simulate the visual processing needed in order to recognize words. The connections among the layers enable more complex behavior to be derived from the system of simple primitive processors. PDP models were prevalent in cognitive psychology in the 1980s and 1990s. By the early 2000s the emphasis had shifted to include neuroimaging experiments to test aspects of connectionist models and PDP simulations of how a system degrades after brain injury.

Hybrid Models

An excellent example of a connectionist inspired model of thinking and reasoning is the Learning and Inference with Schemas and Analogies (LISA) model by Hummel and Keith Holyoak (1997). The connectionist principles can be highly effective for building up toward complex representations such as words or concepts. Reasoning involves building those complex representations into even more complicated entities and matching patterns at a hierarchical scale. The LISA model uses a connectionist style architecture to simulate knowledge at the level of simple semantic information, sub-propositional (language based) representations, and propositional representations (of the sort used in analogies and schemas). The combination of propositional representations (made up of words) and the use of connectionist principles enabled this model to simulate basic results in human cognition using a modeling process that was compatible with features of the human nervous system (Krawczyk, Holyoak, & Hummel, 2004).

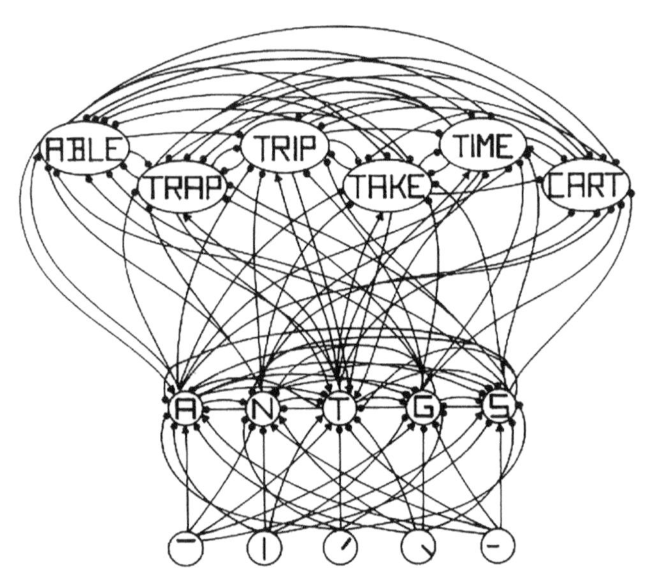

FIGURE 2.22 A schematic example of a PDP connectionist model. In the model, lines are represented in one layer. These pass activation up to a letter recognition layer. Lastly, a word recognition layer receives activation from the relevant letters one layer down. *Rumelhart, D. E., McClelland, J. L., & PDP Research Group (Eds.) (1986). Parallel distributed processing (vol. 1). Cambridge, MA: M.I.T. Press, A Bradford Book. Psychological Review, 88(5), 375–407. Reprinted (or adapted) with permission.*

DEVELOPMENTAL PSYCHOLOGY OF REASONING

Studies of Infant Cognition

There are important contributions to the study of reasoning that can be traced to early infancy and childhood. The work of Jean Piaget (1986–1980) is particularly noteworthy in this area. Piaget will be discussed in much more detail in Chapter 5 when we cover the development of reasoning. For Piaget, the child was seen as an active learner. Piaget specified a series of cognitive stages that children pass through from the preschool years through age 12. These stages traced the competencies of children as they aged. The child was originally seen as being limited to sensory and motor processing in infancy through age two. By the early preschool period Piaget claimed that children begin to exhibit reasoning about the world around them, but that they are limited to a self-centered viewpoint. As the elementary school years begin, Piaget argued that children move toward a phase in which they are capable of more abstract and symbolic thinking. Lastly, Piaget claimed children moved toward a period of formalized cognitive operations. This period includes the development of skills that enable young people to test hypotheses about the world around them and carry out principled deductive and inductive reasoning.

Piaget worked on many notable features of development. He explained the hypothesis that children move away from egocentric thinking as they age. Another major contribution by Piaget was the idea that children move from perceptual to abstract representations as they age. The stages proposed by Piaget are somewhat approximate and much work has been carried out in the subsequent decades testing the predictions of Piaget's developmental stages. Piaget had a long and productive career testing his ideas experimentally, and he also mentored numerous individuals who would go on to become leading developmental psychologists themselves. For several decades researchers have continued to test Piaget's ideas. Many of his proposals about child development have held up under decades of experimental scrutiny. His influence on the cognitive development of thinking and reasoning is far-reaching.

NEUROSCIENCE OF REASONING

The Emergence of Cognitive Neuroscience (1990–Present)

The formation of the field of cognitive neuroscience has led to new perspectives on the study of reasoning that were primarily driven by the availability of new methods. The term *cognitive neuroscience* was coined around 1980 by Michael Gazzaniga, a neuroscientist, and cognitive psychologist George Miller who was known for his work on memory. Gazzaniga is widely known for the study of the cognitive effects of split-brain surgery, in which the two hemispheres of the brain are disconnected in order to control epilepsy. The split-brain surgery afforded researchers an experimental preparation in which they could examine the largely independent contributions of the two cerebral hemispheres. We will discuss the work of Gazzaniga on split-brain patients in greater detail in Chapter 8.

The development of neuroimaging methods enabled new ways to study a range of cognitive phenomena including reasoning, decision making, social cognition, and deduction. The first functional imaging experiments were conducted using a technique called positron emission tomography (PET). PET scanning involves the injection of a radioactive tracer into a person's bloodstream. The participant is then placed within a scanner that is able to detect the presence of the tracer in the brain. As neurons become active, there is an associated increase in glucose metabolism that results in greater amounts of tracer being taken up into different areas of the brain. Attention psychologist and neuroscientist John Duncan et al. (2000) published an intriguing functional neuroimaging study of reasoning and intelligence. In the study, Duncan et al. had presented participants with problems associated with high degrees of fluid reasoning engagement, resulting in lower accuracy and higher response times, and ones that were simpler pattern matching problems. These researchers were able to isolate a larger degree of brain activity in the left middle frontal gyrus associated with more effortful reasoning performance. PET enabled researchers to begin to map the cognitive processes involved in reasoning and other cognitive phenomena to their associated brain areas. This is an ongoing effort that continues to become more sophisticated as new methodologies emerge. We will discuss this technique in further detail in Chapter 3.

By the middle of the 1990s magnetic resonance imaging (MRI) had largely replaced PET as a preferred method for functional imaging. Functional MRI (fMRI) scanning allowed researchers to carry out experiments that addressed locating the cortical areas that are associated with behavior and mental processing. For the study of reasoning, the technology was initially useful for specifying the neural correlates of executive functions in working memory (D'Esposito et al., 1995). In the late 1990s researchers began to apply neuroimaging toward understanding fluid reasoning abilities and the neural basis of intelligence. This is one of the major topics discussed in Chapter 3. Functional MRI enabled investigators to better determine what regions were most associated with different forms of reasoning. Soon researchers were able to carry out neuroimaging studies to compare different forms of reasoning including deductive and inductive reasoning (Goel & Dolan, 2000), analogical reasoning (Wharton et al., 2000), relational reasoning (Prabhakaran, Smith, Desmond, Glover, & Gabrieli, 1997), and spatial reasoning (Vartanian & Goel, 2005). The capability of neuroimaging was obvious and the technology proved useful in adding another form of analysis toward categorizing and understanding the processes involved in reasoning.

The limitations of neuroimaging are also evident after over two decades of functional MRI studies of reasoning. While fMRI allows a reasonably fast sampling rate, it relies largely upon blood oxygen changes for its marker of neural activity. Blood travels slower (on the order of seconds) than neurons can produce action potentials (on the order of milliseconds), thus there is a delay in the response relative to the actual speed of neural transmission. Additionally, fMRI studies provide limited access to the mental representations that are being processed within a cortical area. We still do not understand the neural codes that are being transmitted to other brain areas when an area of the brain is active in a neuroimaging study. Thus, the fundamental representational units of a reasoning task are still largely unspecified. Additionally, neuroimaging tends to lead to localizationist theories, as some regions of interest appear as active in a particular task. The emerging picture facilitated by fMRI studies is that widespread networks are involved in complex cognitive tasks such as reasoning and deciding. While fMRI can also reflect the activity of networks associated with reasoning (Shokri-Kojori, Motes, Rypma, & Krawczyk, 2012), the precise way that regions of interest interact with larger neural networks remains a major target of current and future investigation.

SUMMARY

The study of reasoning has a long and fascinating history. We have examined many of the approaches that have led to our current understanding of this field. In summary, there was remarkable progress made in the 20th century. This moved the field from a philosophical area of inquiry in which the only experiments were primarily involving intuition and introspection about one's mental processes. The methods of experimental psychology enabled researchers to examine the role of context in reasoning accuracy and speed. Additionally, these methods enabled researchers to find consistencies across people forming principles of reasoning.

Psychology continues to participate strongly in the multidisciplinary field of reasoning research. As we will see in the coming chapters, important contributions to reasoning research are being made from neuroscience, comparative psychology, cognitive development, aging research, neurology, psychiatry, and behavioral economics. Social psychologists, as well as technological innovators, including game developers, computer modelers, and software developers are playing increasingly large roles in this research area. Reasoning has come a long way and has a bright and exciting future, thanks in large part to the continuing development of methods and technological advances.

END-OF-CHAPTER THOUGHT QUESTIONS

1. Psychological inquiry made remarkable progress in the 20th century. Had philosophers begun to use experimental methods earlier on, how much progress do you think they would have made prior to advances in physiology?
2. This field moved from being primarily a philosophical area of inquiry to a psychological one. Is there still a strong role for philosophy in the study of reasoning? How are these disciplines related?
3. The methods of experimental psychology enabled researchers to examine the role of context in reasoning accuracy and speed. How far can the field of reasoning get based only upon these behavioral methods? What other data may be important?
4. Early studies of reasoning were carried out in Europe by the Gestalt school of psychology. These researchers focused on the role of insight in problem solving. How much progress have we made in understanding insight since those early times?
5. In the first half of the 20th century behaviorism dominated psychological inquiry in the United States. How would you study reasoning using behaviorist methods of stimulus–response observation?
6. Early studies of deductive reasoning were carried out in the early 20th century. Why do you think these studies managed to emerge despite moving strongly against the restrictions of behaviorism?
7. Modeling cognitive phenomena requires a sensitivity to both the architecture of the nervous system and the interactivity of information within the mind. How do cognitive models relate to the advancing field of AI?
8. Technological advances in brain imaging opened up a wide array of new avenues for linking neuroscience principles with reasoning research. What are some of the limitations of MRI as a tool for studying the nervous system?

References

Baddeley, A. D. (2000). The episodic buffer: A new component of working memory? *Trends in Cognitive Sciences*, *4*, 417–423.

Bartlett, F. C. (1932). *Remembering: A study in experimental and social psychology*. Cambridge: University Press.

Bechara, A., Tranel, D., Damasio, H., & Damasio, A. R. (1996). Failure to respond autonomically to anticipated future outcomes following damage to prefrontal cortex. *Cerebral Cortex*, *6*, 215–225.

Bergman, E. T., & Roediger, H. L., III (1999). Can Bartlett's repeated reproduction experiments be replicated? *Memory & Cognition*, *27*, 937–947.

Best, J. B. (1989). *Cognitive psychology*. St. Paul, MN: Wadsworth Publishing Company.

Broadbent, D. E. (1958). *Perception and communication*. New York: Oxford University Press.

Broca, P. (1861). Remarques sur le siège de la faculté du langage articulé, suivies d'une observation d'aphémie (perte de la parole). *Bulletins de la Société d'anatomie (Paris), 2e serie*, *6*, 330–357.

Broca, P. (1865). Sur le siege de la faculte du langage articule. *Bulletin de la Societe d'anthropologie*, *6*, 337–393.

D'Esposito, M., Detre, J. A., Alsop, D. C., Shin, R. K., Atlas, S., & Grossman, M. (1995). The neural basis of the central executive system of working memory. *Nature*, *378*, 279–281.

Dronkers, N. F., Plaisant, O., Iba-Zizen, M. T., & Cabanis, E. A. (2007). Paul Broca's historic cases: High resolution MR imaging of the brains of Leborgne and Lelong. *Brain*, *130*, 1432–1441. http://dx.doi.org/10.1093/brain/awm042.

Duncan, J., Seitz, R. J., Kolodny, J., Bor, D., Herzog, H., Ahmed, A., et al. (2000). A neural basis for general intelligence. *Science*, *289*, 457–460.

Duncker, K. (1945). On problem solving. *Psychological Monographs*, *58* American Psychological Association.

Ehrlich, K., Mani, K., & Johnson-Laird, P. N. (1979). *Mental models of spatial relations*. University of Sussex: Laboratory of Experimental Psychology.

Flynn, J. R. (1987). Massive IQ gains in 14 nations: What IQ tests really measure. *Psychological Bulletin*, *101*, 171–191. http://dx.doi.org/10.1037/0033-2909.101.2.171.

Goel, V., & Dolan, R. J. (2000). Anatomical segregation of component processes in an inductive inference task. *Journal of Cognitive Neuroscience*, *12*, 110–119.

Haberlandt, K. (1994). *Cognitive psychology*. Needham Heights, NJ: Allyn & Bacon.

Hummel, J. E., & Holyoak, K. J. (1997). Distributed representations of structure: A theory of analogical access and mapping. *Psychological Review*, *104*, 427–466.

James, W. (1890). *The principles of psychology*. New York: H. Holt and Company.

Janis, I. L., & Frick, F. (1943). The relationship between attitudes toward conclusions and errors in judging logical validity of syllogisms. *Journal of Experimental Psychology, 33*, 73–77.

Johnson-Laird, P. N. (1983). *Mental models: Towards a cognitive science of language, inference, and consciousness*. Cambridge, MA: Harvard University Press.

Kahneman, D., Knetsch, J. L., & Thaler, R. H. (1990). Experimental tests of the endowment effect and the Coase theorem. *Journal of Political Economy, 98* (6), 1325–1348.

Krawczyk, D. C., Holyoak, K. J., & Hummel, J. E. (2004). Structural constraints and object similarity in analogical mapping and inference. *Thinking and Reasoning, 10*, 85–104.

Lovejoy, C. O. (1981). The origin of man. *Science, 211*, 341–350.

Miller, G. (1956). The magical number seven, plus or minus two: Some limits on our capacity for processing information. *Psychological Review, 63*, 81–97.

Neisser, U. (1967). *Cognitive psychology*. New York: Appleton-Century-Crofts.

Newell, A. (1980). *Duncker on thinking: An inquiry into progress in cognition*.

Newell, A. (1985). Duncker on thinking: An inquiry into progress in cognition. In S. Koch, & D. E. Leary (Eds.), *A Century of Psychology as Science* (pp. 392–419). New York: McGraw–Hill.

Newell, A., Shaw, J. C., & Simon, H. A. (1958). Elements of a theory of human problem solving. *Psychological Review, 65*, 151–166.

Nickerson, R. S. (1990). William James on reasoning. *Psychological Science, 1*, 167–171.

O'Driscoll, K., & Leach, J. P. (1998). "No longer Gage": An iron-bar through the head. Early observations of personality change after injury to the prefrontal cortex. *British Medical Journal, 317*, 1673–1674.

Popper, K. R. (1959). *The logic of scientific discovery*. London, UK: Hutchinson.

Prabhakaran, V., Smith, J. A., Desmond, J. E., Glover, G. H., & Gabrieli, J. D. (1997). Neural substrates of fluid reasoning: An fMRI study of neocortical activation during performance of the Raven's progressive matrices test. *Cognitive Psychology, 33*, 43–63.

Roediger, H. L. (2000). Sir Frederic Charles Bartlett: Experimental and applied psychologist. In G. A. Kimble, & M. Wertheimer (Eds.), *Portraits of pioneers in psychology* (Vol. IV) (pp. 149–161). Mahwah, NJ: Erlbaum.

Rumelhart, D. E., & McClelland, J. L. (1986). *Parallel distributed processing: Explorations in the microstructure of cognition. Volume 1: Foundations*. Cambridge, MA: MIT Press.

Selfridge, O. G. (1959). Pandemonium: A paradigm for learning. In *Symposium on the mechanisation of thought processes*. London, England: Her Majesty's Stationery Office.

Sharps, M. J., & Wertheimer, M. (2000). Gestalt perspectives on cognitive science and on experimental psychology. *Review of General Psychology, 4*, 315–336.

Shokri-Kojori, E., Motes, M. A., Rypma, B., & Krawczyk, D. C. (2012). The network architecture of cortical processing in visuo-spatial reasoning. *Scientific Reports, 2*, 411.

Skinner, B. F. (1957). *Verbal behavior*. New York: Appleton-Century-Crofts.

Sperling, G. (1960). The information available in brief visual presentations. *Psychological Monographs: General and Applied, 74*, 1–29.

Strickberger, M. (1996). *Evolution* (2nd ed.). Burlington, MA: Jones & Bartlett Publishers.

Tversky, A., & Kahneman, D. (1981). The framing of decisions and the psychology of choice. *Science, 21*, 453–458.

Vartanian, O., & Goel, V. (2005). Right ventral lateral prefrontal cortex mediates hypothesis generation in an unconstrained anagram task. *NeuroImage, 27*, 927–933.

Wallas, G. (1926). *The art of thought*. New York: Harcourt Brace Jovanovic.

Waugh, N. C., & Norman, D. A. (1965). Primary memory. *Psychological Review, 72*, 89–104.

Wharton, C. M., Grafman, J., Flitman, S. S., Hansen, E. K., Brauner, J., Marks, A., et al. (2000). Toward neuroanatomical models of analogy: A positron emission tomography study of analogical mapping. *Cognitive Psychology, 40*, 173–197.

Wilkins, M. C. (1928). The effect of changed material on ability to do formal syllogistic reasoning. *Archives of Psychology, 102*, 1–83.

Woodworth, R. S., & Sells, S. B. (1935). An atmosphere effect in formal syllogistic reasoning. *Journal of Experimental Psychology, 18*, 451–460.

Further Reading

Johnson-Laird, P. N. (1979). Will there be any neat solutions to small problems in cognitive science? *Cognitive Science, 3*, 173–176.

CHAPTER

3

The Neuroscience of Reasoning

KEY THEMES

- Species that possess large amounts of cortex and especially association cortex tend to show advanced reasoning skills.
- The relationship of large brains to complex behaviors may be driven by whether an organism is a predator. Predators need a wide array of behaviors in order to locate and subdue prey.
- Functional neuroimaging involves scanning the brains of participants as they carry out reasoning tasks. This results in the ability to map brain regions that support reasoning.
- Many of the early neuroimaging studies of reasoning suggested that the prefrontal cortex (PFC) was particularly important.
- Relational reasoning and analogical reasoning involve areas of the left PFC. This finding has been indicated by both neuroimaging and electrophysiological studies.
- Neural network studies indicate that the frontal lobes may serve a coordinating or control function. This area may integrate wide-scale activity across the brain, rather than operating as a relational module.
- The materials used in an experiment have a large impact on the results of reasoning studies. Materials invoke our semantic memories, which are supported by temporal lobe regions.
- The integrated sets of information that we can use in reasoning are called schemas or scripts.
- Brain network interconnectivity looks to be an especially promising area toward capturing the complexity of neural processing in reasoning and may further clarify some of the roles of specific brain areas that have been linked to reasoning in various forms.

Reasoning
http://dx.doi.org/10.1016/B978-0-12-809285-9.00003-X

OUTLINE

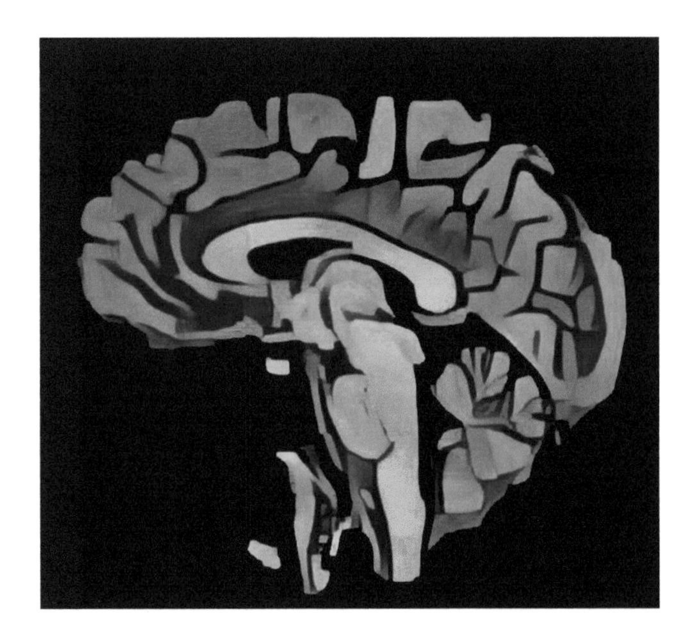

THE NEUROSCIENCE OF REASONING

Introduction

The quest to understand the neural basis of reasoning is a relatively new and complicated endeavor. Researchers working within this subfield of the discipline wish to determine how the cognitive operations of reasoning are carried out within the brain. Achieving this goal would provide some much needed clarity about what categories of reasoning are sufficiently different that they have clear biological differences and rely upon different brain systems. More importantly, a better understanding of how the brain carries out reasoning could guide medical practitioners toward targeting key processes that become impaired in brain injury and dementia patients. A greater understanding of the brain mechanisms for reasoning could also benefit educators as they strive to provide the most fertile learning environments possible at the most developmentally appropriate time.

The history of philosophical inquiry into the nature of reasoning has frequently touched upon this very topic. In the 1600s Rene Descartes famously grappled with mind-body questions. Descartes maintained that reason was separate from the body and that it was the perfect product of the mind. His premises led to the conclusion that the mind is not constrained by biology and that we do not benefit from biological investigations of the mind. By the late 1800s Charles Darwin articulated a different position suggesting that human reasoning and mental abilities exist on a continuum with the abilities of other species and that biological traits have very much to do with reasoning abilities (Fig. 3.1). While this was a highly controversial position in the late 1800s, science would gradually begin to widely adopt this view, which paved the way for educated societies to view other species in a more ethically humane way. Through the 1900s many academic disciplines expanded dramatically including the fields of biology, neuroscience, psychology, neurology, psychiatry, physiology, and anthropology. This expansion has enabled us to begin to integrate across many of these disciplines by gathering evidence

FIGURE 3.1 French philosopher Rene Descartes (1596–1650) and British naturalist Charles Darwin (1809–1882) articulated opposing positions separated by several hundred years regarding the nature of the relationship between the mind and body. *From Wikimedia Commons.*

from a variety of techniques from several fields of study. This integrative process is helping us to define the ability to reason and develop plans to facilitate reasoning skills.

One of the most practical applications for a neuroscience-based understanding of reasoning is applying that knowledge toward understanding disorders of thought and reason. We hope to provide better behavioral and biological markers with the potential to help diagnose and treat brain injuries and neurological conditions. This is an ambitious goal, but one that the modern cross-disciplinary emphasis may one day make possible. My own career has been very much shaped by the convergence of numerous subdisciplines on the topic of neuroscience and cognition. I was trained in a psychology department studying cognition and neuroscience. I then completed my training at both a psychology department and an academic neuroscience center. I am currently a faculty member in a school of behavioral and brain sciences and a psychiatry department. I am also affiliated with a university neuroscience center, as well as a medical brain imaging research center. This lack of organizational boundaries can at times be complicated, but ultimately offers excellent opportunities to study reasoning using a variety of tools and methods in the contexts of experimental psychology and neuroscience. The influence of numerous brain-related disciplines interacting may enable us to help patients to regain reasoning abilities through targeted medical and rehabilitative interventions.

A good place to begin our inquiry into the neuroscience of reasoning is considering the differences between people and animals in reasoning ability. While we will focus more deeply on this topic in Chapter 4, it is important to make some essential points before we begin. When we consider humans in the context of other species the definition of reasoning begins to take shape and possibly change. This comparison forces us to clarify what type of behaviors should be considered *reasoning* and opens up a variety of interesting questions. Is it reasoning when birds flock together and migrate? Is it reasoning when an ant colony works as a group to find food and build a home? Is it reasoning when a trout deftly eats your bait while avoiding getting caught on your fishing hook? Are these instead instances of *learning*? Might some of these behaviors be driven by *instinct* due to the strong genetic guidance? Alternatively, must an organism be creative in a behavior in order for it to count as reasoning, or can reasoning be merely an elaborate form of imitation drawing from the abilities of one's family members? To answer these types of questions, we must consider how the behavior is represented at the level of the brain. This leads us to a separate, but related issue regarding the role of brain size, capacity, and the characteristics of an organism's nervous system. These critical features are intertwined with the reasoning abilities of an organism.

Differences in Brain Capacity

Both the size and characteristics of a nervous system are important determinants of reasoning abilities. As a general principle, larger brains usually lead to more complex behavior, and some of this complex behavior might be defined as reasoning. A challenge we quickly run up against is arriving at a precise definition of reasoning in other species. Reasoning is clearly a human construct that we have applied to several of our own abilities, but how should we define the abilities of animals? We will revisit this topic in the next chapter when we cover comparative reasoning abilities in depth, but for now it may be sufficient to acknowledge some broad trends about the links between reasoning and the brain.

Overall brain size is an important indicator of both brain complexity and behavioral complexity. Many of the largest animals on Earth have large brains. Many of these are mammals that have relatively complex brains with pronounced anatomical and physiological similarities that can be observed across species. We can also observe this trend based on evidence from extinct species. Several species of dinosaurs were enormous by modern day standards, but their heads were small relative to the size of their bodies. Estimates of brain size in some of these species indicate that they had very small brains compared to their body size. The 30 ft long 3.5-ton stegosaurus famously had a brain about the size of a walnut, while the much smaller Troodon had a relatively large brain with a larger proportion of cerebrum, the area comprised of the cortex (Fig. 3.2). The behavioral traits of these animals can only be inferred from the fossil record, but paleontologists suggest that the Troodon was a fast hunter with stereoscopic vision, while the stegosaurus was likely a slower moving grazing animal. Large modern animals can be compared in terms of brain size as well, with whales, dolphins, and elephants emerging as some of the most intelligent, large-bodied, large-brained species. All of these features tend to result in more complex behaviors.

Another strong predictor of reasoning ability and brain size is whether an animal is a herbivore or predator. Herbivores include many large animals such as giraffes, buffalo, bison, moose, elk, and zebras. Despite the large body size of these animals, none of them have especially large brains. Meanwhile, carnivores such as cheetahs, bobcats, and hyenas are much smaller animals that have proportionally large brains. Not only do the predators tend to have larger brains, they are often capable of a richer set of diverse behaviors compared to grazing herbivores. Perhaps these differences can be attributed to the ways these animals obtain food. The daily activities of herbivores include eating, migrating, sleeping, and monitoring for the possibility of danger. If no danger presents itself, they have a relatively peaceful day. By contrast, predators must identify, capture, and subdue prey animals by overcoming their defenses. Prey animals may also be spread over vast territory and may be hidden. These additional pressures likely drive the evolution of enhanced brain capacity in predators.

The proportion of cortical mass to brain size is strongly linked to reasoning abilities. Humans have a remarkably large proportion of cortex relative to body size. Similarly, large cortical mass is a prominent feature of cats, dogs, and many other predatory species. Predators tend to have binocular vision as well, which enables them to better estimate depth cues in the environment. Such abilities are made possible by complex nervous systems that have the processing capacity necessary to model an accurate representation of the visual landscape. Similarly, dolphins and killer whales use sound to augment their sensory representations of the oceanic world in which they live. Predators that must track and pursue single individual prey animals often require sophisticated sensory abilities, and these sensory capacities tend to accompany complex brains. It is also important to note that many species of birds are capable of behaviors that many would classify as involving reasoning skills. Such behaviors include the use of tools, a level of sensitivity to the context, and the ability to process abstract relations among objects. Birds have physically small brains compared to large mammals, but their brain size is proportionally large relative to their small bodies. The proportional size

FIGURE 3.2 We can infer the complexity of behavior in extinct species by considering clues in the fossil record. The troodon was a dinosaur and likely a clever hunter with a relatively large brain. By contrast the stegosaur was a relatively large animal that had a very small brain. *By Reid, Iain James (Drawn by hand) [CC BY-SA 3.0 (http://creativecommons.org/licenses/by-sa/3.0), via Wikimedia Commons.*

between the brain and the body is often a better indicator of how sophisticated the reasoning abilities and complex behaviors of an organism are.

ANATOMICAL CONSIDERATIONS

The Importance of Association Cortex for Higher Cognitive Functions

The cortex, or outer surface of the brain, contains the cell bodies of neurons, which are the primary computational units within the brain. The overall amount of cortex is often related to the complexity of behavior that we observe in a given species. More processing power made possible by greater amounts of cortical tissue could be an important feature of the brain that relates to the complexity of reasoning abilities; however, there are likely some other key biological factors that bear on an individual's ability to reason.

Cortical areas rarely act as isolated modules or units. Instead regions of the cortex tend to communicate in networks. Primate memory physiologist Joaquin Fuster has referred to these cortical networks as *cognits* (Fuster, 2006). The idea that cortical areas are critically interconnected becomes apparent when one considers the structure and functions of single neurons. Most neurons will have numerous dendrites, which are the projections that enable them to receive chemical or electrical information from other neurons. Neurons also typically have axons, which enable the flow of electrochemical reactivity across space within the brain. Lastly, the terminal buttons projecting off of the axon allow neurons to communicate with one another. Thus, it should not be surprising that the types of connections possessed by neurons or cortical areas will lead to large-scale network interactions and to interactions of more localized circuits (Fig. 3.3). It is likely that those network interactions that can span across the hemispheres, from front-to-back, or up-to-down, are especially important for integrating information needed for complex reasoning skills.

In addition to the mass and connectivity of the cortex, the structure of the brain is also important to consider. Some cortical areas such as the primary visual cortex, located in the occipital lobe, receive relatively unprocessed information from the retinas. The neurons within this area of the brain are tuned for detecting edges, features, and differences in lighting. A similar arrangement is found within Heschl's gyrus on the top of the temporal lobes, which is the primary cortical area associated with processing sound. The neurons here, like those in the visual cortex, are finely tuned to respond to different incoming sound frequencies transduced into electrochemical information by the inner ear. The primary motor cortex, which lies just forward of the central sulcus, contains neurons that directly link to the outgoing axons that extend out to the periphery of the body. These neurons are predominantly tuned for sending out motor signals. These primary sensory and motor areas are not limited only to these functions, but a large percentage of the neurons in those areas are heavily involved in sensory and motor functions. These areas may participate in activities that we would call reasoning, but they probably play limited roles bound to information either entering or exiting the brain having to do with the format or modality of the reasoning problem.

Some of the important neurons that are involved in reasoning are those that comprise the *heteromodal cortex*. This term is used to describe the cortex that does not serve one particular function, but rather participates in

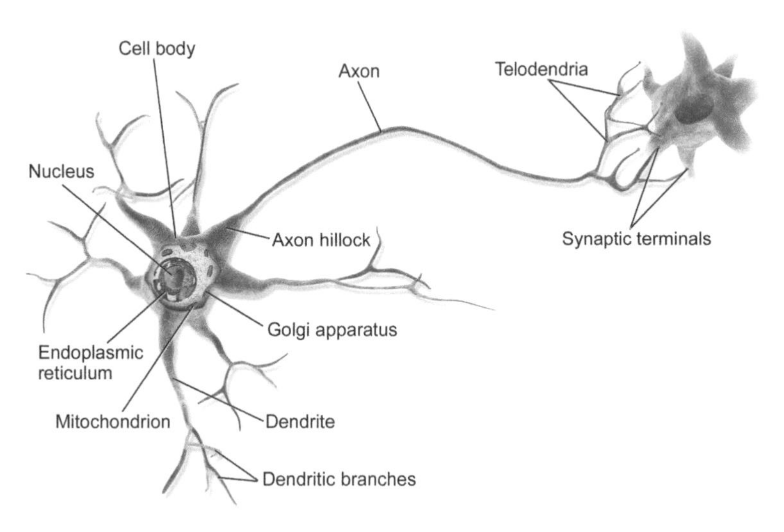

FIGURE 3.3 The structure of a neuron includes the axon, dendrites, cell body, and synaptic terminals. The dendrites, axons and terminal buttons enable neurons to connect to form large-scale and small-scale networks. *CC BY-SA 3.0, from Wikimedia Commons.*

ARCHECORTEX – Evolutionarily older areas of the brain including the brainstem and limbic system

COGNIT – A cortical network for cognition

CORPUS CALLOSUM – A band of white matter connecting the two hemispheres of the brain

GRAY MATTER – The cortex of the brain comprised of cell bodies of neurons and gial cells

HETEROMODAL CORTEX – Cortical areas that can specialize in numerous functions

NEOCORTEX – Evolutionarily newer areas of the brain including the prefrontal cortex

NEURON – The electrically active cells of the brain

WHITE MATTER – The axons of the brain that relay electrical signals to other cells

FIGURE 3.4 Neuroanatomical terms help us to describe aspects of the nervous system.

many functions. Such areas are also called association cortex, as they tend to both receive information from many brain areas and project information to many other areas. Association cortex, or heteromodal cortex, may be in an ideal position to participate in guiding our reasoning functions, as these areas can both receive and relay larger integrative aspects of information processing (Fig. 3.4).

The Expansion of the Association Cortex in Predators

Brains and nervous systems have expanded as organisms have evolved new behaviors that are better tuned to enhance their survival skills. Some of the most widely shared anatomical areas of the brain across species are referred to as *archecortex*, meaning "old cortex." Areas of the archecortex include the limbic system structures, such as the hippocampus and amygdala, the brainstem nuclei, thalamus, and hypothalamus related to regulation of body states. Archecortical areas tend to be involved in primary survival functions. These include regulating the heart rate and respiratory system, coordinating sensory information, and allowing for emotional reactivity toward environmental variables. Archecortical areas tend to be similar across a variety of species. For example, small mammals such as mice have remarkably similar hippocampi, amygdalae, and thalami compared to primates and larger mammals. All of these areas appear to have similar functional roles across a variety of mammals, and they are also located at centralized positions often beneath the later-evolved areas of the cortex.

Association cortex is often referred to as the *neocortex* or "new cortex" due to its relatively late appearance in the evolutionary history of organisms. The neocortex comprises mostly of the outer covering of the brain in large and complex mammals including primates, elephants, dolphins, and whales. Unlike the archecortex, the neocortex tends to lack clearly defined single functions, or even well-defined sets of functions. Rather, neocortical areas appear to become involved in many diverse functions that are carried out by an organism. Such functions may be so difficult to specify that we simply refer to the supporting brain areas as "integrators" that receive highly processed information from other areas of the nervous system. These integrator regions perform processing functions, and then route their outputs to other areas of the nervous system that are downstream and closer to the ultimate action or output of behavior by the organism. The neocortex includes most of the frontal lobes in primates, including the prefrontal cortex (PFC), which is dramatically expanded within the primates. Anatomists currently consider humans and chimpanzees to possess the largest PFC proportionally to body size within the mammals. The PFC is highly important for reasoning functions as we will see in this chapter.

Another important neocortical area is the parietal cortex. The parietal lobes are densely interconnected with the PFC, as well as with the temporal lobes, and occipital cortex. Functions of the parietal cortex include a role in calculating numerical quantities, locating objects in space, coordinating grasp for grasping objects, orienting our hands, and visualizing spatial layouts. The PFC and parietal cortex also show a high degree of correlation. Indeed the frontal-parietal network has been identified as one of the most commonly active task-based networks within the brain through studies of functional brain connectivity. As we will describe in this chapter, the parietal cortex is frequently co-active with the PFC in carrying out reasoning tasks.

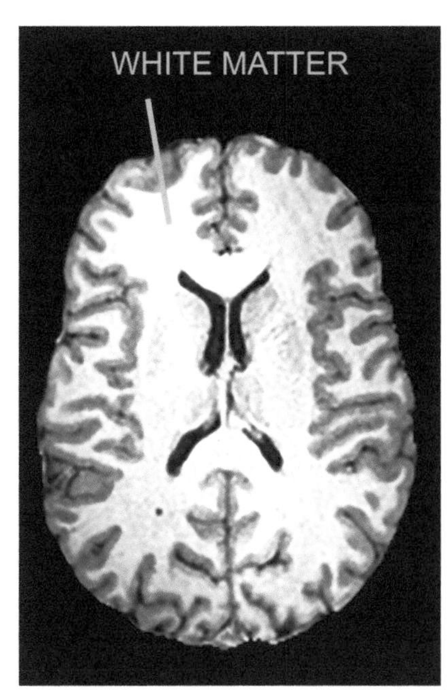

FIGURE 3.5 The neurons of the brain are densely interconnected through bundles of axons that are collectively known as tracts or fasciculi. These are commonly known as the white matter of the brain and this tissue images white in MRI scans.

Communication Across Cortical Areas

Cortical association areas are densely interconnected with the rest of the brain. Many of these connections form the large crossing fiber bands of the corpus callosum, the wide-ranging longitudinal fasciculi connecting frontal, occipital, and parietal cortex, and the U-fiber bands which allow multiple cortical regions to communicate (Fig. 3.5). Even in cases where there is not a single tract or bundle of axons connecting the cortical areas, communication can still be achieved through synapses at the thalamus, which is comprised of a set of nuclei within the midbrain that has a characteristic density of incoming and outgoing fiber tracts reaching across vast cortical territory. The thalamus is sometimes considered to be a relay station capable of routing messages from distant cortical areas to other areas through synapses within its subnuclei. My esteemed neuroanatomy instructor Arnold B. Scheibel would refer to the thalamus as the "gateway to the cortex."

NEUROSCIENCE OF RELATIONAL REASONING

From Hypotheses to Neural Network Models and Experiments

In the 1980s researchers became increasingly interested in modeling cognitive operations in ways that make use of the principles of neural systems. For example, the connectionist parallel-distributed processing (PDP) models pioneered by David Rumelhart and James McClelland (1986) were capable of modeling a range of complex cognitive abilities including perception, reading, and reasoning. These models were termed "biologically plausible," as they featured a series of primitive processing units in the form of positively and negatively weighted units. These units that would share connections with other layered units and simulated energy (representing neurochemically based signaling) were transferred through the connections to produce more sophisticated output from the models.

By the 1990s researchers had become increasingly interested in how actual nervous systems functioned and fewer papers appeared reporting comparisons between human behavior and the connectionist architectures. Some researchers began to describe the approach as being "neurally inspired" rather than "neurally plausible," as no connectionist model was able to lay claim to realizing the goal of a faithful reproduction of an actual nervous system. All such models were extremely oversimplified at the level of connectivity. This led them to be taken less seriously by the research community that was building what are known as "neuronal models." A neuronal model is not aimed at simulating cognitive phenomena, but rather it is meant to model simple neural circuits with high fidelity. Such models might be used to reflect the cell interactions that make up a reflex response in an animal with a parsimonious nervous system. These much simpler neural circuit models could more accurately be characterized as being "neurally plausible."

In the middle of the 1990s several new types of neuroimaging tools were becoming available. Among these were positron emission tomography (PET) and functional magnetic resonance imaging (fMRI) (refer to Box 3.1). These techniques are capable of imaging brain tissue as a person performs a sensory, motor, or cognitive task while having their brain scanned. PET and fMRI allowed researchers to plot functional activity onto anatomical scans, thus mapping out and isolating key areas that were involved in different types of tasks at the scale of the entire brain.

In 1995, Nina Robin and Keith Holyoak published a book chapter in which they hypothesized that the PFC was likely to be an important area for carrying out the operations of human reasoning. Robin and Holyoak described lines of evidence from brain organization, including the fact that the frontal lobes were highly interconnected with many other brain areas, that patients with frontal damage would exhibit challenges in maintaining items in working memory, avoiding distractions, and had difficulties with focusing attention. This was one of the first reports in the reasoning literature articulating the possibility that the frontal lobes may be the most critical region of the brain for relational reasoning or integrating information across dimensions. With this hypothesis, the new functional brain imaging methods could be deployed alongside more traditional investigations of cell-recordings in nonhuman primates, as well as patient studies of individuals with brain damage to test whether the PFC was central to relational reasoning.

Investigating Reasoning Ability in the Brain

One of the experimental tools of choice for evaluations of relational reasoning is the Raven's matrices task (Raven, 1938, 1960). In the task, a person sees a series of abstract visual objects arranged in a three-by-three matrix with one item missing. The goal of the participant is to determine what object or visual pattern best fits into the missing space by evaluating the patterns contained within the other places in the matrix (Fig. 3.6). The simplest matrices problems involve basic perceptual matching, in which the item to complete the problem matches based on appearance alone. More complex Raven's matrix problems require the participant to evaluate the different types of changes that occur either across the rows or down the columns and select the object that best matches the pattern of changes or combinations of

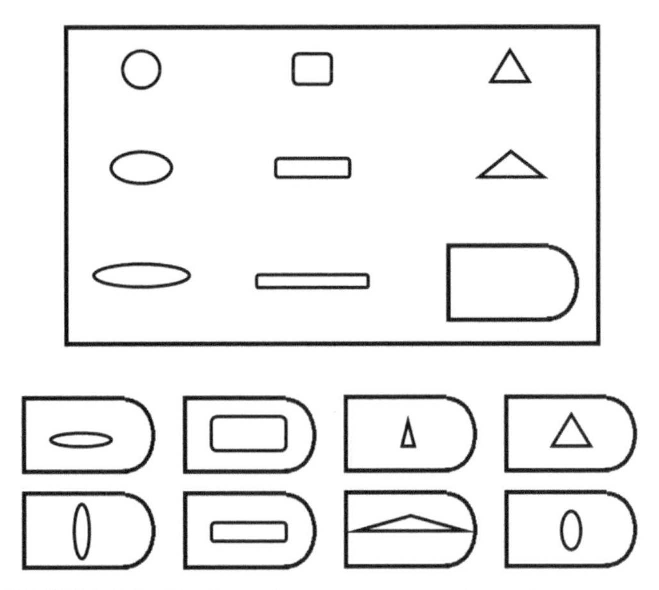

FIGURE 3.6 The Raven's progressive matrices task presents a series of abstract geometric shapes. The goal of the participant is to find the missing piece that best completes the pattern present in the matrix from a set of answer choices.

BOX 3.1

FUNCTIONAL MAGNETIC RESONANCE IMAGING

Magnetic resonance imaging (MRI) is a technique that allows researchers to view internal tissue using the principles of nuclear magnetic resonance. MRI scanners are large devices containing magnetic gradient devices that can generate a magnetic field many times that of the Earth's magnetic field. Magnetic energy can be deposited into the human body using a send and receive coil, which is a sophisticated device enabling both the deposition of energy into a person and the reception of information about the tissue. Researchers can create a picture of any tissue within the body provided that the tissue has composition differences. For example, bone and muscle image vary differently on an anatomical MRI, which is a static picture showing human anatomical structures. Functional MRI uses the same device and the same principles to image the differences between brain activity recorded under experimental conditions.

MRI was first used to image a human in July of 1977 in the physics lab of Raymond Damadian. Early scanners were used to image anatomy at relatively crude levels. The first MRI image was taken of the human chest. It took researchers approximately 5h to generate the first image, which was crude by modern standards. MRI technology improved to the point of becoming a dominant tool for the diagnosis of both disease and injury by the 1980s. In 1990, Seiji Ogawa first demonstrated that MRI could be used to track the activity associated with brain function, not simply structure.

Tracking functional activity was made possible by several key properties of nuclear magnetic resonance and brain physiology. First, neurons produce action potentials at a baseline rate when a particular set of neurons are operating at a resting level. When the brain responds to cognitive or motor demands, the associated neurons tend to activate above their baseline levels producing more action potentials per second. Oxygen in the immediate surrounding areas begins to be depleted in the wake of these more active neurons leading to the presence of more deoxyhemoglobin (deoxygenated blood) relative to the baseline state. In response to this decrease in oxygen, the cardiovascular system delivers an even larger supply of oxyhemoglobin, or oxygenated blood, to the area around the activated neurons. This change in the proportions of local oxyhemoglobin and deoxyhemoglobin leads to a signal that can then be tracked using MRI. The specific MR signal that occurs in this case is called the blood oxygen level dependent (BOLD) signal, and it has been the primary measure used in most of the fMRI studies conducted over the past 20 years. Functional MRI data consist of a series of voxels, which are three-dimensional pixels. There are thousands of voxels acquired across the whole brain during an fMRI experiment, usually obtaining a new reading of the activation (through BOLD difference) every 1 to 2s.

BOX 3.1 *(cont'd)*

When people design fMRI tasks, they typically include one or more control conditions in which some aspect of interest is carefully controlled for. An example of this occurs with fMRI studies of working memory in which a set of letters to be kept track of is presented to a person in the MRI scanner. In addition to a letter memory condition, researchers will often include a control condition in which letters are shown with a delay along with a test item, but without the memory demand. This enables the researcher to perform a statistical comparison between the memory condition and the control (non-memory) condition. Whichever voxels are more active during the memory condition over the control condition are considered to be relevant to the operations of working memory.

Functional MRI scans often consist of activation maps. These maps are color coded to reflect the degree of average BOLD signal contrast between conditions within the voxels of the brain. Typically, densely clustered groups of neurons will become active together and this will lead to a set of grouped clusters of voxels that are considered to be active over their baseline state. The maps of lower resolution voxels (typically 3 mm cubed) can then be plotted onto a higher resolution structural MRI image of the brain allowing the researcher to view the areas of activity accompanying a task (Fig. 3.7).

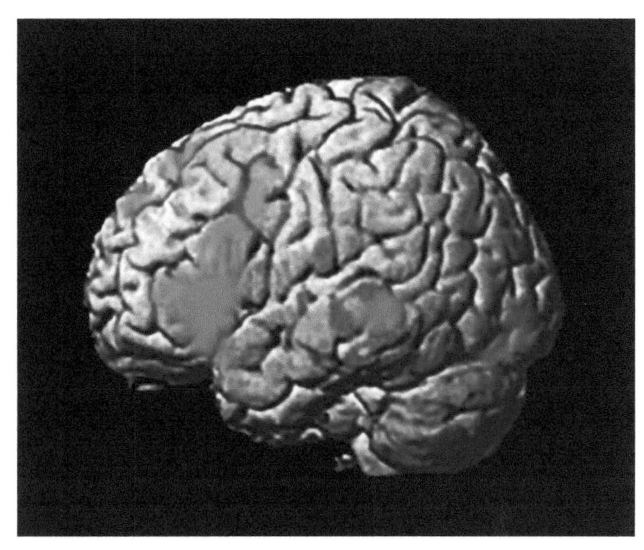

FIGURE 3.7 A color coded map showing fMRI data plotted in red on an anatomical brain scan shown in gray.

changes across the other items. This is a classic relational reasoning task, as it demands evaluation of a series of relationships among the items. The Raven's matrix task is considered to be a culture-fair test, as it lacks any specific identifiable objects. Each figure is a novel geometric shape or pattern. This means that the individualized knowledge, or semantic memory of the individual, is unlikely to affect performance. This task has also been considered to be a test of fluid general intelligence.

Localization of Reasoning and the Prefrontal Cortex

Prabhakaran, Smith, Desmond, Glover, and Gabrieli (1997) at Stanford University performed an early neuroimaging study of reasoning. This study was the first of a line of investigations, in which human participants were scanned using either PET or fMRI while they performed a variation of the Raven's progressive matrices task. In this case, Prabhakaran et al. offered participants figural problems, in which participants evaluated a single dimension of change across a matrix in order to choose a match to complete the pattern from among a set of alternatives. They also offered participants analytic matrix problems, in which multiple relations among task elements had to be integrated in order to solve the problems correctly. This analytic condition provided the best neuroimaging test of the hypothesis that Robin and Holyoak had articulated 2 years earlier, which had stated that the PFC would likely be key to the integration of relational information. The study revealed that bilateral frontal areas within the dorsolateral prefrontal cortex (DLPFC) were more active for the analytic problems compared to the figural problems. Additional areas showing greater activation for the analytic problems included the left parietal cortex, temporal lobes, and occipital regions. All of these active areas were consistent with the maintenance and manipulation of working memory, in addition to any cognitive operation that could plausibly be described as reasoning. A question remained as to whether any of these areas were specifically engaged by relational reasoning, or whether they were alternatively markers of working memory and attention processes that increased most in the demanding analytic condition.

Soon additional matrix reasoning studies appeared in the growing functional neuroimaging literature. These studies were designed to further clarify the role of the relational processing regions that had been identified in the Prabhakaran et al. (1997) study. Christoff et al. (2001) further examined the role of the PFC in matrix reasoning. In this study, participants solved Raven's matrices problems that varied in the number of relations that had to be integrated in order to solve the problem. Participants in this study chose one item that best completed a nine-cell matrix. Similar to the results reported by Prabhakaran et al. (1997), the study by Christoff et al. revealed activation in the left DLPFC, left rostrolateral prefrontal cortex (RLPFC), and bilateral inferior frontal gyrus when two-relation problem solving was compared to one-relation problems. One challenge in interpreting the results in this study was the fact that two-relation problems were more difficult and therefore took longer for participants to solve. This could mean that the PFC was not responding to relational reasoning load, but rather the amount of effort or processing time that was needed to accomplish the task. Christoff et al. addressed this challenge by performing a response-time corrected model for their fMRI data. This procedure was better able to equate the conditions in terms of the fMRI response analysis. This more stringent timing-corrected analysis revealed two left-sided areas of activation within the PFC. The left posterior PFC was one of these areas, which extended over the inferior frontal gyrus. The second area was located within the left RLPFC. These areas specifically appeared to be involved in the integration of relations among items.

Christoff et al. (2001) then examined the time-course of activation of both right- and left-sided DLPFC and RLPFC regions. They found that only the left RLPFC showed a clearly differentiated response to the two-relation problems over the one- and zero-relation problems (refer to Fig. 3.8). The other PFC areas were active across all relational conditions. The left RLPFC region appeared to be the most engaged area in response to integrating relational information, a key skill needed to solve the more difficult Raven's matrices problems.

At this point it remained unclear whether the left RLPFC functions as a relational processing module or whether it was responding to some other measure of task complexity. Another possibility is that the left RLPFC was supporting the maintenance of relational information in working memory. To further explore the involvement of the RLPFC in relational reasoning, Kroger et al. (2002) conducted an fMRI study, in which participants solved problems ranging from zero relations up to four relations. The task was more relationally demanding than the prior fMRI studies of relational reasoning. Kroger et al. found similar results to the previous studies with relational visuo-spatial problem solving evoking bilateral PFC activation and bilateral parietal lobe activation. The left PFC was more engaged than the right PFC in this study. Progressively more anterior

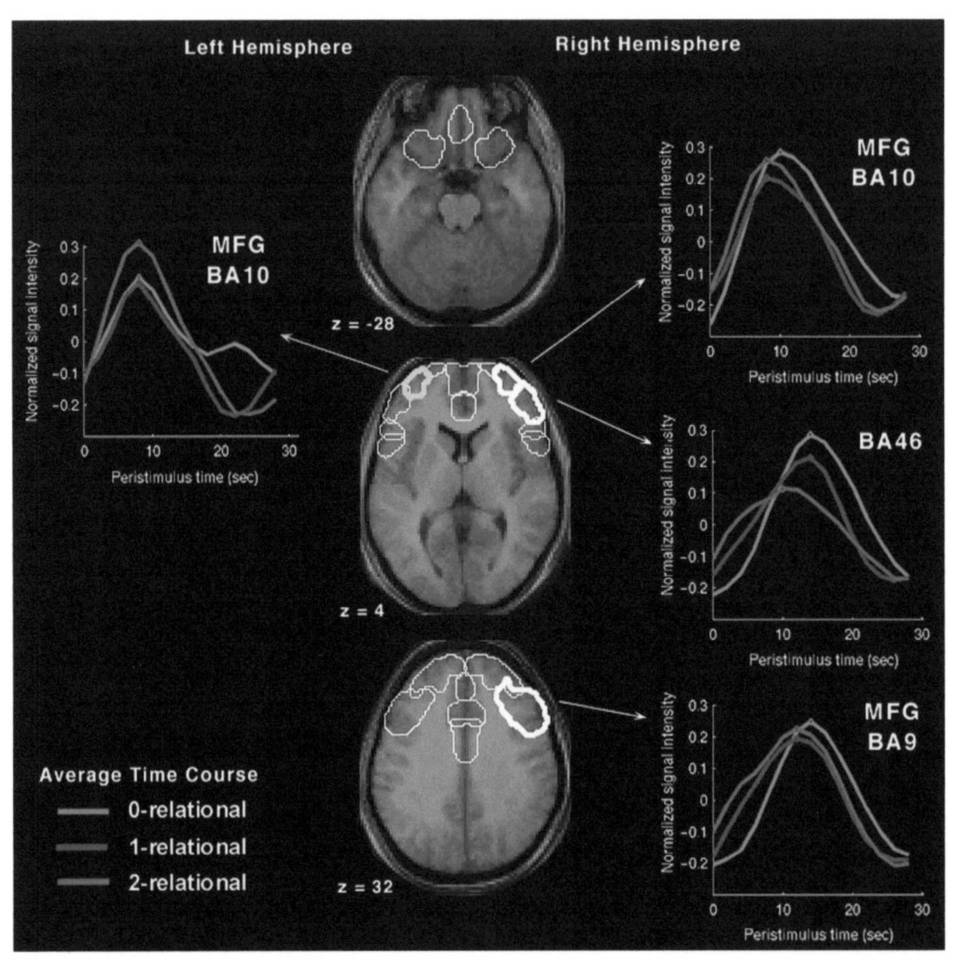

FIGURE 3.8 Christoff et al. (2001) found left PFC activation associated with solving complex relational matrix reasoning problems. *Credit: Christoff, K., Prabhakaran, V., Dorfman, J., Zhao, Z., Kroger, J. K., Holyoak, K. J., et al. (2001). Rostrolateral prefrontal cortex involvement in relational integration during reasoning, NeuroImage, 14, 1136–1149.*

regions of the left PFC became activated at higher relational load levels supporting the role of the left RLPFC in relational reasoning.

A serious experimental issue still remained in determining whether the PFC was responsive to relational processing in matrix reasoning tasks. Most neuroimaging studies require a series of simpler control conditions, which will also evoke brain activation patterns of their own. For example, Kroger et al. (2002) included a three-relation condition that was in effect, a control condition for their four-relation condition. This is possible because the activity evoked in the simpler three-relation condition could be removed by subtracting it out from the commonly active areas present at the four-relation level. A key control condition was included in the study by Kroger et al., and this was represented by matrix reasoning problems that were made more difficult and slower to process, but not by the addition of relational elements. Rather, the researchers added visual noise in the form of distractor items that were not relevant to the task. These extra "noise" items were added to all of the relational elements in the matrix. The noise items did not need to be factored into the solution of the problem, but were nonetheless difficult to ignore. These control conditions added processing time and difficulty to the task without increasing the relational demand on the participant. Participants took longer to solve the problems in the conditions with distracting elements added, and this also led to increases in activation in similar bilateral PFC areas, as well as medial PFC regions. Kroger et al. constructed a brain map in which the activation evoked by relational reasoning was compared with that evoked by problem difficulty that was attributable to distracting elements. There was some overlap among areas within the left and right PFC, but only the relational increase led to left RLPFC activity. Furthermore, larger areas of PFC were activated in response to the increased relational load over the distraction factor. Consensus was beginning to build that relational reasoning demanded left RLPFC activity relative to other forms of executive function demand.

There was also interest in Raven's matrices from the intelligence testing community. This group of scholars considered the Raven's matrices task to be part of the construct *fluid intelligence*. Fluid intelligence was a term coined by Raymond Cattell, who had been influenced by the work of Charles Spearman before him. Fluid intelligence refers to the abilities of an individual that are based on biological factors including the sensory structures, the central nervous system, and heredity (Horn & Cattell, 1966). John Duncan et al. (2000) used PET scanning to evaluate the neural systems involved in fluid general intelligence. They boldly titled their paper *A Neural Basis for General Intelligence* and it appeared in the high-impact journal *Science*. Duncan et al. PET scanned participants while they completed three different tasks that contained a range of difficulty (or loading) on fluid general intelligence. For example, participants solved an abstract figure test that had been developed by Raymond Cattell, along with a complex letter sequence reasoning task and another abstract figure judgment task. The high demand conditions within all three of the tasks were accompanied by higher levels of activation within the left DLPFC within the middle frontal gyrus. Duncan et al. argued that the left DLPFC was likely to be a critical region for supporting the mental processes that are necessary for fluid intelligence (Fig. 3.9).

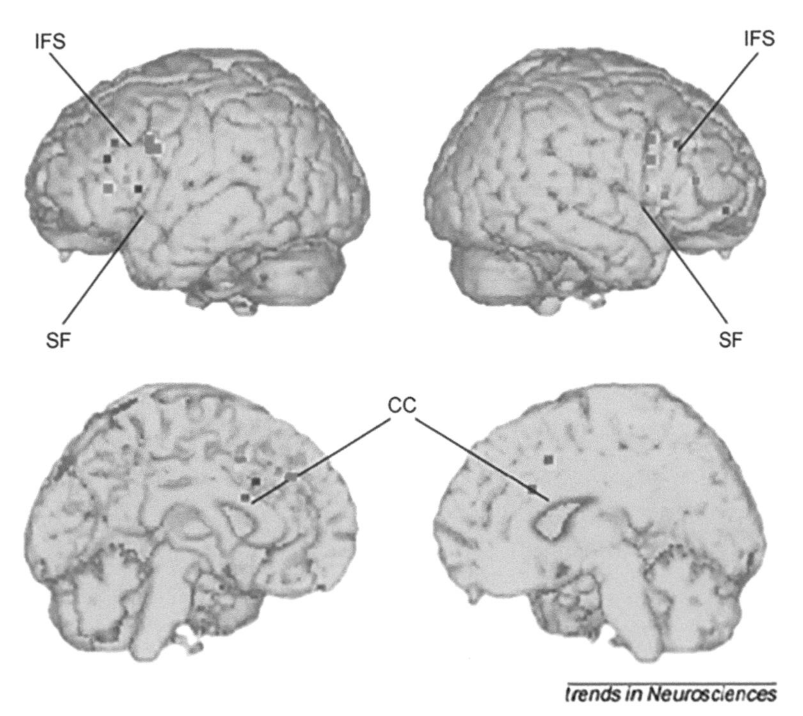

FIGURE 3.9 An influential positron emission tomography (PET) study by Duncan et al. (2000) revealed that a common left frontal region became active across a set of varying fluid reasoning tasks. *Credit: Duncan, J., & Owen, A. M. (October 2000). Common regions of the human frontal lobe recruited by diverse cognitive demands. Trends in Neuroscience, 23(10), 475–483.*

The fluid intelligence paper by Duncan et al. (2000) generated quite a bit of enthusiasm, excitement, and discussion, as this was one of the first studies to provide some clues about the biological basis for the construct of fluid intelligence across multiple task formats within a specific region of the brain. The study also generated a fair amount of controversy as well. Noted intelligence theorist and researcher Robert Sternberg, who is well-known for his triarchic theory that described multiple intelligences, wrote a scathing review of the study referring to it as representing the "*holey* grail of intelligence" (Sternberg, 2000). Sternberg expressed his disappointment in the interpretation of the study recommending that the neuroscience community focus on characterization of the cognitive processes that are supported by the PFC, rather than claiming that a concept as difficult to define and variable as intelligence could be supported by a single region of the brain. This critique noted the similarity of this approach to phrenology, a widely criticized effort to map human abilities to particular areas of the head, which was popular in the early 1800s (Fig. 3.10).

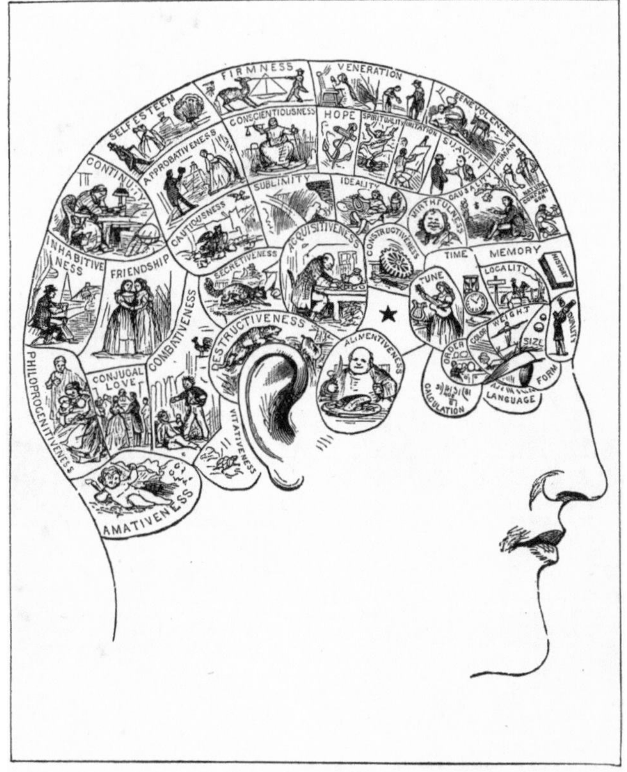

FIGURE 3.10 Phrenology was a movement attempting to link personality and mental capabilities to different regions of the brain. This movement occurred in the 1800s and involved feeling the sizes of areas of the head and concluding that certain propensities were linked to variations in head size. *From Wikimedia Commons.*

Left Prefrontal Cortex and Relational Reasoning With Analogies

The left PFC emerged as important for analogical reasoning very early in the developing neuroimaging literature. Charles Wharton et al. at the National Institutes of Mental Health published one of the first analogical reasoning studies. This was a PET study, in which the analogies consisted of simple shapes that also varied in color and texture. The task was administered in a match-to-sample format. Participants initially encoded a sample set of four shapes followed by a brief delay period. They then had to decide if a second set of four shapes matched the initial sample. In a literal condition, participants simply had to judge from memory whether the two sets of shapes matched. In an analogy condition participants had to judge whether the relations among the four shapes on the test screen matched the relational information present in the sample (refer to Fig. 3.11). Critically, in the analogical match condition the shapes and colors involved in the sample were not the same as those in the match. Rather than match on perceptual features, participants had to understand the similarity between match and sample at a more abstract level. The analogy condition preferentially activated the left parietal cortex, medial frontal cortex, along with a large area within the left PFC. This study indicated a focus of relational processing within the left PFC in an analogical matching condition. The study is reminiscent of the Raven's matrices match studies described earlier in the chapter with some variation in format. Like the matrix reasoning tasks, this analogy task lacked semantic information limiting its ecological validity and generalizability.

A challenge with functional imaging methods are that any inferences one makes are correlational rather than causal. This means researchers cannot infer strong cause-and-effect relationships between cognitive operations and task-evoked brain activation on the basis of those imaging studies alone. There is always the possibility that activity in a given region, such as the left PFC, could be related to some aspect of cognition that accompanies the reasoning construct under examination. For instance, the left PFC may be especially active when people have to maintain many dimensions in working memory within a matrix task. It is also possible that the left PFC is active in reasoning conditions simply because the demand on attention increases in response to the need to notice and consider multiple relational elements. While this second explanation of left PFC involvement would be of great interest to researchers involved in studying reasoning, it would still show that left PFC function in attention increases when reasoning becomes more demanding relative to when it is less demanding. This is not an especially insightful claim about reasoning specifically, and it adds little additional new knowledge about the function

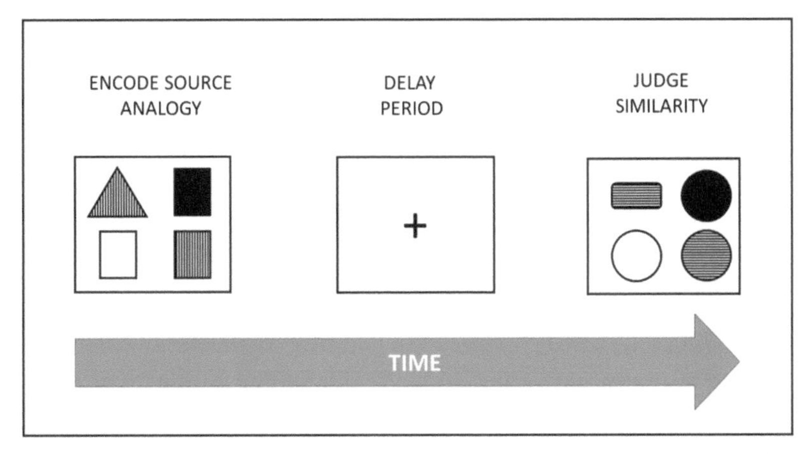

FIGURE 3.11 Wharton et al. (2000) conducted an early analogical reasoning study using pattern stimuli resembling items from matrix reasoning tasks.

of the left PFC. Another challenge of interpretation is that other regions of the brain are also involved in attention, and if the left PFC were in some way deactivated, perhaps these other regions (such as the right PFC) could stand in for the typically active left PFC. The activity in other regions could enable relational reasoning or analogical comparisons to be made all the same. Other methods were needed in order to more firmly establish a role of the left PFC in reasoning. Researchers needed additional methods to establish causal links that would enable a stronger conclusion than the idea that the left PFC is relevant to executive control regions and reasoning operations.

Transcranial magnetic stimulation (TMS) is a method that allows researchers to influence brain function in a safe and temporary way. The method is carried out by positioning a handheld magnetic coil that is roughly the size of a Ping Pong paddle over the scalp of a participant. Pulses of electrical current can be delivered through the coil. Electrical coils will induce magnetic fields when activated, so magnetic energy can be transmitted through the scalp and skull to induce electrical currents within the cortex. TMS pulses are relatively noninvasive. Participants may experience a tapping or tingling sensation around the location of the coil on the scalp, but the method is otherwise safe. TMS may be used to temporarily fatigue neurons in a given area of the brain thereby inducing a temporary reduction in function. Alternatively, TMS can be used in a facilitatory manner to induce an enhancement in brain function, though again this has only a temporary effect. Typically, researchers will deliver TMS while a participant performs a cognitive or perceptual task on a computer. Participant's heads are usually restrained using a metal frame with a chinrest, similar to those used in optometry offices. Many researchers will use MRI scans of the participants' brain in order to better target a particular area of the cortex. TMS is limited by its inability to effectively reach deep brain structures such as the amygdala and hippocampus, both of which are located within the interior portion of the temporal lobes.

Not long after the Wharton et al. (2000) study was published, Charles Wharton, Jordan Grafman, et al. conducted a TMS study of analogical reasoning with Babak Boroojerdi at the Cognitive Neuroscience Section of the National Institute of Neurological Disorders and Stroke. Boroojerdi et al. (2001) presented participants with the same type of analogical relation stimuli previously used in the Wharton et al. (2000) study. The analogies were presented in a sequential match-to-sample format in which a source picture was initially shown to the participants followed by the target picture, which had to be matched to the source. In an analogical condition, participants had to match the target to the sample based on the relations (color, pattern, or shape) among the four figures (see Fig. 3.11). In a literal comparison control condition, participants had to match the target to the sample based on perceptual similarity, thereby controlling for most aspects except relational reasoning. The task was also run with both target and sample pictures being presented simultaneously to allow for a comparison that reduced working memory demand on the participants. Facilitatory repetitive transcranial magnetic stimulation (rTMS) was delivered as participants performed all four conditions (sequential analogy, sequential literal match, simultaneous analogy, and simultaneous literal match). The rTMS was applied over the left PFC, as that region had been activated by analogical reasoning in the Wharton et al. (2000) PET study. Boroojerdi et al. also applied rTMS over the right PFC to evaluate its role in reasoning, as well as regions of the motor cortex as a control condition. Lastly, sham rTMS was simulated with the coil directed away from the participants' scalp, while these same task conditions as a procedural control condition.

Results of the study by Boroojerdi et al. (2001) further confirmed an important role for the left PFC in relational analogical reasoning. The application of rTMS over the left PFC resulted in faster performance for the analogy trials (by 100–200 ms) presented both sequentially and simultaneously. The speed increase did not occur during the sham conditions. No speed advantage occurred when rTMS was applied over the right PFC, suggesting that there was indeed a lateralization present for relational reasoning, as had been suggested by the earlier PET and fMRI studies. Results from the trials in which left motor cortex received rTMS were less clear, as speed advantage was also present for both simultaneous and sequential analogy conditions for the left motor rTMS condition, along with a speed advantage for the literal match condition when presented under sequential conditions (Fig. 3.12). This last finding with rTMS applied to the left motor cortex further supported the left frontal dominance in processing analogical relations, but also suggested that some of the speed decrease may be linked to a motor speed advantage rather than a purely cognitive processing speed advantage.

Up to this point in the literature there was a growing focus on the role of the PFC in reasoning. Over the next several years this focus gravitated toward the left PFC and then the left rostral PFC. There was some concern that the contributions of cognitive neuroscience were becoming limited to merely localizing all activity down to the functions of single brain areas. The focus on PFC localization was complicated by the fact that matrix reasoning and fluid intelligence tests are specifically designed to minimize (or exclude) the contributions of semantic memory or prior knowledge. It became clear that future cognitive neuroscience efforts would have to address the role of prior knowledge and how areas supporting semantic memory interacted with attention and working memory in order to facilitate human reasoning.

Integrating Relational and Semantic Information

Verbal analogies are a commonly used tool for investigating intelligence. Verbal analogies typically take the form of four-term "A is to B, as C is to D" comparisons. Such examples have been used in developmental investigations of children's knowledge and abilities.

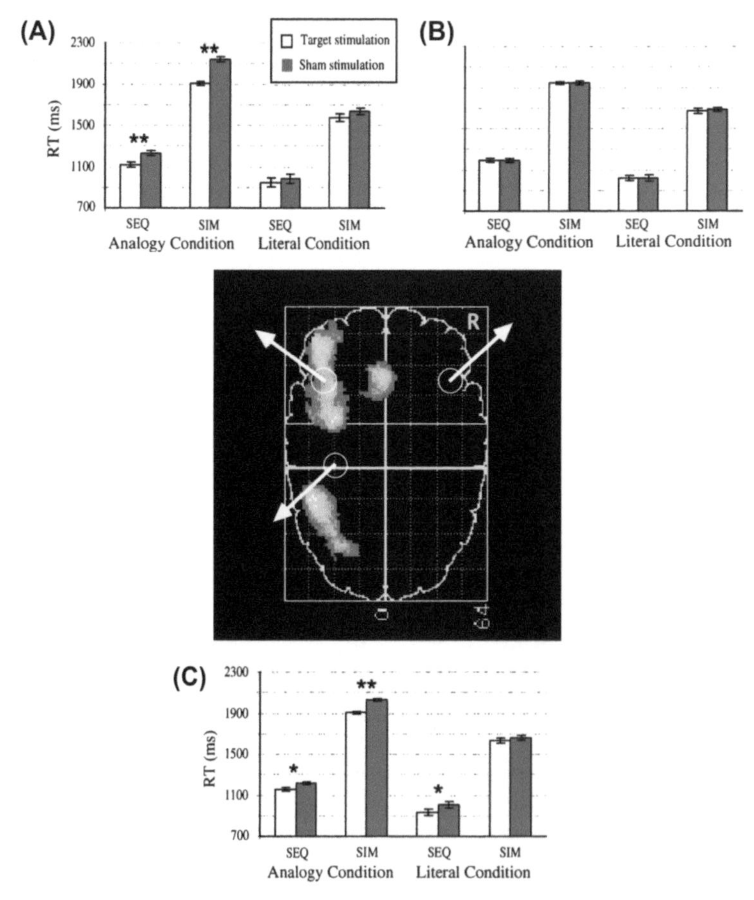

FIGURE 3.12 Boroojerdi et al. (2001) used transcranial magnetic stimulation (TMS) to induce facilitatory input over the left (A) and right PFC (B) and the left (C) and right motor cortex during a reasoning task. The results indicated that the left, but not right, PFC enhanced speed of processing during analogical reasoning. *Credit: Boroojerdi, B., Phipps, M., Kopylev, L., Wharton, C. M., Cohen, L. G., & Grafman, J. (February 27, 2001). Enhancing analogic reasoning with rTMS over the left prefrontal cortex, Neurology, 56(4), 526–528. http://dx.doi.org/10.1212/WNL.56.4.526:1526-632X.*

Robert Morrison et al. (2004) published a paper describing deficits in verbal four-term reasoning experienced by participants who had acquired frontal lobe damage through fronto-temporal lobar degeneration, a form of dementia. These individuals had difficulty solving for the fourth term. For example, participants were asked to complete the analogy "*Play* is to *Game*, as *Give* is to ?" and the answer choices *Party* and *Take* were offered. Individuals with frontal damage had difficulties selecting *Party* as the correct answer, especially when the alternative, such as *Take*, was highly associated with the third term, *Give* in the current example. This finding suggested that in addition to relational processing, verbal or semantic relatedness was also a necessary skill in order to reason analogically. We will discuss this study further in Chapter 7 on disorders of reasoning.

Shortly after the Morrison et al. (2004) study appeared, Bunge, Wendelken, Badre, and Wagner (2005) published one of the first neuroimaging studies to investigate verbal analogical reasoning with four-term items. This study compared brain activation when participants evaluated analogical relationships compared to other situations when they simply evaluated how two words were related. For example, participants evaluated the analogical problem *Bouquet* is to *Flower*, as *Chain* is to *Link*. In this case, all relational information must be considered in order to evaluate whether this is a valid analogy. Meanwhile, other sets of four words were presented in a similar format such as *Note* and *Scale* followed by a second unrelated pair, such as *Rain* and *Drought*. Participants judged whether rain and drought were related. Findings indicated that the judgments of word relatedness involved the left inferior frontal gyrus, a region also known as the left ventro-lateral prefrontal cortex (VLPFC). Meanwhile, the comparison between analogies and evaluations of semantic-relatedness indicated that additional left PFC areas were preferentially active in response to the relational evaluations. These areas were more anterior to the VLPFC and supported the earlier relational integration studies that had been carried out using nonlinguistic materials. The possibility that different left hemisphere specialized areas of the frontal cortex supported word retrieval and relational integration was introduced.

Additional research on the neural correlates of verbal four-term analogical reasoning appeared with additional considerations about task demands and more specificity within the brain. Green, Fugelsang, Kraemer, Shamosh, and Dunbar (2006) asked participants to evaluate four-term verbal analogies and found that a left frontopolar cortical region (an area similar to the RLPFC) was active when participants had to evaluate the validity of an analogy, such as *Planet* is to *Sun*, as *Electron* is to *Nucleus*. This analogical evaluation condition was compared with a categorization condition, in which two word pairs had to simply be evaluated regarding the presence of independent semantic relations among the word pairs. For example, participants would indicate that a true categorical relationship was present between *Cow* and *Milk* and between *Duck* and *Water*, but there was no analogical relationship formed. The subtraction of semantic word-pair evaluations revealed only this single left frontopolar cortex location of activation associated with the evaluation of the relations between analogies. The results of this study largely converged with those of the study by Bunge et al. (2005) in the localization of a left PFC site of activation associated with relational comparisons in analogies that occurred over and above the activation associated with simpler semantic association evaluations.

Studies that include semantic knowledge, or word relatedness comparisons, along with relational processing may be more effective in simulating real-world reasoning. This may be especially the case when comparing the work on four-term analogies to the earlier Raven's progressive matrices fMRI studies, or those that involved evaluations of visual and spatial dimensions in simple analogical pairings. Both types of studies converged in isolating a left anterior PFC regional activation that escalated with relational processing demands. Only the semantic analogical reasoning tasks were capable of differentiating the left PFC areas that are associated with evaluations of semantic association from those that specifically process relational comparisons. While the work of Bunge et al. (2005) and Green et al. (2006) included more semantic information than several of the prior relational reasoning studies, it remained unclear to what degree other factors in semantic knowledge influenced the brain activity associated with relational comparisons.

Researchers had described the interesting properties of distant, or remote, analogies for several years prior to the emergence of functional neuroimaging studies. The term *remote analogical reminding* had been used to describe instances in life when two quite different situations could be seen as abstractly similar based on overall relationships that were analogous, despite great differences in the surface similarities of these situations (Gentner, 1983; Wharton, Holyoak, & Lange, 1996). An example can be found in fables. In the next chapter we will discuss one of Aesop's famous fables. In the fable, a thirsty crow confronts a pitcher of water in which the water level is too low to drink from. The crow solves the problem by adding several stones in order to raise the water level in the pitcher to the point where he can drink. This situation is analogous to a modern day person who finds herself unable to complete a marathon due to being out of shape. She may work persistently running on a daily basis and adding more stamina each time, until at last she can complete the race and experience a feeling of accomplishment. The relationships between the crow and the runner are apparent when one considers the system of relations among the elements. The crow matches to the runner, the quenching of thirst corresponds to the completion of the marathon, and the adding of stones matches the sequence of effort applied in daily running practice. Some within the research community even argued that the more remote an analogy was in terms of domains, the more interesting and satisfying the situations are to compare.

The next fMRI study from Green, Kraemer, Fugelsang, Gray, and Dunbar (2009) investigated the effect of remoteness, or distance, on the brain activity associated with analogical reasoning. These researchers were interested in capturing some elements of creativity and innovation that are characteristic of analogical reasoning. This is difficult to achieve using fMRI, as one must use numerous trials that are repetitive, short in duration, and relatively simple. A clever way to address the remoteness of relational comparisons is to consider word association values. Over the years cognitive and language research labs had quantified the association values between words. This can be done in a variety of ways by local co-occurrence in texts, or by having large numbers of participants generate associations to words. Green et al. used a technique called latent semantic analysis to calculate the similarity between the meanings of words within a high dimensional semantic space (Landauer & Dumais, 1997). The results of the latent semantic analysis enabled Green et al. to experimentally vary the distance among the analogies that were presented in their study using fMRI. As in the prior study (Green et al., 2006) four-term analogies were presented in which participants evaluated the validity of an analogy; however, in this study the distance of the analogical relationships was varied. In highly associated within-domain analogies, the word pairs presented were relatively close in semantic distance. For instance, in the analogy *Nose* is to *Scent*, as *Tongue* is to *Taste*, all of the items link strongly to parts of the face. In distant cross-domain analogies, comparisons were made between word pairs that had low association values, meaning that they do not tend to be strongly associated semantically. An example of such an analogy is *Nose* is to *Scent*, as *Antenna* is to *Signal*. This second example may appear to be a more clever analogy, as the distance highlights the relevant relation (sensory device and what it detects) more clearly when other nonrelational semantic associations are absent. Green et al. interrogated the same left frontopolar PFC region that they had previously identified as being modulated by relational processing when compared to semantic association control conditions. Interestingly, the task-related activity within the left frontopolar PFC increased as the semantic distance among the words increased. This finding indicates that the rostral or frontopolar PFC is responsive to relational information, but that this area is also sensitive to the degree of remoteness between concepts being relationally compared.

The dual role of the left frontopolar PFC in relational integration and in managing remote domains is consistent with much of what has been known about the area from a cell physiology perspective. The PFC is considered to be heteromodal, meaning that this area of the brain is involved in numerous functions. The PFC is also considered to be highly plastic in that PFC cells change their response patterns rapidly in accordance with new task demands. The frontopolar PFC may be considered to be at the top of a hierarchy of information within the brain, serving as a coordinating or switching area to allocate attention toward particular goals or to integrate different types of information that may be recalled in the service of a reasoning task. To properly understand the role of the PFC in reasoning, key questions emerge that earlier functional neuroimaging studies had not yet begun to investigate. Does PFC activity vary depending on the type of reasoning task? What about the format of the task? When do different reasoning demands maximally engage the PFC? Such questions would begin to be addressed as additional labs became involved in the functional neuroscience of reasoning.

The Cognitive Operations of Reasoning

The studies we have reviewed thus far emphasized conditions that emphasized relational integration or managing multiple relations compared to ones that did not. The control conditions in those tasks largely focused on information complexity or visual complexity, but critically the controls did not emphasize any processing of relational or abstract information. A challenge with such studies is that the relational task conditions typically involved several cognitive abilities and those abilities may occur in a sequence over time that is not easily captured during neuroimaging tasks. This is due to the number of consistent and repetitive trials needed to average over blood oxygen signal to create a statistical map of brain activity. In studies conducted without the use of neuroimaging methods, there has often been a strong sense that information relevant to solving reasoning tasks becomes available at different points in time and a variety of cognitive abilities including perception, attention, working memory, and long-term memory are involved in reasoning tasks. These abilities are applied either to infer new knowledge about the world, make a prediction, verify a possibility, or overcome some sort of obstacle. Further investigation of cognitive abilities is a key to understanding the process of reasoning and how it is carried out within the brain. It is also important in this endeavor to specify when and why different cognitive abilities are utilized and for what purpose during the process of reasoning. This type of approach is necessary to understand the temporal dynamics of reasoning and ought to make it possible to avoid overly fixating on single brain areas as "modules" for reasoning. Most would argue that reasoning is too variable and diverse to be fully captured by a single module within the brain.

The act of inference is central to the purpose of making a relational or analogical comparison. Consider the example "Ink is to Quill, as Paint is to Paintbrush." In the analogical reasoning studies we have reviewed thus far in the chapter, participants were typically asked to evaluate whether or not there was an analogical relationship between those word pairs. Participants would likely have evaluated the relationship between the first two words, ink and quill, noting that the quill is used as a writing tool to transmit ink onto paper. The individual would then note that a paintbrush is used to spread paint purposefully onto other surfaces, thus there is a valid analogy present and that this should be verified in the experimental task. Given such instructions, the participants' goal is to simply state whether there is a relational match. This type of task is effective in eliciting rapid evaluations of analogies, which occur by the participant evaluating relations, comparing them, and proceeding to answer whether they are relationally similar. Such cognitive processing may explain why there is a high degree of similarity between many of the matrix relational reasoning tasks and those in which analogies were evaluated, as both tasks emphasize the same process of evaluating relations and then comparing them serially.

In everyday life, we make analogical inferences. These occur when we consider relational similarity between an older source analog from our past and a new situation that is unfamiliar and contain some unknown aspects. In this case, we may wish to use information from the source analog to infer a new conclusion about the new situation. For example, if you are familiar with driving cars, but have never ridden a motorcycle before, you may relate parts of the motorcycle to similar gadgets in your car. You might compare the motorcycle throttle to the gas pedal on your car and infer that it will make the vehicle accelerate in the same way that your car accelerates after you press the pedal. Note that simply comparing two word pairs to verify if they share a relational correspondence does not appear to capture much of the cognitive work that goes into the reasoning process in the motorcycle-car analogy. If we consider this new task goal of inferring new knowledge on the basis of a prior situation, it can change the cognitive operations a great deal and indeed the task itself becomes different in some key aspects.

In 2010 my own lab published a study that investigated four-term analogical reasoning as it occurred over time, with an eye toward understanding when frontal, parietal, and temporal lobe regions became active in the reasoning process. We were particularly interested in the inference stage of processing an analogy, which to that point had not been the subject of much direct investigation in neuroscience studies. A small change in the method from those prior four-term analogy neuroimaging tasks we have reviewed so far made this possible. In our experiment, we initially presented participants with the first pair of the analogy, for example, *spyglass* and *ship*. These objects were presented in picture form, which offered the participants an opportunity to consider the pair, visualize them, and relate them to one another. We next presented a picture of only one element of the second relational pair. For instance, participants saw an image of a *periscope*. We instructed our participants to imagine what would go with the periscope in order to complete the same relational association that had been illustrated by the spyglass and ship. Lastly, we presented either a picture of a valid item to complete the analogy, such as a *submarine*, or a picture of an unrelated item that did not satisfy the analogy, such as an *elephant*. Participants indicated with a button press whether or not the new item did indeed complete a valid analogy or not. We included two control conditions that were presented in an identical manner to the analogy condition, but with different instructions. In a semantic control, we simply asked participants to infer a semantic associate of the third item. For example, they could generate the concept of a sailor to go with the periscope, as this item did not need to share relational similarity with the source pair. In a perceptual condition, we simply asked participants to think of an item that looks similar to the third item in the analogy. For instance, they could think of a flashlight or a piece of drain pipe as looking similar to the periscope. Again, there was no need to consider similarity to the source pair in this condition. Refer to Fig. 3.13 for an example of the task design.

We gathered fMRI data from all phases of this task. This enabled us to ask a new set of questions: what areas of the brain become active during the evaluation of a source pair? What areas activate during the process of analogical inference? How are these areas of activity similar or different to the situation in which one evaluates a full analogy when the final item has been presented? By presenting analogies in this sequence and by spacing all of these task situations out in time, we hoped to better appreciate what different cognitive demands were associated to activity in what areas of the brain.

Let's reconsider the earlier example involving the pen-paintbrush analogy to better illustrate the difference in this analogical reasoning study compared to the prior ones. To infer something about the paintbrush that would complete a relational correspondence, we must initially appreciate that a quill is dipped in the ink with the goal of applying the ink to write or draw. The thinker is likely

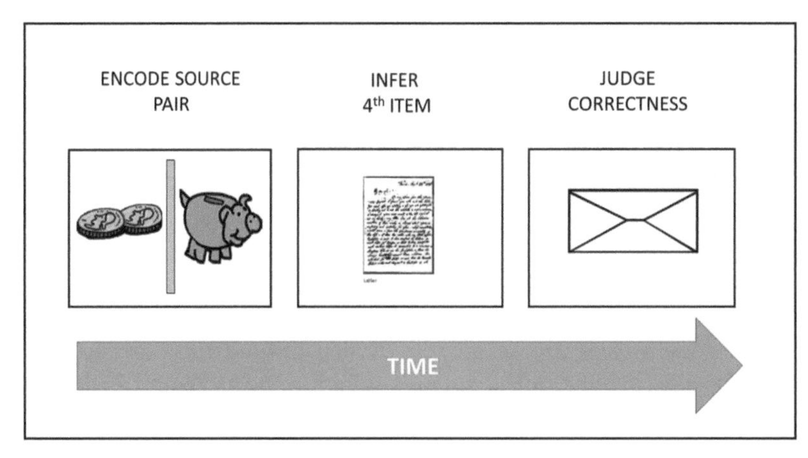

FIGURE 3.13 Krawczyk et al. (2010) evaluated four-term analogy problems as presented over time. This design allowed comparisons of the brain activation associated with encoding relational pairs, inferring a matching relational paired item, and verifying the solution.

to settle upon a dominant association that links these two items, in this case focusing on their functions. Next, the thinker is presented with the paintbrush, at which point he must perform a constrained memory search to isolate the function of the paintbrush and generate an item that is functionally equivalent to the role that the inkbottle plays in the earlier source analog. Lastly, when a paint can is presented, he or she must either note that it agrees with the generated item, or is capable of fulfilling the role of containing the paint that would be needed to allow the paintbrush to function. The cognitive processes present across the sequence are different at each point in the task and it is still much simpler than what we often do during the course of everyday analogical reasoning. To understand the brain processes that are evoked during different aspects of reasoning, it is necessary to go to the trouble of trying to isolate the different steps in a reasoning task and their associated cognitive processes. Only then can we begin to grapple with which functions are needed for reasoning and how these functions are correlated with brain activity.

The results of the Krawczyk, McClelland, Donovan, Tillman, and Maguire (2010) study helped to clarify the importance and functional significance of many of the brain areas noted in prior matrix reasoning and analogical relational reasoning studies. Fig. 3.14 shows a series of the major task-related areas of the frontal lobes: the left DLPFC, Left inferior frontal gyrus (LIFG), medial PFC, left middle frontal gyrus, posterior medial PFC, and right DLPFC. Note that several of these areas, especially the left DLPFC and LIFG, have already been discussed in association with evaluations of semantic relatedness in analogical reasoning tasks and in processing relational information. In addition to these areas, we found numerous other frontal areas including those on the medial surface of the PFC and on the right side as being strongly relevant in this analogical reasoning task. In each major area we plotted activation levels at the encoding period, in which the first two items in the analogy were evaluated, the mapping and inference period, in which a new item was generated in response to the presentation of the third picture, and a response period in which the participant indicated whether or not the final item shared analogical, semantic or perceptual similarity with the third item. We observed some very similar trends across all of the PFC areas involved in the task. The analogical reasoning condition evoked the greatest activity in the encoding period and in the inference period. Forming analogies evoked much more activity within all of the frontal areas except the right DLPFC during the encoding period. This makes sense, as participants knew that they would not need to consider that first pair during the semantic and perceptual conditions. At inference, the left DLPFC, LIFG, left MFG, and medial PFC all showed greatest activation levels for the analogical condition over the others. Several reverse trends were evident at the response period, when analogies typically evoked the lowest levels of activity relative to both the semantic and perceptual conditions. This was accompanied by slightly quicker and more accurate responses during the analogy condition. This pattern is consistent with the possibility that participants had more work to do during this last period evaluating a potentially new fourth item that they had not generated on their own previously for the semantic and perceptual conditions, while the analogical matches were more constrained and therefore easier to evaluate.

My colleagues and I followed up the Krawczyk et al. (2010) fMRI study with a second study that used event-related potential (ERP) recordings in order to better understand the temporal or timing dynamics of the cognitive processes that are involved in four-term analogical reasoning. ERP is a technique used to record electrical potentials off the scalp using electrodes placed on

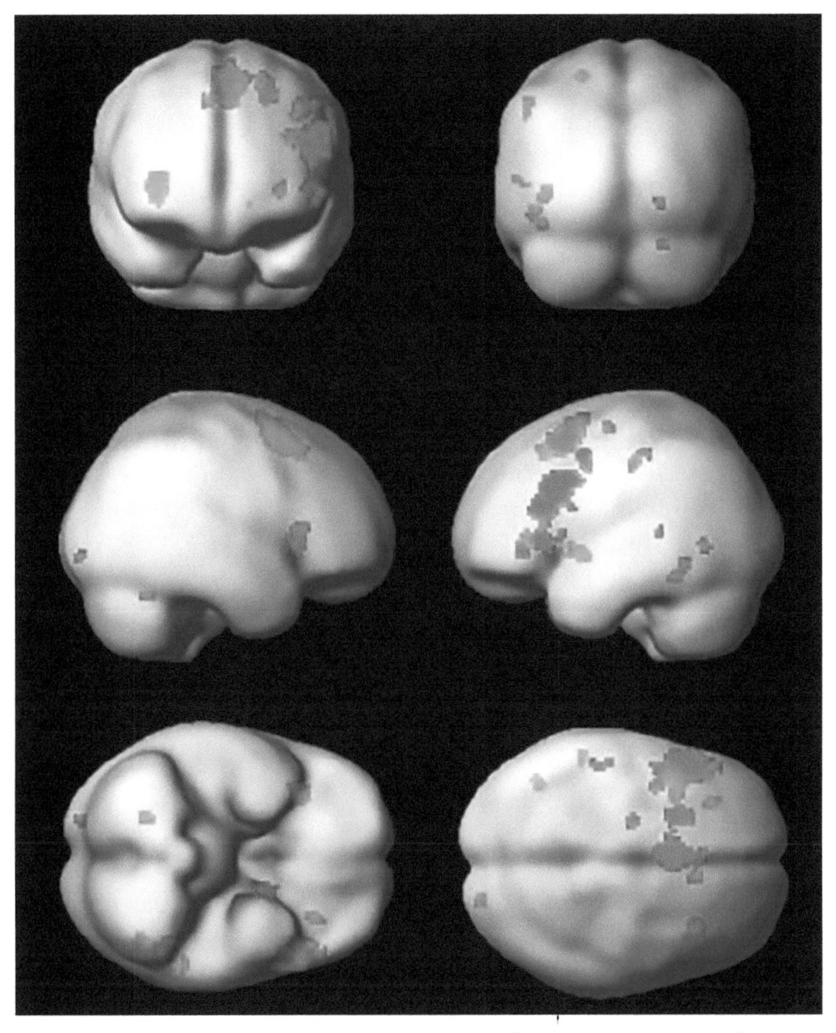

FIGURE 3.14 Results from the analogical reasoning study by Krawczyk et al. (2010) indicated strong left PFC involvement at the time of relational encoding relative to the other task periods. Inferring a new item that would complete a pair sharing relational similarity involved both left frontal and posterior areas and the right DLPFC.

the head. Typically, all of the electrodes are arranged in a mesh and positioned within a cap that enables recording from most of the head simultaneously. ERPs are derived from a recording technique termed electroencephalogram (EEG), which dates back to Hans Berger's pioneering work in the 1920s. Unlike fMRI, which depends upon changes in blood oxygen in order to generate its signal, EEG signals are generated directly from the electrical activity of the brain. Thus, ERPs are much better suited to answering questions about the timing of events within the brain relative to fMRI. The challenge with ERP data is that they cannot be directly traced to a particular brain region. This can be understood by considering the spatial layout of the EEG cap. Typical electrode placements occur centimeters apart and are simultaneously recording signal that is generated from many thousands of neurons that are active many millimeters below the scalp and are passing through the meninges (the outer coverings of the brain), the skull, and hair. What ERPs do very well is provide excellent timing information.

For our analogical reasoning study using ERPs (Maguire, McClelland, Donovan, Tillman, & Krawczyk, 2012), we used nearly an identical experimental setup for this experiment again comparing analogical inferences and verification to the generation and evaluation of semantic associations, as well as perceptually similar objects. During the encoding period, two images, A and B, were presented simultaneously (e.g., spyglass: ship) for 2 s. After a 1-s fixation period, the mapping and inference period included presentation of the C item for 3 s (e.g., periscope). After another 1-s fixation break, the response period included presentation of either the correct target D item (e.g., submarine) or a false item that did not complete the analogy. This period also lasted 2 s. We analyzed data collected from an EEG cap fitted with 64 electrodes while participants evaluated the analogy, semantic and perceptual problems. Data analysis used a Principle Components Analysis to address the locations relevant to analogical encoding and those associated with mapping and inference processes. Additionally, we wished to harness the ERP methods' superior temporal resolution to better understand the timing of these task events in analogical reasoning. Results indicated that left frontal electrodes showed the strongest response to the analogical encoding period. In this case a significant positive EEG waveform was present for encoding of analogical relations relative to the semantic condition beginning 500 milliseconds (ms) after the stimuli were presented and continuing through 1200 ms after stimulus presentation. Meanwhile, a different pair of spatial and temporal patterns were detected supporting the mapping and inference stage of the task. In this case, the left frontal electrodes yielded a negative response, 400–600 ms, post-stimulus that was significantly different for analogies over the perceptual condition. A second site of activity was from the posterior electrodes and was a positive waveform, 700–900 ms, post-stimulus onset.

When considered together the fMRI study by Krawczyk et al. (2010) and the ERP study by Maguire et al. (2012) tell a more complete story than one would achieve with either method alone and reveal some striking features about the neural basis for analogical reasoning. First, the encoding of a relational pair preferentially engaged several left-lateralized PFC regions including the DLPFC, IFG, and middle frontal gyrus in the fMRI task over the other conditions. Additional encoding related areas included the medial PFC and right DLPFC. The ERP study indicated a left frontal electrode site, which contributed a negative waveform, 500–1200 ms, into the encoding period. When considered together, we found that the left PFC is important for the evaluation of relational information in analogical comparisons and that it may also be involved in the maintenance of relational information, but additional PFC regions are also important for the encoding of relational information and these include the right dorsal and midline areas of the PFC. Secondly, the mapping and inference phase of the tasks indicated again that the left PFC was important and potentially somewhat dominant over the right PFC. The positive ERP waveform for the encoding period followed by the negative waveform for mapping and inference are consistent with different cognitive processes driving these effects.

An overall understanding of brain responses evoked by a reasoning task requires a variety of methods that can be applied to the task and an appreciation of the fact that several different cognitive processes are likely to be engaged in any reasoning task. In the domain of analogical reasoning, the initial results from fMRI indicated a left-sided dominance and an almost modular focus on the rostrolateral or frontopolar regions as being critical for relational integration or comparison. Later studies indicated that semantic distance was also a factor that modulated the left frontopolar cortex along with relational integration. When considered in isolation, the encoding of a relational pair is supported by numerous PFC regions with a left-sided dominance overall. The process of inferring a new item that would complete a pair sharing relational similarity involved both left frontal and posterior areas and also showed some degree of modulation in the right DLPFC.

Note that in nearly all of the studies we have reviewed in this section involved four-term analogies. Reasoning in everyday life is much more fluid, less repetitive, and may involve more diverse and complex material. We will discuss these differences in future chapters that delve further into the types of challenges that we actually reason about in everyday life. It is worth remembering that neuroscience studies must be precise about timing and certainty of what task the participant is engaging in at a given moment. Only under those conditions can researchers capture meaningful differences associating cognitive events to activity within the brain.

FUNCTIONAL NEUROANATOMY OF KNOWLEDGE

The Temporal Lobes and Networks for Concepts

To this point in the chapter we have focused to a large degree on the role of the PFC in reasoning. While findings suggest that the PFC is probably the most relevant brain region for functions such as integrating information, dealing with novel situations, and managing our working memory and emotions, it is not the only area important for reasoning. There are many situations in everyday life in which our reasoning does not heavily involve integration or working memory.

Consider a situation in which you are on a date at a nice restaurant. You need to balance your ordering, manners, timing of actions, and track and contribute to the conversations. Having a successful date can be considered a reasoning problem. The date could go poorly for a variety of reasons, but it will surely go disastrously if you are unable to organize your behavior in the restaurant setting. The way to approach this challenge is to use your prior knowledge and avoid overly relying on working memory to get you through the date. You will almost surely do worse in your attempts to appear smooth and sophisticated if you are unfamiliar with restaurant dining, wine, menus, and general protocol. We call this type of knowledge a *schema*, which is a connected set of background information that will apply most of the time. Schemas are developed by observing and being part of many instances of an event or situation. In the case of a restaurant schema, we are likely to have a general sense of what takes place at most restaurants on most date settings. You will expect to arrive on time and possibly have to wait if you have not made a reservation. You will expect to be seated and provided with water and menus. You will expect to have time to study the menu and will have to balance this activity with tracking and keeping up your part of the conversation. You will also expect the food to arrive within a reasonable timeframe and be given an opportunity to consider dessert before the bill is brought. These features of restaurants occur in a particular order and schemas that have this property are often called *scripts*. The reasoning problem of carrying off a successful date, or at least not a disastrous date, probably relies much more on our schemas and scripts than it does on working memory and multitasking. Reliance on the schema will help free up your other mental abilities which you will likely need in order to focus on your partner, observe their needs and actions, and carry on a relevant conversation. You will also need to free your attentional resources for the all-important act of listening. The more you have to use your valuable attention and working memory skills to simply track and predict the mechanics and timing of the meal, the worse your date will likely go.

Schemas and scripts rely mostly upon the cortex of the temporal lobes. The idea of schemas dates back to the work of Sir Frederic Bartlett, a pioneering British experimental psychologist. One of Bartlett's classic works was a study on the "War of the Ghosts," a story about Native American tribal warfare (1932). This story is provided in Box 3.2. The "War of the Ghosts" included elements common to many tribal battle situations, such as rowing a canoe, shooting arrows, and invading a village. The story also included a variety of unusual elements, such as mystical occurrences; being wounded, but feeling no pain; and the notion that the battle had occurred among ghosts. Bartlett asked people to retell the story at a later time, which ranged from just minutes to years later. Findings indicated that when people were asked later on about the story, they would tend to recall a distorted version of the story and the distortion increased along with the passage of time since the original reading of the events. While people will retell the story in slightly different ways, there appear to be some consistent principles by which we distort memories for stories like this.

Frequently people will remember the core elements during a retelling of the "War of the Ghosts," while failing to include many of the specific details. In other words, people tend to recall a gist of the story. Examples include the idea that there were two young men; they went down river, a battle occurred and one was shot by an arrow. The specific details of the story; however, are often lost in the retelling. People are much less likely to accurately recall the name of the village (Egulac), what the young men were initially hunting (seals), or what the young men did when they heard war-cries (hid behind a log). Over time, people would also begin to merge their retelling of the story with their own background knowledge. In other words, they recalled those details that were consistent with their schemas about tribal warfare. Along with this merging of information, people tended to leave out details that were not schema-consistent. As time passed, a greater proportion of the reported story elements were schema-consistent, and many fewer unusual details were reported. At the latest points in time the story was also edited quite a lot suggesting that many details either become lost or simply are less accessible over time.

The work of Bartlett (1932) on the "War of the Ghosts" recall tells us quite a bit about the reconstructive nature of memory and this in turn is informative about how we reason based on schemas, expectations, and gist (Fig. 3.15). First, the changes in recall of the story tell us that memory for details changes over time. This observation is consistent with the notion that memories are reconstructed each time we recall them and represents a fundamental point about how we see the world. The presence or absence of cues is critical to determining whether we will recall a memory. Some cues may evoke memory for certain details, while other cues will evoke other details. Second, the retelling of a story indicates that we tend to build up a general sense of what was important to us about the situation over time. Rarely do we need to extensively focus on all aspects of a narrative or a situation. Our own version of the memory will vary from others, as observed in the "War of the Ghosts" retelling. This is shaped initially by the limitations of perception and attention determining how much information we can perceive at any given moment. The specific aspects of a situation that we attend to will further delineate what we will recall later. Given that both perception and attention are limited, we are already bound to lose many details of any occurrence. Lastly, the specific cues we happen to encode that associate with a memory further shape what we will recall later.

The act of recalling a memory is also heavily influenced by internal factors, and this brings us back to the concept of schemas and scripts. While we have relatively less control of what we can actually see, hear, touch, or smell at any given moment, we have a voluntary ability to attend to some things and not others. This has been compared to a signal-to-noise problem, in which some aspects of a situation will be amplified over other aspects and accompanying this amplification is often an elevation of brain activity within our cortex. The fact that we are able to move our focus of attention is another reason that some of our memories will be different than others. What guides our attention may often be what is relevant to us and this is dependent upon our background knowledge to a large

BOX 3.2

THE "WAR OF THE GHOSTS"

One night two young men from Egulac went down to the river to hunt seals and while they were there it became foggy and calm. Then they heard war-cries, and they thought: "Maybe this is a war-party.". They escaped to the shore, and hid behind a log. Now canoes came up, and they heard the noise of paddles, and saw one canoe coming up to them. There were five men in the canoe, and they said:

"What do you think? We wish to take you along. We are going up the river to make war on the people."

One of the young men said, ""I have no arrows."

"Arrows are in the canoe," they said.

"I will not go along. I might be killed. My relatives do not know where I have gone. But you," he said, turning to the other, "may go with them."

So one of the young men went, but the other returned home.

And the warriors went on up the river to a town on the other side of Kalama. The people came down to the water and they began to fight, and many were killed. But presently the young man heard one of the warriors say, "Quick, let us go home: that Indian has been hit." Now he thought: "Oh, they are ghosts." He did not feel sick, but they said he had been shot.

So the canoes went back to Egulac and the young man went ashore to his house and made a fire. And he met everybody and said: "Behold I accompanied the ghosts, and we went to fight. Many of our fellows were killed, and many of those who attacked us were killed. They said I was hit, and I did not feel sick."

He told it all, and then he became quiet. When the sun rose he fell down. Something black came out of his mouth. His face became contorted. The people jumped up and cried.

He was dead.

FIGURE 3.15 Sir Frederick Bartlett (1886–1969) is well-remembered for his pioneering work on schemas in semantic memory. *Courtesy of the Department of Psychology Archive, University of Cambridge.*

degree. This is perhaps most apparent when we develop a working model of something with our minds and then proceed to test our assumptions by reading more information, talking to others, or seeking out new experiences that will inform us. The influence of both attention and semantic memory can be seen from this example. Our reasoning processes involve language and symbolic coding of information in our networks of semantic memory.

DEDUCTION, INDUCTION, AND THE BRAIN

A Strong Philosophical Distinction

Deductive and inductive reasoning are often compared and contrasted. Fundamentally, deductive reasoning guarantees the validity of a conclusion, provided that the premises are true. For example, if the premises state that all planets are round and that the Earth is a planet, it is valid to conclude that the Earth is round. Induction is an argument from a specific example toward a general rule. Induction is different than deduction, as inductive reasoning does not guarantee a valid conclusion, only a probable conclusion based on experience. For example, if we have heard that both the brown recluse spider and the black widow spider are poisonous, we may be tempted to conclude that all spiders are poisonous and should be avoided. This could be true or possibly false, based on our degree of experience with spiders. We will consider this distinction in greater detail in Chapter 9.

Philosophers have noted the strong distinction between deductive and inductive reasoning. Aristotle wrote about the distinction between inductive and deductive arguments, emphasizing the power of deduction primarily. Later, David Hume emphasized the shortcomings of inductive reasoning, as it was prone to errors due to the need to infer a global rule based on the experience of only a subset of instances. In purely academic terms, there is a strong distinction between deductive and inductive inferences. This can translate into very real consequences in everyday life when we may make errors due to inductive inferences, while deductive logic will always give us a valid conclusion, provided the premises are valid.

Evaluating the differences between these styles of reasoning has been a long-standing goal of psychologists, who are often most interested in the cognitive processes that occur during inductive inference and how these may differ from those engaged during deductive inference. Adding further clarity between these processes at a biological level became possible due to the emergence of functional imaging techniques, notably fMRI and EEG. These techniques have also offered an additional method to investigate the various conditions that are possible within deduction or induction, such as comparisons of deductive reasoning with realistic content compared to pure deduction based on novel and arbitrary rules.

Evidence From Neuroimaging

Some of the first studies to be conducted comparing deductive and inductive inference used the PET methodology discussed earlier in this chapter. Goel, Gold, Kapur, and Houle (1997) published a PET imaging study with the intriguing line "The seats of reason?" in the title. In this study, Goel et al. asked participants to solve classic deductive reasoning problems such as "All men are mortal, Socrates is a man; therefore, Socrates is mortal." In response to these deductive statements, participants were asked to indicate whether the first two statements (all men are mortal and Socrates is a man) entailed the third statement (Socrates is mortal). Participants were also asked to evaluate inductive statements such as: "Socrates is a cat; Socrates has 32 teeth; therefore, all cats have 32 teeth." In the inductive inference condition participants were prompted to indicate whether the third statement was plausible given the first two. Both deductive and inductive reasoning activated areas of the occipital, frontal, and temporal cortex. Goel et al. performed a subtraction analysis, in which the inductive and deductive reasoning conditions were directly compared. Deductive reasoning trials were subtracted (or contrasted) with those involved in inductive reasoning, and this analysis revealed one area more strongly activated by induction, the left superior frontal gyrus on the inside or medial wall of the left frontal lobe. This finding indicated that a potential difference between induction and deduction was associated with the functions of the left PFC. Note that this is a different region than had been observed in other relational reasoning studies around this same period, such as those of Wharton et al. (2000) and Duncan et al. (2000). The region that Goel et al. had identified was more medial than those other studies. It should be noted that there was not a processing load manipulation in the Goel study, as deductive and inductive reasoning about verbal statements are more difficult to vary with different levels of complexity or processing load in the manner that can be done for processing relational aspects of geometric figures discussed earlier in the chapter.

Goel, Gold, Kapur, and Houle (1998) conducted a second PET imaging study, in which they sought to better understand the possible links between spatial cognition and deductive reasoning. In this study, participants solved three types of deductive reasoning tasks (categorical syllogisms, three-term spatial items, and three-term nonspatial relational items) while they underwent PET imaging. All of these deduction conditions showed similar activation patterns when compared to a baseline condition that simply required people to judge the meaning of different statements. In all three cases, deductive reasoning led to activation of a large section of the left lateral PFC in the middle frontal gyrus. Additional areas of activation included the inferior and superior temporal gyri and the left cingulate cortex. Similar regions of the left PFC and temporal cortex appear to be relevant in other types of reasoning problems, notably the solution of analogical and relational reasoning problems.

The content of the problems can dramatically influence deductive reasoning. Goel et al. next conducted an fMRI study of deductive reasoning in which they manipulated the level of realistic content contained within deductive reasoning problems. All deductive syllogisms take a form that can be generically expressed as statements lacking semantic content. For example, these can be stated as "All A's are B's, all B's are C's; therefore all A's are C's." People tend to reason differently about these conditional syllogisms than they might about realistic deductive arguments that contain semantic information. Deductive syllogisms that are content-rich can take the form "All lions are cats, all cats are mammals; therefore, all lions are mammals." Such statements may be faster and easier to process for people if the content is consistent with what they already know. If however, the content goes against what is already known, evaluating these kinds of statements can actually prove to be more difficult for people. Imagine a syllogism such as the following: "All lions are bears, all bears are lizards; therefore all lions are lizards." While this last statement is deductively valid, it does not accurately fit people's background knowledge about categorizing animals. In the fMRI study by Goel, Buchel, Frith, and Dolan (2000), participants evaluated deductive arguments lacking content ("all P are B, all B are C; all P are C"), congruent syllogisms with content that fit with people's knowledge ("all poodles are pets, all pets have names; all poodles have names"), and incongruent syllogisms that were at odds with people's knowledge ("all poodles are pets, all pets are vicious; all poodles are vicious"). In this study, activation for the deductive reasoning conditions over baseline revealed strong bilateral activation in bilateral fusiform gyrus, the left parietal cortex, left middle temporal gyrus, bilateral inferior frontal gyrus, and the basal ganglia bilaterally. These regions are broadly consistent with the earlier PET study by Goel et al. (1998), but fMRI revealed greater right frontal, occipital, and basal ganglia activation that had not previously been detected using PET. This difference may be related to the superior spatial and temporal resolution of fMRI over PET imaging for conducting functional studies. Syllogisms with and without semantic content yielded broadly similar patterns of activation with some differences. Notably, a conjunction analysis was used to identify areas of activation common to both conditions. These included the left inferior PFC, bilateral fusiform gyri, bilateral basal ganglia, and the right cerebellum. There were differences observed with subtractions of the content minus no content conditions leading to activation of the left temporal cortex, left inferior PFC, and bilateral occipital cortex. The reverse subtraction of no content conditions minus content conditions showed greater activation within bilateral occipital, parietal, frontal, and precentral gyrus. This suggests that there are some key processing differences that can be detected within several cortical areas when people reason about information they know about compared to when they evaluate deductive logic in purely formal and abstract terms. Lastly, Goel et al. reported greater activity within two bilateral PFC regions when participants evaluated syllogisms congruent with their world knowledge relative to ones that went against what they know.

These initial studies were successful at identifying the major sites of activation that are associated with deductive and inductive reasoning. They also provided some clues regarding the possible differences between formats of deductive reasoning and the differences between inductive and deductive inferences. Unlike the early studies of relational reasoning that we covered in earlier sections within this chapter, these syllogistic reasoning studies evoked broader and somewhat less-specific activation across the bilateral frontal lobes, along with other brain areas including the occipital cortex, and sometimes parietal and temporal regions. The cognitive operations underlying deductive and inductive forms of reasoning may be broadly similar and commonly involve working memory, semantic memory, and decision-making processes.

In summary, deductive and inductive inference appeared to show largely similar activation of several cortical areas. Among these regions is a left-lateralized set of areas involved in reasoning about real-world situations and content along with a right-lateralized focus for general deduction about arbitrary premises. Other research groups had similarly reported frontal, parietal, and occipital involvement in deductive reasoning about three-term series problems presented as syllogisms (Knauff, Mulack, Kassubek, Salih, & Greenlee, 2002) and for solving deductive reasoning problems requiring the evaluation of conditionals and relational arguments, which is consistent with the engagement of visuo-spatial processes in reasoning (Knauff, et al., 2002). Fangmeier, Knauff, Ruff, and Sloutsky (2006) reported largely bilateral activation for a visuo-spatial deduction task, which indicated the possibility that the right-lateralized activity that had been observed in prior studies was potentially less clear due to visual and spatial processing. More work was needed to further specify which cognitive processes were linked to which brain areas in inductive and deductive conditions.

A Focus on Process-Related Areas in Deduction and Induction

Further research led to further discoveries about the core cognitive mechanisms of deductive and inductive reasoning. This later set of studies was also able to further map cognitive mechanisms onto the emerging brain activation data that were increasingly being collected. To further understand the relevance of the evoked activation in frontal, temporal, and occipital areas and across hemispheres, it became necessary to carry out experiments that enabled more precise mappings between specific points in the reasoning process and areas of activation. This approach is similar to the process investigators took to more precisely understand stages of analogical reasoning and the activity elicited in those tasks. Diana Rodriguez-Moreno and Joy Hirsch (2009) published the results of a multistage fMRI study that helped to clarify the likely associations between brain activity and specific parts of the deductive process. Participants in this study saw an instruction at the start of each trial indicating whether it was a deductive reasoning trial or a control trial that simply required participants to verify if a particular word appeared in the problem. This control condition enabled the researchers to estimate activation levels that were associated with text processing, attention, and judgment that were not specific to deductive reasoning specifically. After the instruction cue, participants were shown a premise either as text on a screen or auditorily through headphones in order to better understand the role of language processing modality on brain activation. An example premise would be "All politicians recycle glass bottles." Next a second statement was presented, for example, "People who recycle glass bottles like wildlife." A short delay period followed this statement enabling the researchers to separate activation

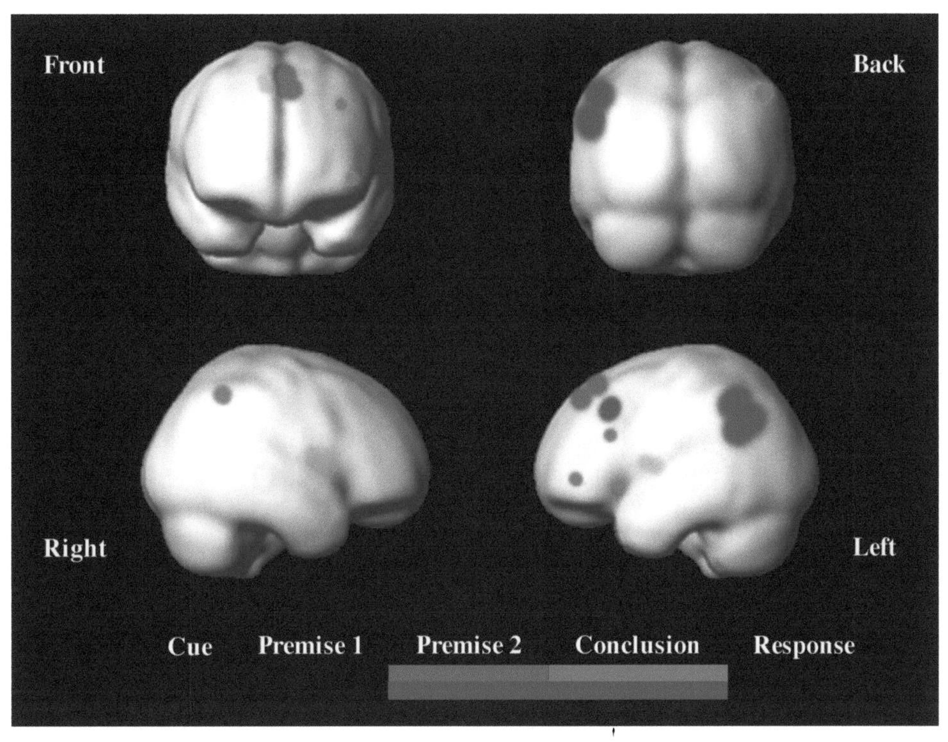

FIGURE 3.16 Rodriguez-Moreno and Hirsch (2009) reported a shift of activation from superior frontal and parietal regions during the second premise of a deductive reasoning task (presented in red) to areas of the left frontal lobe, bilateral parietal and caudate nucleus during the conclusion (presented in blue). Some of these areas showed sustained activity through the second sentence and conclusion (presented in purple). *From Rodriguez-Moreno, D., & Hirsch, J. (March 2009). The dynamics of deductive reasoning: An fMRI investigation, Neuropsychologia, 47(4), 949–961.*

from the premises to that evoked by the conclusion. Conclusions were presented as a third statement such as "All politicians like wildlife." Lastly, there was a word presented for use in the control condition. The detail of this design enabled a variety of important comparisons that helped to isolate the areas of activation that are specific to the reasoning process, while excluding other active areas that are related to other cognitive or sensory processes.

Rodriguez-Moreno and Hirsch (2009) reported a set of areas that were active in response to deductive reasoning demands when presented through auditory or written format. These included the DLPFC, VLPFC, frontopolar PFC, supplementary motor areas, posterior parietal cortex, inferior parietal lobules, and the caudate nuclei. Several of these same regions have been previously discussed in association with relational reasoning and with induction. In addition to the identification of areas involved in reasoning, Rodriguez-Moreno and Hirsch (2009) were able to specify at what points during the reasoning process brain areas became engaged. In the case of the task instruction cue and the first premise there were no differences for deduction than for the control task requiring verification of whether a word had been present in the premises. Additionally, no differences were observed between deduction and the control task for the first premise. Only at the point when the second premise was presented did the deductive and control conditions diverge, with deduction being associated with increases in activation within the left PFC, left parietal cortex, and angular gyrus. The authors described this period of the deductive task as being related to premise integration and conclusion generation. Similarly, the deduction task differed from the control task during the presentation of the conclusions at the point when participants would have had to evaluate the validity of the full syllogism. This comparison revealed activation within the left middle and superior frontal gyri, the right medial frontal gyrus, bilateral parietal cortex, and bilateral caudate bilaterally. Note that left frontal regions were activated during both the integration of premises and the evaluation of a conclusion, as was the parietal cortex (Fig. 3.16). Lastly, Rodriguez-Moreno and Hirsch's study of deduction compared the differences that format had in evoked brain activation. When presented in an auditory format, greater numbers of active voxels emerged for deduction relative to the presentation by text. An additional observation was that while similar frontal, parietal, and caudate areas were active from the period after the presentation of the second premise until the conclusion, the auditory condition tended to have a slightly delayed time course of activation.

The shift in research was moving toward investigation of stages in reasoning. The stages in deduction were also tested by Fangmeier et al. (2006), who evaluated simpler deductive conditions represented by letters. In this task, single letters were presented to participants in the following manner:

Premise 1: V X
Premise 2: X Z
Conclusion: V Z

Participants were able to evaluate these premises by considering them in terms of the sentences "V is to the left of X," "X is to the left of Z," and asked to evaluate whether the conclusion "V is to the left of Z" is valid. In this case, the comparison yields a valid conclusion. Fangmeier et al. were also able to evaluate the stages of deduction in this paradigm. The presentation of the first premise was limited to activation within temporal and occipital cortex, possibly linked to recognition and memory. The second premise activated the anterior PFC in a manner consistent with the prior results that we have discussed on relational reasoning in which integration of information is frequently marked by greater levels of activation within the frontopolar PFC. The final conclusion evaluation, or validation phase, was associated with greater activation within the PFC and lateral parietal areas, similar to those results reported by Rodriguez-Moreno and Hirsch (2009).

While the specific locations of activation differ, many of these same frontal and parietal regions were found to be involved in the stages of analogical reasoning reported by Krawczyk et al. (2010), which included the integration of two relations and the generation of a term to complete the problem. Both types of reasoning involved the encoding of two items or premises, integration of information across time, and the generation of a conclusion followed by verifying whether a final piece of information fit with the problem solution generated.

Work by Cocchi et al. (2014) tested this relationship directly by varying relational complexity conditions within a test of deductive reasoning. The task was a variation of the classic Wason card selection task (Wason, 1966) in which individuals are presented with a deductive reasoning rule such as "If a card has a vowel on one side, then it must have an even number on the other side." When people solve the Wason task, they are presented with numbers and letters and have to determine whether the card would need to be flipped over to see if the number or letter on the back side of the card could disprove the rule presented. For example, if a card has a letter A on one side, it would need to be flipped to see if it indeed had an even number on the back side. If it does not, then this would violate the rule. In Cocchi et al. experiment they also introduced a relational complexity variable such that some cards required greater amounts of relational information that would be necessary to determine whether the card flip would disprove the premise.

Functional MRI results from Cocchi et al. (2014) largely replicated those of Fangmeier et al. (2006) and those of Rodriguez-Moreno and Hirsch (2009) indicating strong DLPFC activation along with parietal activation among other areas in the premise evaluation phase, followed by DLPFC, frontopolar PFC, striatum, thalamus, and insula activity during the card decision phases, when deductive reasoning was maximally engaged. Additionally, Cocchi et al. reported increases in these PFC regions when relational complexity was increased. In this study, we see a pattern of convergence in which deduction and relational processing are considered as being governed by some of the same brain regions. This type of study indicates that the field had begun to move past investigations in which different types of reasoning were always predicted to engage different neural systems. It remains an open question to what degree there is overlap among the neural systems supporting different types of reasoning. Accomplishing this objective will require further research. Some clues to the resolution of the overlap are presented next in the following section on neural networks.

NEURAL NETWORKS OF REASONING

A Shift Toward Neural Networks

To this point in the chapter we have been focusing largely on the locations of activity associated with different aspects of reasoning and how these can be related back to cognitive events. Many of the neuroimaging studies of relational reasoning have been focused primarily on identifying which areas are most essential for discriminating between purely relational comparisons and those that involve semantic or perceptual similarity. Such comparisons lead us to conclude that the most critical area for relational reasoning is likely the left anterior PFC, which has been called either the rostrolateral PFC (Wendelken, Nakhabenko, Donohue, Carter, & Bunge, 2008) or the frontopolar PFC (Green et al., 2006). This regional focus should not be interpreted to mean that the left anterior PFC is a module purely for reasoning, or that it functions in isolation by carrying out the full extent of reasoning functions and passing that information on to other regions. If this were the case, then damage to the left anterior PFC could be predicted to lead to a catastrophic inability to perform relational reasoning tasks. This is not the case, as we will discuss in Chapter 7 covering disorders of reasoning. Frontotemporal lobar degeneration (FTLD) is a form of dementia that typically causes frontal lobe atrophy and loss of function. Individuals who develop frontal-variant FTLD often lose functionality in the anterior-most regions of the PFC first and progressively lose more frontal cortical function as the disease progresses. These patients do not typically reach a neurological clinic until their behavior becomes sufficiently disorganized that it cannot be ignored or attributed to simple changes in personality or lifestyle. At such a time, most patients will have lost dramatically more cortical function than just the anterior PFC, and some patients who are actively being treated for frontal-variant FTLD do not express a complete deficit in analogical reasoning, but rather differences in speed or reductions in accuracy (Morrison et al., 2004).

To understand reasoning in the brain, we must consider the myriad of brain regions that have been discussed as being activated specifically in association with both relational reasoning and semantic processing. Additionally, we must consider the connectivity within white matter that enables these brain regions to connect and share information.

A very useful technique has emerged called functional connectivity. This technique was originally carried out in resting-state studies. A resting-state fMRI scan consists of scanning a person while they are awake and thinking to themselves internally, rather than performing a task. Some resting-state scans require participants to keep their eyes open and perhaps focus on a single point on a screen, while other investigators have simply let participants think internally. What emerges from resting-state scans is a consistent feature of neural systems. Areas will tend to correlate with one another, which is an indication that neurons within these areas are likely communicating as their blood-oxygen demands rise and fall together. This finding was originally reported by Biswal, Yetkin, Haughton, and Hyde (1995). This group observed strong correlations between the right and left motor cortices. The areas of the brain that are actively correlated or connected are not random. There is a characteristic pattern for resting-state activity in a set of areas that are clustered toward the midline of the brain between the two hemispheres. These midline areas have been called the default mode network (Gusnard & Raichle, 2001), which includes areas such as the medial PFC and precuneus, a region in the middle of the parietal lobe. Other rest-state functional connectivity studies have identified other sets of commonly active neural regions that show spontaneous and persistent correlations. Among those most relevant to reasoning functions is the task-network or fronto-parietal control network, which is frequently engaged when someone exercises his or her working memory or cognitive control abilities. Similarly the cingulo-opercular network, including regions within the cingulate cortex and the frontal operculum, is thought to play a supporting role in directing attention and sustaining task focus.

Neural network approaches have been shown to be highly informative for reasoning studies providing a clear way to understand how cortical areas interact, rather than promoting the further search for isolated cognitive modules. Task-based connectivity was carried out assessing the correlations between areas within the deductive reasoning Wason card task carried out by Cocchi et al. (2014). In this study, the task-related set of areas, including the frontopolar PFC, DLPFC, anterior cingulate cortex, parietal and occipital lobes, and the anterior insula, was active during the deductive reasoning conditions. Cocchi et al. (2014) analyzed the correlations among these areas finding that there was greater functional connectivity among the areas when relational complexity increased. The areas that were active during this deduction task overlapped with the fronto-parietal control network and the cingulo-opercular network that we had discussed earlier. Both of these networks are strongly active during cognitive control tasks. Cocchi et al. (2014) suggested that the enhanced connectivity of these network areas may indicate that the DLPFC and frontopolar PFC serve as important nodes in the neural networks that carry out reasoning functions. Further work will be needed to better delineate whether there are regions that do act in a more modular way despite the widespread connectivity.

SUMMARY

The series of studies on the neuroscience of reasoning covered in this chapter suggest a variety of summary points. These studies also point to several unresolved questions that will guide future research directions.

Species that possess large amounts of cortex and especially association cortex tend to show advanced reasoning skills. These species include primates, dolphins, and whales, as well as elephants and certain birds. The relationship of large brains to complex behaviors may be driven by whether an organism is a predator. Predators must possess a wide array of behaviors in order to locate and subdue prey.

Many of the early fMRI studies of reasoning suggested that the PFC was particularly important. Additionally, we have noted that the frontopolar PFC and left DLPFC have important roles based on imaging and electrophysiological studies. Some network studies indicate that the frontopolar PFC has been reported as an active component of numerous reasoning tasks across several domains, including visuo-spatial relational reasoning, analogical relational reasoning, deductive and inductive reasoning. This indicates that the PFC may serve a coordinating or control function, possibly integrating wide-scale activity across the brain, rather than operating as a relational module. We have also noted the impact of materials on reasoning studies in the brain. Materials invoke our semantic memories, which are supported by temporal lobe regions. The integrated sets of information that we can use in reasoning are called schemas or scripts. Brain network interconnectivity looks to be an especially promising area toward capturing the complexity of neural processing in reasoning and may further clarify some of the roles of specific brain areas that have been linked to reasoning in various forms.

END-OF-CHAPTER THOUGHT QUESTIONS

1. A large amount of cortex is associated with advanced reasoning skills. Can you think of some exceptions to this tendency?
2. The relationship of large brains to complex behaviors may be driven by whether an organism is a predator. What specific behaviors of predators may have driven their brains to develop?
3. Functional neuroimaging provided the ability to map brain regions that support reasoning. What are some of the benefits and limitations of this type of study?
4. Many of the early neuroimaging studies of reasoning suggested that the prefrontal cortex was particularly important. Is this area developed for reasoning or for other cognitive processes, such as memory and attention?
5. Neural network studies indicate that the frontal lobes may serve a coordinating or control function. How might we test this possibility?
6. The materials used in an experiment have a large impact on the results of reasoning studies. Materials invoke our semantic memories, which are supported by temporal lobe regions. Can you think of some ways to test this?
7. The integrated sets of information that we can use in reasoning are called schemas or scripts. How could you adapt the script or schema concept into a testable neuroimaging study?
8. Brain network interconnectivity looks to be an especially promising area for reasoning. Are all types of reasoning supported by networks, or do some types appear to rely upon regional brain activity?

References

Bartlett, F. C. (1932). *Remembering: An experimental and social study.* Cambridge: Cambridge University.

Biswal, B., Yetkin, F. Z., Haughton, V. M., & Hyde, J. S. (1995). Functional connectivity in the motor cortex of resting human brain using echo-planar MRI. *Magnetic Resonance Medicine, 34*, 537–541.

Boroojerdi, B., Phipps, M., Kopylev, L., Wharton, C. M., Cohen, L. G., & Grafman, J. (February 27, 2001). Enhancing analogic reasoning with rTMS over the left prefrontal cortex. *Neurology, 56*(4), 526–528. http://dx.doi.org/10.1212/WNL.56.4.526:1526-632X.

Bunge, S. A., Wendelken, C., Badre, D., & Wagner, A. D. (2005). Analogical reasoning and prefrontal cortex: Evidence for separable retrieval and integration mechanisms. *Cerebral Cortex, 15*, 239–249.

Christoff, K., Prabhakaran, V., Dorfman, J., Zhao, Z., Kroger, J. K., Holyoak, K. J., et al. (2001). Rostrolateral prefrontal cortex involvement in relational integration during reasoning. *NeuroImage, 14*, 1136–1149.

Cocchi, L., Halford, G. S., Zalesky, A., Harding, I. H., Ramm, B. J., Cutmore, T., et al. (2014). Complexity in relational processing predicts changes in functional brain network dynamics. *Cerebral Cortex, 24*, 2283–2296. http://dx.doi.org/10.1093/cercor/bht075. Epub April 5, 2013.

Duncan, J., Seitz, R. J., Kolodny, J., Bor, D., Herzog, H., Ahmed, A., et al. (2000). A neural basis for general intelligence. *Science, 289*, 457–460.

Fangmeier, T., Knauff, M., Ruff, C. C., & Sloutsky, V. (2006). fMRI evidence for a three-stage model of deductive reasoning. *Journal of Cognitive Neuroscience, 18*, 320–334. http://dx.doi.org/10.1162/jocn.2006.18.3.320.

Fuster, J. M. (2006). The cognit: A network model of cortical representation. *International Journal of Psychophysiology, 60*, 125–132.

Gentner, D. (1983). Structure-mapping: A theoretical framework for analogy. *Cognitive Science, 7*, 155–170.

Goel, V., Buchel, C., Frith, C., & Dolan, R.J. (2000). Dissociation of mechanisms underlying syllogistic reasoning. *Neuroimage, 12*(5), 504–514.

Goel, V., Gold, B., Kapur, S., & Houle, S. (1997). The seats of reason: A localization study of deductive & inductive reasoning using PET (O15) blood flow technique. *NeuroReport, 8*(5), 1305–1310.

Goel, V., Gold, B., Kapur, S., & Houle, S. (1998). Neuroanatomical correlates of human reasoning. *Journal of Cognitive Neuroscience, 10*(3), 293–302.

Green, A. E., Fugelsang, J. A., Kraemer, D. J., Shamosh, N. A., & Dunbar, K. N. (2006). Frontopolar cortex mediates abstract integration in analogy. *Brain Research, 1096*, 125–137.

Green, A. E., Kraemer, D. J., Fugelsang, J. A., Gray, J. R., & Dunbar, K. N. (2009). Connecting long distance: Semantic distance in analogical reasoning modulates frontopolar cortex activity. *Cerebral Cortex, 20*, 70–76. http://dx.doi.org/10.1093/cercor/bhp081.

Gusnard, D. A., & Raichle, M. E. (2001). Searching for a baseline: Functional imaging and the resting human brain. *Nature Reviews Neuroscience, 2*, 685–694.

Horn, J. L., & Cattell, R. B. (1966). Refinement and test of the theory of fluid and crystallized general intelligences. *Journal of Educational Psychology, 57*, 253–270. http://dx.doi.org/10.1037/h0023816.

Knauff, M., Mulack, T., Kassubek, J., Salih, H. R., & Greenlee, M. W. (2002). Spatial imagery in deductive reasoning: A functional MRI study. *Cognitive Brain Research, 13*, 203–212.

Krawczyk, D. C., McClelland, M. M., Donovan, C. M., Tillman, G. D., & Maguire, M. J. (2010). An fMRI investigation of cognitive stages in reasoning by analogy. *Brain Research, 1342*, 63–73.

Kroger, J. K., Saab, F. W., Fales, C. L., Bookheimer, S. Y., Cohen, M. S., & Holyoak, K. J. (2002). Recruitment of anterior dorsolateral prefrontal cortex in human reasoning: A parametric study of relational complexity. *Cerebral Cortex, 12*, 477–485.

Landauer, T. K., & Dumais, S. T. (1997). A solution to Plato's problem: The latent semantic analysis theory of acquisition, induction, and representation of knowledge. *Psychological Review, 104*, 211–240.

Maguire, M. J., McClelland, M. M., Donovan, C. E., Tillman, G. D., & Krawczyk, D. C. (2012). Tracking cognitive stages in analogical reasoning with event-related potentials. *Journal of Experimental Psychology: Memory, Learning and Cognition, 38*, 273–281.

Morrison, R. G., Krawczyk, D. C., Holyoak, K. J., Hummel, J. E., Chow, T. W., Miller, B. L., et al. (2004). A neurocomputational model of analogical reasoning and its breakdown in frontotemporal lobar degeneration. *Journal of Cognitive Neuroscience, 16*, 260–271.

Prabhakaran, V., Smith, J. A., Desmond, J. E., Glover, G. H., & Gabrieli, J. D. (1997). Neural substrates of fluid reasoning: An fMRI study of neocortical activation during performance of the Raven's progressive matrices test. *Cognitive Psychology, 33*, 43–63.

Raven, J. C. (1938). *Progressive matrices: A perceptual test of intelligence, 1938, sets A, B, C, D, and E*. London: H. K. Lewis.

Raven, J. C. (1960). *Standard progressive matrices: Sets A, B, C, D, & E*. London: H.K. Lewis & Co.

Robin, N., & Holyoak, K. J. (1995). Relational complexity and the functions of prefrontal cortex. In M. S. Gazzaniga (Ed.), *The cognitive neurosciences* (pp. 987–997). Cambridge, MA: MIT Press.

Rodriguez-Moreno, D., & Hirsch, J. (March 2009). The dynamics of deductive reasoning: An fMRI investigation. *Neuropsychologia, 47*(4), 949–961. PMID:18835284.

Rumelhart, D., & McClelland, J. L. (1986). *Parallel distributed processing: Explorations in the microstructure of cognition*. Cambridge: MIT Press.

Sternberg, R. J. (2000). The holey grail of general intelligence. *Science, 289*, 399–401.

Wason, P. C. (1966). Reasoning. In B. Foss (Ed.), *New horizons in psychology* (pp. 135–151). Harmondsworth: Penguin Books.

Wendelken, C., Nakhabenko, D., Donohue, S. E., Carter, C. S., & Bunge, S. A. (2008). "Brain is to thought as stomach is to ??": Investigating the role of rostrolateral prefrontal cortex in relational reasoning. *Journal of Cognitive Neuroscience, 20*, 682–693.

Wharton, C. M., Grafman, J., Flitman, S. S., Hansen, E. K., Brauner, J., Marks, A., et al. (2000). Toward neuroanatomical models of analogy: A positron emission tomography study of analogical mapping. *Cognitive Psychology, 40*, 173–197.

Wharton, C. M., Holyoak, K. J., & Lange, T. E. (1996). Remote analogical reminding. *Memory & Cognition, 24*, 629–643.

Further Reading

Bartlett, F. C. (1920). Some experiments on the reproduction of folk stories. *Folk-lore, 31*, 30–47.

Ferrer, E., O'Hare, E. D., & Bunge, S. A. (2009). Fluid reasoning and the developing brain. *Frontiers in Human Neuroscience, 3*, 46–51.

CHAPTER

4

Comparative Reasoning: A Cross-Species Perspective

Daniel Krawczyk, Aaron Blaisdell

KEY THEMES

- Field research can assess reasoning abilities in animals through observing the variety of ways that an animal solves problems. These problems often consist of obtaining food through foraging or hunting. Problem solving can also be expressed by how an animal avoids predators or danger.
- More complex reasoning abilities tend to occur in species that have large brains with proportionally large amounts of cortex. The prefrontal cortex supports numerous cognitive processes and may integrate information from diverse sources within the brain.
- Field studies of animal behavior provide a window into a species behavior in their natural environment. Field studies have the unfortunate property of not allowing control of extraneous factors which can prevent strong conclusions.
- Laboratory studies of animal reasoning provide a strong degree of experimental control. Such studies tend to have challenges in replicating the problem solving that a species actually carries out in its natural environment. A balance between field research and laboratory-based studies is likely to maximize the advantages and minimize the disadvantages of these two approaches.
- Classical conditioning, as started by behaviorist psychologists in the 20th century, provides an associative learning-based account of many reasoning behaviors. These behaviors can be linked to striatal dopaminergic neurons present in the basal ganglia.

Reasoning
http://dx.doi.org/10.1016/B978-0-12-809285-9.00004-1

- Imitation is a highly successful strategy for animals to learn and apply solutions that they observe in other members of their species. Associative learning may be a mechanism that enables imitative behaviors to occur.
- Theory of mind and causal reasoning studies indicate that animal cognition may be more sophisticated than would be expected based on the formation of simple associations.
- Animals are capable of tool use, as chimpanzees and corvids use both sticks and stones to reach inaccessible foods, to displace water, and for display purposes. Additional studies of tool use confirm that orca whales use water as a tool and humpback whales use air bubbles as tools for hunting fish.
- Animals are capable of relational reasoning and symbol trained chimpanzees have been able to solve simple analogies. These studies suggest that background knowledge is critical to the capacities that an animal is capable of displaying and that specialized training can allow animals to outperform their fellow untrained species members.

OUTLINE

BIOLOGICAL DIFFERENCES BETWEEN ANIMALS AND HUMANS

Introduction

For most of human history we have considered ourselves unique relative to other forms of life on the planet. This may stem from the fact that we tend to use symbols in our thinking, such as language, numbers, and pictures. We develop the ability to use these symbols to represent objects in the world at a young age. Further, our conscious recollections and impressions of the world around us are captured effectively by words and images. Contrast this with the fact that we have considerably less access to the perceptual processes occurring within our minds. As we gain experience with people, places, and situations, we become experts at different skills and go about our daily routines accomplishing things with increasing automaticity, thereby freeing our conscious minds to wander, form impressions, comment on our circumstances, and generally maintain an ongoing inner monologue.

It is tempting to assume that other species do not have these same abilities. When our dogs leap around and bark at squirrels in the backyard trees, we may react by scolding them in as obvious and demonstrative a manner possible, rather than calmly and delicately discussing with them how we do not enjoy the barking sound and that they would win more friends in the neighborhood if they would kindly stop this behavior. Dogs and other animals do not behave as if they grasp complex language in its full richness, so we often use stimulus-response learning methods in order to guide their behavior.

This chapter explores evidence about the mental lives of other species, their capabilities, tendencies, and how these may be similar and different relative to our own. As research delves deeper into the psychology and neuroscience of behavior, we have begun to realize that animal behavior is highly sophisticated and in many instances outpaces our own abilities. For example, we have learned that some species use sound in much more sophisticated ways than humans for locating other animals and in evaluating aspects of their environments. Our growing appreciation for the mental capabilities of other species has coincided with an ever-greater consideration for the ethical treatment of other species. Those species that we judge to possess the most evident levels of consciousness by displaying the highest sophistication of behavior are usually those that are afforded the most protection and value by people.

Evolutionary Considerations

Before we delve into the fascinating studies and observations of animal reasoning, it is worth considering some evolutionary theory. The ongoing effort of cladistics, a branch of biology seeking to classify species into sensible family-relatedness trees, has been in a constant state of change for several decades primarily due to the rapid pace of new knowledge about the genetic similarity among species. While biologists had once relied upon morphology, or the structure of two species, as a basis for similarity comparisons, now most consider genetic-relatedness to be the strongest factor linking different species together. Genes are a good starting point, but we then must also consider the environmental demands placed on a species to truly evaluate its reasoning abilities.

Taking an evolutionary perspective on reasoning ability forces us to consider the causes for particular traits that different organisms possess. For example, when we consider the keen visual acuity and depth perception of many predators, such as cats and dogs, we can interpret this in light of the fact that these animals must be able to detect and pursue prey animals. Similarly, when we consider the reasoning abilities of other species, it is worth considering their particular lifestyle, how they find food, and how they avoid danger. In highly social species there are also strong family or group demands that influence the reasoning strategies used by these species. Evolutionary pressures place survival needs as prioritized behaviors, and these are likely to play a strong role in determining the reasoning capacities of different species.

FIGURE 4.1 The great white shark is a formidable predator that relies on a successful body plan built for speed and ambush hunting techniques. *From Shutterstock.com*

Highly developed reasoning abilities in other species are marked by the ability of an animal to obtain food, avoid danger, or generally interact with one another in flexible ways that take the context of the situation into account. Simple strategies for survival are observed in crocodiles and sharks, both large predators that specialize in rapidly biting and devouring food. These species use what biologists call *stereotyped responses*, or action patterns, that are highly predictable across circumstances, highly similar across individuals, and largely insensitive to prey type and other environmental variables. Such animals are termed to be evolutionarily old, meaning that their body shapes, size, defenses, and likely their behaviors, have remained consistent for millions of years. The fossils of ancient sharks and crocodilians appear remarkably similar to their modern day ancestors and include braincases (the area of the skull containing the brain) that are comparable to the shark and crocodile of modern times. Similar brain size coupled with other characteristics including strong jaws, jagged teeth, and powerful bodies suggest that the behavior of the ancient megalodon shark (*Carcharocles megalodon*) was likely to have been very similar to its more recent ancestor, the great white (*Carcharodon carcharias*). Observations of the brains of great whites and other modern sharks tell us that these species possess an extensive limbic cortex, an area of the brain specialized for olfaction, memory, and survival responses such as the reflexive "fight or flight" response (Fig. 4.1). In evaluating the reasoning abilities of other species, we can consider the size and capacity of the brain as key indicators of the level of reasoning ability that the animals are likely capable of achieving.

Differences in Brain Capacity

Species capable of flexible and complex behaviors typically have greater amounts of cortex, or brain matter (in the case of birds) that is more densely interconnected. There is a division between animals with brains having smooth cortical surfaces, including mice, rats, and crocodiles and animals that possess such large amounts of cortex that it must be folded in order to fit within the braincase of the skull. Species with large folded cortex are termed *gyrencephalic*, which refers to the folds of the cortex, termed gyri. Gyrencephalic animals include all of the primates, dolphins, and elephants. When considering our great white shark from the prior section, it reacts in highly similar ways when the scent of blood is in the water. It hunts by ambushing unsuspecting prey, and its manner of testing the prey for suitable eating is often to take a test bite of it. In sum, it hunts in a relatively predictable manner that is relatively inflexible regarding prey species and situation. Compare this with the behavior of the killer whale, *Orcinus orca*, a similar-sized predator also widely inhabiting the world's oceans. Orcas specialize in feeding upon different types of prey across the world. Orcas in Finland prefer herring; those in the waters off Vancouver Island eat salmon, while others miles off the shore of California feed upon other large sea mammals. Further, the orca is not limited to one method of hunting. Orcas hunt alone or in groups sometimes using coordinated efforts to produce waves to force seals off ice flows. In other orca populations, individuals have been observed to launch themselves onto the sand, beaching themselves in shallow water in order to capture seal pups on shore. This diversity of prey and variety of specialized methods for obtaining it is reminiscent of primate behavior. The common link may be rooted in brain capacity.

Whales and primates are similar in having large amounts of cortex to support their complex behavior. Orca whales have an enormous 5600 g brain, 1000 g larger than the brain of the much bigger humpback whale, and orcas possess more cortex than even the elephant (Manger, Prowse, Haagensen, & Hemingway, 2012). By comparison, the great white, while very similar in body size to the orca, has much smaller brain weighing 22.85 g with considerably less cortical mass. As mentioned earlier, much of the brain mass of the great white is devoted to smell (Yopak, Lisney, & Collin, 2015). Aside from hunting methods, the lifestyle of the killer whale is also more complex than that of the great white, as orcas typically live in family groups that travel together, hunt together, and actively raise their young, unlike sharks. Further, the orca has a highly sophisticated vocalization ability making a range of calls and frequently communicating with it's group (pod) members. The sonar abilities of orcas and dolphins are renowned and are influenced by the large amounts of cortex that this species has. Thus, brain capacity maps onto behavioral complexity in terms of both amount of cortex and how specialized that cortex is.

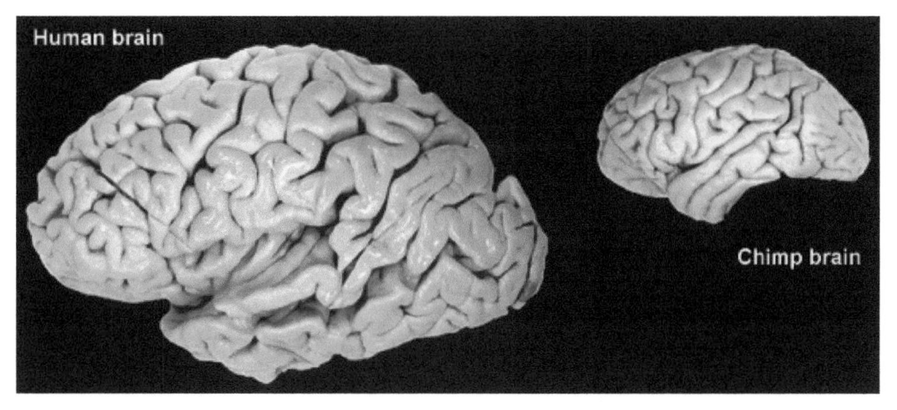

FIGURE 4.2 The size of the cortex in humans and chimpanzees. Both species have large amounts of cortical surface as indicated by the large degree of gyri (bulges) and sulci (folds) present in the cortex. Chimpanzees possess a large amount of frontal cortex, which is termed "association cortex" as it is involved in integrating information from many other sources within the cortex. *From National Geographic: Human and chimp brain to scale, by Todd Preuss, Yerkes Primate Research Center.*

Primates have large brains with large amounts of cortex. Humans lead the way in terms of brain size followed closely by chimpanzees, *Pan troglodytes* (refer to Fig. 4.2). The cortex is comprised of cell bodies, which are the source of electrical impulses enabling the processing power of the brain. Humans have an unusually large amount of prefrontal cortex (PFC) termed neocortex due to its late evolutionary status relative to other areas of the brain. The PFC is centrally involved in numerous brain networks important for attention, memory, and action. Our large brains with comparatively large PFC are linked to our reasoning and symbolic processing capability. Similarly, the chimpanzee is also known to have a large brain, extensive cortex, and disproportionately high PFC mass. Accompanying this high degree of cortical processing power is the sophisticated behavior of chimps. They live in hierarchical social groups, communicate in flexible ways, and hunt a variety of prey. Chimps are attentive parents and even use rudimentary tools. We will discuss more of these capabilities later in this chapter. The common ground between chimps and humans breaks down when we consider language and capacity for symbol use, in which humans appear to stand alone.

COMPARATIVE BIOLOGY

What Can We Learn From Animal Studies?

Animal research has produced vastly important knowledge that we have used to help treat diseases and understand anatomy and physiology, and it has even assisted us in delivering therapies and emotional support for individuals. While, these are all excellent achievements, animal research tends to occupy a controversial place for many people. We must always consider whether animals in research are being treated ethically. Is the knowledge gained worth the sacrifices made by research subjects, in terms of their freedom over their behavior, living in captivity, and being removed from their habitat in nature? These questions have led to reductions in funding for certain types of studies, and in the United States, research with nonhuman primates has dropped dramatically. Nonetheless, we have learned and will continue to learn much about human reasoning capacity by comparison to that of other species.

Comparing Problem Solving Across Species

Reasoning is a challenging area of study in which to make meaningful comparisons between humans and other species. Much of the time, we engage similar learning mechanisms and even similar brain regions relative to other complex animals including most mammals. When we solve problems, we tend to rely heavily on past experience and animals do this as well. Animals rely on brain regions such as the hippocampus to navigate and remember locations and objects and we do this as well. A major shift occurs when we consider how humans cope with new problems that do not fit prior circumstances particularly well. This may be where humans and other species most clearly diverge. New problems demand trial-and-error learning. Such situations place added demands on our working memory abilities rather than our long-term memories and require greater levels of attention. Humans may be unique in that we often verbalize to ourselves how we might go about solving a new problem. We are also likely to engage in planning that can extend far out in time from the current situation. Some of the most innovative human solutions occur because we can factor time into completing a task. We see little evidence of these abilities for most other species, and their behavior suggests that they do not understand the complex mechanics of tools, use the symbol system of language, or predict likely outcomes into the extended future.

A very promising area of research comparing problem solving in animals and humans is the work of Samuel Deadwyler et al. on neural pattern analysis for solving delayed non-match-to-sample problems. This work features a problem-solving task in which the subject is initially shown a cued location and then after a variable time delay has to choose the non-cued location to obtain a reward. Researchers record from around 30 neurons in the CA1 subfield of hippocampus, which then sends that information to the CA3 hippocampal subfield leading to an eventual behavioral output. A fascinating thing happened when the researchers "implanted" a successful non-match CA1 neural pattern from a donor rat into the CA3 subfield of a less-trained rat. Remarkably, the recipient rat was much quicker at learning the non-match task with longer delays. This is one of the only instances in which neural patterns from one animal have been successfully used to train a different one. The same principle appears to apply in PFC areas, as demonstrated by Deadwyler et al. (Opris, Hampson, Gerhardt, Berger, & Deadwyler, 2012) in nonhuman primates, in which multicellular neural patterns predict behaviors. This technique has been tried in humans on the same non-match-to-sample task in an individual undergoing neurosurgery. The key point of these studies for our purposes is that the brain generates and makes use of patterns that have been established over time. Neural ensembles can modify their firing properties to represent information and the specific patterns of those ensembles enable an organism to respond predictably and accurately. This is likely the basis of much of the problem solving that animals engage in and likely underlies a surprising amount of human behavior as well.

How to Test the Reasoning Abilities of Animals

Reasoning has been challenging to study in nonhuman subjects because humans frequently have difficulty in engaging animals in tasks that the animal finds motivating. Typically, laboratory-based animal cognition tasks involve some type of arbitrary rule or association that must be learned in order for the subject to receive a food or liquid reward. Other tasks use avoidance of punishments in the form of shocks delivered to the paws or puffs of air delivered to the eye to motivate behavior. These conditions enable experimental control, but are less similar to the environmental context that the species finds itself in naturally. Animals are often highly specialized and have neural networks that are tuned to operate on certain key problems in their natural environments.

Field Studies of Animal Behavior

Field studies have been a major contributor to our knowledge about the reasoning abilities of other species. This is especially the case for studies of large mammals with complex gyrencephalic brains such as elephants, dolphins, whales, and other primates. Much can be learned by careful and sustained observation of animals interacting within their environments. This approach avoids the ethical challenges that are associated with maintaining captive animals and requiring them to perform various actions for rewards and to avoid punishments. Field studies also free the researcher from having to devise situations and contexts that are representative of the natural environment of the animal.

An example of the value of field studies comes from the work of stress neurobiologist Robert Sapolsky. In order to understand the influences of stress on primate health, Sapolsky conducted field research for years in Kenya observing and recording the daily lives of a baboon troop. The observations of the baboons' reasoning abilities are documented in his excellent book, *A Primates Memoir* (2005). In nonhuman primate groups every individual is highly sensitive to where each one ranks in the dominance hierarchy of the troop. Dominance hierarchies determine who is able to get first crack at food, when and with whom one mates, as well as how the troop will migrate. Dominance hierarchies would initially appear to dictate much of how the troop's daily activities unfold; however, Sapolsky noted numerous complexities in the behavior of the individuals and surprising aspects of the baboon troops' actions. The hierarchy certainly influenced the social lives of these primates by providing an overall structure guiding the movements of the group and goings on within the troop. In addition, individuals looked out for their own interests, snuck food when higher ranking animals were not looking, and were capable of collaborating and forming alliances that facilitated the self-interest of certain individuals. In sum, these primates demonstrated strong social reasoning abilities operating within groups and exhibited clever skills to outwit their fellow troop members and defend their own keep. Many of Sapolsky's observations were possible because of his willingness to commit to spending long periods in the natural environment of these animals and deliberately not intervening in the behaviors of the baboon subjects.

Laboratory-Based Studies of Animal Reasoning

Within the vast literature describing animal research studies, there exists only a subset of reports on what we might properly call *animal reasoning*. Again, this may be due to the brain capacity differences among species.

Much of the research investigating behavior has emphasized learning and the way in which animals respond to either rewards or punishments. This is certainly complex behavior and brain systems are highly adaptive across species to enable animals to thrive in a variety of ecosystems. Species such as sea slugs, hermit crabs, mice, rats, and rabbits have all shown strong abilities to predict danger, avoid punishment, adapt to new types of threats and seek new types of rewards. Many genetic "knockout," or "knock-in" species have been bred with deactivated or activated genetic factors in order to evaluate biochemical influences on the cellular systems that support behavior. This type of learning research has had a reduced impact on the subject of reasoning. Perhaps this is because it can be difficult to define what we really mean by reasoning. When it comes to reasoning, the research community has tended to study those species with large brains and complex social lives.

A guiding question in animal reasoning research is how to categorize reasoning abilities in animals. The research community has not clearly organized a taxonomy of animal reasoning styles in the same way that we have for humans. This lack of clear definitions has resulted in many studies focusing on whether animals are capable of various forms of human reasoning, which may miss critical details related to problem solving that occurs in the wild.

Large primate facilities are a major contributor to animal reasoning research. Of particular note are facilities that house colonies or troops of animals including higher primates such as chimpanzees. Such facilities include the Max Planck in Germany, the Yerkes National Primate Research Center associated with Emory University and Georgia State University, the New Iberia Research Center at the University of Louisiana, and the Primate Research Institute at Kyoto University in Japan. These centers have maintained a clear focus on problem solving and often include both tool-use opportunities and social context manipulations to evaluate the flexibility of primate behavior. An example of such work was carried out by Hanus, Mendes, Tennie, and Call (2011) at the Max Planck Institute in Leipzig, Germany. The researchers presented chimpanzees and gorillas with a peanut that was placed at the bottom of a long, clear tube. The primates were not able to reach the nut using their fingers and could not move the tube. Critically, the chimpanzees and gorillas had access to water and some of the chimpanzees developed a solution in which they used water as a tool by transferring the water to the tube by taking it into their mouths and spitting it into the tube. As the tube filled, the peanut floated toward the top opening enabling the animal to obtain the reward. This solution is both novel and somewhat surprising, as chimps are unlikely to have any prior memories of similar situations to draw from. Gorillas did not succeed at this task and neither did orangutans tested in a second experiment. The utilization of water as a tool to obtain the peanut appears to be a clever solution arrived at through reasoning, not unlike the way humans think. In fact, it is so similar to a human solution that such a problem was actually described by the writer Aesop in one of his famous fables. In the 2000 year old tale, a crow adds stones as tools in order to raise the water level in a pitcher to reach the liquid in order to quench its thirst (Fig. 4.3).

FIGURE 4.3 Crows are intelligent birds and can demonstrate some surprising reasoning skills, both in and out of laboratory environments. *From aesops fables Milo winter 1919 ill schutblad a; photo by Jan Willemsen. CC BY-NC-SA 2.0*

In a case of life mimicking art, Bird and Emery (2009) demonstrated that rooks, *Corvus frugilegus*, a form of crow perform much the way Aesop's fable predicted (see Box 4.1). These birds were able to obtain a floating worm by adding stones to a tube of water (Fig. 4.4). Two of the rooks were able to solve the problem on their first try and the birds were able to discriminate between water and sawdust, only performing the stone solution for the water trials. On top of this, the rooks also learned to utilize larger stones in order to speed up the process. Orangutans in a prior study by Joseph Call's group (Hanus et al., 2011) had previously solved the water tube task.

In all instances of the water tube-reasoning task, we are left to wonder whether the animals were using a learned strategy based on observation or whether they were visualizing a solution in advance and then applying it. If the latter is true, it would suggest that these successful animals have a greater sense of causal attribution and can think flexibly about the possible uses of

BOX 4.1

MIND OVER MATTER

In one of the famous fables of Aesop, a thirsty crow comes upon a pitcher that had once held water. Upon inspection, the crow noticed that there was some remaining water in the bottom of the pitcher and out of his reach. After trying unsuccessfully to tip the pitcher over, the crow realized that he could add some pebbles to the pitcher to displace the water. The crow was able to get a drink at last after depositing enough stones in the pitcher. This fable tells us of the value of using one's mind to solve a problem.

Birds can do remarkable things in the domain of reasoning. In a direct application of Aesop's fable, researchers have devised an interesting line of experiments. Sarah Jelbert et al. tested whether the Caledonian crows actually do use stones or other objects as tools to retrieve food from a liquid-filled container. In the studies, researchers devised a method in which they suspend a floating worm at the top of water contained within a clear tube. Jelbert, Taylor, Cheke, Clayton, and Gray (2014) tested whether Caledonian crows were capable of comprehending the situation and retrieving the highly desirable worm building off a prior Aesop's Fable Task that had been devised by Bird and Emery (2009) to test cognition in rooks. The Caledonian crow, which inhabits the Pacific island of New Caledonia, makes an interesting test subject, as these birds are observed in the wild using small hooks as tools fashioned out of small sticks or pieces of leaf. The birds are able to use the hooks to extract tasty insects from logs and trees. This ability is uncommon in natural environments, and it has drawn comparisons to chimpanzee behavior.

Jelbert et al. (2014) gave the crows several tasks, which included: (1) water-filled tubes compared to sand-filled tubes, (2) sinking objects compared to floating objects, (3) solid compared to hollow objects, (4) narrow or wide tubes, and (5) high or low water levels presented in the narrow and the wide tubes. The birds were able to demonstrate the power of their minds by solving the Aesop's fable task by dropping objects into the containers to raise the water level to a degree that allowed them to obtain a small floating cube of meat attached to a cork. The researchers observed several interesting things. The crows tended to drop solid objects (rubber erasers) into a water-filled tube rather than one filled with sand. This suggests that they have some understanding of the context of the task and the properties enabling displacement of the surrounding matter. The crows preferred solid objects that sink rather than hollow objects that float. The birds also preferred to work with a tube that contained more water than one with less, conserving effort in the quest to obtain the reward. There were some limits to the crow's performance, as they appeared to be insensitive to the width of the tube, failing to recognize its relevance to the task. Cheke et al. suggested that the crows demonstrated a reasoning capability similar to that of children in the 5–7 year age span.

Another bird that can pass the water tube task is the great-tailed grackle (Logan, 2016). Wild grackles were able to pass this task in a lab. Logan showed that four out of six grackles were efficient at problem solving.

FIGURE 4.4 The crow is a capable problem solver. The tube contains a reward floating on the surface of the water, but the water level is too low for the crow to reach the reward. These crows are able to make use of stones as tools to displace water in order to raise the water level to a point at which the reward can be reached. *By Leigh Ann Cox, Artist.*

objects in the environment. It is worth noting that Hanus et al. (2011) also tested children on the water tube task. In this task, the children were provided with a water pitcher and were given an opportunity to water plants with the pitcher prior to the water tube experiment. Approximately 8% of four-year-olds solved the problem, while 58% of eight-year-olds were able to solve the problem. Another variable in the experiment was whether the children observed the peanut initially floating in a small amount of water that was not sufficient to make the nut reachable, or whether the peanut was initially sitting at the bottom of a dry tube at the beginning of the experiment. Children at age eight were more successful if they initially encountered the floating peanut. This feature of the experimental setup suggests that it was helpful to highlight the fact that the nut can float. This same clue was also provided to the chimpanzees, but it did not affect their performance.

Functional fixedness is a term that describes a tendency to view an object as having only one function. For example, a set of house keys can be used for scraping, cutting, and making a jingling noise, in addition to their clearest function of opening locks; yet, we may tend not to think of our keys as allowing these possibilities due to the very strong association that keys have with lock opening. Functional fixedness may have limited the ability of both the apes and the children tested on the water tube task. Such a possibility appeared likely, as two different chimpanzee groups had varying levels of success at the task. A chimp group in Leipzig was not initially able to solve the task, while one in Ngamba was successful. A particular Leipzig chimp named Frodo appeared to suffer from functional fixedness in this task. Frodo was able to solve the water tube task successfully retrieving the peanut by using water from a variety of sources as tool. Frodo could gather water from hose, from a bottle held near his cage, and from a variety of dispensers. Oddly, Frodo consistently failed at the water tube task when the only water available was contained in his water dispenser that was used on a daily basis for drinking. It is plausible that overtraining in drinking from that particular dispenser led Frodo to perceive his own dispenser as being limited exclusively as providing water to solve the problem of thirst. Such a conclusion was supported in another experiment by Hanus et al. (2011), in which they introduced new water dispensers for the Leipzig chimps. This change of presentation led to a significantly greater ability of the Leipzig chimps to solve the task. This experiment, perhaps more than the others, lends some insight into the mental representations that these chimpanzees have about the water tube situation task. When the setup is overly familiar, participants may fail to explore and therefore miss the alternative function that a water dispenser can play in helping to solve the peanut task. A similar situation may have occurred in the case of the children who failed to solve the task, as the water pitcher had been used for watering plants prior to the water tube task, potentially biasing children away from seeing the pitcher as providing a solution.

The Tension Between Experimental Control and Ecological Validity

There are clear trade-offs that have to be made when studying animal reasoning. One of the most important goals is to establish a proper balance between experimental control and ecological validity. Experimental control in the context of animal studies refers to the researchers' ability to isolate a particular type of behavior by seeking to minimize environmental influences. Such influences can include the presence of other members of the species, complexities of the environment, which can include distracting items, uncontrolled noises, sights, or smells, along with carryover effects from other prior behaviors or recent activities. For example, it would be undesirable for an experimenter to spend a considerable amount of time trying to corner and isolate a rodent just prior to enlisting its participation in a timed maze running task, as the animal may become anxious and fatigued before it has begun the experiment leading to overly long maze completion times. Additionally, it may be undesirable to test multiple animals in a fixed order in the same maze directly one after another, as the rodents may smell leftover food or the scent of the prior rodents and behave differently than they would otherwise. Alternatively, well-controlled laboratory conditions that minimize extraneous factors may also result in behavior that is not representative of what an animal will do in its natural environment. In this case, the experimental setup may look so unlike the animal's familiar habitat or life in the wild that there is little chance that the reasoning ability captured would be relevant in the daily life of the animal or that it would generalize to other members of the species. An example here would include cases in which a chimpanzee is tested on tool use with only complicated manmade objects at its disposal. Provided the chimpanzee were able to perform some action with the tools, it is very unlikely that any such scenario would inform us about the problem-solving tendencies of chimpanzees that occur in their natural habitat.

There is not a clear answer as to how to optimize the balance between experimental control and ecological validity (Fig. 4.5). The researcher must make a determination as to how much a behavior resembles what may occur in the natural environment of the animal and seek to control for as many possible variables as

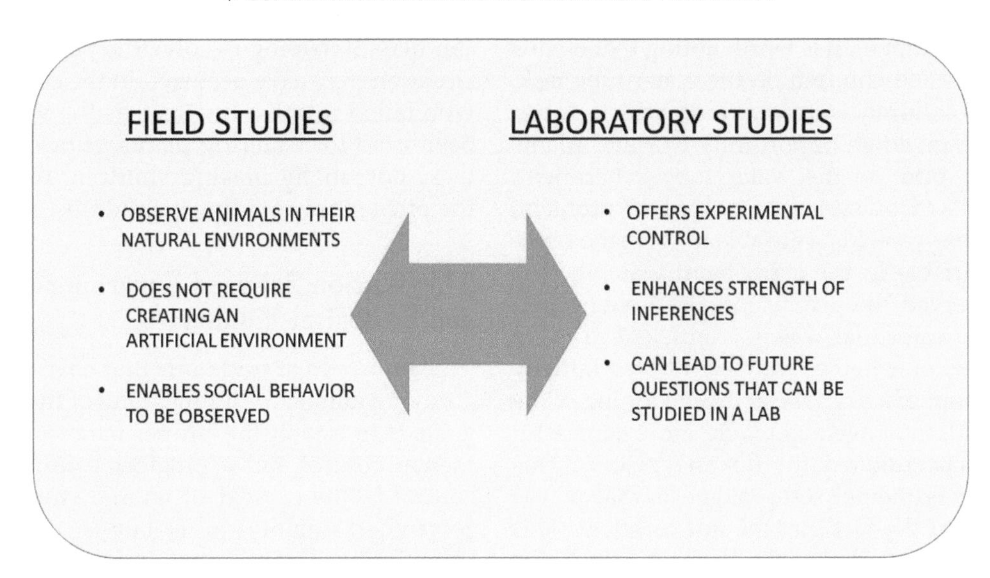

FIGURE 4.5 Researchers much consider many factors when they balance experimental control with ecological validity in animal research.

can be reasonably achieved. As a result, we are likely to continue to need both field studies of animal reasoning, as well as more controlled laboratory-based studies, in order to further our knowledge of the reasoning ability of animals. Future research may seek to balance these through the creation of experimental setups within research facilities that house large colonies of primates. Abilities to investigate reasoning in other species with high brain capacity including elephants, dolphins, and whales appears to be less promising due to the challenges associated with housing and replicating environments that resemble the natural habitats of these large animals. Additional complexities arise with studying cetaceans in field research; as such, animals are unpredictable in their migration, difficult to track, capable of rapid movement, and require extensive allocations of time, effort, and expense to track and observe. Lastly, treating large, intelligent mammals ethically in captivity has become an increasingly widely appreciated challenge. These factors result in a greater knowledge base having been achieved for the study of primates over other large-brained species.

FIGURE 4.6 Ivan Pavlov (1849–1946) is one of the most famous early experimental psychologists and behaviorists. He famously studied the behavior of dogs in classical conditioning research. *Credit: Wellcome Library, London Ivan Petrovich Pavlov. Photograph by Deschiens.*

ASSOCIATIVE LEARNING AS A BASIS FOR ANIMAL REASONING

Behaviorism and Classical Conditioning

One of the most widely studied areas of animal research is associative learning, which investigates the methods by which animals pair objects, sights, sounds, or other cues with outcomes, typically rewards or punishments. Such studies extend back to the early days of experimental psychology. Indeed, most early psychologists including Wilhelm Wundt, William James, and Hermann Ebbinghaus were considered associationists, as they were keenly interested in the study of how things came to be paired together in the mind.

Perhaps the most famous early associationist studies were those of Russian psychologist Ivan Pavlov (Pavlov & Anrep, 2003), who is widely known for his classical conditioning research performed with dogs (Fig. 4.6). In 1914, Pavlov paired certain tones (termed the *conditioned stimulus*) with the delivery of a reward in the form of meat powder and observed a salivation response from the hungry dogs. Eventually Pavlov was able to evoke the salivating response to the tone alone, thus the salivation became a *conditioned response* that had come to predict the delivery of a reward. This powerful demonstration of learning led to more than a century of additional research into how animals come to associate stimuli such as sounds, sights, and situations with the prediction of either rewards or punishments. Such studies have been highly informative about both behavior and the neural systems that enable these memories to be encoded, stored, and extinguished. Perhaps the best-characterized learning system in the brain relies upon the neural circuitry of the cerebellum and was mapped out in exquisite detail by Richard Thompson et al., who investigated the mapping between conditioned tones and eye-blink responses in rabbits that came to expect the delivery of a puff of air (McCormick & Thompson, 1984). For the purposes of understanding reasoning in animals, the classical conditioning research of behaviorists, such as Pavlov, B. F. Skinner, and later researchers, in learning and memory is extremely powerful.

A clarity that emerged from classical conditioning research is that associative learning applies to species with very simple nervous systems, such as the sea slug, *Aplysia*, across the spectrum of complexity up to and including large and socially complex mammals. The ability to adapt to new predictors for desirable and undesirable outcomes is an important process that may provide a key building block for the higher-order system-level associations that are characteristic of human reasoning. This technique potentially applies to other higher primates as we will describe later in this chapter.

Dopaminergic Systems and Predicting Rewards

Until now the neural systems supporting classical conditioning phenomena were not well understood at the level of neuronal signaling. Neuroanatomical regions such as the amygdala, the hippocampus, and the cerebellum were known to be involved in the conditioning process, and damage to these structures through neurological injury or surgery was demonstrated to abolish associative learning. A landmark finding linking the midbrain dopaminergic system to learning was established by Schultz, Dayan, and Montague (1997). In this study, Schultz et al. evaluated a conditioned learning setup in which monkeys were trained to associate a tone with a food reward. Schultz recorded from neurons that fire on the basis of the neurotransmitter dopamine, previously established to be sensitive to rewards. These neurons were recorded with an electrode in the striatal region of the basal ganglia. The basal ganglia receives a wide array of information from other parts of the brain and also sends processed information to other brain areas which are associated with performing actions, including the PFC.

The striatal dopamine neurons of the basal ganglia (Fig. 4.7) receive several critical categories of information needed to enable learning to occur. These include sensory stimuli, such as sights and sounds, information about the quality of a reward, information about the relative timing of sensory stimuli and the delivery of a reward, and their responses can be modified flexibly and rapidly. The neurons that were recorded in the Schultz study initially became active in response to the cue and then activated again seconds later to the delivery of the reward. Over time, the neurons adapted to respond only at the time that the cue was presented. In other words,

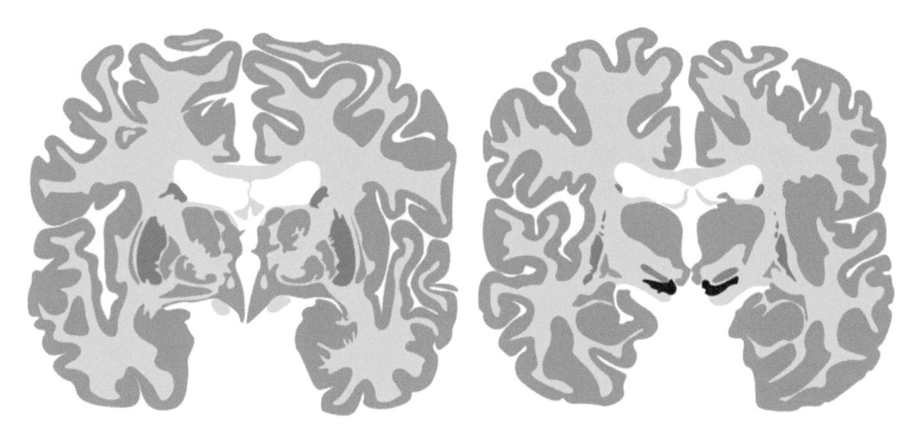

FIGURE 4.7 There are neurons within the basal ganglia (highlighted areas) that are sensitive to reward signals and also signal the nervous system when a prediction error has been made. Such cells are an important clue about how organisms receive signals that adjustments in behavior are needed. *Image by Mikael Häggström and Andrew Gillies – Creative Commons Attribution ShareAlike + GFDL license, CC BY-SA 3.0,* https://commons.wikimedia.org/w/index.php?curid=10301415.

these neuronal responses became adapted to predict upcoming rewards.

Observation of the reward cue-prediction sensitive neurons led to a second interesting effect as well. When the expected reward that had been predicted by the cue was withheld and not delivered, the activity of the striatal dopamine neurons reduced. This finding becomes important in the context of cued behavior. Since these striatal neurons communicate reward associative information to the PFC and other action-related brain regions, the increases and the decreases in responsiveness of these cells appear to signal to an organism, whether to maintain a particular action or to withhold it and explore other possibilities, as the previously learned behavior now appears to be no longer profitable. Similar responses in the nucleus accumbens and human striatum have been observed in response to monetary rewards, juice, tastes, and odors (Berridge, 2007).

The importance of this dopaminergic reward prediction neural system is that it is well-positioned to enable rapid, dynamic, and flexible responding to occur. This response has clear value in a variety of situations including foraging for food and exploring alternative strategies or response options. This same system has been investigated in the context of pathological gambling, addiction, and substance abuse, all of which represent deficits of human reasoning and decision making.

ASSOCIATION AS A BASIS FOR PROBLEM SOLVING

The Clever Hans Phenomenon

In Germany in the early 20th century, a horse called Clever Hans became famous for his feats of intelligent behavior (De Sio & Marazia, 2014). Hans' owner Wilhelm von Osten, a former schoolteacher, worked with Hans to enable the horse to answer questions using stamps of his hoof. Hans was capable of stamping out an appropriate number of hoof beats in response to a question about the number of letters in a person's name, even a half hour after he had been introduced to them. Hans could also indicate the colors of objects by stamping out previously stated codes with his hoof; he could even calculate fractions accurately, as indicated by hoof stamps. Bystanders were amazed and remarkably there was serious talk that Clever Hans indicated that horses could possess reasoning skills equal to those of humans! A special commission of avowed skeptics evaluated Clever Hans in 1904 and concluded that there was no trickery responsible for Hans' performance and that horse was not merely responding to "training" from his owner. Wilhelm von Osten also professed to believe fully in the horses mathematical, deductive, and observational skills. Incredibly, Hans and Von Osten inspired a rather large community of people that it was necessary reevaluate the intelligence of animals altogether amid reports surfacing in esteemed news sources acclaiming the impressive deeds of this brilliant horse. A notable exception was skeptic, Oscar Pfungst, who remained unconvinced.

The Clever Hans act inspired one of the first rigorous investigations into what have become known as *experimenter expectancy effects*. These represent situations in which an experimental subject performs actions consistent with the wishes of the experimenter. Experimenter expectancy effects typically occur when a research participant detects subtle cues that have been provided by the experimenter. These cues may be unintentional and thus very difficult for the experimenter to avoid providing. Experimenter expectancy effects can completely disrupt the validity of experiments whether the cues are provided explicitly or implicitly. Oscar Pfungst decided to carry out a more thorough examination of Clever Hans' mental feats. By isolating Hans from his trainer, Von Osten, Pfungst confirmed that the horse's apparently remarkable reasoning skills were actually responses to subtle changes in posture and expression that were being provided by Wilhelm von Osten. For example, when Clever Hans was asked to divide 20 by 5, he would begin stamping after the question was asked and then stop his stamps when he detected very slight cues such as gasps of breath, subtle head nods, or expressions of emotion by Von Osten. The "Clever Hans" effect refers to cases where an animal's behavior comes under the control of inadvertent cues given off by the experimenter's own behavior rather than from the experimental task. In many cases, the animal in question has learned to associate an arbitrary command or cue with a behavior. This is both an example of behaviorist learning and the influence of a human selectively interpreting what he or she wishes to see from an animal. In other words, people being unwilling or unable to look for disconfirming evidence, a bias blind spot that plagues human reasoning that we will cover in greater detail in Chapter 11, covering decision making and abductive reasoning.

Problem Solving in the Wild

To understand the reasoning abilities of animals, it is necessary to observe and record the skills, strategies, and solutions that are employed in the animals' natural environment. Such studies are often challenging to carry out, as observation requires a set of skills and concerns that are quite different from those in laboratory environments. Field research typically requires spending considerable time observing animals in their natural habitats, being careful not to influence their behavior by one's presence. Additional challenges of interpretation arise, as one cannot leverage the control of the laboratory in the outdoor environment. Such studies tend to be carried out by biologists, who specialize in either ethology, neurobiology, or ecology. These approaches can be particularly useful when they are supplemented by more targeted laboratory-based studies that complement the field research by shedding additional light on challenging questions of interpretation.

Reasoning skills become important when considering food choices made by different species in their natural environments. Optimality theory has been one of the guiding strategies that helps us to better understand feeding behavior (Richardson & Verbeek, 1986). This theory considers feeding behavior in terms of both calories obtained and costs to obtain those calories. There are typically a variety of food items available to both predators and foraging animals, but potential foods vary in nutritional value. To maximize benefits, animals must also factor in the costs that go along with obtaining possible food items. In the case of foraging animals, this may chiefly consist of avoiding dangers that are presented in the form of poisons or avoiding natural defenses that are presented by plants or animal food sources. For predators, this may also include a calculation of how much energy will be required to catch a prey animal, subdue it, or to circumvent a prey animal's defenses. We can assume that most species do not explicitly calculate optimal strategies, but rather their neurobiology becomes tuned toward beneficial strategies over time (this would include both the evolutionary time during which an animal develops and the lifespan of the individual). This neural tuning toward optimal foraging may derive from the associative learning mechanisms that we reviewed earlier in this section of the chapter. Neural tuning for optimality may also include the actions of specialized neural circuits that are specifically guided to compute particular aspects of behaviors or strategies. Examples of this latter case can be observed in the insect world, in which organisms with relatively simple nervous systems carry out highly stereotyped and rigid behaviors that are nonetheless extremely well-matched to accomplish a particular task. For instance, bees are well-known for their "dances" that include particular flight patterns and bodily movements. These dances are heavily genetically guided, carried out across individuals, and capable of relaying key information about food quality and distance that must be navigated in order to obtain the food.

Optimality theory can be observed as a guiding principle in solving foraging problems. An example of this can be seen in crows living in the Pacific Northwest. These birds feed on littleneck clams that the crows dig from their burrows. The clams must be opened by lifting them into the air and dropping them from a sufficient distance that they will crack on the rocks below. Critically, the largest clams that would provide the greatest number of calories also require the most energy to transport and are less likely to crack on the rocks than other sizes of clams. Meanwhile, the smallest clams are easily transported and most likely to break open when dropped, but also provide the least caloric benefit. In order to solve this clam problem, the crows have to be selective. They cannot predict in advance what size clam they will dig up initially, but once they obtain a clam and assess its size, they then must make a decision about whether to fly it up to a height and drop it (Fig. 4.8). The flight expends valuable energy in the process. Alternatively, they can simply abandon a big-sized clam and put forth additional effort in locating a more suitable sized clam that is more likely to break open when dropped. Observation of these crows in their natural habitat indicates that they tend to gravitate toward selecting middle-sized clams for dropping (Richardson & Verbeek, 1986). Such mid-sized clams represent the trade-off between obtaining a maximal caloric value, while minimizing the costly possibility that the clam will fail to break open when dropped. Richardson and Verbeek were able to generate a predictive model that quantified this trade-off by plotting clam size to the percentage of time such clams were eaten. The crows may use a reinforcement learning mechanism that is informed by trial-and-error in order to achieve this successful clam-eating strategy.

FIGURE 4.8 A crow must expend the energy by flying high enough to break open a clam. *By Myburgh* – http://en.wikipedia.org/wiki/Image:Pied_crow.jpg, *CC BY-SA 3.0,* https://commons.wikimedia.org/w/index.php?curid=766186.

FIGURE 4.9 Lion prides are capable of cooperative hunting. Group hunting allows these animals to subdue prey animals that would be too large for a single animal alone to successfully hunt. *From Shutterstock.com.*

Another interesting example of problem solving in the wild has been observed in lion prides. Lion prides are organized primarily around the lionesses who spend much of their time raising cubs, carrying out much of the hunting, and foraging in coordinated groups (Fig. 4.9). Lions can go for many days without eating, but must be careful about their prey selection factoring in both the energy that will need to be spent to capture prey and how risky it will be to subdue the prey. Some evidence suggests that lions can be quite successful in terms of their overall number of calories obtained per day when hunting alone or in pairs (Packer, 1986). Why then do lions often hunt in groups and live in sometimes large and highly social environments? Does this lifestyle benefit their survival, or do many mouths to feed actually hurt a lion's probability of survival?

One possible reason for group hunting in lions is that the effort expended to obtain the calories is likely much higher the fewer lions are involved in a hunt. There are several species of large prey that lions will attack, including hippos, giraffes, and wildebeest, that would likely not be suitable for any single individual or even a small group of juveniles to subdue. Common lion prey animals also include the Cape buffalo, which can weigh up to a ton and has large horns and a highly irritable disposition. A small group of lions can potentially take down a Cape buffalo, but it is a risky proposition even for several lions. Group hunting enables lions to potentially attack any prey animal that they encounter, which would vastly reduce the search time and travel effort involved in selecting only the smaller prey animals. Again, it is not clear to what degree lions possess awareness of these strategies. It may be that group hunting also provides numerous other benefits including companionship, greater protection for the young, and protection for the individual in cases of injury or disease. Optimality theory helps us to understand the behavior of a variety of animals as they factor in risks and benefits to solve their feeding challenges. Next, we move on to primates and the use of innovative strategies carried out by individuals and groups.

Tool Use and the Transmission of Hunting Techniques Within Species

For many years, it was believed that tool use was one of the hallmarks of human intelligence and superior reasoning ability that separated us from other species, including other nonhuman primates. This notion came into question when it was observed that chimpanzees in the wild regularly use tools (Whiten et al., 1999). Such tools include stones, which can be used to break open nuts; leaves, which can be used as sponges to absorb and consume water; and sticks, which prove very handy for extracting ants or termites from their inaccessible mounds. The stick technique is possibly the most impressive tool-use case for chimps in the wild. If the chimpanzee tries to grab ants with its hand, the ants will bite it. Many chimps have been observed to find a long stick, place it into the anthill, and wait to extract it along with a variety of ants that can then be licked off the stick. It is an effective and clever technique, but how does this hunting behavior develop? The technique may have initially been happened upon by chance through trial-and-error and then formalized through observation and repetition by other chimpanzees. One way to address this question is to observe whether young and inexperienced chimpanzees use the stick hunting technique and how they perform it. Young chimps have been observed to experiment by sticking their hand into the anthill repeatedly and thereby receive repeated nasty bites. These young individuals also tend to observe successful adults using the stick technique. It may be a combination of trial-and-error learning and observation that leads ultimately to success.

A follow-up laboratory study of chimpanzee behavior was carried out by Tomasello, Davis-Dasilva, Camak, and Bard (1987) in order to clarify the possible mechanisms for learning in chimpanzee tool use. In the study, experimenters placed a food item out of reach and the chimpanzees were given a T-bar that could be used to reach the food. The critical question was whether young chimpanzees would be more readily able to learn to use the T-bar as a tool if they were exposed to an adult chimpanzee who demonstrated this behavior compared to older chimps and those that did not observe the adult demonstrator. The study team found that young chimpanzees did use the T-bar both faster and more effectively when they had previously witnessed the adult using it. Meanwhile, the older chimpanzees were not as readily able to benefit from the modeled tool use. This finding suggests that young chimpanzees may be particularly receptive to copying the strategies that they see modeled by older animals. This situation also indicates a form of *cultural transmission* in which animals can learn reasoning techniques from their local group, but these do not apply more widely to other animals who either have not hit upon a particular problem-solving strategy or who have not had others model the behavior.

In some instances, biologists refer to imitation of a behavior by other members of the group as cultural transmission (Whiten et al., 1999). There is some debate among individuals in the scholarly community about what actually constitutes a *culture*. Anthropology has tended to take the term culture very seriously and restrict it to groups of human individuals only. The field of biology has focused on a less-restrictive definition that is mostly informed by the means of learning a particular behavior, in this case tool use. The two dominant possibilities are first that the behavior is genetically transmitted to offspring and relatives in the group, while the second is that the behavior is learned through imitation with older, more experienced individuals modeling a process that is then noticed and repeated by the young. Whiten et al. (1999) reported on the behavior of seven different chimpanzee groups that were both genetically and geographically isolated from one another. Such dispersion would likely ensure that the behaviors present in each group must have developed independently and therefore be more representative of the capacity of the species than studies that are restricted to local groups only. The researchers categorized behaviors according to whether they were present in each of the particular groups, how commonly the behavior was observed in each group, and whether the absence of a behavior in a group could be explained by some features of the local environment, such as a lack of appropriate materials with which to fashion a given tool. The results indicated some very strong similarities among the seven chimp groups regarding stick use. Mostly all of the groups used sticks for grooming, displays, drumming, and investigation. Strong similarities were also observed among the groups in terms of the use of a bunch of leaves as a sponge.

These results suggest that some behaviors are either genetically transmitted, or they may just simply be common behaviors that most individuals of a particular species will happen upon and then transmit through modeling to the rest of the group, either intentionally or unintentionally. Strikingly, there was also a variety of tool-use behaviors that were either completely distinct to one of the seven groups, or were highly uncommon in most of the other groups. Such behaviors included club use, termite fishing, extracting bone marrow using a pick, and using wood as a hammer along with a stone as an anvil for breaking open nuts. These actions not only represent impressive reasoning strategies, but also suggest that they are likely culturally transmitted among the individuals in the group, as they are so particular to certain groups only. In large-brained primates imitation and social learning are likely major driving forces in food acquisition methods and other sophisticated problem-solving behaviors that involve strategies or use of environmental materials.

BOX 4.2

WOLFGANG KOHLER AND THE MENTALITY OF APES

Wolfgang Kohler (1887–1967) was a noted Gestalt psychologist who was also an early comparative scientist carrying out the first well-documented studies of chimpanzee cognition and behavior. Kohler spent an extended visit at the Anthropoid Station in Tenerife, a Spanish island off the coast of Morocco during the time of World War I. The island had a population of nine research chimpanzees whom Kohler was able to study during the time he was there. Kohler's research consisted of providing the chimpanzees with a variety of "objectives." An objective was typically a piece of fruit placed in variety of out-of-reach locations or behind obstacles. The experiments either occurred within an enclosed pen or in an outdoor area called the "playground." The chimpanzees were typically provided with a variety of objects to use in their attempts to reach their objective and obtain the food. For instance, Kohler placed food on pulley out of reach of the chimpanzees. Often, several of the chimps were allowed to work on getting the food at a time. At their disposal were wooden boxes that could be stacked to reach high objectives, poles that could be used for climbing, and sticks for reaching both up to heights and around obstacles. The chimps were observed to use a variety of methods, sometimes employing several simultaneously. Sticks were common implements that the chimps used to reach to greater heights. The "jumping stick" method involved using a long stick to vault up to a greater height to grab fruit, or in some cases shimmying up the stick when it was placed on one end like a stilt. Sticks were often used as reaching aids to reach objectives, much as they are in the wild when chimps excavate termites and ants. In some cases the chimps would work in groups of two or three and combine methods by using stacked wood boxes as platforms and then use a stick to reach fruits fastened high overhead (Fig. 4.10).

One of the most famous examples of chimpanzee reasoning occurred when Kohler's star chimp, Sultan, managed to assemble a long pole out of two hollow poles with one fitting inside the other. This "double-stick" of Sultan's could then be used to reach fruit that either stick alone was too short to reach. It is worth considering how Sultan reached the double-stick solution. The two fitting poles were deliberately

Continued

BOX 4.2 (*cont'd*)

FIGURE 4.10 Wolfgang Kohler's famous chimpanzees in action. (A) Chica makes use of a pole as a "jumping stick" to obtain bananas that were suspended out of reach. (B) Sultan makes the "double-stick" by inserting a thin stick inside a larger hollow stick in order to construct a pole with a longer reach. (C) Several chimpanzees attempt to obtain suspended bananas using a combination of crates and boxes. (D) An unsteady tower is created in order to reach the suspended fruit. *All images courtesy of The Mentality of Apes. By Internet Archive Book Images [No restrictions], via Wikimedia Commons.*

provided to Sultan, and he had failed to reach the fruit with either alone. The insight of assembling the double-stick appeared to occur by chance, as he was handling the two sticks while seated on a box. At some point, he arranged the two sticks end-to-end and appeared to realize the possibility that they could be joined together. Sultan very quickly began using the double-stick in food reaching. Thus, an accidental solution emerged, and Sultan gained an appreciation of what the new tool would be capable of achieving. Once mastered, Sultan would reassemble the double-stick with great speed when a situation presented itself for its use.

In another case, a wound coil could be used as a reaching tool by unwinding it. The chimpanzees did not initially grasp the idea that they could unwind the coil in a particular manner. Rather, they were mostly observed to fumble with the coil and occasionally unwind it in a haphazard way often attempting to use it before it was fully unwound and optimized for the task. In this way, the chimpanzees appeared to hit a wall in their reasoning process. They could happen upon solutions to improve reaching, but did not appear to think through the problems in a systematic way. Thus, they may have grasped only the concept of a reaching tool but failed to demonstrate the ability to plan as to how to chain sequences together to solve problems.

Web clips of Kohler's chimps in action are available today and strikingly, when the chimps stacked boxes and crates to try to reach food overhead, they often assembled their towers without regard for the center of gravity and attempted to impulsively climb clearly unstable structures. Limitations of planning were evident in these attempts. Nonetheless, Kohler's chimpanzees are classic animal reasoning studies and appear to show that capabilities beyond simple mimicry and associative learning operate in these primates. Kohler and the Gestalt school of psychology were highly interested in insight. The chimps were credited with insight in some of their efforts as they genuinely were observed to pause and then rapidly implement a solution that they had apparently thought through to some degree.

The term culture has also been applied to whales and dolphins, another large-brained group of mammals (Rendell & Whitehead, 2001). Humpback whales have been observed to use "bubble-nets" in which a group of whales will corral fish by blowing columns of bubbles and then gulping them down once the fish are tightly grouped. Sperm whales have been observed to form defensive "rosette" formations with their lethal tails facing outward in a circle to defend their young from killer whale attacks. To assemble and maintain the rosette, these sperm whales must be able to communicate the presence of danger, group the calves in the center near their noses, and then complete the coordinated spatial pattern together. Any nonconforming whales would undermine the security of the whole formation.

Killer whales provide some of the most impressive examples of group problem solving and even techniques that use water as a kind of tool. Groups of killer whales have been observed to cooperatively swim toward an ice flow on which a seal (a potential meal for the orcas) was positioned. The orcas surface and dive right under the ice flow causing a wave to slosh the seal over the back of the flow to where it could be caught. This technique was originally observed in 1979 and thought to be a single incident, but was later verified to be a regular hunting method applied by pods of killer whales (Visser et al., 2008). Such coordination may be possible due to this species' elaborate acoustic ability and a massive cortical surface specialized for sound processing. In another interesting example, orcas off the coast of Argentina have become specialists in hunting seal pups that hide out on the beach. To solve this problem, these 20,000-pound 20-ft long whales will race full force at the shore beaching themselves in the very shallow water and grabbing an unsuspecting seal pup in their jaws. The whales must then rely on some degree of luck and wave movement as they wriggle themselves back into the water. Interestingly, orcas that hunt using the beaching method also appear to encourage their young in a manner of teaching. Orca calves have been observed being pushed toward shore and learning the technique of wriggling back out to sea. Such observations suggest that active teaching may go on in this species. Accompanying this possible teaching is the additional observation that sometimes orcas will catch seals only to return them unharmed to the shore or back onto an ice flow using their massive jaws. In this way orcas appear to "play with their food" basically offering calves opportunities to catch prey under controlled conditions (Fig. 4.11).

Such instances have been termed *stimulus enhancement*. This term refers to a situation in which an action is performed and other members of the species observe a successful hunt and then imitate it. These situations present challenges for interpretation, since it is often unclear how much awareness animals have for the full consequences and planning of actions. Alternatively, they may simply rely on focusing on one element of a successful strategy and repeating it without fully grasping the broad implications of their actions. This situation is also frequently encountered when interpreting the chimpanzee research. Do chimps and whales really understand plans that involve sets of interactions that depend upon one another? Reasoning about cause and effect is key in achieving true understanding under these conditions. We next turn our attention to inferences about causality in the following section on animal causal reasoning.

CAUSAL REASONING IN NONHUMANS

Moving Beyond Arbitrary Associations

The ability to reason about cause and effect may be a cornerstone of human cognition. Cause is perhaps most critical when considering situations in which someone does some action that leads to a poor outcome. Did he push the victim of the fall causing her to topple out of the window? Did his brain tumor cause him to act in this aggressive manner? Will incarcerating him cause him to rethink his actions and refrain from this type of behavior in the future? Causality is key to establishing who did what to whom, as well as being central to many solutions to human problems. For example, are heightened carbon emissions a cause of global warming? Does human activity cause an increase in carbon emissions? If so, what can people do to cause global warming to slow down? We will discuss these types of questions further in Chapter 8.

There is growing interest in how causes are attributed in the brain and what this may say about the mechanisms and the limitations of how causal reasoning operates.

FIGURE 4.11 Killer whales *Orcinus orca* are clever problem solvers that have innovated their own hunting techniques in different locations around the world. *From Shutterstock.com*

Some of these studies in humans were discussed in the last chapter describing functional imaging studies that reveal differences in parietal, motor, and PFC in response to causal attribution, but what about nonhumans? Is there evidence for establishing cause in other species? What might their behavior indicate about causal attributions? Why would animals need to consider cause?

In the previous section, we considered many diverse solutions that primates and cetaceans use in order to obtain food. Cause can certainly be studied in the context of food acquisition, but the bedrock of human reasoning has much to do with our ability to anticipate causes and effects that do not directly benefit us in concrete, observable ways. The ability to attribute cause has allowed us to diagnose and treat diseases, engineer automobiles and airplanes, as well as plan dinner parties, and navigate international business deals. Animal cognition appears to have relatively severe limitations when it comes to attributing causality and predicting sophisticated effects from their associated causes. This may be the reason that there is not a dog government or a chimpanzee space program.

Much of the learning research conducted in nonhuman species has focused on the way in which animals encode arbitrary associations. This was the cornerstone of the behaviorist movement in American psychology during the 1900s and offered an excellent degree of experimental control; as such, studies removed the influence of prior associations that an animal may have learned previously, which would be unknown to the experimenter. Many species are quite capable of taking note of sensory cues, maintaining them, and pairing them with other actions that follow. The real trick of establishing cause has to do with the ability to integrate across a range of diverse information to assemble knowledge (Box 4.3).

Predictions in Rodents

Rats are highly capable foragers. They are able to move about their environment to find food. Rats are led by a keen sense of smell, vision, and hearing and can avoid threats and overcome obstacles along the way. Rats are also very clever experimenters as any homeowner can tell you who has had the experience of finding a rat infestation in their garage or attic. Rats can gnaw through boxes, scurry along beams, and generally go undetected as they solve the problem of how to obtain consistent food and avoid danger.

Much work has been conducted on associative learning in rodents dating back to the behaviorist tradition in American psychology. Aaron Blaisdell et al. began to conduct some influential experiments evaluating whether rats are capable of inferring causes. Causal reasoning is clearly a hallmark of human cognition, but in a relatively simple mammalian forager, could one find evidence that rats work out what cause and effect to occur out in the world? Unlike basic association learning, in which two arbitrary items come to be paired in the mind, causal learning has a direction. Causes hold a special status that differentiates them from their effects. The structure of a causal situation can be modeled and imagined, at least for humans, and causes and effects have their own cues.

Blaisdell and Waldmann (2012) described a situation in which an individual who is naïve to trends in meteorology observes a drop in the reading on a barometer followed by rainy weather. How would such a person attribute cause and effect in this situation? Is the barometer the cause of rainy weather? Alternatively, could an independent (and in this case unobservable) event, such as atmospheric pressure, be a common cause for both effects: changes in the barometer and in the weather? A key question is what is a cause and what is an effect in this arrangement? One can answer these questions by tampering with a barometer and making it read out a lower number. If no rain follows, then this rules out the barometer as the cause of the weather, and instead supports the common cause account. Addressing questions about cause and effect necessitates conducting a test by intervening with an action. Performing this test would likely inform one about the cause of the rain.

Blaisdell, Sawa, Leising, and Waldmann (2006) reported on an influential finding simulating this situation in rats with a few modifications. Rather than atmospheric pressure, the experimenters used a light that would appear within the animal testing box. Instead of rain, the light preceded the occurrence of a tone and instead of a low barometer reading, the rats observed a delivery of sucrose solution. In a training phase, rats were exposed to light followed by the tone. They were also exposed to light followed by the sucrose. The question then is what would rats do later on when tested? Would they behave as if they expect sucrose delivery when the light comes on by performing a nose poke at the location of delivery? The results indicate that they do nose poke at the location indicating that they had learned that light is followed by sucrose. Do they also perform nose pokes when the tone is presented at test? Again, the answer is yes. The rodents in this case are behaving as the hypothetical naïve person would if they believe that both the atmospheric pressure drop and the barometer numerical drop both are associated with rain and therefore they put on a raincoat when they observe either the drop in pressure or the drop in barometer reading alone. In this instance, it is not clear whether causal knowledge has been inferred by the rat (or person), or if the subject's behavior reflects merely an associative chain.

Blaisdell et al. then added another interesting twist to the experiment. Rats were given the same training consisting of the paired light and tone along with the paired light and sucrose. Next, after a delay, rats received a

BOX 4.3

THE AQUATIC ESCAPE ARTIST

At an undisclosed aquarium, fish were mysteriously disappearing in the night. So, the story goes, aquarium staff placed a security camera running in the exhibit area overnight only to discover that the culprit was a clever octopus that could escape from its tank from the top and make its way around the other exhibits dining as it went. The story has become a bit of an urban legend at aquariums with the specific details and camera footage not being readily available, but it does point to the surprising intelligence of several species of octopus. These animals are highly capable at escaping from predators armed with both speed and a supply of ink acting as nature's smokescreen.

In 2012 at the Monterey Bay Aquarium in California, an employee was doing her rounds in the shale reef exhibit and saw what she thought was a banana peel lying on the ground. Further inspection revealed that this object was in fact a young red octopus (*Octopus rubescens*) out for a 3 a.m. stroll around the aquarium grounds. The octopus quickly dove out of sight in a cloud of ink after being returned to the exhibit. In fact, there were no red octopus exhibits at that particular aquarium making this not only an escape artist, but also a likely stow away. Aquarium staff concluded that the octopus likely entered the aquarium attached to a rock or sponge as a tiny juvenile. The octopus appeared to have made a life for itself feeding off crabs and other invertebrates, hunting nocturnally and living hidden among the rocks and kelp of the exhibit.

Octopuses are invertebrates most closely related to cuttlefish and squid, making them somewhat unlikely candidates to be credited with a high degree of intelligence. That stated, octopuses are very bright animals as evidenced by their rapid learning and memory capacity along with a wide-ranging repertoire of behaviors. These versatile mollusks are able to walk, swim, explore, reach, probe, dig, collect stones, open clams, and change color to camouflage themselves. Further, the octopus has excellent vision and a remarkable ability to sense the environment around it.

Clues about the source of these abilities of the octopus may be found in the organization of its nervous system. Like vertebrates, the octopus possesses a brain organized into several distinct lobes with the vertical lobe being similar to the hippocampus in vertebrates with both structures being rich in numbers of densely interconnected neurons that play a key role in learning and memory. Furthermore, the octopus has an unusually large number of neurons relative to other mollusks and other invertebrates more broadly. This complex nervous system is capable of processing vast quantities of information about the organism's environment, as the octopus has the most sophisticated eyes of the mollusks resembling a camera system, like those of many vertebrates. Octopuses also have numerous chemical and touch sensitive cells around their bodies. Unlike their distant vertebrate cousins ranking high on the animal intelligence spectrum, the octopus has a more distributed nervous system with the eight arms having considerable autonomous control that can function without significant top-down influence from the central nervous system. These features may allow the amazing octopus to become the escape artist and chameleon that it so often appears to be in aquarium environments, which they can adapt to remarkably quickly due to their strong learning abilities (see Fig. 4.12).

FIGURE 4.12 The red octopus is a capable problem solver able to crawl, hide, swim, and squeeze itself through tiny openings. *Credit: Photography by Linda M. Drew.*

test condition in which the tone occurs. Again, the rats should behave as if tone will lead to food and they will nose poke to demonstrate this. Now an additional factor is added in the form of a lever not previously encountered. When the rat pressed the lever, the tone appeared. The rats explored this option by pressing the lever and receiving only a tone. These rats rapidly stopped nose poking to the tone. This suggests that during initial training, rats had formed a causal model with the light as a common cause of both tone and sucrose. When the rat intervened to make the tone happen, it reasoned as if it, and not the light, caused the tone at test, and thus they did not expect the light *or its other effect, the sucrose*, to be present! This is analogous to the hypothetical person trying to figure out cause in the meteorology problem mentioned earlier. Such a situation would occur in which the person experiences dropping atmospheric pressure coupled with rain and experiences atmospheric pressure dropping associated with a lower barometer reading. Then later on, this person would be offered the chance to press a small button that makes the reading on a barometer drop. Pressing the button would fail to predict subsequent rain as an effect. Such a person would then limit taking a raincoat (nose poking for rats) to only those situations in which the atmospheric pressure actually drops (the light activates in the rat experiment), but not for those situations in which the barometer drops following the intervention (the presentation of the tone produced by the lever press for the rat). Associative learning alone cannot account for the results of this experiment. Rather, some degree of causal knowledge appears to have been rapidly acquired when the rats were permitted to carry out their own small experiment by pressing the novel lever at the test period. There is little doubt that rats do run into limitations, as their behavior in the natural environment does not appear as sophisticated as that of larger mammals. The finding is impressive nonetheless and suggests that there is more to the reasoning process than the formation of simple arbitrary associations, even in mammals with small brains and comparatively little cortical mass relative to primates and cetaceans.

Do Non-Human Primates Understand Cause and Effect?

People are very good at evaluating cause and effect in some cases and are skilled at generating stories about what appeared to have caused something while being wrong the whole time. Likely, our extended cortex allows us to move beyond associations and generate causes that are removed in space and time. Large cortical surface and expansive frontal lobes may also allow chimpanzees to understand cause.

For an interesting series of examples, we can consider the behavior of Sarah, a language-symbol trained chimpanzee who was trained in the laboratory of David Premack. Sarah was trained in an era in which several research labs were focused on the project of training chimpanzees to use language. During the process of training Sarah, the researchers were able to carry out a wide variety of interesting experiments intended to characterize the abilities of the chimpanzee in terms of social understanding and causal reasoning. Sarah was able to show some rather impressive abilities. For example, she was trained to understand pencils and erasers, as well as their effects of making marks on paper or removing marks. Sarah was able to appropriately respond to problems that required her to interpret what caused a particular effect. For example, she could correctly indicate that a pencil was the object that causally linked blank paper to marked paper. She could also perform the reverse and indicate that an eraser was needed to link marked paper to blank paper.

In another stronger test of causality, Premack and Premack (1983) described situations in which they offered Sarah videos of problem scenarios to watch and interpret. In the videos, human actors were faced with problem scenarios that modeled the famous chimpanzee experiments performed by Wolfgang Kohler on the island of Tenerife (Box 4.2). For example, a human enclosed within a cage was shown trying to reach some desirable food that was placed out of reach. In another video, a crate was needed as a platform to reach food placed high up, but the crate was weighed down by cement blocks. Impressively, Sarah could watch these videos and then select an appropriate still photograph depicting the solution, for the most part. The task is interesting, as Sarah, unlike Kohler's chimps, was asked not to solve the task herself and receive a food reward, but rather to perceive a situation experienced by a human and be able to perspective-take and recognize a static representation of the solution. To select a picture showing the solution, Sarah was given a closed envelope containing two alternative images, the correct solution and a distractor item. Sarah was able to recognize the predicament faced by the human and consistently solve these problems. This is impressive and especially interesting for the cement block problem, as Sarah would have needed to realize that the human actor's struggle to move the crate was in fact caused by the extra weight placed on the object. This was one of the few problems that Sarah failed to solve correctly.

Going a step further, Premack et al. (1983) tested Sarah's ability to understand the functions of different objects, thereby providing additional insight into this symbol-trained chimpanzee's causal reasoning skills. An additional set of videos was created in which a human actor struggled with several dilemmas. In one video, the actor struggled inside a locked cage. Sarah was offered either a picture of a key or a twisted key that would not function. In another video, the actor struggled clasping at his chest in an attempt to stay warm, while looking over at an unlighted heater. Sarah was offered pictures of a lit heater or an unlit heater. Similar problems involved viewing an actor trying to wash a dirty floor with a hose that was not connected and the actor attempting to operate a phonograph that was not plugged in. Sarah was excellent at these problems, missing only the key problem. In performing these problems, Sarah had to be able to appreciate that she was viewing a dilemma and appropriately be able to assess the cause of that dilemma.

The finding that Sarah was able to perform effectively on these problems indicates that she showed an elaborate ability to evaluate cause. Selecting the solution to the problem required perspective taking, evaluating the function of the object that the actor was attempting to use, seeing the cause for the failure that the human was experiencing, and then noticing what item would enable the object to correctly function. Note that the experiment was cleverly designed to avoid simple object matching owing to the fact that the incorrect alternative items were in fact the same object, but missing a key functional attribute (such as the unlit heater wick). If Sarah were solving these problems by a simpler associative learning strategy, she would likely not have had a basis for choosing the specific version of the object that would solve the dilemma faced by the human.

Comparisons of children and chimpanzees provide further insight. Premack and Premack (1983) also tested three-and-one-half year old children on the video selection tasks that Sarah was successful at solving. In around half of the cases, the children failed to solve these and simply selected an object that was associated with the situation, rather than the one that would cause the situation to be solved successfully for the actor. This finding indicates that Sarah, unlike these young children, was able to reliably perform a relatively sophisticated form of causal reasoning. While Sarah appeared to have outperformed young children on this task, she is unlikely to have been able to approach the performance of children even a few years older. By age five, most children possess such a mastery of language and knowledge of labels for emotional states that they would likely have been able to quickly and accurately notice and describe the appropriate responses. By age 10, children would be further capable of writing symbolic poetry or making up songs and metaphors to describe the actors' modeled predicaments.

While primates and rodents show abilities that suggest that they have some ability to understand cause and effect, these species do not appear to reach the degree of causal attribution that humans are capable of performing. In the case of human causal reasoning, we are able to string together chains of hypothetical causes and their effects almost endlessly in order to predict many future states of the world. In a study by Povinelli and Dunphy-Lelii (2001), young children and chimpanzees were given the opportunity to stand a wooden L-shaped block on end. Later, on test trials, a visually identical L-block was offered, but it contained a weight that prevented it from being stood up as the other one had. The children instantly recognized that the block was different and spontaneously turned it over to investigate what might be different about it. Meanwhile the chimpanzees repeatedly attempted to stand the weighted block on end without apparent interest in the cause of this new blocks' instability (Fig. 4.13). People routinely infer causes that are hidden. Examples include gravity, fate, magnetic attraction, and even supernatural intervention. The reasoning process of animals appears to be more limited and one of the reasons may be due to their limitations of social perspective taking, which we turn to in the next section (Box 4.4).

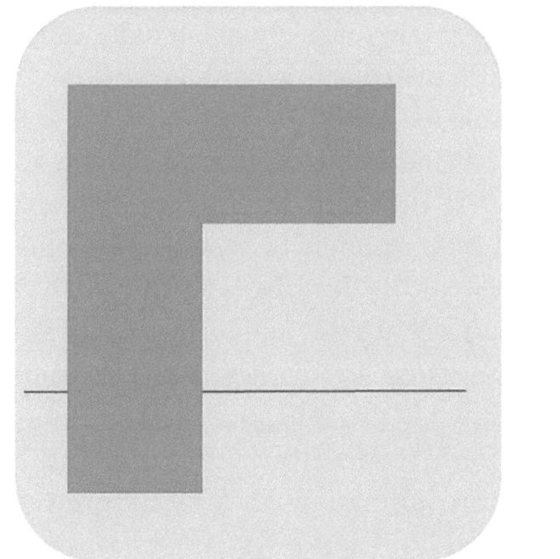

FIGURE 4.13 A clear difference can be observed between children and nonhuman primates when an L-shaped block has been weighed to no longer stand on end. The children in this case quickly investigate what is causing the block to fall, while chimpanzees do not. *Credit: Povinelli, D. J., & Dunphy-Lelii, S. (2001). Do chimpanzees seek explanations? Preliminary comparative investigations. Canadian Journal of Experimental Psychology/Revue canadienne de psychologie expérimentale, 55(2), 185.*

BOX 4.4

A CHIMPANZEE GOES TO COLUMBIA

In the late 1950s, there was a vigorous debate about the nature of language acquisition. Prominent behaviorists including B. F. Skinner maintained that all of language could be explained by behaviorist learning principles. This would include instances in which behavior would be shaped by parent or caregiver approval in response to different utterances made by a small child and going through this sequence repeatedly would eventually yield the full richness of human language. The behaviorists were opposed by a growing group of young linguists and cognitive psychologists who would eventually introduce a nativist view into the field, maintaining that language is instead guided by genetic tendencies that were further informed by environmental influences. Linguist Noam Chomsky was at the center of this debate, which came to represent a growing split in the field of psychology that forever changed how language was viewed. The behaviorist account of language acquisition culminated with Skinner's (1957) book entitled *Verbal Behavior*. The reply to this work from the growing cognitive and linguistic community came in the form of a scathing review by Noam Chomsky entitled *A Review of B. F. Skinner's Verbal Behavior* (1959).

A strong test of the hypothesis that language competence could be attained through the principles of behaviorist conditioning occurred when researchers attempted to train chimpanzees to acquire language in much the same manner that young children do. Chimpanzees are highly social and share a large degree of genetic similarity to humans, but critically do not show evidence of any type of communication in their natural environment that approximates human language. Several research labs became involved in trying to train chimpanzees or gorillas to use American Sign Language (ASL). Other groups, such as that led by David Premack, used symbol systems in which arbitrary plastic tokens were used in order to represent words. Sarah, the chimpanzee who performed so many causal and analogical reasoning feats was trained using such a system.

One of the most interesting cases of chimpanzee language training occurred in the laboratory of Herbert Terrace in New York City at Columbia University in the 1970s. Terrace had been a student of the strident behaviorist B. F. Skinner and was therefore in the unique position of having an opportunity to directly test his advisor's claims about the potential power of behaviorist training being capable of enabling a chimpanzee to communicate complex and abstract sentences using ASL. In order to conduct the experiments, Terrace obtained a 2-week-old male chimpanzee whom he named Nim Chimpsky, poking fun at linguist Noam Chomsky one of Skinner's major rivals in the language debates surrounding the book *Verbal Behavior* and the explanatory power of behaviorist principles to generate language in all of its complexity (Fig. 4.14).

Nim began his life in New York in 1973 living in a brownstone residence of an acquaintance of Terrace who was raising several children. Male chimps are highly territorial and concerned with the dominance hierarchy of the group they are in, whether composed of other chimpanzees or humans. The ongoing challenges of managing Nim's behavior led to his move to his second residence in New York, a large mansion north of Manhattan that had been acquired by Columbia. Nim spent his days being trained in ASL at the Terrace lab at Columbia and remained at his elaborate New York residence, the rest of the time being raised by a rotating group of students and research assistants from Terrace's lab for the next 4 years. Nim was successful in learning numerous words in ASL, but his major utterances proved to be requests for food, treats, and other self-focused needs. It is worth considering that he was after all a very young primate.

Children typically speak their first words around 12 months of age and begin to acquire language in their second year. Most babies will initially comprehend linguistic information in their first year and then launch into language production with little coaching or direct encouragement needed. From ages 2 to 3 years, small children build their vocabularies into the thousands of words and rapidly progress from two to three-word utterances to almost limitless multi-word phrases. Nim was decidedly less successful in language usage reaching perhaps 120 words by age four and was not convincingly able to move past utterances of two to three words. Further, Nim never convincingly mastered grammar, the rules by which humans assemble language to convey statements interpretable to others.

What had gone wrong? Nim was receiving a somewhat different treatment of language, as he was constantly spoken to and was being directly instructed in ASL at the Terrace laboratory. Thus, Nim ought to have had a chance at mastering ASL if such an ability is within the grasp of his species. Eventually Nim appeared to view his trainers and caregivers as possible threats to his dominance within the group and became progressively more unmanageable as he grew in size and strength. The Nim Chimpsky project was terminated in 1977 with Nim having achieved only a modest vocabulary that lacked grammar and was limited for the most part to

Continued

BOX 4.4 *(cont'd)*

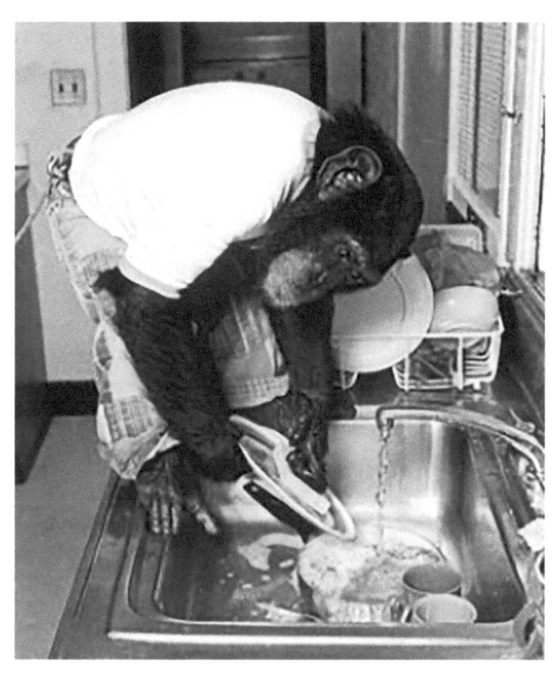

FIGURE 4.14 Nim Chimpsky was a young male chimp trained in the use of American Sign Language. The Nim Chimpsky language-learning project was originally intended to provide an example of how language could be learned exclusively through the principles of behaviorist associative learning. After several years, Nim was determined to have learned approximately 125 words and failed to show evidence of mastering grammar. Herbert Terrace, the lead investigator of the project deemed the behaviorist methods to have failed to produce language. *Credit: Photography by Herbert Terrace. From Wikimedia commons.*

simple requests or demands. Herbert Terrace himself was led to declare that the project had been a failure and that Nim did not vindicate B. F. Skinner's ideas on language acquisition. Rather, the Nim Chimpsky study appeared to have supported the types of things Noam Chomsky had been claiming, namely that language is unique to humans, makes use of both genetic and environmental guidance, and that other species are simply not wired to acquire grammar and language in the manner that humans do. While some other primates have attained larger vocabularies than Nim, few show competence that approaches the capability of an average three-year-old human. Chimpanzee language experiments have also tended to fail in that these animals do not show creative use of the language, which is common in small children.

SOCIAL COGNITION

Testing for a Concept of Self in Animals

Theory of mind is a term describing the capacity to attribute states of mind, or intentions to oneself and to others. It originates from philosophy of mind, a category of inquiry dealing with the ability to read the intentions of other beings. Theory of mind has been widely adopted by developmental psychology to describe the capacity of small children to show empathy or understand others. Theory of mind has also been used to characterize the differences in the perception of others observed in individuals with autism or schizophrenia. Such individuals often show difficulties in comprehending the likely views, intentions, or thoughts of other people, thereby making social interactions more difficult.

For an individual to have a theory of mind, he or she must appreciate that there is a difference between oneself and others. Such an appreciation could then lead to the ability of an individual to simulate or imagine that the way that he or she sees the world is similar to the way that another individual views the world. Alternatively, one may use evidence gained from social interactions in order to conclude that another individual is perceiving things differently. A concept of self has been tested for in numerous species using a technique that has become known as the "mirror test," which was originally carried

FIGURE 4.15 Passing the mirror test is considered to be an indication that an individual has a sense of self. This is a gateway skill leading to elaborated social reasoning abilities. *From Shutterstock.com*

out by Gordon Gallup in 1970 to determine whether chimpanzees had the ability to differentiate themselves from other individuals. Gallup initially presented the mirror test to two chimpanzees (1970). The mirror was introduced and a variety of behaviors were noted such as making threat gestures and faces. The key test occurred when Gallup placed a mark on the brow ridge of the chimps. When provided a mirror, they were able to scratch at their own bodies, rather than attempting to investigate the mark on the strange chimpanzee in the mirror (Fig. 4.15). Since that original test, several other species including elephants and magpies, a type of crow, have been able to pass the mirror test. In one of the more interesting variations, captive dolphins were tested after a trainer drew tattoo-like patterns of lines and shapes on their backs. After a mirror was placed outside the glass of their tank, the dolphins passed the mirror test by swimming over to the mirror and twisting and turning at angles that would have allowed them to view their newly decorated bodies (Tschudin, Call, Dunbar, Harris, & van der Elst, 2001). This rigorous test included "sham" markings made using a non-marking pen. The sham markings ensured that the dolphins were not responding to mere touch sensations, but rather they appeared to be interested in viewing the new visual markings that appeared on their bodies. Having an appreciation of oneself is an important stepping-stone on the way to being able to understand others and reason about their intentions. While the mirror test offers an interesting clue about the ability of an organism, it may not be sufficient to adequately determine whether an animal has a self-concept. The fact remains that we cannot fully appreciate how the animal is responding in this particular case, or what cues may influence this task.

Evidence for Theory of Mind in Chimpanzees, Elephants, and Cetaceans

Theory of mind is a concept that derives from philosophy describing the notion that one has an appreciation of others possessing a mind. Woodruff and Premack first applied this term to chimpanzees in 1978. The ability to infer other minds would enable the inference of mental states by analogy to one's own intentions and knowledge. Simple tests of theory of mind involve asking whether chimpanzees would be able to discover the location of hidden food based on following the gaze of a human. In several tests, it appeared that chimps were able to follow the gaze of a human to learn where food had been hidden (Povinelli, Bierschwale, & Cech, 1999). Other tests have been devised in which chimpanzees beg for food from a human experimenter. The chimpanzees in this case make an open hand gesture in effect requesting food from a human. In a clever test by Povinelli, Eddy, Hobson, and Tomasello (1996), chimpanzees were confronted by two human experimenters, one of whom wore a dark blindfold over her eyes and the other wore a similar cover over her mouth. The chimpanzees did not preferentially beg for food from the experimenter whose eyes were uncovered. This suggests that chimpanzees have some limited form of theory of mind, as they do not appear to be able to imagine that the experimenter seeing them was key to the success of the food request. This insensitivity to sight suggests that chimpanzees do not have a complete theory of mind. As described by Vonk and Povinelli (2006) chimpanzees will respond appropriately when one experimenter is facing toward them and one facing away. In these cases, they will tend to request food from the human who is facing them. This may indicate that they are able to problem solve based on visual cues. Facing toward the chimpanzee may be an obvious enough cue that the animal is able to decide appropriately from whom food is more likely to be available under those conditions, but this would not equate to a full ability to perspective-take when conditions are more subtle.

For further evidence of theory of mind in chimpanzees, we return to Sarah, the highly capable chimpanzee who was trained to process symbols for objects by David Premack et al. In their 1983 book *The Mind of an Ape*, Ann and David Premack described reasoning scenarios that were acted out by a human performer attempting to solve some of the famous problems presented to the chimps of Tenerife by Wolfgang Kohler decades earlier (Box 4.2). We reviewed some of the findings associated with these videos in the earlier section on causal reasoning, and we now return to this task to discuss Sarah's Theory of Mind. In some of the videos Sarah was presented with actors performing the problem-solving situations of the Kohler chimps and the experimenters varied whether or not Sarah liked the actor or not. In some cases her trainer Keith, whom she liked, was shown acting out the attempts to reach out-of-range bananas, while in other videos the actor was an individual she did not like. After watching each video Sarah was asked to respond by selecting a picture that showed the solution to the situation, for example, the actor standing on a box to reach the previously inaccessible bananas. As alternatives, Sarah was offered a picture showing a minor catastrophe that did not solve the problem, for example, the actor was shown tripping over the box and failing to achieve the objective and an irrelevant situation that did not show a solution to the problem. Impressively, Sarah was able to solve these problems for the human actor most of the time, apparently by taking their perspective. Even more impressively, Sarah tended to select the solution for her friend Keith, while selecting the minor catastrophe outcome for the actor whom she disliked! In other words, Sarah was able to select the future state of the world that she would most like to see happen factoring in the character of the actor. This finding further supports the idea that Sarah was not simply associating the videos with situations she has seen before, or even inferring what she might have done in the place of the particular actor. Rather, these cases appear to suggest that Sarah understands the dilemma faced by the human actor, attributes cause to the failure, and can perform a mental simulation of a future state of affairs, namely that the actor succeeds in obtaining the food, or fails by having the solution go awry. Furthermore, Sarah appears to have been able to modulate her preference for the future state based on how she feels about the past behavior of the individual actor in the situation.

People are capable of almost limitless simulation with numerous embedded causal inferences in social situations. Does Sarah show a similar ability as a chimpanzee? David Premack et al. (1983) provide a least one additional modification to the video simulation task that indicates that the answer is no. The chimpanzee eventually runs up against limitations. In a clever modification of the likable or dislikable actor task, Sarah was offered a version in which she would have to take the perspective of another chimpanzee named Gussie, as she watched either Keith the trainer, or a less-likable fellow, grappling with the inaccessible banana scenarios. In the modified perspective-taking version, Sarah was shown a video of Gussie watching the human actor video. In other words, Sarah watched Gussie watching a human actor trying to obtain the out-of-reach fruit. In this case, the video was stopped after Gussie was offered the pictures showing a successful outcome, a minor calamity, or an irrelevant outcome. If Sarah could simulate Gussie's thinking as she thought about trainer Keith's problem, it would indicate that chimpanzees are able to take a dual-perspective understanding, what one individual is thinking about another individual's thoughts. To show this kind of responding, Sarah was offered pictures in which Gussie was shown selecting an embedded picture of Keith succeeding or the disliked actor failing. In all of her attempts, Sarah was not able to reliably demonstrate comprehension of Gussie's intentions. Thus, chimpanzees, while impressive, do not appear to have a full grasp of understanding others states of mind as they pertain to the actions of yet a third individual.

RELATIONAL REASONING IN ANIMALS

The Role of Symbols in Relational Thinking

A core feature of human reasoning ability is our rampant tendency to think in symbols. Language is one of the most obvious types of symbolic cognition that characterizes humans and this ability allows us to make very complicated situations that are quite removed in space and time into concrete concepts that we can share with one another simply by using words. While words are core symbolic representations that are used in human reasoning, other species that do not share our verbal capabilities, either at the level of speech or at the core level of linguistic competency in the brain, may also use symbols in reasoning. We turn now to considering tasks that are intended to measure the abstractness of representation that an organism can reason about. This ability is tested by analyzing an individual's capacity to perceive and reason about the relations among objects. Relational correspondences, or matches, will emerge later in the analogical reasoning chapter and the capacity to re-code objects and situations in terms of symbols is extremely helpful in relational reasoning and may actually be necessary if an organism is going to transcend mere perceptual matching in order to reason about systems of items or objects and how they interact. Such an ability may be the precursor to a broader understanding of the world and an ability to abstract across situations to make new predictions about the world.

The match-to-sample problem is an excellent tool for assessing the ability of an organism to evaluate similarities among objects. When extended to include comparisons of relations among objects, the match-to-sample becomes very much a proper reasoning task. A match-to-sample task begins with a simple condition in which one object is shown, a ball for instance. Next, a pair of objects is added with one being a duplicate of the first item, the ball, and the other being a new item, a block. The subject has to select the item in the pairing that matches the original ball. In the case of animal research, a food reward is typically given for correct performance. The match-to-sample task is able to be solved by many species including pigeons, rats, monkeys, and chimpanzees. It amounts to being a perceptual matching task rather

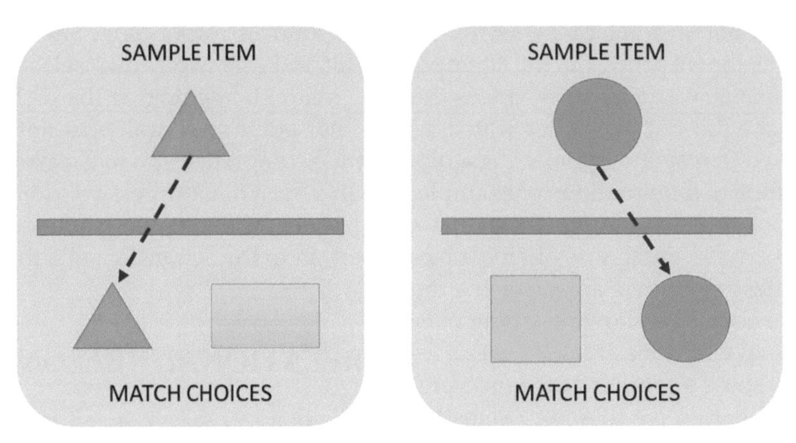

FIGURE 4.16 Variations of the "match-to-sample" problem.

than a true reasoning task, as the subject can get the correct answer by attending to the features of the item and repeating those features in the new pairing.

The match-to-sample task can be made more complex in a couple of straightforward ways. By taking away the initial object, the ball, for a short period before the presentation of the selection objects. Such a situation has become known as a delayed match-to-sample task. The delayed match-to-sample problem is more difficult than the original problem because the subject must now maintain a representation in memory of what the match object was in order to perform well. This type of task has been used for many years to evaluate working memory abilities across species. The match-to-sample has been modified to become the *non*-match-to-sample task by requiring the animal to select the item that is not the same as the original, which would make the block the correct answer in our original problem. The non-match-to-sample task becomes more difficult due to an increase in abstraction. When asked to select the item that is not the same as the sample, the subject has to both encode the perceptual elements of the sample, but also enact a non-perceptual rule such as "choose the object that was not already seen." These difficulty factors can be combined into a delayed non-match-to-sample task. Despite these challenges rats, mice, and monkeys can be trained to select the nonmatching item even after delays of up to 20 s. As mentioned earlier, the patterns of neural firing within hippocampus and PFC are likely responsible for the maintenance of the rule and the objects specific to any given trial. Refer to Fig. 4.16.

The Relational Match-to-Sample Problem

Things get more complex when additional objects are added into the match-to-sample task. In a relational match-to-sample task two items are presented in the sample phase. These may be two objects such as a pen and a stapler, or they may be two pens, the key being that they are either the same object or two different objects. Alternatively, relational match-to-sample tasks have been presented in the form of colored squares. These can be easily manipulated to form sameness and difference relations. The match items for such tasks consist of four new items that are separated into pairs. Critically, the paired items must all be new and therefore not perceptually identical to either item presented in the sample phase. The matching items must also differ in their sameness property, such that one pair will consist of two identical objects, while the other pair will consist of two different objects. In the example problem in which the sample consisted of two pens, the match options might be two oranges or a chess piece and a thimble. The correct answer in that case would be the two oranges on the basis that these objects are the same, as were the two pens. Critically, the problem is solved based on the sameness relationship between the object pairs, rather than on the sameness of any one object in the sample to any one object in the match options. Relational match-to-sample problems can also match on differences. For example, if a sample included a pen and a stapler and the match alternatives were again two oranges or a chess piece and a thimble, in this case the chess piece and thimble would be the correct choice on the basis of differences between paired items. Many years of describing this task to confused students has taught me that pictures make the relational match-to-sample task much easier to comprehend and I therefore refer you to Fig. 4.17.

Animals have been demonstrated to be capable of solving the relational match-to-sample task. One of the most notable examples was Sarah, the chimpanzee who had received specialized training to communicate by symbols in the lab of David Premack. Sarah was trained in an era in which several research labs were focused on the project of training chimpanzees to use language. In

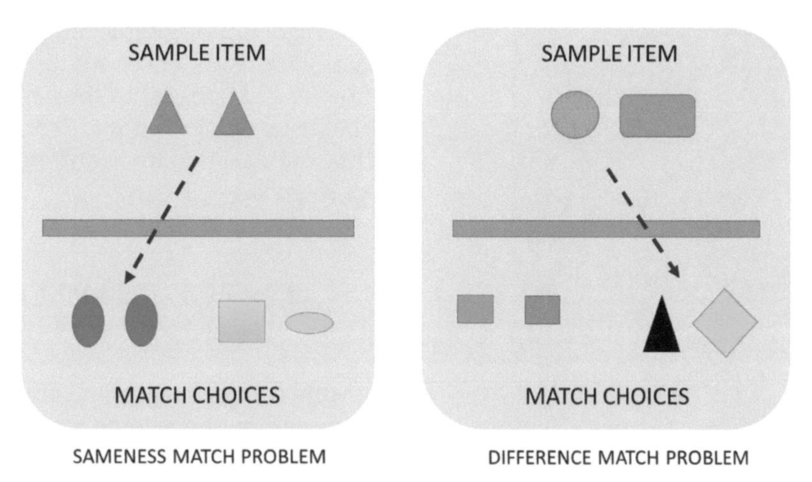

FIGURE 4.17 The "relational match-to-sample" problem. The sameness match version is depicted on the left with the difference map on the right. The *dotted arrow* indicates the correct answer.

the studies by Premack, Sarah was able to work with plastic tokens that represented words. In this way Sarah could be evaluated for her linguistic competence as she could string symbols together to form phrases, or in the case of reasoning, she was able to attempt answers to problems such as the relational match-to-sample task using the plastic tokens.

Sarah was initially tested on the simple match-to-sample problem using actual physical objects (Premack & Premack, 1983). The objects were placed in front of Sarah in a triangular pattern with a sample item at the top and two other items forming the base of the triangle. For example, if an apple were the sample it would be placed above the match items, such as a banana and an apple, which would be placed an equal distance from the sample apple. Sarah was instructed to make the match decisions through a process whereby the experimenter would move the fruit under the appropriate match. This was relatively easy for Sarah to notice and she was quickly solving the match-to-sample problem by the movement of the matching item to the position directly below the sample. A critical test was then performed in which Sarah had to transfer the match-to-sample knowledge to a new set of items. She was able to do this readily and without many trials, suggesting that she had some sense of matching the conceptual rules of the task (choose the object that is the same) rather than merely matching based on the perceptual features of the apple with those of the second apple.

Proportional relational problems were constructed to further assess Sarah's abilities. These problems consisted of tubes of liquid that were one-quarter filled as a sample. Sarah was able to correctly match one-quarter of a wooden disk as a correct answer compared to a three-quarter-sized disk. This problem suggests a relatively high degree of abstraction, as the chimpanzee was matching across states of matter and the sample and incorrect alternative were far more similar to one another than either was to the liquid in the tube. Sarah was further tested on proportional analogies comparing apples to bananas. In these cases, she was able to complete problems such as full banana is to one half of a banana as full apple is to one-half of an apple. Lastly, Sarah was able to perform other visual analogies consisting of dimensions such as shape, color, and presence of a dot on the shape. Experimenters constructed paper shapes and Sarah was able to attain an accuracy of around 80% on these problems. Thus, she appeared to be a highly capable relational problem solver, which may have stemmed from her extensive training in symbolic knowledge. Later efforts cast some doubt on the relational abilities of other primates (Penn, Holyoak, & Povinelli, 2008), but other research suggested that even non-symbol trained primates are capable of relational reasoning (Fagot, Wasserman, & Young, 2001), although this discrimination may be solved by a lower-level perceptual mechanism, such as display entropy.

SUMMARY

Animals are capable of highly interesting behavior. Over the past century there has been a wide array of fascinating studies examining questions about animal reasoning capabilities. Some general trends about nervous systems involvement indicates that more complex reasoning abilities tend to occur in species that have large brains with proportionally large amounts of cortex (Fig. 4.18). The PFC supports numerous cognitive processes and may integrate information from diverse sources within the brain.

FIGURE 4.18 Cats are predators with sophisticated brains and an interesting range of capabilities. *Benjamin the Cat. Image courtesy of David Martinez.*

To learn the most we can about animal reasoning, researchers must conduct both field studies and laboratory studies. Field studies give insight into the capacity of animals acting within their natural environment. Field studies do not allow control of extraneous factors and this can limit the conclusions that are able to be drawn. Laboratory studies of animal reasoning provide experimental control. Such studies tend to have challenges in replicating the problem solving that a species actually carries out in its natural environment. On balance, laboratory tasks have excelled at revealing surprising features of animal behavior.

Classical conditioning provides an associative learning-based account of many reasoning behaviors. These behaviors can be linked to striatal dopaminergic neurons present in the basal ganglia. Learning can lead an animal to accomplish imitation. This is a highly successful strategy for animals to learn and apply solutions that they observe in other members of their species. Associative learning may be a mechanism that enables imitative behaviors.

Other interesting capabilities are also possible in mammals and other large-brained species. Theory of mind and causal reasoning studies indicate that animal cognition may be more sophisticated than would be expected based on the formation of simple associations. Animals are capable of some types of tool use, as chimpanzees and corvids use both sticks and stones to reach inaccessible foods, to displace water, and for display purposes. Orca whales use water as a tool. Animals are capable of accomplishing certain relational reasoning tasks as well. These studies suggest that background knowledge is critical to the capacities that an animal is capable of displaying and that specialized training can allow animals to outperform their fellow untrained species members.

END-OF-CHAPTER THOUGHT QUESTIONS

1. More complex reasoning abilities tend to occur in species that have large brains with proportionally large amounts of cortex. Can you think of any exceptions to this observation?
2. Field studies of animal behavior provide a window into a species behavior in their natural environment. What factors make field studies difficult to conduct on animal reasoning?
3. Laboratory studies of animal reasoning provide a strong degree of experimental control. Does too much control make the situation inappropriate for the animal, as it is not similar to the animal's natural environment?
4. How much of animal reasoning is due to sophisticated associative learning?
5. Imitation is a highly successful strategy for animals to learn and apply solutions that they observe in other members of their species. Is it fair to call animal groups that share strategies "cultures"? Why or why not?
6. Theory of mind and causal reasoning studies indicate that animal cognition may be sophisticated. What are some limitations of perspective-taking that animals run into?
7. Animals are capable of tool use. What are some of the most impressive instances of tool use in animals? What are some of the limits of animal cognition that prevent the building of more complex tools?
8. Can animals be credited with truly "reasoning" after what you have read in this chapter? Why or why not?

References

Berridge, K. C. (2007). The debate over dopamine's role in reward: The case for incentive salience. *Psychopharmacology, 191*, 391–431.

Bird, C. D., & Emery, N. J. (2009). Rooks use stones to raise the water level to reach a floating worm. *Current Biology, 19*, 1410–1414.

Blaisdell, A. P., Sawa, K., Leising, K. J., & Waldmann, M. S. (2006). Causal reasoning in rats. *Science, 311*, 1020–1022.

Blaisdell, A. P., & Waldmann, M. R. (2012). Rational rats: Causal inference and representation. In E. A. Wasserman, & T. R. Zentall (Eds.), *Handbook of comparative cognition* (pp. 175–198). Oxford: Oxford University Press.

Chomsky, N. (1959). A review of Skinner's verbal behavior. In L. A. Jakobovits, & M. S. Miron (Eds.), *Readings in the psychology of language* (pp. 142–143). (Prentice-Hall, Inc., 1967.

De Sio, F., & Marazia, C. (2014). Clever Hans and his effects: Karl Krall and the origins of experimental parapsychology in Germany. *Studies in History and Philosophy of Biological and Biomedical Sciences*, *48*, 94e102.

Fagot, J., Wasserman, E. A., & Young, M. E. (2001). Discriminating the relation between relations: the role of entropy in abstract conceptualization by baboons (*Papio papio*) and humans (*Homo sapiens*). *Journal of Experimental Psychology: Animal Behavior Processes*, *27*(4), 316.

Gallup, G., Jr. (1970). Chimpanzees: Self recognition. *Science*, *167*, 86–87.

Hanus, D., Mendes, N., Tennie, C., & Call, J. (2011). Comparing the performances of apes (*Gorilla gorilla*, *Pan troglodytes*, *Pongo pygmaeus*) and human children (*Homo sapiens*) in the floating peanut task. *PLoS One*, *6*, e19555.

Jelbert, S. A., Taylor, A. H., Cheke, L. G., Clayton, N. S., & Gray, R. D. (2014). Using the Aesop's fable paradigm to investigate causal understanding of water displacement by New Caledonian crows. *PLoS One*, *9*, e92895.

Logan, C. J. (2016). Behavioral flexibility and problem solving in an invasive bird. *PeerJ*, *4*, e1975.

Manger, P. R., Prowse, M., Haagensen, M., & Hemingway, J. (2012). Quantitative analysis of neocortical gyrencephaly in African elephants (*Loxodonta africana*) and six species of cetaceans: Comparison with other mammals. *Journal of Comparative Neurology*, *520*, 2430–2439.

McCormick, D. A., & Thompson, R. F. (1984). Cerebellum: Essential involvement in the classically conditioned eyelid response. *Science*, *223*, 296–299.

Opris, I., Hampson, R. E., Gerhardt, G. A., Berger, T. W., & Deadwyler, S. A. (2012). Columnar processing in primate prefrontal cortex: Evidence for executive control microcircuits. *Journal of Cognitive Neuroscience*, *24*, 2334–2347.

Packer, C. (1986). The ecology of sociality in felids. *Ecological Aspects of Social Evolution*, 429–451.

Pavlov, I. P., & Anrep, G. V. (2003). *Conditioned reflexes*. Courier Corporation.

Penn, D. C., Holyoak, K. J., & Povinelli, D. J. (2008). Darwin's mistake: Explaining the discontinuity between human and nonhuman minds. *Behavioral and Brain Sciences*, *31*(2), 109–130.

Povinelli, D. J., Bierschwale, D. T., & Cech, C. G. (1999). Comprehension of seeing as a referential act in young children, but not juvenile chimpanzees. *British Journal of Developmental Psychology*, *17*(1), 37–60.

Povinelli, D. J., & Dunphy-Lelii, S. (2001). Do chimpanzees seek explanations? Preliminary comparative investigations. *Canadian Journal of Experimental Psychology/Revue canadienne de psychologie expérimentale*, *55*(2), 185.

Povinelli, D. J., Eddy, T. J., Hobson, R. P., & Tomasello, M. (1996). What young chimpanzees know about seeing. *Monographs of the society for research in child development* i-189.

Premack, D., & Premack, A. J. (1983). *The mind of an ape* (1st ed.). New York: Norton.

Rendell, L., & Whitehead, H. (2001). Culture in whales and dolphins. *Behavioral and Brain Sciences*, *24*, 309–382.

Richardson, H., & Verbeek, N. A. M. (1986). Diet selection and optimization by northwestern crows on Japanese littleneck clams. *Ecology*, *67*, 1219–1226.

Schultz, W., Dayan, P., & Montague, P. R. (1997). A neural substrate of prediction and reward. *Science*, *275*, 1593–1599.

Skinner, B. F. (1957). *Verbal behavior*. Prentice-Hall.

Tomasello, M., Davis-Dasilva, M., Camak, L., & Bard, K. (1987). Observational learning of tool-use by young chimpanzees. *Human Evolution*, *2*, 175–183.

Tschudin, A., Call, J., Dunbar, R. I., Harris, G., & van der Elst, C. (2001). Comprehension of signs by dolphins (*Tursiops truncatus*). *Journal of Comparative Psychology*, *115*(1), 100.

Visser, I. N., Smith, T. G., Bullock, I. D., Green, G., Carlsson, O. G. L., & Imberti, S. (2008. Vonk, J., & Povinelli, D. J. (2006). Comparative cognition: Experimental explorations of animal intelligence). Antarctic peninsula killer whales (*Orcinus orca*) hunt seals and a penguin on floating ice. *Marine Mammal Science*, *11*.

Vonk, J., & Povinelli, D. J. (2006). *Comparative cognition: Experimental explorations of animal intelligence*.

Whiten, A., Goodall, J., McGrew, W. C., Nishida, T., Reynolds, V., Sugiyama, Y., et al. (1999). Cultures in chimpanzees. *Nature*, *399*, 682–685.

Yopak, K. E., Lisney, T. J., & Collin, S. P. (2015). Not all sharks are "swimming noses": Variation in olfactory bulb size in cartilaginous fishes. *Brain Structure and Function*, *220*, 1127–1143.

Further Reading

Holyoak, K. J., & Thagard, P. (1995). *Mental leaps: Analogy in creative thought*. Cambridge, MA: MIT Press.

Ko¨hler, W., & Winter, E. (1925). *The mentality of apes*. London: K. Paul, Trench, Trubner & Co.

Mendes, N., Hanus, D., & Call, J. (2007). Raising the level: Orangutans use water as a tool. *Biology Letters*, *3*, 453–455.

Sapolsky, R. M. (2001). *A primates memoir*. New York: Scribner.

CHAPTER

5

Reasoning Origins: Human Development During Childhood

KEY THEMES

- The developmental process is remarkably dynamic as children move from infancy up through adolescence.
- Development involves both biological and environmental influences. These factors interact throughout human development.
- Children begin with a process of cortical thickening as large numbers of synaptic connections are formed. From age three onward, the cortex undergoes a tuning process as some synaptic connections strengthen and others weaken.
- Jean Piaget described a series of stages of cognitive development that have formed the basis for a large body of research evaluating children's thinking and reasoning.
- The context of a problem becomes a significant factor in determining how children will reason and developmental reasoning studies require sensitivity toward making the experimental stimuli understandable and interesting to the child.
- Children exhibit competency in causal reasoning and learning from a very young age.
- Relational and analogical reasoning grow during the elementary school years and are supported by increases in cognitive control and decreases in impulsivity.
- Decision making is guided by heuristics or mental shortcuts. Children make use of these shortcuts from a relatively young age.
- Moral thinking progresses during childhood and incorporates increasingly complex and abstract content over the period of development.

Reasoning
http://dx.doi.org/10.1016/B978-0-12-809285-9.00005-3

OUTLINE

REASONING ORIGINS

Introduction

Reasoning abilities dramatically increase from birth through the childhood years. In this chapter we will explore the origins of reasoning as observed through the capabilities of young children as they progress toward adolescence. There are special challenges in measuring the reasoning skills of children, and in some cases novel methods must be used. Children see the world so differently than adults that many contributions come from labs that specialize in the study of very young children. Throughout the chapter you will note the importance of providing the child with a context that they can understand, and one that does not overly lead them toward a particular behavior.

We will begin the chapter by describing the rapidly emerging abilities of very young children. There is a remarkable degree of advancement, which occurs in the infant mind and the cognitive abilities of young children are equally remarkable to observe. We will describe some of the cognitive advances of children as they progress through the first several years of life and move through the preschool years to the elementary school years. There are clear characteristics present in children's reasoning abilities at these early ages. Their nervous systems and experience both enable their abilities and limit them compared to older people. Children's brains also show changes and growth that support their rapidly developing capacities. The noted developmental psychologist Jean Piaget described a series of cognitive stages that children progress through during childhood. Piaget's focus was on the thinking and reasoning skills of the child. This makes his work highly relevant to this chapter, as well as a standard reference point for many developmental researchers in the past century. Much of the work in childhood reasoning was carried out in order to test and improve upon ideas that originated with Piaget.

We will then move into discussing the reasoning capacities of children as their abilities emerge and change through the elementary school years. Some of these skills are present from a very young age including reasoning about cause and the ability to notice that other people may not share their perspective. Other skills develop later including relational thinking, as observed in analogical reasoning, or intelligence test performance. In addition, children can make use of heuristics for decision making from a surprisingly early age and exhibit some fascinating skills in deductive and inductive reasoning. Lastly, we will discuss the emergence of moral thinking and how children evaluate the actions of others, their intentions, and emotional information.

A few key points should be made as we examine the reasoning abilities of children. First, children are highly dynamic in how they examine the environment around them. They are not passive observers, but rather interact extensively with the people, places, objects, and things surrounding them. Just ask anyone who has ever spent time teaching preschoolers and they can confirm this. Reasoning is heavily influenced both by contextual factors in the environment, background knowledge about the world, and cognitive abilities (Fig. 5.1). As children seek out different environments, they gain new knowledge and build autobiographical memories. Lastly, their abilities in working memory, attention, and metacognition grow. Reasoning performance is not determined entirely by biological development. The use of instruction and strategies impacts how children perceive a reasoning task and the manner in which a child is taught about their goal state can affect their performance.

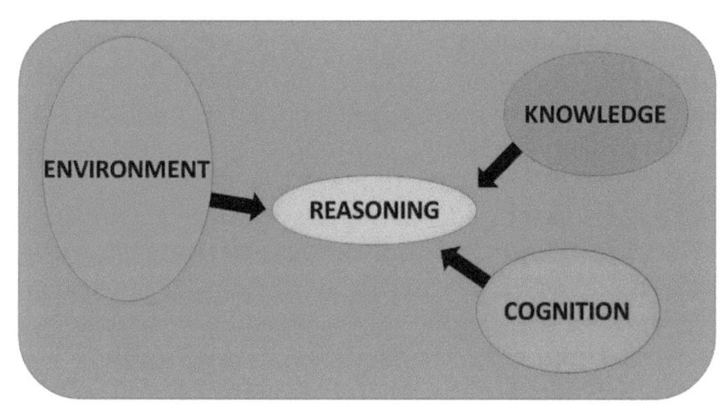

FIGURE 5.1 Children's reasoning receives influences from several key factors. Among these are contextual factors in the environment, background knowledge, and cognitive abilities.

ASSESSING COGNITIVE ABILITIES

Nature Versus Nurture

There is an almost constant and dynamic interaction between biological factors and social or contextual factors in infant and child development. At a casual glance, infancy and childhood may initially appear to be heavily driven by biological change. The genetics and temperament of an infant are strongly influenced by the biological factors present at birth. Nonetheless there is a great deal of environmentally driven change occurring as well. This can be observed in the case of identical twins who are biologically the same but will already show some variations simply based on their shared in utero environment. One twin may be slightly larger than the other. There can already be some differences in temperament that may be attributable to how the children were oriented in the womb and how nutrients were shared. Generally speaking, reasoning skills utilize both our biological guidance toward developing intellectual capacities and the acquisition of knowledge based on experience. Both biological and environmental factors constantly interact dynamically to support the changes we observe in children's reasoning abilities over the course of childhood. While some factors are more genetically guided, others are more environmentally driven.

The Dynamic Nature of Brain Development

The brain undergoes radical changes in the period from infancy through childhood progressing through to adolescence. This dynamic set of changes mirrors the equally remarkable set of cascading changes that occur in behavior, cognition, and reasoning as children grow and learn about the world. When scientists discuss brain development, they often use the term "wiring" to indicate that much of the neurological change that is observed in the developing brain has to do with the connections that are being formed. When children learn, make discoveries, connect new concepts, or undergo some strategic difference in the way that they think about the world around them, there is also likely to be a large degree of connectivity change within the

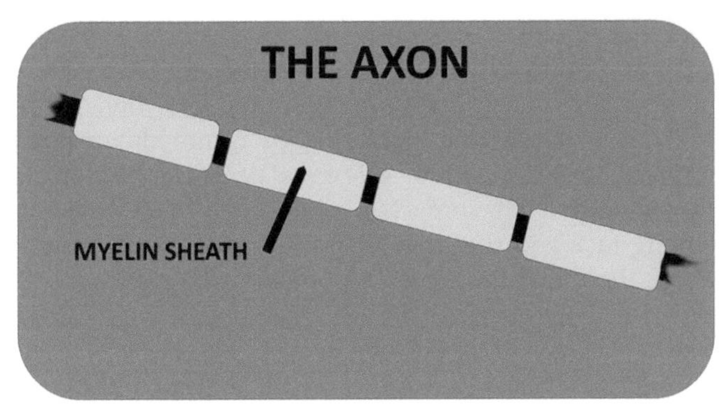

FIGURE 5.2 The axons of the brain are covered with a sheath of myelin, which enables rapid transduction of electrical currents from one cell body to another.

brain. As we discussed in Chapter 3, many of the cells within the brain are neurons, which are the electrically active cells capable of sending coded messages in the form of changes in rates of action potentials. There are also a myriad of glial cells, which supply the neurons with metabolic and nutritional support. Glial cells also provide structural and functional support to the brain. The myelin-covered axons of the neurons are analogous to electrical wiring, as they enable electrical communication between neurons and populations of neurons (Fig. 5.2). Perhaps a better analogy for the axons is that they act like telephone lines enabling communication among brain areas in the form of coded messages. It is this connectivity that changes so dramatically as children age and develops different reasoning styles and capabilities.

Synaptogenesis and Pruning

Children gain the foundation for their reasoning capacity from the biological substrates specified by their genes and how genes interact with the environment. Early animal studies involving rodents showed that rats that were provided with rich environments involving wheels to run on and playmates developed quicker capacities when tested on transfer tasks such as running mazes. Furthermore, these enriched rats had thicker cortices to support their advanced capability over rats that had grown up in a deprived environment. This result indicates that environmental deprivation can disrupt the potential for richer cognitive development that would likely have come about through stimulating play and exploration (Gopnik, Meltzoff, & Kuhl, 1999).

Neurological development occurs with biologically programmed cells that set about multiplying and aligning in order to provide the basic wiring to support brain functions. When this process is occurring, it becomes perhaps equally important for the emerging neurocognitive pathways of the brain to receive environmental input to further guide and refine the pathways that are formed. This was elegantly demonstrated by pioneering neuroscientists David Hubel and Thorsten Wiesel (1960). These researchers surgically closed one eye in a baby cat and left the other alone to receive all of the cat's visual input from the environment. When examining the outcome of this sensory deprivation by opening the chronically closed eye, they discovered that the cat had effectively become blind in the previously closed side. Interestingly, the structure of the chronically closed eye looked to be normal, but the neural inputs within the brain had effectively failed to set up in a way that enabled sight. This study demonstrates the importance of environmental inputs on the development of the brain.

How does the brain tune its wiring in response to environmental stimulation? One important aspect of this process is called synaptogenesis, which refers to the formation of new synapses connecting the electrically active cells within the brain. Experience helps to shape which neurons form synapses and added repetitions of specific experiences strengthen and facilitate these synaptic connections over time. In effect, this process enables the brain to wire itself and tune its connections in order to respond effectively to learned experience. The young child's brain thickens dramatically as cells proliferate from approximately 2500 synapses per neuron to a massive 15,000 synapses per neuron by age three, potentially more synapses than an adult brain has.

The cognitive advances from age three onward in reasoning ability are profound. Children begin to add more complex and nuanced behaviors into their cognitive repertoire. These include the ability to represent increasingly complex visual and spatial information enabling for the relational reasoning ability that is often the focus of intelligence tests. These abilities grow alongside an increasing understanding of cause, consequences, and moral sensibility. The emergence of these abilities is accompanied by a steep drop in synaptic connectivity. This unintuitive "less-is-more" phenomenon comes about as the less frequently stimulated synapses receive decreasing support from the body and are gradually eliminated. The selective elimination of the synapses that are less guided by behavior has been termed *pruning*, as it mimics a gardener selectively eliminating certain less-desirable branches of a tree.

Neural synaptic pruning is a dynamic process and occurs at different rates in different areas of the brain. A landmark magnetic resonance imaging (MRI) study conducted by Nitin Gogtay et al. (2004) at the National Institutes of Health plotted the developmental time course of cortical volumes and color-coded the results as children progress from age five through adulthood.

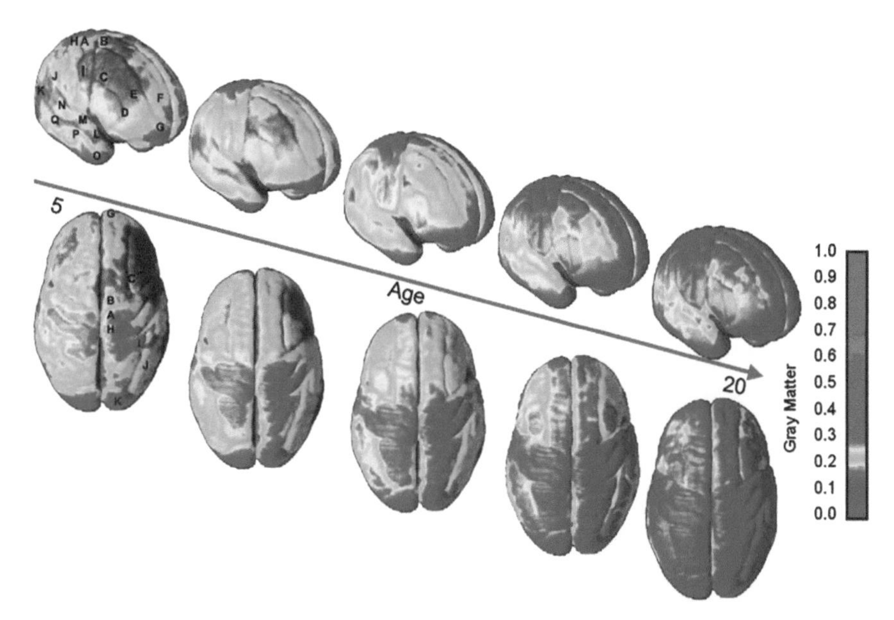

FIGURE 5.3 Gogtay et al. (2004) plotted the developmental time course of cortical volume in the years between childhood and adulthood. Notice that cortex with higher gray matter volume is shown in the warm colors, while the more mature (lower volume) cortex is depicted in cool colors. The side bar shows a color representation in units of gray matter volume. Note that the frontal and temporal lobes are some of the last regions to mature and these regions support reasoning functions. *From Gogtay, N., Giedd, J. N., Lusk, L., Hayashi, K. M., Greenstein, D., Vaituzis, A. C., et al. (2004). Dynamic mapping of human cortical development during childhood through early adulthood. Proceedings of the National Academy of Sciences, 101, 8174–8179. Copyright (2004) National Academy of Sciences, U.S.A.*

In Fig. 5.3, you can see the changes that occur within the brain with thicker cortex being colored either red or bright green and thinner cortex appearing in blue tones. As shown in Fig. 5.3, the brain undergoes a pronounced decrease in cortical volume over the course of childhood through adolescence. The favored interpretation for this pattern is that much of the thinning of the regions is related to the neural circuits of the cortex maturing and dropping off less-used connections. The striking end result of this process is that the brain becomes more fine-tuned toward supporting practiced behaviors as we move through these years.

As we discussed previously in Chapter 3, some of the most critical brain areas important for reasoning lie within the temporal cortex, which supports semantic knowledge, and areas of the frontal and parietal cortex, which support working memory and abstract integration of information. Note in Fig. 5.3 that temporal lobe and frontal lobe areas are some of the last to mature during adolescence. These regions retain a higher volume level that endures from childhood for a longer period of time than the other areas of the brain, such as the sensory and motor areas of the cortex. This neural maturation process mirrors the reasoning skill acquisition and development that is supported by the temporal and frontal areas. We see a relatively thinned and streamlined brain with connections that are strengthened toward maximal support of goal-directed behavior and inhibitory control only late in the adolescent brain and into adulthood (Fig. 5.4).

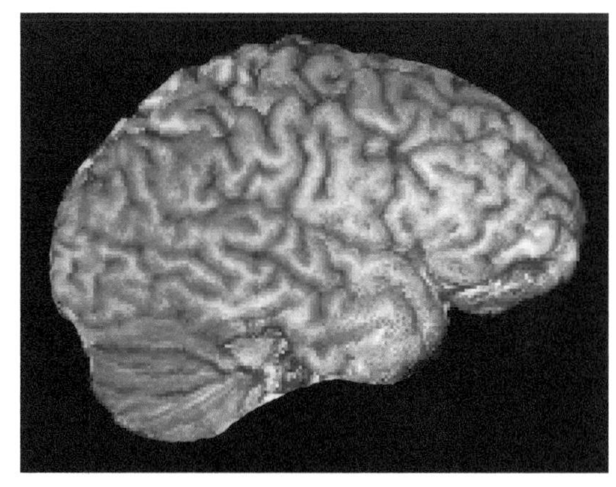

FIGURE 5.4 The adult brain is densely interconnected with a cortical surface that has become tuned to support the most practiced behaviors. Our brains continue to change in a variety of ways as we move through adulthood and there is still an active debate in neuroscience over the average age at which adults achieve a fully mature cortex.

DEVELOPMENT OF REASONING IN CHILDHOOD

The Developmental Stages of Jean Piaget

Jean Piaget (1896–1980) made long lasting contributions to the field of developmental psychology (Fig. 5.5). He is perhaps best known for proposing a set of stages of development that children progress through from birth to early adolescence. The focus of Piaget's stages was on the ability of a child to perceive the world, think, and

FIGURE 5.5 Jean Piaget (1896–1980) made strong and enduring contributions to the field of developmental psychology and influenced a generation of researchers. *By Unidentified (Ensian published by University of Michigan)–1968 Michiganensian, p. 91, Public Domain, Wikimedia commons.*

reason. He placed a strong emphasis on both how children think about objects and also how they think about other people's perceptions, or theory of mind. While the emphasis on developmental stages suggests strong divisions between different periods, development actually proceeds in a continuous manner. This means that the precise ages at which each stage might occur are subject to some debate. Despite the issues researchers have had with the rigidity of these stages, Piaget's stages have served as an effective guide for much of the research on cognitive development, especially as it is applied toward the acquisition of reasoning skills.

Piaget suggested several key characteristics of cognitive development. He focused on structures and how they are modified. The term *structure* applied to knowledge structures, or in other words, sets of information and knowledge (Piaget & Inhelder, 1969). The structures of thought are built through experience and can be called to mind when needed and relied upon when a problem solution can be retrieved from memory. Two other terms apply to the use of structures. *Adaptation* describes a situation in which a child is confronted with new information that they have no experience with, but that can be used to influence their existing knowledge structures. Adaptation may take two forms: *assimilation* and *accommodation*. Assimilation occurs when a child learns new additional information and integrates this information into his or her existing knowledge structures. For example, a young child who has had the experience of using a key to open his front door has accumulated some knowledge about door locks. This knowledge is specific to one door and his knowledge of locking doors in general is almost surely incomplete. Let's imagine that he now sees his mother open a padlock on the back gate using a smaller key. The movement of the key is similar and the mechanics of the lock appear to operate quite like that of the front door with which he is familiar. This new lock can be understood by assimilation. The knowledge that smaller locks are fit with smaller keys and that the movement of the key and the lock are highly similar to what has been done previously with the front door lock. The child will likely assimilate the new information about the padlock. Note that assimilation does not dramatically alter the existing knowledge structure. Rather, there is an incremental change in knowledge. The movement is the same, as are the key and lock device.

Accommodation represents another situation, which occurs when the child must adapt to new knowledge. Let's now imagine now that the boy has the opportunity to see his grandparents open their front door using a keypad that replaces the use of a key. They punch in a set of buttons and the door lock opens. This is clearly very different than using the key, which the child was familiar with. The child can accommodate to this new information by modifying his knowledge structure about door locks. This keypad is different than the key. It involves no tool and the movements involved in the opening of the lock are very different as well. The same knowledge structure about locks can be maintained and used in this case, but it requires an adjustment to the knowledge structure itself, rather than a simple assimilation of new information into an older schema. The child will now be better equipped to understand combination locks and even passwords for computer-based accounts (Fig. 5.6).

According to Piaget, the term operations described the use of logic or mental representation. An operation became available only after the child passed preschool age and could then reason using logical operators and think about abstract information. In modern terms, an operation would involve the use of information from the past that has been assimilated or undergone accommodation to allow the child to reason in a manner similar to adult capabilities. The development of operations became a major milestone for Piaget in the growing capabilities of the child.

Piaget proposed that children move through four major stages of development from basic sensory and motor movement up to abstract thinking. These stages are termed *sensorimotor development*, *preoperational period*, *concrete operations*, and *formal operations*. There are rough age guidelines as to when these stages occur and Piaget

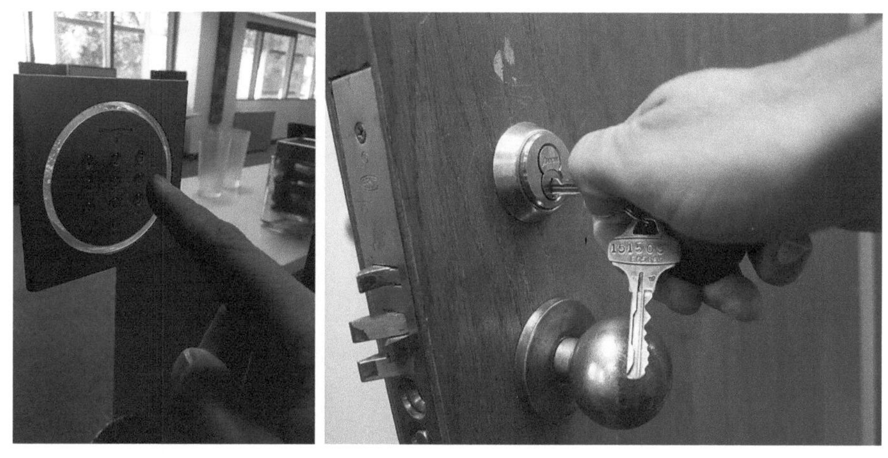

FIGURE 5.6 According to Jean Piaget, accommodation and assimilation processes are used to understand new knowledge based on what is already known. The example of door locks demonstrates the idea that variety can be understood by modification of existing knowledge.

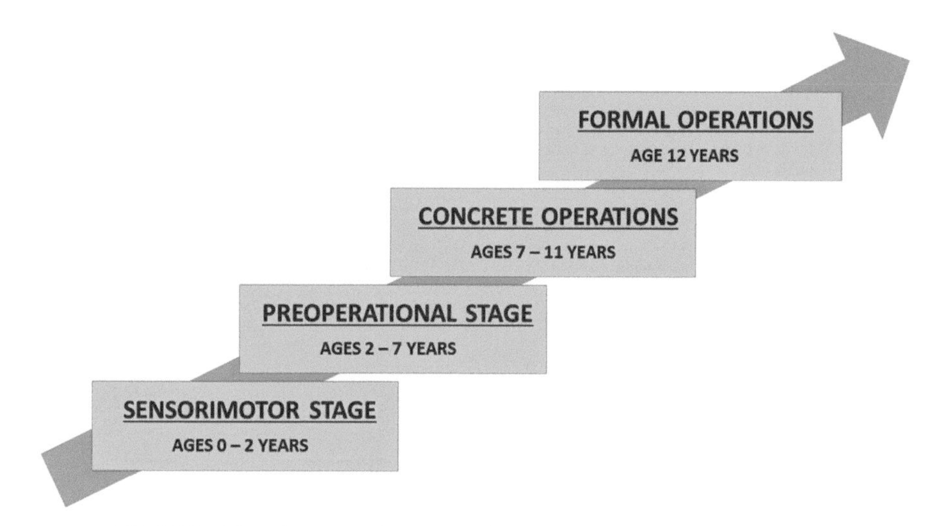

FIGURE 5.7 The stages of cognitive development proposed by Jean Piaget.

indicated several key features of each stage that are characteristic of children who occupy that period of development (Fig. 5.7). According to Piaget, the age at which the child occupies each of the stages is somewhat less important than the sequence of the stages. He maintained that some children are more precocious than others and may advance more rapidly to the next stage. Piaget maintained that the sequence of the stages was set, as each progressive stage arose from the capacities developed in the previous stage. For this reason it was not possible for a child to skip over any of the stages (Piaget, 1967).

The Sensorimotor Period: 0–2 Years of Age

Piaget's first stage, sensorimotor development, takes place from birth to around age two. As you can surmise from the title, this stage emphasizes basic sensory and motor development. The child is essentially functioning based on his or her direct actions and their outcomes within the environment. The child does not yet demonstrate reliable object permanence at this stage. This means that when an object is visually occluded, the child no longer acts as if the object exists. It is a case of "out of sight, out of mind" for the child in the sensorimotor stage. Additionally, a child will tend to think on the basis of what has become habitual for him or her. For instance, if a child is regularly able to find her favorite toy bear next to her high chair, she will go to look for it there when it is not available. This may be the case even if she has seen the toy bear placed out of sight behind a chair. The child does not show strong evidence of mental representations at this stage, but is nonetheless able to interact with the world in a purposeful way.

The Preoperational Period: 2–7 Years of Age

The next Piagetian stage following sensorimotor operations is the preoperational stage. The child occupies this stage from approximately ages 2–7 years, in other words, during the preschool years through kindergarten. The child begins to show evidence of symbolic thinking during this

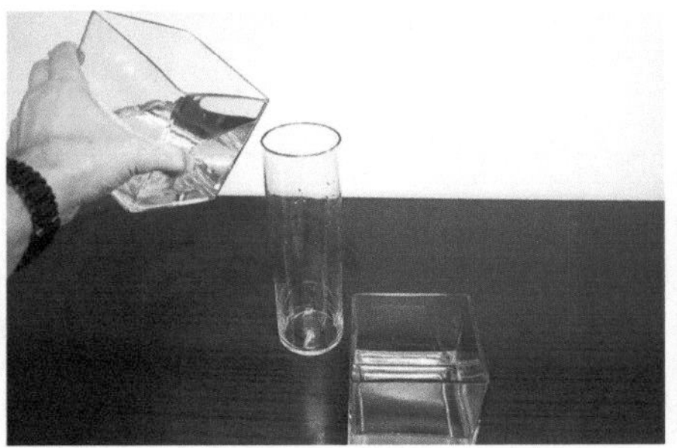

FIGURE 5.8 Piaget's (1965) conservation of liquid task. The same amount of water is displayed in the two glasses. Typically, a child will observe that the two glasses of water are indeed the same. The water from one of the glasses is poured into a taller, but thinner glass. This makes the water level considerably higher in the tall glass. Conservation is achieved when a child states correctly that the tall, thin glass still contains the same amount of liquid.

stage. One of the most obvious signs of the child's newly developed symbolic capacity comes in the form of language use. The child clearly links abstract representations (e.g., the word "ball") to the spherical objects he sees in his environment. During this period of language acquisition, the child will demonstrate the ability to use mental imagery in the form of imaginative play or thinking about impossible situations. The child will also show evidence of being led by perceptual appearances, rather than logical and abstract thinking. For example, children at this stage will tend to answer questions based on their intuition, rather than any sort of deliberative strategy for thinking. One such example involves the concept of *conservation*.

Piaget was interested whether the preoperational child was able to represent quantity in a symbolic or abstract way, as this would demonstrate a clear advance over the reactive cognitive tendencies that characterized the child's development in the previous sensorimotor stage. The conservation concept involves a child understanding that an object or substance remains the same despite changes in the direct perceptual information that is provided. An excellent example of conservation involves assessing the child's beliefs about the transformation of a liquid. For Piaget (1964), an understanding of the conservation of a material was essential to the ability to demonstrate concrete logical abilities. Piaget's conservation of liquid task is now often considered to be a classic in the literature and one of the most reliable and elegant demonstrations of a child's logical thinking ability. Children in this task are shown two identical glasses of liquid initially. The child is asked if there is the same amount of water in the two glasses, or whether one glass has more water. Typically, a child will observe that the two glasses of water are indeed the same. Next, the water from one of the two glasses is poured into a taller, but thinner glass. This makes the water level considerably higher in the tall glass than it had appeared earlier when it had been in the shorter glass (Fig. 5.8). Now the child is asked again whether the amount of water is the same or different in these two glasses. A

FIGURE 5.9 Flavell's (1963) "falling sticks" task requires relational thinking. The sticks must be considered in relation to one another and in relation to a sequence unfolding over a period of time.

child at the preoperational stage (around age 5 years) will now usually state that the amount of water is different. Despite having seen the water transferred, the child does not demonstrate the ability to realize that there has only been a change in appearance, rather than an actual substantive change in the amount of liquid. The change in perceptual inputs amounts to a change in the actual substance itself for a child in this preoperational stage.

Another perceptual error documented by the developmental psychologist John Flavell (1963) gives a similar clue about children in the preoperational period depending upon the perceptual inputs over abstract knowledge in their predictions about the world. Children in this experiment were shown a series of sticks that are meant to show the movements through time of a bar that falls from a vertical, upright position to a horizontal one. The task is for the child to arrange the sticks in the proper sequence in order to demonstrate the specified transformation (Fig. 5.9). This task appears to be an easy one for anyone who can access the concept of a falling object. You might even imagine a series of rapidly shot photos depicting an object toppling over onto its side. You may consider an analogy to the arrangement of dominos that will topple over in succession as the first in a line is tapped out of balance. Children younger than age six in the study by Flavell found this stick arrangement task to be extremely difficult and often placed the sticks in somewhat random arrangements. These children at the preoperational stage appeared to treat each stick as an independent object and failed to sequence them appropriately. Flavell's "falling sticks" task represents a case of relational thinking, in which the sticks must be considered in relation to one another and in relation to a succession occurring in time. Children at the preoperational stage appeared to treat this task as independent still images, rather than integrating them into a situation involving the toppling over of objects.

Another hallmark characteristic of a child at the preoperational stage of thinking is egocentrism. This refers to the fact that the child will consider other people to have access to the same information that they have available to them. For example, a child may talk to his long distance grandparents on the phone and act out a situation in which a toy bear drives a convertible car with his toys. While doing this, the child may assume that his grandparents can see what he is doing and that they will find it funny that the bear is driving the car. This indicates that the child thinks that the perceptual information must be available to others since it is available to him. In another common example, small children will charmingly try to hide from an adult by covering their eyes while sitting in plain view. The child thinks that the adult cannot see her because she can no longer see the adult. These examples are somewhat like the conservation principle in that a change in perceptual information with visual input either appearing or disappearing is sufficient to lead the child to assume that the same perceptual information can be immediately appreciated by others since it is subjectively there or not there for the child. This represents a sort of logical violation in which the child cannot represent what is occurring for other people. Such perceptual mistakes will often lead to difficulties with theory of mind concepts or mentalization errors about the nature of other people, as the child will not be able to adequately represent what is likely to be known or unknown by others. In other words, the child will have difficulties with perspective taking and may therefore inappropriately judge the actions of others.

Piaget and Inhelder (1967) described what is perhaps one of the strongest demonstrations of the egocentrism construct. In this task children were shown a model of three miniature mountains with one mountain between the two others displayed on a square platform. The mountains overlapped to some degree and each had a characteristic object positioned on the top (a cross, a snowy peak, or a miniature house) (Fig. 5.10). The child was shown a doll that was placed at a position next to the child or across from him or her. The child was asked to choose from a series of pictures of the mountains shown from different perspectives to indicate how the doll saw the mountains. Children at the preoperational stage below the age of seven reportedly failed to differentiate between their own perspective and that of the doll. For example, if the child was looking at the mountain with the cross to the far left, the snow-capped peak in the middle, and the mountain with the house to the far right, he or she would select that image for the doll regardless of where the doll was actually situated. This task appeared to show that the child is incapable of taking the perspective of the doll and would therefore be unlikely to be able to take the perspective of other people in general. The child at this preoperational stage would be apt to consider whatever they see to be equivalent to what others are also looking at.

The Concrete Operations Period: 7–11 Years of Age

According to Piaget at approximately age seven children move toward a more mature period of thinking. This period is marked by a shift in abilities from direct perception toward the use of operations, though these are still limited. The age range for children in the concrete operations period places them solidly within the elementary school grade levels. This is a period during which children gain much of their early formal schooling.

Children at the concrete operations stage are capable of passing several of the benchmark tests that are challenging for those at the earlier stages. Notably, children can pass the conservation test by age seven; however, they still display some clear limitations on the ability to do so. Piaget described the performance of children transitioning between the sensorimotor and concrete operations stages as understanding conservation of liquid when the perceptual differences between variously shaped glasses lead to relatively small differences between the liquid levels. In these cases the child can pass this liquid conservation test, but they will revert back to preoperational performance claiming that there are different amounts of water in the glasses when the perceptual differences are large. This would be the case if a very tall and thin beaker were used for transfer of the water with a very high water level (Bjorklund, 2005). Children who fully enter into a concrete operations period are capable of understanding the conservation of liquid, mass, and numbers. For Piaget, the transition period between preoperational and concrete operations functionality moved in three phases with the child demonstrating no conservation, intermediate levels, and then full conservation (Brainerd & Brainerd, 1972).

During the concrete operations period children's thinking becomes less egocentric, thus improving their mentalization, or perspective-taking abilities. At around age eight children gain competence at the "three

FIGURE 5.10 The "three mountains" task requires children to describe the scene from another point of view. Children can pass this task when they accurately visualize how the mountains are arranged from a viewpoint that differs from their own viewpoint (Piaget & Inhelder, 1956).

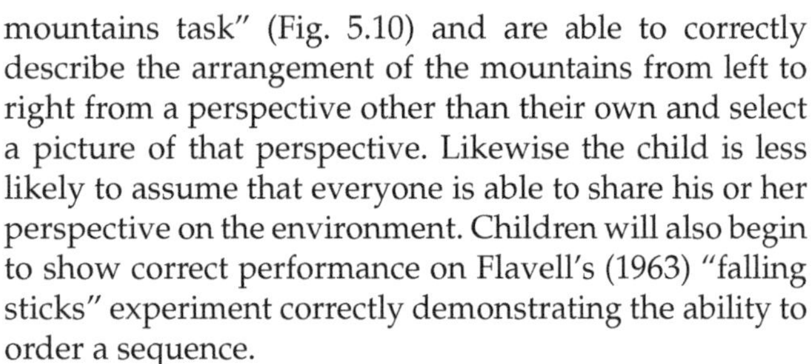

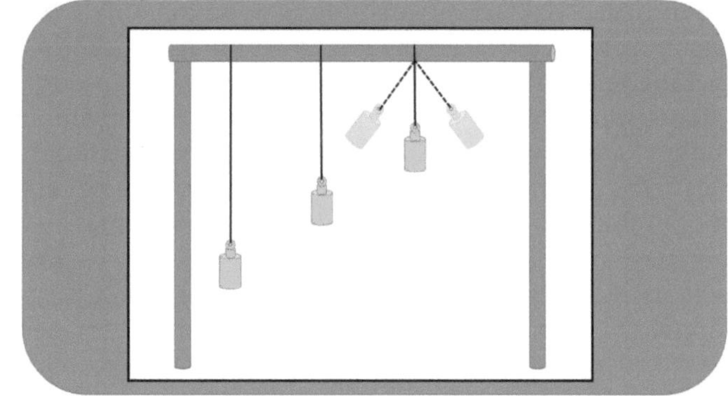

FIGURE 5.11 The pendulum task requires children to test hypothesis in order to isolate which factor is most important in determining the rate at which a pendulum swings. Children at the concrete operations stage cannot reliably reason this way.

mountains task" (Fig. 5.10) and are able to correctly describe the arrangement of the mountains from left to right from a perspective other than their own and select a picture of that perspective. Likewise the child is less likely to assume that everyone is able to share his or her perspective on the environment. Children will also begin to show correct performance on Flavell's (1963) "falling sticks" experiment correctly demonstrating the ability to order a sequence.

Another important marker for Piaget's concrete operations period is the ability for a child to solve transitive inference problems. For example, a child was shown a series of relational pairings between individuals. Let's imagine that a child is told that Anne is taller than Barb. She is next told that Barb is taller than Claire. Will the child be accurate in answering the question "who is taller, Anne or Claire?." Piaget and Inhelder (1967) indicated that children at the concrete operations stage are able to solve for these relational comparisons that they had not previously been shown explicitly before. This type of solution involves both working memory and possibly the ability to use mental imagery to imagine the relative sizes of the individuals described.

Children are not fully abstract in their thinking at the concrete operations stage. According to Piaget (Inhelder & Piaget, 1958) the child is still not able to think entirely abstractly about some areas, such as scientific inductive reasoning. The "pendulum problem" is one of the last frontiers for a child moving through the stages of cognitive development. The pendulum problem is pictured in Fig. 5.11 and involves three different lengths of string to which multiple weights can be attached in order to make a weighted pendulum that swings back and forth. The child is asked to determine which factor, or set of factors, determines the rate of swing that the pendulum attains, or in other words how quickly the pendulum swings. The child in this experiment is presented with the option to vary the height from which the pendulum weight is dropped and to manipulate the amount of force that can be applied to the pendulum in addition to using multiple levels of weights and lengths of string. At this point the child is free to experiment by adding or subtracting weight to the pendulum, varying the length of string, and altering the manner by which the pendulum starts off. To answer this challenging problem, a child needs to form a hypothesis and systematically test for each of the possible variables in order to isolate the one key factor that solves the problem. You may have already guessed that the answer to this task is that the length of the string is the key factor. A longer string will result in a slower swinging pendulum (think of a slow Foucault's pendulum with a very long cable that you may have seen swinging in a museum), while a shorter string yields a very rapidly oscillating pendulum (as you can observe by watching a small desktop pendulum set). Children at the concrete operations stage are not reliably able to test a hypothesis in order to arrive at the correct solution. While they may make sensible moves in the task and successfully evaluate the effects of weight, force, height, and string length, they will rarely arrive at the single solution. This suggests that children at concrete operations are unable to be systematic and test a model scientifically. The child is rarely able to demonstrate true inductive reasoning by hypothesis testing until they pass on to the formal operations period.

The Formal Operations Period: 12 Years of Age and Older

At around the age of 12, Piaget claimed that children reach a stage at which they can use operations formally in a manner similar to an adult. This occurs around the time of transition from elementary to middle school. At this point, children can reliably pass tests avoiding egocentric thinking. They can pass relational transitivity tasks making inferences about comparisons that they did not witness explicitly. The child can now demonstrate inductive reasoning skills by passing the "pendulum" task that requires them to formulate and test a hypothesis through experimentation. At this stage, the child also gains the ability to reflect on his or her own thinking and to imagine hypothetical possibilities reasoning propositionally. A formal operational thinker is now capable of solving problems using a wide array of logical considerations. These include the ability to solve class inclusion problems of the type frequently tested in deductive reasoning tasks of the following form:

All A's are B's
Some B's are C's
Therefore, all A's are definitely C's

This type of problem can be solved using mental imagery, or more formal analysis by diagrams. We will discuss many more examples of propositional logic and syllogisms later in this chapter and revisit these abilities in Chapter 9.

Evaluating Piaget's Theory: A Perspective From Decades Later

Over the past 50 years numerous researchers have sought to test the validity of Piaget's ideas. Tests have not always supported Piaget's claims, but it is clear that his work has had an incredible impact upon how we view reasoning in the developing child. Piaget's work has also contributed strongly to the experimental methods that are employed in the field and many of his tasks are considered to be classics.

The "three mountains task" originally reported by Piaget and Inhelder (1958) has been widely interpreted as a test of egocentrism (refer back to Fig. 5.10). There was some debate about the actual age of acquisition for this type of perspective-taking ability. Some researchers noted that the "three mountains" task is overly challenging for children, as they are rarely asked to take perspectives in this way. Many children display signs of stronger perspective taking than Piaget originally credited children with at the ages corresponding to the preoperational stage. The original estimate by Piaget and Inhelder was that children are egocentric on this task until around age seven. This prediction has not held up under further experimental testing under different conditions. For instance, if a child is shown a card that has different images on each side, such as a joker and a king, children as young as two-and-one-half years of age can correctly indicate that if they are seeing the king that another individual facing them would be seeing the image of the joker. This demonstration suggests that children have awareness that other individuals have their own perspectives, but that the visual complexity, including the number of objects present may determine their success in describing the perspective of another person (Bjorklund, 2005).

Another challenge for Piaget's original concept of conservation has been raised in the intervening years since the stages were originally proposed. Several research studies conducted in the 1970s demonstrated that children can be instructed on the conservation of liquids or solids. Brainerd and Allen (1971) and Field (1987) demonstrated that children as young as three can show the ability to understand conservation. This indicates that children at the earliest period of the sensorimotor stage are actually capable of thinking this way, which would represent operational thinking according to Piaget. Field (1987) indicated that approximately three quarters of the studies performed in the 1960s through the 1980s had been able to show conservation abilities present in preschoolers casting further doubt on the strong conservation position proposed by Piaget originally.

Lastly, competence on the classic transitive inference task indicating the arrival of concrete operations may actually be solved at an earlier period of development than Piaget had proposed. Trabasso, Riley, and Wilson (1975) conducted a series of studies in which children evaluated transitive relationships using physical sticks of different colors. Children participating in these studies were successively trained on the relations between several pairs of sticks. For instance sticks A longer than B, stick B longer than C, and stick C longer than D (Fig. 5.12). Children as young as 4 years of age, solidly within the preschool preoperational years were then capable of making the correct inference when comparing the key pairs that had not previously been directly presented together (A and D, for example). This indicates that children at the preoperational period are likely capable of richer relational reasoning abilities than originally claimed by Piaget.

CAUSAL REASONING IN CHILDREN

Causal reasoning is among the most critical skills for a child to understand the workings of the world around them. Making sense of situations and events requires consideration of the objects involved, as well as the relations between objects. Over time children will notice that objects will interact with one another in regular ways. A child observing his shadow projected onto the side of a building will quickly realize that moving his arms will lead to a mirrored movement by his shadow. He may then become curious and investigate the mechanisms

FIGURE 5.12 Transitive inferences can be made about hypothetical heights of individuals. Young children often struggle to make such inferences accurately. Meanwhile, children can sometimes solve these problems when presented in the form of physical sticks (Trabasso, 1975).

of this relationship. My own son noticed his shadow around the age of two and pointed out that it was "a picture" of himself. This statement captures the child's understanding of the relations between himself and the shadow and implies some level of causality. He realizes that he has the ability to influence his shadow, rather than the other way around.

Understanding causal relations becomes a critical tool for enabling learning. When two things occur in a sequence, children are quick to try to work out their relationship. When a causal link has been established, its consequences are clear to the child. At that point the child has established a key regularity about the world. The child will next be able to further investigate, possibly through carrying out experiments in order to determine the mechanisms that allow one thing to cause another. The ability to establish causal connections between events leads to making predictions about what will happen in future situations. This ability can also lead to the formation of schemas that include explanations about the world. One of the keys to human causal reasoning is the ability to conduct experiments in order to test the mechanisms that we hypothesize to exist when one thing appears to cause another.

Children exhibit causal reasoning abilities from a young age. Early theorists, including Piaget, claimed that children did not possess causal reasoning abilities. Researchers in developmental psychology have demonstrated that children do understand principles of causality. These include temporal priority, spatial priority, and contingency (Bullock, Gelman, & Baillargeon, 1982). There has been some debate about whether children express a fully formed ability to reason about cause. There has also been considerable debate about what leads children to infer that a causal connection can be made between events.

For many years, dating back to David Hume's work on associationism, scholars have theorized that repeated observations about two occurrences will lead a reasoner to infer a causal connection if there is enough regularity between the two events. The philosopher J. L. Mackie (1974) questioned this view by noting that cause does not simply reduce to the repeated co-occurrence of events, but rather that there is also information present in what is not observed to happen. In this manner, our causal thinking involves conducting thought experiments, in which we imagine situations that we did not have the opportunity to directly observe. These cases that lack a direct observation can be especially informative about how children determine cause. Such cases have provided a basis of many experiments.

Research on causal reasoning has frequently focused on the impact of unobserved situations. For instance, if the engine dies in a car and it pulls off to the side of the road, then we can infer that the vehicle would have continued on if it not run out of gas, or experienced a problem with the spark plug. This assessment of cause involves a counterfactual judgment. In other words, we reason about an alternative state of affairs that is not what we witnessed, but one that would have occurred if we imagine the assumed cause to be removed (such as running out of gas or a mechanical failure). Harris, German, and Mills (1996) conducted a study of counterfactual thinking in children to determine the role of this process in their causal reasoning. Preschool children ages 3 to 5 years were tested on a counterfactual thinking task. In this task they observed a doll carry out an action. They were then asked what would have happened if a different situation occurred. For example, children were shown a clean floor and a doll with dirty shoes was demonstrated walking on the floor resulting in dirty footprints. The children were asked what would have happened had the doll done something different, such as removing her shoes prior to entering the room. The procedure used by Harris et al. is summarized in the text below:

> One day, the floor is nice and clean (the experimenter pointed to a square of white plastic). But guess what? Carol comes home and she doesn't take her shoes off. (The experimenter brought a doll to the edge of the floor). She comes inside and makes the floor all dirty with her shoes. (The experimenter then made the doll walk across the floor, leaving dirty footprints).

The preschoolers were asked a pair of factual questions about the situation at the present time and an earlier time, followed by a critical test question, "What if Carol had taken her shoes off, would the floor be dirty?". Results indicated that children were significantly less accurate on the counterfactual test questions compared to questions on which they were asked to reiterate what the current situation was. The children were still accurate much of the time though, which indicated that they possessed the ability to reason using counterfactual information. Even children as young as 2 years old can use mental imagery to determine the consequences of actions such as predicting what would occur if something were to be painted or poured (Harris, Kavanaugh, Wellman, & Hickling, 1993). Counterfactual thinking requires the child to use knowledge about past situations to infer new information about causes that can be applied to a current situation. Evidence suggests that this ability develops at a young age and is used throughout childhood.

Another relevant aspect of causal reasoning in children involves analyzing the rules that a child needs in order to determine causes. The simplest rule-based account of causal inference makes use of "if-then" statements. For example, a child could learn that they can expect hot water to pour out of a faucet if they turn the handle on the left side. This is a straightforward situation involving a direct cause linking the left handle to the production of hot water. The child can perform such "if-then" causal reasoning on a variety of problems by age three. More complex inferences are needed in daily life, in which outcomes are not so predictable or direct. More complex and less-direct situations are common in today's technologically based societies, where causes may not be straightforward and the linkage between an action and an outcome can be indirect.

At what age do children exhibit reasoning about conditional possibilities that exceed the inferences that can be made by "if-then" rules? Frye, Zelazo, Brooks, and Samuels (1996) examined an interesting case in which children at age 3, 4, and 5 years observed a ball that moved down a covered ramp. In this experiment, the ramp had two input holes on the top and two output holes on the bottom. When a marble was placed on the left input hole it typically fell out directly below it from the left output hole under the ramp. Similarly, the marble could be placed into the right input hole and it would typically fall directly out from the right output hole beneath the input hole.

Frye et al. (1996) next introduced a more complex condition about a rule to be evaluated. They introduced a light in the experiment that informed the child whether a different situation would occur (Fig. 5.13). When the light was illuminated a marble placed into either input hole would be shunted across the ramp and would fall out of the output hole on the opposite side. The experimenters explicitly informed the children about how the ramp operated and included information about the light and its effect on what the marble would do. Children were asked to predict where the ball would emerge from the ramp with the light on and the light off, which signified the different conditions. Children were tested in both configurations (with the light on and off) and the ramp always functioned as they had been taught. The four- and five-year-old children were able to correctly adjust their predictions when the light turned on or off. Three-year-olds had difficulty with this task and were not reliably able to alter their predictions about the marble based on the conditional light. Most of the three-year-olds also tended to predict that the marble would always emerge from the output hole directly under the input hole. Predicting the operations of this "two-input, two-output" ramp required keeping multiple "if-then" rules active at a time and modulating a prediction based on the state of the light. Correctly predicting the outcome of this task requires a more complex representation such as "If the light is on and the marble is put in the left, then it will roll across to the right." Such a situation is termed a *conjoint conditional* and may be beyond the grasp of children aged three and under. One challenge with this study was the fact that three-year-olds may simply be unable to switch strategies once they have committed to one particular prediction. It is also possible that the young children were not able to overcome their tendency to always predict that the marble would fall from the output hole directly below its point of input.

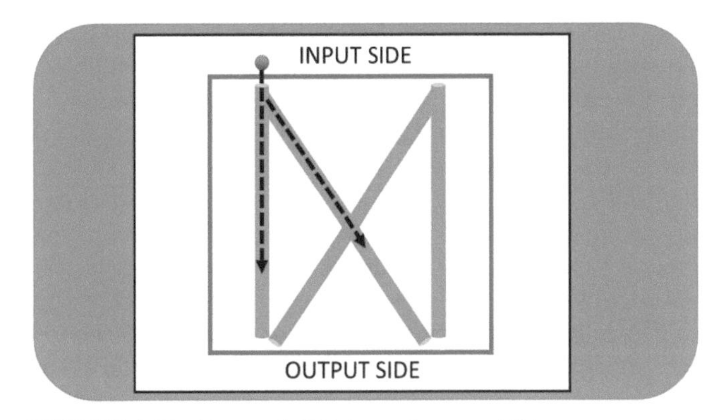

FIGURE 5.13 Frye et al. (1996) tested young children on a complex rule condition using a ramp and ball setup. They introduced a light in their experiment and when illuminated, it signaled that a marble placed into either input hole would unexpectedly cross the ramp and would fall out of the output hole on the opposite side. The ramp was covered to block the child's view of the ball's movement.

Frye et al. (1996) described a second study involving a modified ramp that clarified the causal inference performance of the three-year-olds. In this experiment, the same type of ramp setup was presented, but this time there was only one input hole and two output holes. The light signified that the marble would roll down the ramp and emerge at the opposite end from where it had been put in. Alternatively, the marble would simply drop out directly below the input hole if the light was turned off. Notice that this task can be solved with a simple "if-then" rule of the form "if the light is on, then the ball will come out the opposite side" and "if the light is off, then the ball will come out the same side." To solve this prediction task there is no need for any additional relational factors that include the starting point of the marble. The results demonstrated that children at age three are able to solve the ramp task with the one-to-two hole condition, presumably using the simpler "if-then" rule. These children were also able to overcome their tendency to assume the marble would drop from the output hole directly below the input hole that it was placed into. This suggests that three-year-old children in the prior study (Frye et al., 1996) were overwhelmed by the complexity of the secondary factor (input hole location) and could not solve the conjoint conditional.

Do Young Children Use the Markov Assumption to Determine Cause?

The *Markov assumption* is important for inferring a cause in situations that describe a chain of events. This assumption can be used in order to distinguish between situations that involve mere associations among events and those that contain causal relations (Pearl, 2000). The Markov assumption can be applied to a series of events as follows: the evaluation of any one event is independent; in other words, it does not depend any other event, except in the cases of other events that the event causes, or that cause it directly. For example, consider an unfortunate situation in which you drop your drink, which lands on your laptop, which then causes an electrical short in the computer. In a causal model of this series depicted in Fig. 5.14, dependent events are Event 1, dropping the drink and Event 3, the laptop shorts out. The Markov assumption holds that Event 1 and Event 3 are independent events, but are conditional upon Event 2, the drink landing on the laptop. Event 3 does not in turn lead to any other event, and Event 2 is the only cause of Event 3. So, if someone wishes to predict the possibility of Event 3 (the laptop shorts out) and he or she knows about Event 2 (the drink spilling on the laptop), then having any additional knowledge about Event 1 does not matter. Event 1 only matters to the occurrence of Event 3 through the influence of Event 2.

Any two events may co-occur because there is a causal relation between them. This can take the form that Event 1 may cause Event 2, or Event 2 may cause Event 1. Another possibility exists. There may be a third event, Event 3 that acts to cause Event 1 and Event 2. Gopnik, Sobel, Schulz, and Glymour (2001) evaluated children's abilities to distinguish among these different types of causal arrangements. Gopnik et al. provided children

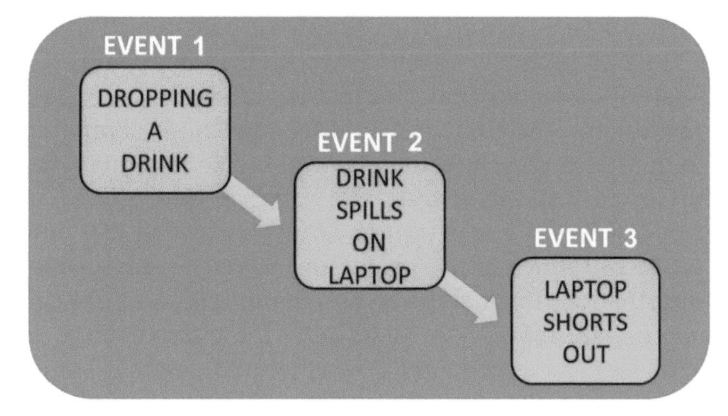

FIGURE 5.14 A causal model can be used to illustrate the Markov assumption. Event 1 and Event 3 are independent events, but are conditional upon Event 2.

with a story that they were being given a device called a blicket detector, which would light up and play music in the presence of certain objects called "blickets." Every time an object activated the blicket detector, it was to be called a "blicket." After children learned to label objects as blickets that activated the machine, they were given a critical experimental trial in which Item A activated the blicket detector, while Item B did not. Next, Item A and Item B together activated the blicket detector multiple times together. Gopnik et al. were curious about what children would make of this situation. Critically would children label Item B as a blicket, given the fact that Item B alone failed to activate the blicket detector, but did activate the detector when it appeared along with Item A? Children as young as two-and-one-half years old labeled Item A as being a blicket, but not Item B. These results indicate that children are able to distinguish between the ability of different items to cause an effect based on their properties. This type of inference can be made purely based on association, as Items A and B possess varying strengths of association to the blicket detector, with Item A having a greater tendency to activate the machine.

Taking this a step further enables us to consider how the child actually draws a causal inference among the items and their ability to activate the blicket detector. There are two other important situations that Gopnik et al. (2001) exposed the children to. These different objects possess different strengths of association, with Item A having the strongest causal relationship to activating the blicket detector. Meanwhile, Item B is associated with activating the detector but to a weaker degree, as it was never shown to have activated the detector directly when on its own without Item A. In an experiment that helps to determine the causal relations between items, children are shown a situation involving the blicket detector that is at first ambiguous about which item (A or B) caused the detector to activate. The children are then provided with new evidence that could resolve that ambiguity that existed in the first situation.

This kind of inference is called a *retrospective inference*, as it is able to clarify something about a scenario that was observed in the past. This new opportunity for information could allow a child to work backward to determine how to interpret the initially ambiguous situation that he or she had previously seen.

Sobel, Tenenbaum, and Gopnik (2004) conducted an experiment to test retrospective causal inferences with preschoolers. In one condition, three- and four-year-olds were shown a situation in which Item A and Item B activated the blicket detector when they were together. Next, the children saw that Item A on its own did not activate the blicket detector. The children inferred that Item B must be a blicket, as it had to be the item that had caused the detector activate in the first (A and B) situation. The children also inferred that Item A was not a blicket. Notice that the children had never directly seen Item B activate the detector on its own. This is critical, as the inference is based on an unseen, but suspected, case in which B would activate the detector if it were to be presented on its own. When the preschool aged children were asked to take an action to try to activate the blicket detector, they only put Item B with the detector, not both Item A and Item B. In a second key condition the preschoolers saw two new objects, Items C and D, activate the blicket detector when they were exposed to it together. Next, the children saw Item C activate the detector on its own. In this case, the children made a surprising inference that is inconsistent with that made in the first condition (in which the children had labeled Item B to be a blicket without seeing it activate the detector directly). In this second condition the preschoolers inferred that Item C was a blicket, but not Item D. Notice that there is a problem with this inference from a purely associationist account. The associative strength of evidence between Item B and the blicket detector in condition one was identical to the evidence presented about Item D and the blicket detector in condition two. If Item B is a blicket, Item D should be as well according to strength of association argument. When children saw Item A fail to activate the detector, then there should be no influence of the causal status of Item B, as Item B was not present when Item A was exposed to the machine (on its own). Children should therefore treat Item B and Item D the same purely based on their association values, as these were identical (activating the blicket detector with a paired item, but never being shown the direct situation in which those items were exposed to the detector alone). This finding demonstrates that children as young as three are capable of taking into account causal information, rather than mere association values between related (or unrelated) events, as these children were able to infer that it must have been Item B that caused the blicket detector to activate, since they later saw that Item A was incapable of activating it alone.

In another experiment eight-month-olds were tested on this same blicket experiment to determine whether causal reasoning occurs in infants this young. Sobel and Kirkham (2006) had to modify the blicket detector task dramatically in order to test these infants, since children so young would be unable to comprehend the instructions. Rather, these experimenters presented the infants with Items A and B in the form of boxes appearing on a screen and instead of a blicket detector, consequence C or consequence D (presented on different sides of a computer screen) appeared to be linked to Item A or B alone or when presented in unison. Just as the children in the earlier experiments, the infants showed evidence (based on where they looked for the outcome) that they expected B to cause a certain outcome in one case, but not in another despite never having seen the clear linkage between Item B and a given outcome. This suggests that children even as young as 8 months are able to attribute cause and do so based on an inference process not purely based on association strength. In this study, Sobel and Kirkham showed evidence that children this young can already make retrospective causal inferences. Impressive indeed and this result is a strong indication of how active and dynamic an information processor a young child is, even in the preverbal first year of life.

The concept of the blicket detector as used by Gopnik et al. (2001) has been used in several follow-up experiments to further investigate the role of association strength in determining how children reason about causes in very basic direct cause situations. For example, Sobel et al. (2004) used the blicket detector experimental cover story with a modification. In this case the occurrence of blickets varied from many to few. The children viewed the outcomes of 12 identical items that were individually presented to the blicket detector. Some children saw 10 of the items activate the detector and pass the test to be considered blickets. Other children saw only two of the items activate the detector. The children then evaluated Item A and Item B, as had been carried out previously in the original Gopnik et al. (2001) study. The exposure to the few or many blicket population changed the results of children's causal inferences about Items A and B. Children who observed the large number of blickets (10) determined that Item B should be considered to be a blicket. Meanwhile, those children who had been presented with rare blickets (two) determined that Item B was not a blicket. Children in the study by Sobel et al. made use of information about the population level of blickets to determine the causal strength of each object's ability to activate the blicket detector. This set of results is difficult to reconcile with a pure association model of causal inference once again. It appears that children can infer cause under these conditions.

DEDUCTION AND INDUCTION IN CHILDREN

Deduction and induction are two of the core categories in reasoning dating to the times of classical Greek philosophy. The core difference between deduction and induction is linked to the validity of the inferences made. Deductive reasoning will reliably lead to a valid conclusion or inference, provided that the premises of the situation are valid. Inductive reasoning may lead to a valid inference, but there is always some degree of uncertainty that accompanies induction. Deduction moves in a direction in which we are provided with general rules about how the world works and we have to look narrowly to determine whether a specific case fits the general rule. Induction, by contrast moves from a smaller number of specific cases toward inducing a general rule that would likely apply to a larger set of possibilities.

In this section of the chapter we will examine the developmental processes that enable deduction and induction.

Deductive Reasoning in Children

Deductive reasoning is frequently studied using three-term series problems. These problems follow a formula in which a thinker is supplied with two propositions and then asked a question. For example, "If Ben is better than Joe, and Bill is worse than Joe, then who is best?" In this case, we consider the relational statements in order determining that Ben leads the group followed by Joe and lastly Bill, so Ben is our answer.

This three-term series task has been studied for many years in the context of child development. One of the earliest findings reported by Burt (1919) and by Piaget (1921) was that children interpret comparative statements according to judgments of membership. Piaget (1921) found that 9- and 10-year-old children were not reliably able to solve problems such as

> Edith is fairer than Suzanne;
> Edith is darker than Lili.
> Which of the three has the darkest hair?

In this case, the child must consider the relations between the different women with regard to hair color and reach the conclusion that Suzanne has the darkest hair. Piaget interpreted the difficulty that children have with this problem as being related to category membership. He claimed that children will interpret both Edith and Suzanne as being fair-haired due to the appearance of the word "fairer" in the first line, while they will interpret both Edith and Lili as being dark-haired based on their proximity to the word "darker" in the second line. This tendency to reason based on category membership rather than relational statements leads children to make this error.

Other problems can occur for deduction on the basis of how the people involved in the relation are interpreted. For example, Margaret Donaldson (1963) indicated that children ages 10–12 make errors on this type of problem on the basis of contradictions. In the women's hair color example, children using the categorization strategy placed Edith and Suzanne into a fair-haired category and Edith and Lili into a dark-haired category, with the contradiction that Edith is both fair-haired and dark-haired. Donaldson also noted that some children often interpret this situation as implying that there are actually two different women named Edith being described, one with lighter hair and one with darker hair. Donaldson also noted that children have difficulty with interpretation of the relational comparisons based on the language used to describe the relations among different people.

A very common type of problem used to evaluate deductive reasoning is the syllogism. Syllogisms in deductive reasoning can be generically expressed as statements involving hypothetical or arbitrary cases. For example, syllogisms can be stated as "all A's are B's, all B's are C's; therefore all A's are C's." Hawkins, Pea, Glick, and Scribner (1984) investigated the ability of children at ages 4 and 5 years as they performed syllogistic reasoning under several conditions of interest. Hawkins et al. noted three potential explanations why young children might struggle on deductive reasoning tasks. First, young children may lack the necessary ability to deductively reason in formal syllogistic problems. Second, children's performance may be hindered by the truth values contained within a problem. This is a challenge with syllogisms, as valid deductions need not be true in the real world and children may prefer real-world truth to formally valid deduction. Thirdly, young children may not fully understand what experimenters are asking in tasks that are meant to assess their reasoning abilities. These three challenges essentially have to do with (1) problem complexity, (2) problem content, and (3) task conditions. Hawkins et al. examined the effects of these three challenges on the reasoning abilities of children.

In the task by Hawkins et al. (1984), children ranging from age four to five were asked to solve deductive syllogisms varying in problem complexity, type of content, and task conditions (Fig. 5.15). These variables were operationalized as follows:

Problem content: Children performed three different types of problems: (1) fantasy problems focused on premises describing fictional creatures, (2) congruent problems with premises that fit with real-world

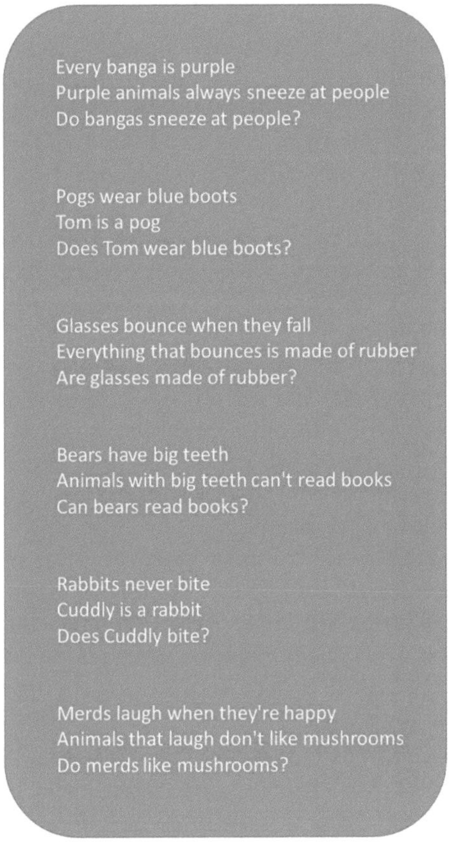

Every banga is purple
Purple animals always sneeze at people
Do bangas sneeze at people?

Pogs wear blue boots
Tom is a pog
Does Tom wear blue boots?

Glasses bounce when they fall
Everything that bounces is made of rubber
Are glasses made of rubber?

Bears have big teeth
Animals with big teeth can't read books
Can bears read books?

Rabbits never bite
Cuddly is a rabbit
Does Cuddly bite?

Merds laugh when they're happy
Animals that laugh don't like mushrooms
Do merds like mushrooms?

FIGURE 5.15 Hawkins et al. (1984) presented deductive reasoning problems in the formats indicated.

situations or knowledge, and (3) incongruent problems with premises that were not possible based on real-world knowledge.

Complexity: Quantifiers such as "some" and "all" were made implicit rather than explicit to vary problem simplicity, and negative premises were included in order to make problems more complex (based on original findings by Wilkins, 1928).

Task conditions: the different syllogism types were presented in different orderings.

The children in this study solved sets of deductive reasoning problems among three categories: fantasy, congruent, and incongruent. These problems were all presented in a format that allowed children to respond with a simple "yes" or "no" answer, but they were also offered the opportunity to justify their answers. Results indicated that the young children were capable of solving deductive reasoning problems. Evidence also indicated that the children were highly sensitive to the varying conditions. The children had difficulty with problems that were inconsistent with factual information about the real world. They were also influenced by the task conditions. The fantasy problems were removed from real-world knowledge altogether, and this enabled better reasoning performance by the children. The tendency for knowledge about the world to influence how we answer deductive reasoning problems is not unique to children, but the tendency for children to have difficulty separating the problems from real situations was a strong predictor of performance. The task conditions affected children in another way with ordering of the problems being a factor. Performance was best when fantasy-based problems were presented first. The lack of influence of real-world knowledge appeared to give the children their best opportunity to succeed. Meanwhile, information that either fits with world knowledge or opposes it placed children into a task mindset mentality that disrupted their abilities to reason independently about the different conditions.

The study by Hawkins et al. (1984) indicates that there are several factors that affect children's reasoning. Among these are the influence of previous knowledge and an inability to shift strategies once the children started down a particular problem-solving path. Lastly, a social factor appeared relevant, as children suspected the experimenters of trickery when the initial problems countered real-world circumstances. The children had particular difficulty with that set of problems (Hawkins et al., 1984). It remained unclear from this study whether the influence of world knowledge was a primary factor disrupting children's deductive reasoning about verbal syllogisms, or whether it was the task context that guided them toward using ineffective reasoning strategies.

Children are heavily influenced by real-world knowledge, and counterfactual descriptions present particular challenges for younger children. Does the context of the problem allow them to overcome this tendency? Follow-up research by Dias and Harris (1988) evaluated the role of task context in deductive reasoning. Specifically, these authors further evaluated the reasons that children show evidence of effective deduction when real-world information does not contradict valid conclusions. They tested participants aged 5 and 6 years old on syllogisms presented in a question format as follows:

All hyenas laugh
Rex is a hyena
Does Rex laugh?

The children answered with a "yes" or "no" statement after repeating back the syllogism contents so that the experimenters could be sure that they had been understood. This experiment featured three key conditions: syllogisms that are consistent with world knowledge (e.g., "all cats meow, Rex is a cat, does Rex meow?"), syllogisms that contradict real-world knowledge ("all cats bark, Rex is a cat, does Rex bark?"), and syllogisms that had contained information that the children were not previously familiar with, such as the question about hyenas mentioned above. Of particular interest was the condition containing syllogisms contrary to world knowledge.

In order to test for the influence of task context, Dias and Harris (1988) included a novel condition in their experiments. They presented the syllogisms both in a standard exclusively verbal format and in a visual/verbal format that was carried out using props. For instance, an experimenter would describe the premises aloud and also act them out by making a toy cat bark or meow. The results indicated better performance for the visual/verbal condition over the verbal only. Most intriguingly, the children were excellent at the syllogisms that ran counter to world knowledge, provided that they had watched the premises acted out by the experimental props. This study indicates that the problem context itself is the reason that children around age five have difficulties with deductive reasoning. It appeared that the purely abstract verbal descriptions cued responses to the syllogisms that were based on background knowledge rather than an evaluation of the premises. This study indicates that children at age five are capable deductive thinkers, but they are more vulnerable to experiencing interference from their background knowledge in standard problem formats.

Inductive Reasoning in Children

Inductive reasoning is an extremely valuable tool for young children. As we discussed in the previous section, the influence of world knowledge on children's ability to evaluate deductive arguments is dramatic. Acquiring world knowledge requires regular use of induction, as children use examples in order to develop rules and make sense of the regularities that they observe in their environments. Unlike deductive reasoning, inductive reasoning does not guarantee a valid conclusion, but inducing rules is perhaps more important to developing a working knowledge of the world than using the process of deduction. There are numerous forms of inductive reasoning. In this section we will focus on category-based induction.

An important form of inductive reasoning occurs when we consider a variety of instances, form a category of information based on those instances, and then proceed to fit other new information into that category. For example, a child may see several rabbits hopping around inside a pen and then conclude that all rabbits hop. A more sophisticated inductive inference is possible based on more diverse instances. For example, when a child visits an airplane museum and sees numerous propeller planes, she may infer that all airplanes fly by means of propeller. There are differences between inductive inferences based on biological instances, which are termed *natural kinds*, and inferences based on artificial or manufactured objects, termed *artificial kinds*. Several investigators have focused on the age differences observed in inductive inferences among young children relative to older children. Of particular interest is the age at which children are able to make inferences about more diverse categories. This is important as diverse categories require more consideration of variable information. For example, it is possible to make simple inferences, such as inducing the rule that all rabbits hop on the basis of seeing numerous hopping rabbits. A more complex inference is when a child decides that all fish have a swim bladder on the basis of learning about several fish that have that organ.

There are several common categorization effects that are shown by adults. These include typicality effects (Rosch, 1975) in which a highly typical member of a category is seen to be a "good" example. For instance a golden retriever is a high typical dog, being medium-sized and having a common dog shape, while a pekingese is not a typical dog having a smaller body than average and a very short face (Fig. 5.16). The resulting typicality effect is that the golden retriever will be judged to be a member of the dog category a bit faster than the pekingese. The result also influences inductive reasoning, as more typical members of a category will be judged to have traits that are universal to the category (Osherson, Smith, Wilkie, Lopez, & Shafir, 1990). Similarly, Osherson et al. (1990) demonstrated similarity effects, in which animals that are highly similar to one another are seen to have more influence over the strength of inductive arguments.

The influence of typicality and similarity effects occur in children as young as 5 years old. López, Gelman, Gutheil, and Smith (1992) evaluated these effects in children. To test whether children's inductive inferences are sensitive to the typicality of animals, the investigators showed children pictures of animals and asked them questions about whether another example animal would have that same property as those shown in the pictures. For instance the children were shown pictures of dogs and bats. The children were told that dogs have leukocytes inside and bats have ulnaries inside. Based on these statements, children are more likely to conclude that all animals have leukocytes inside, as dogs are more typical than bats. Children as young as 5 years old also show similarity effects. For instance, when the children considered that a horse has leukocytes inside, they were more likely to judge a donkey to also have leukocytes, rather than a squirrel.

FIGURE 5.16 Typicality judgments (Rosch, 1975) can be made about many categories. (A) Many would consider the golden retriever to be a typical example, as it is medium-sized with relatively common features present in many dogs. (B) Breeds including the pekingese with unusual features including short legs, small stature, and short snouts would be considered by many to be lower in typicality.

Previous research indicates that children under the age of nine are less sensitive than adults to the diversity of a sample tending to generalize similarly in their inductive inferences between diverse and uniform samples. For example, López et al. (1992) asked children ranging in age from 5 to 9 years to evaluate biological properties within diverse pairs of animals (e.g., cats and buffalo have ulnaries inside) and again in relatively uniform pairs (e.g., cows and buffalo have leukocytes inside). The children were not reliably able to apply the diversity principle when asked to consider whether another animal (such as a squirrel) has leukocytes or ulnaries. Other studies indicate that diversity-based reasoning has been found to emerge around age eight and becomes more robust by ages 10–11 years (Rhodes & Liebenson, 2015).

These studies demonstrate that inductive inference is a part of a child's development from a relatively young age, but inductive inferences become more advanced as children move toward the age of 10. Typicality effects and similarity effects emerge relatively early by age 5. This indicates that category knowledge is growing at rapid rate by this point and nuanced distinctions can be made in evaluating the likely properties of new category members based on the similarity to previously learned category members. The sensitivity of a child to the diversity of instances is more difficult to acquire and does not emerge until around age 10.

RELATIONAL REASONING IN CHILDREN

Thinking Relationally

One of the more abstract forms of inductive inference involves drawing a relational correspondence between two objects, places, or things. Early on, children do not show a strong capability to reason by relations. Rather, there is a tendency for young children to pay attention only to objects in an isolated way. In doing so, children will often miss the correspondences that different objects share in a larger context.

One of the hallmarks of relational ability is success on the Raven's progressive matrices task (Raven, 1938, 1960). Raven's matrices consist of a set of abstract objects or patterns. There are typically nine sections in a Raven's problem with the final sector left blank for the reasoned to fill in. Most Raven's problems also offer the participant a set of multiple choice patterns that would complete the matrix. The simplest Raven's problems are pattern match problems of the type shown in Fig. 5.17A. In this problem the thinker simply fills in the missing piece based on appearance or perceptual matching. More difficult Raven's matrices problems characteristically require that multiple different transformations be integrated across the rows and down the columns in order to form an integrated answer (Fig. 5.17B). This second type of problem requires relational integration, a process by which the relationships among multiple objects or patterns need to be considered in the answer to the problem (Robin & Holyoak, 1995). There are several versions of matrix reasoning problems available across a variety of tests. In addition to the Raven's matrices, there is a matrix reasoning task on the Weschsler advanced intelligence scale (WAIS) (Wechsler et al., 2003), which tests perceptual reasoning skills. A test called the Naglieri nonverbal ability test is particularly well-suited to testing young children and is commonly administered around the age four and up (Wechsler et al., 2003Naglieri, 2007).

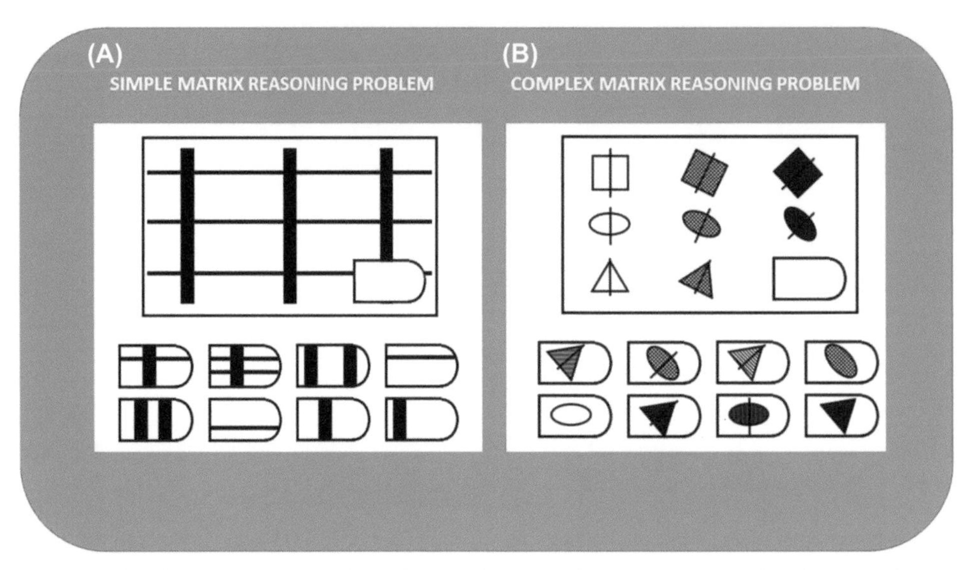

FIGURE 5.17 (A) An example of a simple perceptual matching problem similar to that presented in the Raven's matrices task. (B) A more complex problem involving relational matching.

Relational Shift in Children

There has long been a debate in developmental psychology about the precise timing of when children become capable of relational thinking. Researchers including Jean Piaget (1977a) have maintained that children are not capable of fully abstract thinking until they reach age 10 to 12 (Goswami & Brown, 1989).

Developmental psychologist and cognitive scientist Dedre Gentner has postulated that the absolute age for relational thinking is less important than the child's perception of the task demands and how a situation is described. Gentner (1988) has called the acquisition of relational thinking ability a *relational shift*. As children reach an age at which they can appreciate relational terms, such as *above, on, under, larger*, and *shorter*, they also appear to gain competence in the ability to process and integrate relational information. Gentner has demonstrated that younger children tend toward seeing similarity in the form of object attributes, such as color, shape, and size. This leads the children to match items based on appearance rather than the relational correspondence between items or situations. Older children tend toward the strategy used by adults which centers upon noticing and making use of shared relations among objects. This is the core of Gentner's (1983) structure mapping theory, which has also been implemented in a computational model that is sensitive to both object-based similarity and relational correspondences that can lead to more abstract perception, inference, and problem solving (Falkenhainer, Forbus, & Gentner, 1989).

The hypothesis that relational language is the basis for relational thinking has received support from a study evaluating relational reasoning in adult stroke patients. Patients who had suffered damage to the left hemisphere regions associated with language production and comprehension were evaluated on relational reasoning ability. These patients are termed aphasic, indicating that language deficits are present. Aphasic patients were compared with non-aphasic left hemisphere stroke patients on relational thinking. These patients performed problems from the Raven's colored progressive matrices task, which included both problems that focused on the ability to match visual patterns and relational problems that required the integration of elements in order to solve. Aphasic patients had greater difficulty with relational problems compared to visual pattern match problems, and furthermore they had more difficulty than non-aphasic stroke patients. This result suggests that language is linked to relational thinking, as postulated by Gentner (1988). Voxel-based lesion mapping of neuroimaging data indicated that areas such as the left middle and superior temporal gyri, along with the left inferior parietal lobule, were most linked to relational processing in the Raven's problems, while visual pattern matching problems were most associated with damage to visual cortical posterior left hemisphere areas (Baldo, Bunge, Wilson, & Dronkers, 2010) (Fig. 5.17).

Analogical Reasoning in Children

Analogical reasoning yields a type of inductive inference known as an analogical inference. These types of inferences are induced on the basis of reference to a set of relations shared among a current situation and a previous situation that had been experienced before, or one that is available to compare to. Acquiring relational reasoning ability, as is measured by tasks such as the Raven's matrices, is critical for children to make use of analogies. Analogical reasoning can be a gateway to noticing abstract similarities between situations and a potent tool for new learning and inductive inference. Ultimately, children may build up sets of knowledge in the form of schemas that can be applied widely to a variety of new situations on the basis of analogical information (Hummel & Holyoak, 1997).

When children are able to move from one situation to another noticing relational correspondences then they have taken a strong step toward abstract reasoning. Such abilities may most commonly be observed once children reach age 9 or 10. A rather well-known study investigating the age at which children begin to notice and use abstract similarity involved children applying knowledge about a model room to a situation in a larger room. Developmental psychologists DeLoache and DeMendoza (1987) created a small model of a room complete with small furniture, as one would find in a dollhouse. Children in the experiment watched DeLoache place a small model of a dog that the experimenters referred to as "Little Snoopy" in a particular location within the model room. Children were then shown an actual room within the building that had the same furniture and layout as the small model room had. DeLoache wanted to determine whether children would notice the correspondences between the model room and the actual room. Children aged 2–3 years were tasked with finding "Big Snoopy," which was a larger toy version of the dog in the larger actual room. For example, when the small toy Snoopy had been hidden behind the couch in the model room, a successful child would proceed to the larger version of the couch in the big experimental room and locate the Big Snoopy toy. Three-year-old children were successful at finding Big Snoopy about 85% of the time. By contrast, children as old as two-and-a-half performed at chance, only locating Big Snoopy about 15% of the time. DeLoache also noted that the three-year-olds readily talked about how the big room was just like the model and appeared to understand the relational correspondences between the model and larger rooms (Fig. 5.18). Meanwhile, the two-year-olds typically seemed unable to appreciate the similarities between the model and real rooms, having no idea where to find Big Snoopy. DeLoache noted that both two- and three-year-olds scored above 80% on a memory test of where Little Snoopy had been hidden in the model room. This indicates that the failure of the younger children was due to an inability to notice the relational similarities between the model and larger rooms rather than a failure of memory. Note that age three is a young age for symbolic abstract thinking to occur. This study demonstrates that children as young as three are already capable of symbolic relational thinking provided the task circumstances are appropriate to elicit a relational match.

FIGURE 5.18 In this clever developmental psychology task a child is presented with a model of a room in which a small dog is hidden in a particular location. The child can later use the model as an example for where a larger dog is hidden in an actual room (DeLoache & DeMendoza, 1987).

A more complex analogical reasoning task investigated whether children were capable of showing analogical transfer between abstract stories. Holyoak, Junn, and Billman (1984) investigated children's abilities to use an abstract story as an analog to a current situation. There were both young children (aged 4–7 years) and older children (age 10–12 years) enrolled in this study. Children were initially presented with a story about a genie who had a problem moving from one magic bottle to another. The genie needed to collect his jewels and move them. In one condition the genie used his magic staff to pull the new bottle closer to his old home and then place the jewels into the new home. In another condition the genie was described as using his magic carpet by rolling it up and then transferring the jewels across from the old bottle to the new one using the rolled carpet as a ramp. The children in this study were next asked to solve a problem in which they had to generate as many ways as they could think of to move a set of small balls from one bowl into another. The children were seated and the new bowl to which the balls were to be transferred was not easily within reach of the children. The experimenters provided a set of items for the children to use in order to generate solutions and among these items were a cane for walking, a cardboard tube, and a sheet of poster board. Both the younger and older children could form an analogy between the magic staff and the cane and used the cane to pull the new bowl toward them. A majority of participants in both the younger and older groups applied the staff/cane analogy as a solution to moving the balls. A large number of participants at each age group also used a partial analogy between the rolled up magic carpet that the genie in the story used as a ramp and the cardboard roll. This was not the most relationally

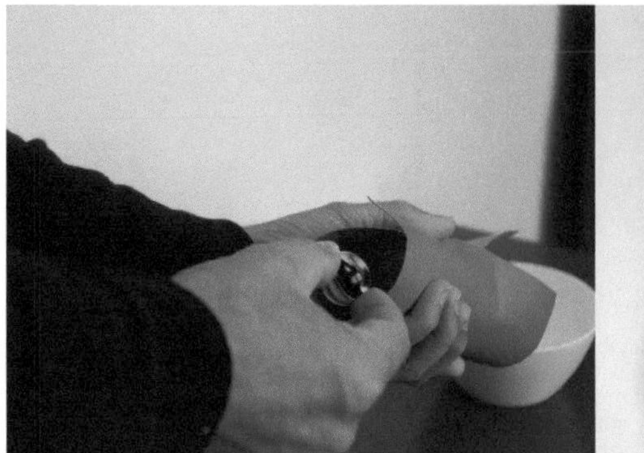
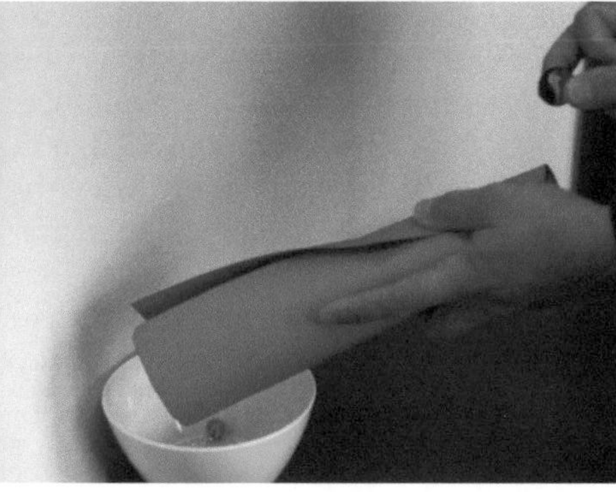

FIGURE 5.19 The genie problem presented children with a story that was analogous to a problem that they would next have a chance to solve. An analogical solution involved rolling a poster board into a tube in order to transfer balls from one bowl to another (Holyoak et al., 1984).

elegant solution, as the child did not have to roll up the cardboard (Fig. 5.19). The interesting result was that only about 10% of the children in the younger age group applied the magic carpet story to the problem by rolling up the poster board and using it as a tube to transfer the balls. By contrast, a majority of the older participants were able to readily make use of the magic carpet/poster board analogy quickly rolling up the poster board as the genie had done. This study indicates that young children around age five or six are capable of using analogies, but they were much more likely to notice and use a concrete action-based analogy such as using the cane to reach with. Young children missed the solution involving the transformation of a flat object into a tubular one, a more complex relational scenario. Why did the young children make use of the analogies to the cane and the cardboard tube, but failed at the rolled tube solution? It may be that the magic carpet and the poster board lacked sufficient similarity for the younger children to notice that these could function similarly. Another possibility is that having to transform the poster board into a tube added complexity to the solution and this blocked the carpet/poster board analogical solution. Not all analogies are equally useful or noticed by children nearing the end of preschool or beginning kindergarten.

Another important factor influencing young children's ability to solve analogies is the type of relations that are used in problems. Standard four-term analogies follow a common format "A is to B, as C is to D." This type of analogy has seen widespread use in tests of intelligence or academic capacity (e.g., Miller's analogies test). Children were thought to be unable to complete these analogies until the formal operations stage theorized by Jean Piaget. Piaget (1977b) presented children with four-term analogy problems such as: *bicycle is to handlebars as ship is to?* A correct D term answer to this problem would be ship's wheel or rudder, but younger children struggled to complete these types of problems frequently generating responses that had some superficial or semantic similarity to the C term (ship). Such answers do not take into account the higher-order relations within the analogy, and Piaget observed that children younger than 12, the age at which they enter into formal operational thinking, tend not to adequately appreciate higher-order relational structure as seen in these analogies.

Developmental psychologist Usha Goswami suspected that perhaps younger children were capable of making relational comparisons in four-term analogy problems. Goswami noted that children from a very young age can appreciate causal relationships that they observe in the world around them. This capability is consistent with some of the observations that we covered in the earlier section of this chapter on the development of causal reasoning in young children. Goswami tested children on four-term analogies using picture items. The problems included analogies with embedded physical causal relations such as *melting, breaking,* or *cutting,* as children between the ages of three and four can reason about these physical causes (Bullock et al., 1982). An example of a causal analogy presented in picture format by Goswami was "playdoh is to cut playdoh, as apple is to cut apple." Alternative incorrect picture choices were included as distractor items along with the correct answer. For example the children saw the following answer choices for the "playdoh/apple analogy": cut apple (correct answer), cut bread (correct relation, but incorrect object), bruised apple (incorrect relation, but correct object), ball (a mere appearance match), and a banana (semantically related object). Note that Goswami also ensured that the problem contents were familiar to the three-, four-, and six-year-old children who were tested. The children did quite well on problems with familiar causal content. Developmental effects were evident within the group data. Six-year-old children were excellent at these analogies performing near ceiling levels. Four-year-old children also did remarkably well with about 90% of children correctly

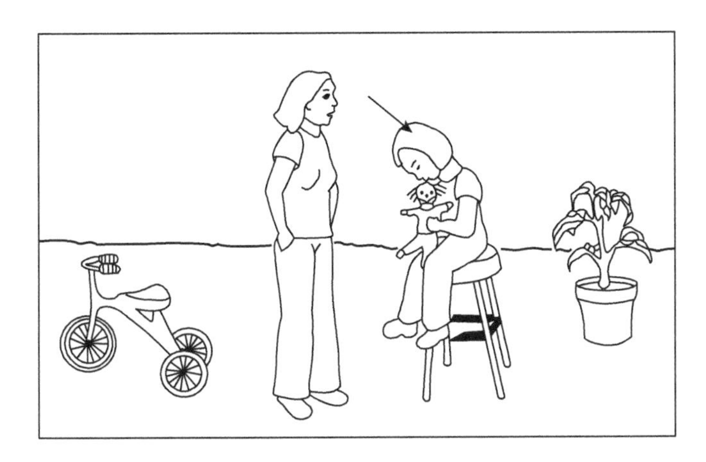

FIGURE 5.20 Scene analogies can be effective ways to test children's relational capabilities (Richland et al., 2006).

solving the analogies when the correct solution has been explained to the child at the completion of each problem. Remarkably, even by age three, about half of the children were able to solve the four-term analogy problems when the experimenter provided explanations. In this study, Goswami demonstrated strong analogical competence in relatively young children, who would still have been in the preoperational stage marked by concrete and egocentric thinking according to Piaget. The analogical competence of the four- and six-year-olds relative to the three-year-olds may be related to the older children's enhanced ability to make inferences about the physical causes within these age ranges.

Scene analogy studies can take into account more than four relational terms. These types of tasks have been effectively used for evaluations of relational over object similarity (Markman & Gentner, 1993). Richland, Morrison, and Holyoak (2006) developed a set of scene analogy problems that could be used to evaluate both mere appearance matches and relational analogical answers. Richland et al. tested children at ages 5–6 years and children aged 9–10 years on problems that involved line drawings of a scene (Fig. 5.20). The children looked at the first scene such as a girl kissing a doll while in the presence of three other objects, people, or things. Notice in Fig. 5.20 that there is an arrow pointing to the girl in this problem. Children were asked to select an item that best matched the girl in a second scene. The child next looked at the second scene in which was presented a similar situation. In this case the child saw a woman kissing a dog while a girl looked on. The question posed by the researchers is which type of match would the children prefer? Would they match analogically by picking the woman (kissing the dog) as a match to the girl who was kissing the doll, or would they select a mere appearance match such as the other little girl who was depicted as an onlooker in the second scene. Richland et al. (2006) demonstrated that younger children ages 3–4 years old tended to select a match based on mere appearance, while older children at ages 9 and 10 understood the relational correspondences between the analogical matching pairs and selected matches based on relations. It appears that even when analogical matches are apparent in problem content, younger children may not select them.

In other conditions within the experiment, Richland et al. varied the number of relations present in the problem. When two relations were present in the problems, children had greater difficulty selecting the analogical response. Also, in some cases mere appearance matches were removed from the problems, which enhanced analogical matching. In this study three factors emerged as important for encouraging abstract higher-order analogical responding: age with older children being more likely to make analogical matches than younger, relational complexity with problems involving two relations discouraging analogical responding over single relational problems, and distraction, with appearance-based matches discouraging analogical responding. While young children can indeed perform analogies, there are some clear contextual or environmental factors that enhance those responses or inhibit an analogical strategy.

The changes in relational thinking that accompany aging are not entirely explained by strategy use or language development. There is also a strong role for brain development in the shift toward greater abstract thinking. Wright, Matlen, Baym, Ferrer, and Bunge (2008) at the University of California, Berkeley, presented evidence for differences in brain development linked to relational reasoning. Both adults aged 19 to 26 and children aged 6 to 13 were tested on simple analogical relations among problems in a functional MRI study. These included analogies such as "chalk is to chalkboard as pencil is to?." Participants selected among four possible answer choices to complete the problem. A distracting answer that did not complete the analogy, such as eraser, was included as an answer choice along with irrelevant pictures and the correct answer, paper. The analogies were shown in picture form. The results indicated that adults solved the analogy problems with a higher degree of accuracy than the children. Neuroimaging results showed that children activated a subset of the same areas that had been activated by the adults when solving analogical reasoning problems. When adults solved analogy problems, activation was observed within bilateral regions of the prefrontal cortex (PFC) including ventrolateral (VLPFC) sectors. These areas are frequently associated with working memory activity (D'Esposito, Detre, Alsop, & Shin, 1994) and have been demonstrated to be some of the later areas of the brain to mature according to patterns of cortical volume (Gogtay et al., 2004). Adults also showed additional analogical reasoning-related activation in the parietal and occipital lobes, along with the basal ganglia. In children, fewer areas were significantly active in association with analogical reasoning. Active regions in the children aged 6 to 13 included bilateral premotor cortex and right occipital lobe, regions that are not commonly associated with reasoning, as we discussed in Chapter 3. Wright et al. indicated that the strong activity for children within the occipital cortex (often associated with visual processing) could be explained by the finding that children took longer to solve the analogy problems than adults and therefore received more visual stimulation within the occipital lobes.

JUDGMENT AND DECISION MAKING IN CHILDREN

How Children Make Decisions

How do children make decisions? A casual observation would suggest that young children are overflowing with emotion and are much more impulsive than adults. Meanwhile older children are often less impulsive, but may still react to situations more emotionally than many adults. In this section, we will discuss some of the evidence for how children carry out decisions. In particular, we will emphasize how children use heuristics, which are helpful rules of thumb that will frequently yield a fast and effective decision. Heuristics have been evaluated extensively in adults (Tversky & Kahneman, 1981), and we will cover much of this territory in Chapter 11 on decision making. A disadvantage of using heuristics is that they can at times trip us up and lead to ineffective decisions, or ones that are overly led by the context of the situation, rather than an effective logical analysis of what is optimal for us to do at a given time. In the following sections we will explore how children go about making decisions with regard to several well-known heuristics that adults frequently employ.

The Representativeness Heuristic in Children

The representativeness heuristic is a commonly observed bias affecting the way in which we judge the likelihood of a particular instance. This bias makes us feel that situations that resemble a frequently observed situation or option are more common. This can trip us up when we have to estimate the chances of a particular outcome, and we overly rely on predictions based on how representative something is rather than a careful and thoughtful analysis of its individual likelihood.

The representativeness heuristic operates when we consider the likelihood of someone holding a particular profession. For example, if we meet a new person and are asked to decide whether he is more likely to be a lawyer or an engineer, we can consider multiple factors about the person and the profession in order to make this decision. We might consider the characteristics of the person and relate them to how often we have experienced those features in people holding a given job. If we are asked to judge whether Linda is an artist or an engineer we may consider things we know about Linda in order to help make this decision. Linda enjoys painting, is fond of nature hikes, holds liberal political views, and is involved in the local music scene. These personal traits probably suggest that Linda is a poet, rather than an engineer. Other information should be considered before we commit to this decision. Engineering is a popular major in college. There are a variety of types of engineers and in most parts of the United States we will probably find many more engineers than poets. This background information is known as the base rate. If there are 2 poets and 20 engineers out of every 100 people, we ought to rethink the likelihood of Linda being a poet. Tversky and Kahneman (1973) demonstrated that adults tend to overweigh the evidence based on personal characteristics and underestimate the base rates of a particular profession within the population.

A study of the representativeness heuristic in children provided evidence about their tendency to weigh decisions based on traits versus base rate information. Jacobs and Potenza (1991) asked children to make judgments about people based on short hypothetical descriptions. In this study, children at ages 6, 8, and 11 years read descriptions similar to that below: *"In Juanita's class 10 girls are trying out to be cheerleaders and 20 are trying out for the band. Juanita is very popular and very pretty. She is always telling jokes and loves to be around people. Do you think Juanita is trying out to be a cheerleader or for the band?"*

Many of the youngest children based their estimations of likelihood on their own criteria ignoring both base rate and personal characteristics. Six-year-olds often used idiosyncratic strategies, such as making a likelihood judgment based on their own personal preferences about what Juanita should try out for.

Older children at the ages of 8 and 11 tended to use the personal characteristics that were described in the stories, such as Juanita's attractiveness and popularity. This study indicates that as children age, they tend to base more of their decisions on the representativeness heuristic, which is similar to what adults often do (Tversky & Kahneman, 1973). Children in this study did not make extensive use of the base rate information about the number of children trying out for either of the activities.

The Role of Availability in Children's Decisions

Adults make decisions using another important heuristic known as availability (Tversky & Kahneman, 1973). When people use the availability heuristic, they base their estimate of the probability of an event or situation on the fluency of how readily examples come to mind (Tversky & Kahneman, 1973). For example, if I were to judge the number of people that use Mac laptops, I may overestimate the number based on the vivid mental image I have of a silver laptop with a glowing apple logo on the back of the screen. This is more readily accessible to me than other laptops that I have seen from a variety of other manufacturers; therefore, I might tend to overestimate the actual percentage of people who use Mac laptops.

Davies and White (1994) conducted a study of the use of the availability heuristic in children at age 7 and at age 10 using a method that had been previously introduced by Tversky and Kahneman (1973). The children in this study were presented with one of two lists of names. One of the lists included the names of 20 women who were not particularly famous and 19 famous men. The gender of names on the other list was reversed with 19 famous women and 20 less-famous men. Children were asked to state whether there were more men's or women's names on the list. Children at both ages 7 and 10 exhibited a bias toward estimating the gender with the more famous names to have been more plentiful. Note that this is actually incorrect, as there were always more of the less-famous names than famous ones. The bias suggests that the availability within memory of the famous names was higher and therefore the children estimated the number of names of a given gender inappropriately. Davies and White also conducted a recall test, which substantiated this tendency, as children could recall more names from the famous list than the less-famous list. Interestingly recall levels had been higher for famous names for a proportion of the 7-year-olds over the 10-year-olds and this equated to a higher use of the availability heuristic within that age group. Overall these results suggest that children by age seven will base their estimates of probability or frequency on their ability to recall examples. Note that this distortion is similar to what adults do.

Anchoring in Children's Decision Making

Another very powerful bias that adults exhibit is known as anchoring. This occurs when people begin a judgment with a quantitative starting point. Wherever the person starts will tend to bias their subsequent judgments about probability. Tversky and Kahneman (1974) originally described this phenomenon in an interesting experiment in which people had to estimate the number of United Nations member states based in Africa. The participants watched as a random number was generated based on the spin of a wheel. Despite the clear arbitrary assignment of this number, it dramatically influenced people's judgments about how many African UN countries there actually are.

Tversky and Kahneman (1974) also demonstrated the effects of anchoring on a task that may be more familiar to school-age children. They offered participants 5 s during which to make an estimate of a multiplication product for a series of problems. The researchers varied the order of presentation of the numbers in the problems. On some trials, the numbers were presented in an ascending order starting with one and ending with eight ($1\times2\times3\times4\times5\times6\times7\times8$). In other cases the number order was reversed and presented from eight descending down to one ($8\times7\times6\times5\times4\times3\times2\times1$). The correct answer for one problem was 40,320. Participants had to quickly produce a ballpark estimate given the time constraints. The differences in ordering dramatically affected people's median estimates for this problem. The ascending sequence (from one-to-eight) resulted in an estimate of 512, while the descending sequence led to an estimate of 2250! Notice the dramatic differences from the correct answer (40,320) indicating that people had likely used heuristics in order to generate their answers. Participants may have only had time to roughly multiply the first three or possibly four numbers, produce a plausible answer to that sequence, and then crudely adjust up from their current anchor point to provide a final answer. Adjustments from a particular anchor are most often too low, as people fail to adequately move their estimation once they have settled on the anchor point.

H. David Smith (1999) tested children on the use of anchoring and adjustment heuristics on a math sequencing problem. The children tested were in either fourth, sixth, or eighth grade and these participant groups were compared to a college student sample. The math sequencing task was based on the original multiplication work that had been performed by Tversky and Kahneman, but was modified to become an addition problem for testing with younger children. The ascending sequence used in this study was $1+2+3+4+5+6+7+8+9$, while the descending sequence presented the same numbers in reverse ($9+8+7+6+5+4+3+2+1$). The correct answer is 45. There was a main effect of sequence across the groups with the ascending sequence resulting in a mean estimate of 40.88 while the descending order yielded 49.37. There were age-related effects as well, with the surprising result that sixth and eighth graders were more accurate than the other groups. College students tended to underestimate the magnitude providing mean estimates in the low thirties with little effect between ascending and descending orderings. Meanwhile the fourth graders, who were approximately age 9–10 years old overestimated from an anchor with their answers being over 50. These children showed an almost seven digit difference between the two orderings (ascending = 51.74 and descending = 58.51). This study suggests that children as young as 9 or 10 years already use heuristics. Furthermore, the use of heuristics is tempered by the facility or practice that one has with the material, as the sixth and eighth grade students may have been most practiced in serial addition at that stage in their schooling.

DEVELOPMENT OF MORAL REASONING

Stages of Moral Development

As children develop cognitively, we also see them develop different social abilities and an understanding of what is right and wrong. Much of child development involves learning by being provided with rules and receiving consequences if those rules are not adhered to. As children develop, their sense of moral justice also changes. In this section, we will discuss the developmental trajectory of changes in moral reasoning for children and how those relate to actual decisions about morality.

Jean Piaget theorized about the development of moral thinking in children in addition to making major contributions to the study of cognition. For Piaget (1932) children largely fell into two major categories of moral ability. From age four through seven he claimed that children exhibit *heteronomous morality*. This term refers to a view that actions and consequences are rigid and unchanging. The heteronomous child sees moral situations as being clearly defined and not under the control of human actions. For children in this stage, the consequences of breaking rules are immediate and are applied automatically. The child in this initial phase of moral thinking may consider it to be worse to damage two windows by accident then to deliberately throw a rock through one window. As the child develops, Piaget claimed that they would enter into a transition period around the ages of 7 to 10, during which they would retain some of the same heteronomous moral thinking along with some new more flexible appreciation of moral consequences. Lastly, around the age of 10 the child was predicted to exhibit *autonomous morality*. At this stage, the child sees moral thinking as a human construct and that people's actions and intentions must be considered when contemplating the violation of a rule. Children in this last stage should be able to understand that it is worse to deliberately destroy property than to accidentally do so. After children reach this stage, they would likely see the value in following rules for the benefits of getting along in society rather than purely to avoid the inevitable punishment that is expected by the heteronomous child when a rule is broken.

One of the most notable descriptions of moral development comes from Lawrence Kohlberg (1958, 1986). Kohlberg based his moral stages of development in part on the existing stages of cognitive development described by Piaget (sensorimotor, preoperational, concrete operations, and formal operations). The stages of moral thinking described by Kohlberg are based on an interview technique that he developed, in which children were asked about 11 different moral dilemmas. The most commonly cited dilemma involved whether it is morally appropriate for a man to steal a drug in order to save his dying wife. The story was presented as follows:

> A woman was near death. There was one drug that the doctors thought might save her. It was a form of radium that a druggist in the same town had recently discovered. The drug was expensive to make, but the druggist was charging ten times what the drug cost him to produce. He paid $200 for the radium and charged $2000 for a small dose of the drug. The sick woman's husband, Heinz, went to everyone he knew to borrow the money, but he could only get together about $1000 which is half of what it cost. He told the druggist that his wife was dying and asked him to sell it cheaper or let him pay later. But the druggist said: "No, I discovered the drug and I'm going to make money from it." So Heinz got desperate and broke into the man's laboratory to steal the drug for his wife. Should Heinz have broken into the laboratory to steal the drug for his wife? Why or why not?

In this example the children would be asked a series of questions probing whether it was appropriate for Heinz to steal the drug, and why it might or might not be acceptable. The justification that the child provides for their answer is the key to this interview. In fact, the reasoning behind the child's decision about the theft decision is more important than the theft decision itself. The answer to this and other dilemmas helped Kohlberg to classify children's moral development into three levels and six stages (two stages per level) (Fig. 5.21).

According to Kohlberg, the first level of moral development is called pre-conventional reasoning, which is divided into two stages. During this period, a child views the moral world as being rigidly defined with actions that are reliably linked to punishments or rewards. The first stage in pre-conventional moral reasoning is heteronomous morality, the same term that Piaget had applied

FIGURE 5.21 Lawrence Kohlberg (1927–1987) described a sequence of stages of moral development. Credit **UAV 605.295.8 (Box 7; Kohlberg), olvwork357137, Harvard University Archives** *Harvard University. News Office. Harvard University Gazette photographs May 23, 1986–August 9, 1991.*

to children's thinking prior to age seven. Kohlberg estimated that children age nine and younger reason predominantly at this level. Children at this stage express a desire to follow rules simply to avoid punishment or gain concrete approval. At this stage, the child is likely to determine that Heinz will be sent to jail if he is caught stealing the drug, so therefore he should not do so. The next pre-conventional stage is known as individualism, instrumental purpose, and exchange. This stage occurs in early adolescence and at this time a child begins to show moral progress in terms of acknowledging that other people should be able to pursue their own interests, that exchanges between people should be fair, and that the golden rule applies with people acting kindly toward others so that the favor will be returned to them. In this stage, a child may justify Heinz's theft of the medicine on the grounds that he had offered the druggist a fair price and that as long as he agreed to give him the money he had raised, then perhaps Heinz would come away without punishment after taking the medicine.

The second Kohlberg moral developmental level is conventional moral reasoning. In this case the child operates according to a set of moral principles that are internal to their own thinking. The child typically observes these standards as exhibited by society.

The third stage of conventional moral reasoning in the adolescent is known as mutual interpersonal expectations, relationships, and interpersonal conformity. The child is now aware that kindness, warmth, and fairness are appropriate moral values to aspire to. The child at stage three might justify Heinz' theft of the medicine by claiming that the druggist was not acting fairly and that Heinz was therefore justified in taking the medicine.

Stage four of the conventional level is called social systems morality, in which the child now begins to adopt the morality that is being shown to him or her by society itself through the laws, rules, customs, and regulations that the child observes within his or her environment. At this stage the child may determine that Heinz should not have taken the drug, as he will experience problems with society and will feel that he has acted in a way that is not just toward the druggist.

The fifth level of Kohlberg's moral judgment scale is the post-conventional stage, which will not be reached even by many adults. This stage relates to values that become abstract and moral guidance is highly internalized. Social contracts and individual rights are important at this stage during which an individual examines the rules of society and makes explicit connections between the laws and the value that they convey upon society. This may involve a realization that following laws makes society function in a more uniform and sensible manner for all of its members. The thinker at this stage may answer that the problem with stealing the medicine is that we cannot simply choose when to apply laws and when not to, or society will become less just overall.

Lastly, the sixth stage of Kohlberg's moral development is called universal ethical principles. This is a stage that not everyone is expected to reach and it involves developing a strong moral compass unique to oneself. At this stage an individual can make up his or her own mind about the moral actions possible based on their conscience. The individual at this stage thinks mostly about human rights and how to enact or follow laws in a way that maximizes the most human benefit. At this last stage an individual will likely justify the theft of the medicine by the logic that human life is valuable and outweighs the profit motive of the druggist in this particular case.

The development of morality is key to a healthy society and our reasoning about social situations, monetary exchanges, and fairness all hinge upon our society raising children who grasp and utilize principles of moral thinking. As we will discuss in Chapter 7, covering disorders of reasoning, there are cases involving psychopathology in which individuals act in inappropriate ways due to failures to reach maturity on moral grounds. The frontal cortex is highly linked to the development of moral knowledge and this region is also important for acting out moral thinking in actual daily life situations (Anderson, Bechara, Damasio, Tranel, & Damasio, 1994).

SUMMARY

The developmental process is remarkably dynamic as children move from infancy up through adolescence. The process is both a biological one and an environmental one with both factors frequently contributing to the output of increasingly sophisticated and abstract reasoning behavior. Children begin with a process of cortical thickening as large numbers of synaptic connections are formed. From age three onward, the cortex undergoes a tuning process as some synaptic connections strengthen and others weaken. The net result of this process is a decrease in cortical volume from age 5 through 20.

Children's thinking is guided by a variety of factors. There are relatively stable and reliable stages that a child will progress through as they develop. The context of a problem becomes a significant factor in determining how children will reason and developmental reasoning studies require sensitivity toward making the experimental stimuli understandable and interesting to the child. Children exhibit some competencies in causal reasoning and learning from a very young age.

Children show increasing reasoning abilities as they develop. Some of the skills such as relational and analogical reasoning grow during the elementary school years and are supported by increases in cognitive control and decreases in impulsivity. The child becomes less concrete in how he or she views and interacts with the world. This increasing abstraction ability encompasses semantic knowledge, deduction, and moral thinking.

END-OF-CHAPTER THOUGHT QUESTIONS

1. Biological and environmental influences interact throughout human development. What are some cases in which each of these influences acts independently?
2. During the tuning process of development, some synaptic connections strengthen and others weaken. What does this suggest about the timing of education?
3. Jean Piaget described a series of stages of cognitive development. Which of these stages seems most accurate today?
4. The context of a problem helps to determine how children will reason. How much do you think the problem context is responsible for children appearing more or less competent than they actually are?
5. Children exhibit competency in causal reasoning and learning from a very young age. Why is this important for survival and development?
6. Relational and analogical reasoning grow during the elementary school years. What relative importance does working memory have in relational reasoning compared to impulse control?
7. Decision making is guided by heuristics or mental shortcuts. Can you think of some examples in which children use literal information over heuristics?
8. Does modern society show sensitivity to the variations in moral thinking that exist? Are there cultural differences in the way that we think about morality?

References

Anderson, S. W., Bechara, A., Damasio, H., Tranel, D., & Damasio, A. R. (1994). Impairment of social and moral behavior related to early damage in human prefrontal cortex. *Nature Neuroscience, 2*, 1032–1037.

Baldo, J. V., Bunge, S. A., Wilson, S. M., & Dronkers, N. F. (2010). Is relational reasoning dependent on language? A voxel-based lesion symptom mapping study. *Brain and Language, 113*, 59–64. http://dx.doi.org/10.1016/j.bandl.2010.01.004.

Bjorklund, D. F. (2005). *Children's thinking: Cognitive development and individual differences* (4th ed.). Belmont, CA: Wadsworth (Earlier editions published in 1989, 1995, and 2000.).

Brainerd, C. J., & Allen, T. W. (1971). Training and transfer of density conservation: Effects of feedback and consecutive similar stimuli. *Child Development, 42*, 693–704.

Brainerd, C. J., & Brainerd, S. H. (1972). Order of acquisition of number and quantity conservation. *Child Development, 43*, 1401–1406.

Bullock, M., Gelman, R., & Baillargeon, R. (1982). The development of causal reasoning. In W. J. Friedman (Ed.), *The developmental psychology of time* (pp. 209–254). New York: Academic Press.

Burt, C. (1919). The development of reasoning in school children. *Journal of Experimental Pedagogy, 5*, 68–77 121–127.

D'Esposito, M., Detre, J. A., Alsop, D. C., & Shin, R. K. (1994). The neural basis of the central executive system of working memory. *Nature, 378*.

Davies, M., & White, P. A. (1994). Use of the availability heuristic by children. *British Journal of Developmental Psychology, 12*(4), 503–505.

DeLoache, J. S., & DeMendoza, O. A. P. (1987). Joint picture book reading of mothers and one-year-old children. *British Journal of Developmental Psychology, 5*, 111–123.

Dias, M. G., & Harris, P. L. (1988). The effect of make-believe on deductive reasoning. *British Journal of Developmental Psychology, 6*, 207–221.

Donaldson, M. A. (1963). *Study of children's thinking*. London: Tavistock.

Falkenhainer, B., Forbus, K. D., & Gentner, D. (1989). The structure-mapping engine: Algorithm and examples. *Artificial Intelligence, 41*, 1–63.

Field, D. (1987). A review of preschool conservation training: An analysis of analyses. *Developmental Review, 7*, 210–251. http://dx.doi.org/10.1016/0273-2297(87)90013-X.

Flavell, J. H. (1963). *The developmental psychology of Jean Piaget*. New York: D. Van Nostrand Company.

Frye, D., Zelazo, P. D., Brooks, P. J., & Samuels, M. C. (1996). Inference and action in early causal reasoning. *Developmental Psychology, 32*, 120–131.

Gentner, D. (1983). Structure-mapping: A theoretical framework for analogy. *Cognitive Science, 7*, 155–170.

Gentner, D. (1988). Metaphor as structure mapping: The relational shift. *Child Development, 59*, 47–59.

Gogtay, N., Giedd, J. N., Lusk, L., Hayashi, K. M., Greenstein, D., Vaituzis, A. C., et al. (2004). Dynamic mapping of human cortical development during childhood through early adulthood. *Proceedings of the National Academy of Sciences, 101*, 8174–8179.

Gopnik, A., Meltzoff, A. N., & Kuhl, P. K. (1999). *The scientist in the crib: Minds, brains and how children learn*. New York: Harper Collins A.

Gopnik, A., Sobel, D., Schulz, L., & Glymour, C. (2001). Causal learning mechanisms in very young children: Two-, three-, and four-year-olds infer causal relations from patterns of variation and covariation. *Developmental Psychology, 37*, 620–629.

Goswami, U., & Brown, A. L. (1989). Melting chocolate and melting snowmen : Analogical reasoning and causal relations. *Cognition, 35*, 69–95.

Harris, P., German, T., & Mills, P. (1996). Children's use of counterfactual thinking in causal reasoning. *Cognition, 61*, 233–259.

Harris, P., Kavanaugh, R., Wellman, H., & Hickling, A. (1993). Young children's understanding of pretense. *Monographs of the Society for Research in Child Development, 58*, I–107. http://dx.doi.org/10.2307/1166074.

Hawkins, J., Pea, R. D., Glick, J., & Scribner, S. (1984). "Merds that laugh don't like mushrooms": Evidence for deductive reasoning by preschoolers. *Developmental Psychology, 20*, 584–594.

Holyoak, K. J., Junn, E. N., & Billman, D. O. (1984). Development of analogical problem solving skill. *Child Development, 55*, 2042–2055.

Hubel, D. H., & Wiesel, T. N. (1960). Effects of monocular deprivation in kittens. *Naunyn-Schmiedebergs Archiv für Experimentelle Pathologie und Pharmakologie, 248*, 492–497. http://dx.doi.org/10.1007/BF00348878. PMID:14316385.

Hummel, J. E., & Holyoak, K. J. (1997). Distributed representations of structure: A theory of analogical access and mapping. *Psychological Review, 104*, 427–466.

Inhelder, B., & Piaget, J. (1958). *The growth of logical thinking from childhood to adolescence: An essay on the construction of formal operational structures (developmental psychology)*. Basic Books.

Jacobs, J. E., & Potenza, M. (1991). The use of judgement heuristics to make social and object decisions: A developmental perspective. *Child Development, 62*(1), 166–178.

Kohlberg, L. (1958). *The development of modes of moral thinking and choice in the years 10 to 16* (Unpublished doctoral dissertation). University of Chicago.

Kohlberg, L. (1986). A current statement on some theoretical issues. In S. Modgil, & C. Modgil (Eds.), *Lawrence Kohlberg: Consensus and controversy* (pp. 485–546). Philadelphia, PA: The Falmer Press.

López, A., Gelman, S. A., Gutheil, G., & Smith, E. E. (1992). The development of category-based induction. *Child Development, 63*, 1070–1090.

Mackie, J. L. (1974). *The cement of the universe: A study of causation*. London: Oxford University Press.

Markman, A. B., & Gentner, D. (1993). Structural alignment during similarity comparisons. *Cognitive Psychology, 25*, 431–467.

Naglieri, J. A. (2007). *Naglieri nonverbal ability test* (2nd ed.). San Antonio, TX: Pearson.

Osherson, D. N., Smith, E. E., Wilkie, O., Lopez, A., & Shafir, E. (1990). Category-based induction. *Psychological Review, 97*, 185–200.

Pearl, J. (2000). *Causality*. Oxford University Press.

Piaget, J. (1921). Une forme verbale de la comparaison chez 1'enfant. *Archives de Psychologie, 18*, 141–172.

Piaget, J. (1932). *The moral judgement of the child*. New York: The Free Press.

Piaget, J. (1964). *The early growth of logic in the child*. London: Routledge and Kegan Paul Ltd.

Piaget, J. (1965). The stages of the intellectual development of the child. *Educational psychology in context: Readings for future teachers*, 98–106.

Piaget, J. (1967). *Six psychological studies*. New York: Random House.

Piaget, J. (1977a). *The grasp of consciousness: Action and concept in the young child*. London: Routledge and Kegan Paul Ltd.

Piaget, J. (1977b). *The development of thought: Equilibration of cognitive structure (Trans A. Rosin)*. Viking.

Piaget, J., & Inhelder, B. (1956). *The child's concept of space*. Routledge & Paul.

Piaget, J., & Inhelder, B. (1958). *The growth of logical thinking from childhood to adolescence*. New York: Basil Books, Inc.

Piaget, J., & Inhelder, B. (1967). *The child's conception of space*. New York: W. W.

Piaget, J., & Inhelder, B. (1969). *The psychology of the child*. New York: Basic Books.

Raven, J. C. (1938). *Progressive matrices: A perceptual test of intelligence, 1938, sets A, B, C, D, and E*. London: H. K. Lewis.

Raven, J. C. (1960). *Standard progressive matrices: Sets A, B, C, D, & E*. London: H.K. Lewis & Co.

Rhodes, M., & Liebenson, P. (2015). Continuity and change in the development of category-based induction: The test case of diversity-based reasoning. *Cognitive Psychology, 82*, 74–95. http://dx.doi.org/10.1016/j.cogpsych.2015.07.003. [1087].

Richland, L. E., Morrison, R. G., & Holyoak, K. J. (2006). Children's development of analogical reasoning: Insights from scene analogy problems. *Journal of Experimental Child Psychology, 94*, 249–271.

Robin, N., & Holyoak, K. J. (1995). Relational complexity and the functions of prefrontal cortex. In M. S. Gazzaniga (Ed.), *The cognitive neurosciences* (pp. 987–997). Cambridge, MA: MIT Press.

Rosch, E. (1975). Cognitive representations of semantic categories. *Journal of Experimental Psychology: General, 104*, 192–233.

Smith, H. D. (1999). Use of the anchoring and adjustment heuristic by children. *Current Psychology: Developmental, Learning, Personality, Social, 18*, 294–300.

Sobel, D. M., & Kirkham, N. Z. (2006). Blickets and babies: The development of causal reasoning in toddlers and infants. *Developmental Psychology, 42*, 1103–1115.

Sobel, D. M., Tenenbaum, J. B., & Gopnik, A. (2004). Children's causal inferences from indirect evidence: Backwards blocking and Bayesian reasoning in preschoolers. *Cognitive Science, 28*, 303–333.

Trabasso, T. (1975). Representation, memory, and reasoning: How do we make transitive inferences? In A. D. Pick (Ed.), *Minnesota symposia on child psychology* (Vol. 9). Minneapolis: University of Minnesota Press.

Trabasso, T., Riley, C. A., & Wilson, E. G. (1975). The representation of linear order and spatial strategies in reasoning. In R. Falmagne (Ed.), *Reasoning: Representation and process*. Hillsdale, JN: Erlbaum.

Tversky, A., & Kahneman, D. (1973). Availability: A heuristic for judging frequency and probability. *Cognitive Psychology, 5*, 677–695.

Tversky, A., & Kahneman, D. (1974). Judgment under uncertainty: Heuristics and biases. *Science, 185*, 1124–1131.

Tversky, A., & Kahneman, D. (1981). The framing of decisions and the psychology of choice. *Science, 21*, 453–458.

Wechsler, D., Kaplan, E., Fein, D., Kramer, J., Morris, R., Delis, D., et al. (2003). *Wechsler intelligence scale for children: Fourth edition (WISC-IV)*. [Assessment instrument]. San Antonio, TX: Pearson.

Wilkins, M. C. (1928). The effect of changed material on ability to do formal syllogistic reasoning. *Archives of Psychology, 102*, 1–83.

Wright, S. B., Matlen, B. J., Baym, C. L., Ferrer, E., & Bunge, S. A. (2008). Neural correlates of fluid reasoning in children and adults. *Frontiers in Human Neuroscience, 28*. http://dx.doi.org/10.3389/neuro.09.008.2007.

CHAPTER

6

Reasoning Over the Lifespan

KEY THEMES

- Reasoning abilities in adolescents may be impacted by biological changes including alterations in hormone levels and neuronal pruning that take place throughout the teenage years.
- The prefrontal cortex is an important area that undergoes maturation during the period of adolescence.
- Memory improvements that take place during the teenage years may be related to enhancements in strategy use and the development of heuristics gained through life experiences.
- Risky decision making may be higher in the teenage years. This may be due to a lack of experience with risks leading teens to rely on calculation of risks rather than heuristics.
- Older adults tend to show decline in a variety of areas including vision, hearing, cardiovascular health, and gray matter thickness.
- Changes in the cognitive abilities of older adults indicate that as people age they see reductions in controlled attention, working memory, speed of processing, and fluid reasoning.
- Controlling interference becomes a challenge for older adults, and this inhibitory control deficit can lead to challenges in reasoning.
- Some older adults remain highly active in their older years and maintain cognitive and brain health. Researchers are seeking to understand the factors that enable improved health through the lifespan.

Reasoning
http://dx.doi.org/10.1016/B978-0-12-809285-9.00006-5

OUTLINE

REASONING ACROSS THE LIFESPAN

Introduction

Reasoning abilities continue to change through the lifespan. We have already discussed the dramatic changes in thinking and reasoning ability that take place in the childhood period. The adolescent brain experiences a wide variety of change as well, as young people move from childhood into adulthood. There is then a long period of relative stability during the twenties through the fifties. During this period we often consider people to have attained adult levels of competency at reasoning. We gain a great wealth of knowledge during these adult years and this impacts our reasoning abilities during these productive years of our lives. It may be our experiences with other people and the world around us that impact our reasoning skills the most in this period of adulthood. Around age 60, cognitive abilities change again and with them we can observe changes in people's reasoning capabilities. This is a complex process involving changes in biology, psychology, and people's occupational and social situations as they age. The brain, cognition, and reasoning abilities specifically all undergo some degree of change in the later years of people's lives.

We begin this chapter by discussing the changes that accompany adolescence. Some of the research in this area directly compares the reasoning performance of adolescents to that of younger children. It may be helpful to refer back to Chapter 5 in order to re-familiarize yourself with some of that developmental research in young children if the need arises. The adolescent brain undergoes a variety of changes. Several of these are biological and relate to changes in hormones and brain maturation. Some of these biological changes are related to cognitive changes in basic processes including attention and memory functions. Many changes can be linked to the maturation of the prefrontal cortex (PFC), a region we discussed in great detail in Chapters 3 and 4.

During adolescence, changes in biology and the brain can impact decision making and risk taking. We will discuss the changes that occur in abstract thinking as young people occupy the developmental stage that Jean Piaget referred to as *formal operations*. This period of development can be directly compared to the reasoning and decision making of an adult. We will focus on how adolescents make decisions about risks. Risk taking in adolescence is widely recognized and often criticized. We will focus on some intriguing research that tries to address the question of why young people tend to take greater risks than they might decide to later on in adulthood.

As adolescents build their knowledge about the world, their reasoning abilities and cognitive tendencies continue to develop. This transition is related to differences in availability of knowledge, as well as a growing awareness about one's own mental abilities. Cognitive psychologists refer to this awareness as *metacognition*, and the way that adolescents approach challenges in reasoning is related to changes in their memory systems and what they think they know versus their actual knowledge. The enhancement in memory for adolescents compared to younger children enables teens to make different types of inferences. We will discuss how adolescents begin to see the world from a broader perspective and how this impacts the types of inferences that they make.

From there we will move into the domain of adult cognition, which will also be the focus of the majority of the chapters that follow. Adults occupy a relatively stable point in cognition. There are not quite so many dramatic biological or neurological changes associated with adulthood as there are in children and adolescents. It may appear somewhat abrupt that we have spent an entire chapter discussing reasoning in childhood, half of a chapter on reasoning and adolescence, and then we finish off with a discussion of the older years. Where you may ask is the information on adult reasoning? Since most research in the mainstream of neuroscience and psychology continues to be conducted with adult participants in their twenties to forties, you can find extensive discussions of reasoning in early and middle adulthood throughout the remaining chapters of the book in which we cover a variety of different types of reasoning.

As adults age, we begin to observe more biological changes in the individual. In addition to physical changes in aging, the brain also begins to show consistent patterns of alterations in older adults. We will discuss the biological change that occurs within the brains of older adults and how these changes link to the differences that we see in cognition. While cognition shows some clear decline in areas such as speed and memory, what is the evidence for changes in people's reasoning? We will focus on a wide array of changes that contribute to differences in reasoning abilities in older adults.

We will conclude the chapter by discussing some of the practical implications of aging on people's reasoning abilities. There are critical areas to consider, and these include how people reason in everyday life in order to live independently and navigate through their day. We will consider the implications of reasoning abilities in older adults for maintaining an active and healthy lifestyle and how older adults respond to new learning in the form of cognitive training interventions. We will also consider the challenges associated with aging on decision making about financial affairs and medication use. One of the clear features of aging is that individuals vary considerably. Some individuals are considered to be healthy agers, as they appear to maintain strong cognitive and reasoning skills late into their lives. Other individuals are less successful agers and experience difficulties with speed, memory, and other physical limitations. There are other individuals who face the challenges of dementia and other brain diseases as they age. This topic will be addressed in Chapter 7 when we discuss disorders of reasoning.

BIOLOGICAL CHANGES IN ADOLESCENCE

Biological Change and Hormones

The predominant changes that occur in adolescence are both physical and psychological. The role of hormones is prominent starting in the early teen years with strong influences over individuals of both genders. The social world of teens is dramatically different from that experienced during childhood. The teenage years can resemble both childhood and adulthood. Individuals struggle during this time to take on different roles in society. At times teens may feel that they are being tugged between immature and frivolous behavior and mature high-stakes responsibilities. Changes in the social environment of teens strongly influence reasoning abilities. Chapter 12 is devoted to a thorough examination of the influences of social behavior on the reasoning process in adults, but we will also discuss some of this material in sections of this chapter as it relates to teenagers.

Physical changes are largely governed by changes in different hormone levels. Hormones are molecular influences that largely are governed by the endocrine system. The relevant hormones are already present in people since birth, but the amounts of hormones change radically at adolescence. Males have more androgens, the most prominent of these being testosterone, while females have more estrogen and progesterone. Each of these hormones acts on different receptors that influence physical changes.

Hormone levels and aspects of adolescent development occur at different rates across the population. This is why there can be such disparities in the height and weight of different people throughout the teenage years. These changes are likely to increase the emphasis on thoughts and comparisons about appearances and body types. The role of cultural factors becomes prominent as well in this case, as people try out different roles, meet changes in expectations, and explore the world around them.

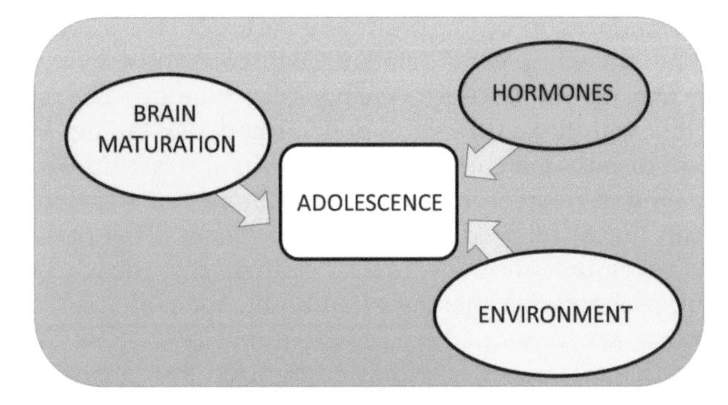

FIGURE 6.1 The changes occurring within adolescence include biological change in the form of altered hormone levels. There are also brain developmental changes, which include the maturation of cortex by strengthening certain pathways through synaptic pruning. Teens undergo changes in cognitive abilities during this period and increase their base of knowledge dramatically. Lastly, the social environment and societal expectations change during the adolescent years.

THE BRAIN AND REASONING FROM ADOLESCENCE TO ADULTHOOD

Cortical Change

The basis for change in reasoning performance over the period from childhood through adolescence is accomplished by increases in societal expectations, psychological factors, including enhanced knowledge and strategy use, as well as biological factors in brain development (Fig. 6.1). As we discussed in Chapter 5, the brain undergoes an extensive pruning process over the course of childhood and adolescence, as dense connectivity within the neural networks of the brain eventually gives way to sparser and more targeted connections. This reduction in diffuse connectivity leads to thinner cortex with more efficient connections present as young people develop through the teenage years.

As we discussed in Chapter 5, Gogtay et al. (2004) plotted the developmental trajectory of cortical volume exhibited from age five through adulthood. Take a moment to turn back to Fig. 5.2. Remember that in this figure thicker cortex is color-coded in either red or green colors, while thinner (more mature) cortex appears in blue tones. Over the course of development toward adulthood the brain undergoes an overall decrease in cortical volume that is related to the pruning process, by which we enhance critical connections and decrease the connections that are nonessential for adult cognition.

Age-Related Changes in Frontal Lobe Network Interconnectivity

In Chapter 5, we discussed a trend in brain development in which the temporal lobe and frontal lobe areas are some of the last to undergo the neural pruning process, typically thinning only at the later periods of adolescence. This neural maturation process is also reflected in the activation patterns of the cortex that are associated with reasoning tasks, as measured using functional magnetic resonance imaging (fMRI). The dorsolateral prefrontal cortex (DLPFC) is strongly connected to the lateral parietal cortex and several other regions known collectively as the *fronto-parietal network* (Laird et al., 2013). This network has been identified as supporting task-based cognitive activity and is involved with the performance of tasks that demand attention and working memory resources. In addition to the DLPFC, the fronto-parietal network is connected to the parietal cortex, the orbitofrontal cortex, regions of the motor cortex, the anterior insula, the ventral occipital cortex, and the frontal eye fields.

Wendelken, Ferrer, Whitaker, and Bunge (2016) at UC Berkeley assessed the degree of correlation of the fronto-parietal cortex in association to reasoning abilities over the course of development in a large longitudinal dataset that included individuals aged six to eighteen. The analysis included the degree of functional connectivity among the cortical network areas of the fronto-parietal network and performance on reasoning measures. Wendelken et al. assessed reasoning performance on three diagnostic measures. These included the block design task and matrix reasoning task from the Wechsler Intelligence Scale for Children–Revised (Wechsler et al., 2003). The block design task requires participants to arrange a set of red and white blocks in the same manner that is presented to them on a set of cards. This task requires spatial reasoning abilities (Fig. 6.2). The matrix reasoning task is similar to Raven's progressive matrices task and measures the relational reasoning abilities, a measure of fluid intelligence. We have discussed this task already in other chapters. A concept formation task was also included, and this assessed the ability to identify and state the rules for concepts in response to examples. All three of these tasks measure slightly different cognitive processes that are all important for reasoning.

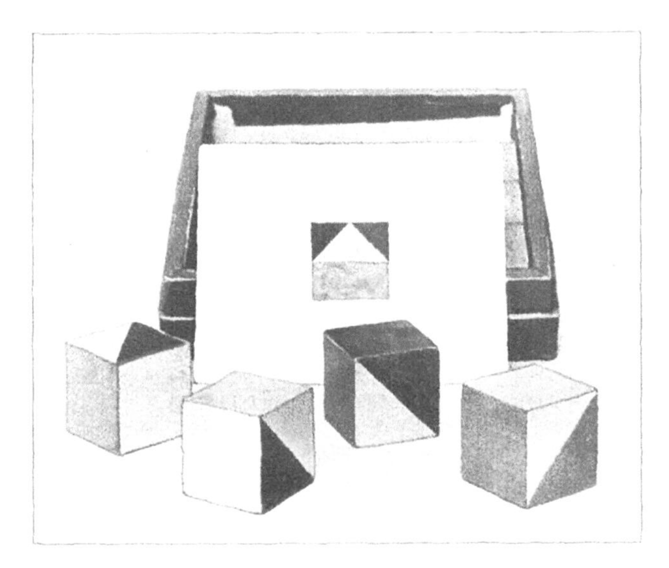

FIGURE 6.2 Historical sketch of the original block design test from 1920. The block design task requires participants to arrange a set of patterned blocks according to a sample as shown here. The block design task requires participants to arrange a set of patterned blocks according to a sample as shown here. *Image by S.C. Wikimedia Commons.*

The results reported by Wendelken et al. (2016) indicated differences within the fronto-parietal regional connectivity that accompany development from childhood through adulthood, as people develop in reasoning ability. The researchers reported two major age-related increases in the connectivity of this network. The major frontal hub of the fronto-parietal cortex, the DLPFC, exhibited differential age-related patterns between the two cerebral hemispheres with a left-hemisphere increase in connectivity with the superior parietal cortex and an increase in connectivity between the right DLPFC and inferior parietal lobe. The rostrolateral PFC (anterior middle frontal gyrus) is often associated with relational reasoning (Vendetti, Matlen, Richland, & Bunge, 2015). This region showed an age-related increase in connectivity with inferior parietal cortex, bilaterally. This connectivity increase was accompanied by another age-related increase in connectivity within the ventrolateral PFC, which showed increased connection to the superior parietal cortex. The age-related change in connectivity was largest between the ages of 10 and 14 reaching into early adolescence. This age pattern was reported for connectivity changes occurring within the PFC, showing both increases and decreases with age. It also occurred for age-related decreases in parietal connectivity and age-related increases in connections between the rostral PFC and parietal cortex.

These findings suggest that the period of life occupying late childhood into early adolescence involves some pronounced changes within the task-based fronto-parietal network. Furthermore, the specific ways that the frontal and parietal regions interconnect are somewhat variable, and this is consistent with our discussion in Chapter 3 that different areas of the PFC and parietal cortex are associated with different types of reasoning abilities.

Wendelken et al. (2016) also reported age-related connectivity differences associated with reasoning. For the youngest children (6–8 years old), there were no significant relationships between connectivity and reasoning and processing speed was the major factor driving reasoning abilities (as measured by the block design, matrix reasoning, and concept formation tasks). In older children aged 9–11 years, reasoning assessments were associated with connectivity increases between the left and right rostral PFC. The connectivity changes in this age group were mostly attributable to increases in working memory ability. Lastly, among adolescents aged 12–18 years they observed that reasoning ability was related to increased functional connectivity between left rostral PFC and inferior parietal cortex. These results indicate that reasoning advances in a predictable pattern associated with changes in the fronto-parietal network. As we will discuss in the upcoming sections there are a range of increases in cognitive abilities in young people that are likely supported by the fronto-parietal network.

Frontal Lobe Changes in Reasoning Capacity With Age

The PFC, parietal, and temporal cortex are important for relational reasoning as we discussed in Chapter 3. One of the key areas associated with relational reasoning is the left rostral PFC. This area has previously been linked to solving relational problems including four-term verbal analogies of the form "A is to B, as C is to D" (Bunge, Wendelken, Badre, & Wagner, 2005). A similar left-sided region of the anterior PFC or frontopolar PFC has been associated with completing four-term verbal analogies with longer-distance semantic category relations as well (Green, Fugelsang, Kraemer, Shamosh, & Dunbar, 2006). Does the activation of the left anterior PFC change as the children age and become more capable at relational reasoning in adolescence?

A study by Ferrer, O'Hare, and Bunge (2009) evaluated the differences between activation for adolescence and children in association with solving four-term picture analogies. Participants viewed a set of items expressing an analogical relationship. For example, participants saw a relationship such as "bee is to hive, as spider is to?". Answer choices were offered and included the correct completion to the analogy ("spider web"), an incorrect semantic associate ("insect"), an incorrect perceptual distractor item ("octopus"), and an irrelevant item, such as a dessert. Vendetti et al. (2015) described the finding in a review article indicating that adolescents aged 14–18 years old showed improved abilities in analogical reasoning

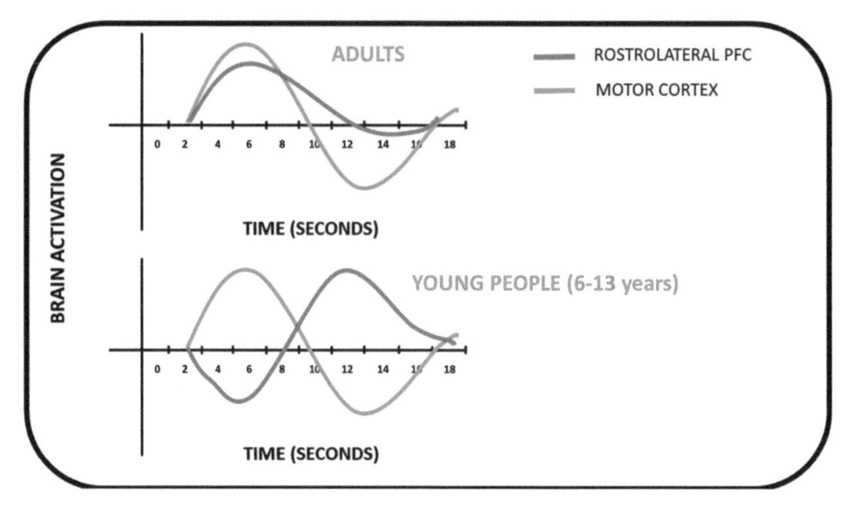

FIGURE 6.3 In adults, the left rostral PFC shows a peak of activation in fMRI during reasoning that nearly coincides with a similar peak in task-related activation of the motor cortex. Meanwhile, in younger children and early adolescents aged 6–13 years the pattern between these two regions varies. In this case, the motor cortex activated in the same manner as it had for adults, but the left rostral PFC activated significantly later. *Based on Ferrer, E., O'Hare, E. D., Bunge, S. A. (2009). Fluid reasoning and the developing brain.* Frontiers in Neuroscience, 3, 3.

compared to the younger children between ages 6 and 9 years. Despite these behavioral differences in reasoning performance, there were relatively few differences in brain activation between these two groups. Ferrer et al. (2009) further analyzed the left rostral PFC in this task evaluating its time course of fMRI activation compared to that of the primary motor cortex. In this analysis, adults showed a strong co-activation of the left rostral PFC and the motor cortex, with both regions peaking at a similar timescale indicating involvement of both areas within the task (Fig. 6.3). This fits with the position that the left rostral PFC is an integration area that is sensitive to relational comparisons in adults (Bunge, Helskog, & Wendelken, 2009). Meanwhile, Ferrer et al. described the activation pattern of children and early adolescence aged 6–13 years. In this case the motor cortex activated in the same manner as it had for adults, but the left rostral PFC activated significantly later. Fig. 6.3 shows an approximate time series between these regional activation patterns. Note that the peak of the left rostral PFC occurs about 1 s after the motor cortex peak in adults, but almost 5 s later in children and young adolescents. Note that the timescale of fMRI is dramatically delayed from the firing of action potentials within neurons. The activity is delayed by several seconds in the case of fMRI. This pattern of difference suggests that while activation associated with analogical reasoning is similar overall between children and adolescents, there are some substantial differences in the timing of activation of the left rostral PFC. This timing difference within the brain may impact analogical reasoning performance, as left rostral PFC is not actively engaged early in order to integrate relational elements in younger people.

COGNITION AND BEHAVIOR IN ADOLESCENTS

Abstract Thinking

As we discussed in Chapter 5, Jean Piaget began a tradition of research targeted at achieving a better characterization of the thinking process from the period between birth and adolescence. Piaget specified that children enter the formal operations stage of thinking at around the age of 12 years. This milestone is marked by clear competencies in passing several of the tests of formal reasoning developed by Piaget. By the period of formal operations, teens are capable of avoiding the egocentrism that is characteristic of young children's thinking. By this point, the teenager can reason both deductively and inductively. They can also succeed on hypothesis testing and inference tasks. Piaget (1971) described the newly acquired ability of *reflective abstraction* during formal operations. Reflective abstraction refers to the idea that adolescents are now capable of taking inputs and performing transformations upon them in order to arrive at new inferences, or possibilities. This capacity has been referred to as *introspection* at other times in psychological history. Notably, philosophers of mind have made heavy use of the introspection method, including David Hume and Rene Descartes, whom we discussed in Chapter 2, focusing on the history of reasoning. Early psychologists Wilhelm Wundt and Hermann Ebbinghaus also made use of introspection in order to examine mental abilities. This set of abstract reasoning abilities forms the basis for much of the work that we will discuss on propositional logic and syllogistic reasoning in Chapter 9.

Teenagers capable of formal operations in thinking possess an increased metacognitive ability, in addition

FIGURE 6.4 Metacognition refers to the ability that we have to think about our own cognition and the contents of our minds. Metacognitive ability increases as children move into the adolescent years.

to purely mechanistic reasoning abilities that involve abstract, or higher-order, constructs. Metacognition refers to our abilities to ponder our own cognitive abilities (Fig. 6.4). An example of this relates to memory. Young children have a lower working memory span, which is related to lower capability in verbal rehearsal at a younger age. Through childhood, both working memory span and long-term memory abilities increase. The ability to self-assess one's own capability increases alongside these more basic abilities. Young children have lower memory capacity in general, but will often estimate that they have a much larger memory ability than they actually do. People in general tend to overestimate their own memory or reasoning capacity, but as we age we tend to become more accurate at judging abilities overall. Formal operations involve metacognitive abilities in judging what we know, what we do not know, and discovering many of our limitations of both cognitive ability and overall knowledge.

Alongside the growth of metacognitive abilities, adolescents maintain some elements of egocentric thinking until they reach adulthood. This is in part due to the changing expectations and new roles that are increasingly expected of our youth. Piaget maintained that adolescents are increasingly concerned with their future in society. They may wonder how they might make changes to the status quo and tend to want to rebel against the institutions that adults maintain. A generation of experimental psychologists followed up on Piaget's original ideas. In the 1970s David Elkind and Robert Bowen (1979) discussed adolescent thinking as *imaginary audience behavior*, in which teenagers tend to think of themselves as being onstage and imagine that they are regularly being evaluated and scrutinized to a greater degree than they are objectively. Elkind and Bowen developed a scale to assess people's willingness to expose different aspects of their thinking to an imagined audience. The imagined audience scale (IAS) consists of two subscales (referred to as the abiding self and the transient self). The abiding self was defined as being a stable construct that includes the characteristics of mental abilities and personality traits that are viewed as being fixed and permanent (or abiding) aspects of the self. By contrast, the transient self consisted of temporary and possibly superficial aspects of oneself, including having mismatched clothing or having made a social gaffe. Results of the IAS in young people revealed that teens in the eighth grade level were significantly less willing than both younger and older individuals to reveal aspects of themselves to an audience. There was also a gender difference in which boys tended to reveal more of their inner thinking to an audience than girls. Elkind also described the egocentric tendencies of teenagers as revealing a personal fable, in which an individual imagines that they are unique and invulnerable (Bjorklund, 1989). This kind of thinking leads us to our next section discussing the riskiness of teenage behavior and the belief that bad things will not happen to oneself, but are instead limited to other people.

Decision Making and Risk in Adolescence

The teenage years are a time when people are known for making irrational, impulsive, and potentially risky decisions. These tendencies may be attributed to surging hormones, impulsivity, and an illusion of invulnerability. All of these factors may be in part biologically driven and based on the changes that individuals undergo during adolescence. Evidence has suggested that the logical reasoning abilities of 15-year-olds are comparable to those of adults. Therefore, adolescents may be as capable as adults are at perceiving risk and their vulnerability toward risks (Reyna & Farley, 2006). Could there be other factors related to the brain and cognition that may lead teens to take exceptional risks?

Decision making researcher Valerie Reyna at Cornell University has proposed another reason that teens expose themselves to risky behavior. Reyna and Brainerd (2011) have described two thinking styles in their *Fuzzy-trace theory*. The essence of Fuzzy-trace theory is that we store many instances of categories of situations over time. This growing base of knowledge is similar to what we have termed *schemas* in other chapters of this book. A schema is a category of knowledge that is built up over experiencing something numerous times. As we accrue experience, we acquire a distilled version of the knowledge for this situation, a *trace* in the terms of Fuzzy-trace theory. The fuzziness comes about by the fact that it is not a literal representation, but rather a gist-based overall representation that gets the job done most of the time. Gist representations are useful in helping us to identify and react to a particular situation, but they lack details. In Chapter 11, we will discuss dual-process models of decision making. Fuzzy-trace theory is one such model. In addition to gist representations, Reyna and Brainerd (2011) maintain that we also store detail-based representations, which are higher fidelity and have clear and detailed sets of knowledge by which we can guide our behavior.

As an example, Fuzzy-trace theory can be applied to a gambling situation in which we may choose to seek or avoid financial risks. Let's imagine that a middle-aged man named Phil is presented with a financial game (Fig. 6.5). Phil can choose Option A and receive a sure gain of $200, or decide to take Option B, a gamble with a 1/3 probability of gaining $600 and a 2/3 probability of gaining nothing. In this instance, Phil sees that he can get $200 for sure with Option A. That sounds pretty good. He also quickly notes that with Option B, there is a pretty high (2/3) chance that he will end up without any money and that immediately seems like it should be avoided. Phil opts to take Option A, gets his $200 and plans to treat his wife to a lovely dinner. According to Fuzzy-trace theory Phil has made use of a gist-based representation in this example. He sees the chance of not gaining any money and finds that choice unappealing opting instead for the lower, but sure payout of $200.

Later that day Phil gets the chance to try his luck again. This time he is given $600 to gamble with and a new choice is offered to him. He can take Option C and receive a sure loss of $400, or he can gamble on Option D, in which he is presented with a 1/3 probability that he will get to keep his entire $600, and a 2/3 probability that he will lose it all and have no money (Fig. 6.6). Now Phil gets uneasy. The threat level of this situation is high. He likes the $600 he has been given and would hate to lose it. The first choice of Option C sounds pretty bad. Losing $400 would be awful. That would be most of the money he has. Option D seems alright. After all, Phil gets to keep all of the $600 if he gets lucky on the gamble. Because he'd hate to take a $400 loss, Phil opts for Option D. Regrettably, his one-out-of-three chance for keeping the money doesn't work out and he loses the money.

In this second scenario, Phil has fallen victim to a phenomenon known as the framing effect (Tversky & Kahneman, 1985). If you consider Option A from the first scenario, Phil stands to gain $200. This is equivalent to Option C, in which Phil gains $200 (after losing $400 of the original money). Option B and Option D are also equivalent in terms of how much money Phil ends up with in both cases. They are framed, or presented, differently. This alteration in framing tends to make people change their minds about a decision. Many people choose to go with the safe choice (Option A) when the decision is framed as a gain and then reverse their preference and choose the risky bet (Option D) when the decision is framed as containing a loss. Fuzzy-trace theory maintains that we have a gist-based representation for financial outcomes. This gist representation leads us to gamble in order to avoid a loss and we will otherwise

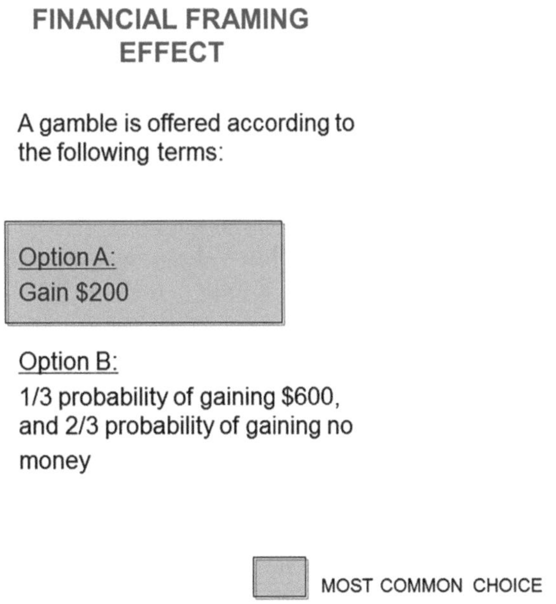

FINANCIAL FRAMING EFFECT

A gamble is offered according to the following terms:

Option A:
Gain $200

Option B:
1/3 probability of gaining $600, and 2/3 probability of gaining no money

FIGURE 6.5 A gambling situation is presented in which there is a chance to take a sure win alongside an opportunity to gamble in order to gain more money with the associated risk that one will gain no money. Many adults will tend to be risk-averse in these cases and choose to take the sure gain, even though it will pay out lower than the gamble.

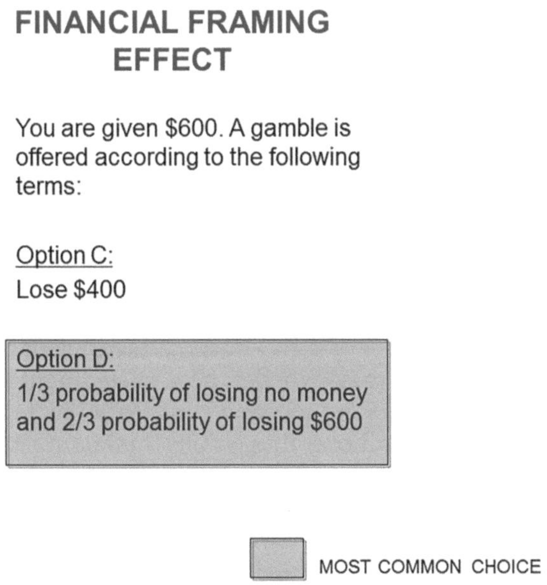

FINANCIAL FRAMING EFFECT

You are given $600. A gamble is offered according to the following terms:

Option C:
Lose $400

Option D:
1/3 probability of losing no money and 2/3 probability of losing $600

MOST COMMON CHOICE

FIGURE 6.6 In another gambling situation in which there is a sure loss offered or the chance to gamble between an even higher loss and no loss at all, people will tend to become riskier and choose to gamble.

avoid risk when we are seeking gains. This prediction fits with Kahneman and Tversky's (1979) observation that losses carry more psychological impact than equivalent gains, making us seek to avoid losses to a greater degree than we will seek gains. This is covered in greater detail in Chapter 11.

What would a risky, thrill-seeking teenager do when faced with these same options? According to Reyna, Wilhelms, McCormick, and Weldon (2015), teenagers are more likely to take risks in the case of seeking out gains. This is especially true in cases in which the gain option has high numbers. Teenager's lack of experience leads them to have fewer representations to guide them in these circumstances. The lack of a strong gist-based representation operating within an adolescent may lead them to rely upon calculation based on the details (in this case the numbers involved in the gamble). The adult simply chooses to take the $200 in Option A, rather than gamble with the money by choosing Option B. Meanwhile, this same adult will usually choose to gamble when faced with a loss. According to Reyna et al. (2015) a teen will tend to focus on the details, as they do not yet have these gist-based heuristics to use. When gain numbers are possibly high (such as an option A with a sure gain of $250, or an option B in which there is a 1/4 probability of gaining $1000 and a 3/4 probability of gaining no money), the teenager may opt to gamble, as they are more focused on the magnitude of the numbers in this case, rather than a general gist-based heuristic to take the sure money and avoid the gamble. In this scenario, teens may choose to gamble in both the gain and the loss conditions, thereby making them more risky than adults, whose gist representations focus on conservatism in decision making most of the time.

Reyna et al. (2015) discussed the role of risk in decision making pointing out that adolescents may operate on the basis of different gist representations than adults. Adults have a lifetime of experience with which to build up gist-based or script-like representations. For example, an adult may have a general gist representation containing the heuristic: "If you have been drinking alcohol, do not get behind the wheel of a car." This gist representation is relatively insensitive to details, such as exactly how much alcohol is over the legal limit. The general rule to simply avoid the behavior regardless of the fine-grained details may hold. Teenagers by contrast, have not had nearly as much time in their lives to amass a large amount of gist representations through experience. This may mean that they will tend to focus on details, including how many drinks one has consumed in the past hour, what the alcohol content of the drinks is, and what their likely food intake has been. While these factors all influence one's blood alcohol level, it can appear to an adult to be unwise to take the risk of driving based on some general calculation of these factors. It is not that the teens in such a case are being deliberately risk seeking or irresponsible, but rather they simply have a different set of knowledge that they are going to apply about what constitutes a risk (Fig. 6.7).

FIGURE 6.7 Teenagers and adults may be guided by different heuristic representations of risk. While an adult may consider any mixture of alcohol and driving to represent an inappropriate risk, a teenager may instead attempt to calculate his or her blood alcohol level and focus on the details of the situation, thereby leading to a greater overall tendency toward risk-taking behavior.

Biological Changes Associated With Risk-Taking Behavior

There are also relationships between hormone levels and risk taking in adolescence. Zdena Op de Macks et al. (2011) investigated the role of hormones and adolescent risk taking in girls aged 11–13 years. The participants took part in a probabilistic decision-making task, in which they could choose to play or pass on each round of the task depending on the risk level and potential reward that could be earned on each decision. Participants underwent fMRI to reveal activation related to risk taking and hormone levels. Op de Macks et al. reported that higher levels of risk taking were observed in association with higher testosterone levels (which are present in females, but to a lower degree than males) among the participants. This increased risk by testosterone association was also linked to greater brain activation within the orbitofrontal cortex. This is the same region that is associated with greater risk-taking behavior in patients who have damage to the area

(Anderson et al., 1994). Meanwhile, higher levels of the hormone estradiol were linked to more conservative behavior on the decision-making task. The association between estradiol and lower riskiness was linked to greater activity within the nucleus accumbens, another region involved to reward processing. These results indicate that there are separate brain regions associated with greater and lower risk-taking levels in adolescent girls. This study also provides an interesting biological link to differences in hormone levels that appear to impact the riskiness of choices made at these age levels.

EXECUTIVE CONTROL AND ADVANCED REASONING SKILLS

Changes in Memory Capacity in Adolescence

Memory capacity is one of the major determinants of reasoning abilities. If we cannot encode the information that is relevant and central to finding a solution, then even if we happen to solve a problem correctly once, this experience will not have a strong bearing on whether we will solve such a problem in the future. Likewise, if we cannot retrieve critical information at the appropriate time, then again, we will be quite limited with regard to using our past knowledge and successes on reasoning tasks in the future.

Memory ability can be measured in a variety of ways. One of the most obvious memory measures linked to reasoning is working memory ability. Working memory is relevant for maintaining information about a particular problem or decision. Both explicit and implicit long-term memory can be relevant for reasoning as well. Implicit information can be most relevant for decision making, as we discussed in Chapter 3, while explicit information in the form of semantic and episodic memory is also very relevant.

Working memory is relevant for maintaining information about a particular problem or decision. This may include premise information, sets of relational comparisons, or facts and figures needed for making a complex decision. Maintenance of information and the ability to manipulate that information, or make transformations on the contents of working memory, are both core capacities that are necessary for reasoning (Fuster, 2005). Both working memory and attention are important for fluid reasoning tasks such as the Raven's progressive matrices or transitive inference tasks. Working memory increases initially in childhood as young children advance in their language abilities. The developing language abilities allow young children to make greater use of verbal rehearsal strategies. Children make little use of rehearsal to enhance working memory at the beginning of grade school, but by the fifth grade as many as 85% of students use rehearsal in order to enhance their memory performance (Flavell, Beach, & Chinsky, 1966). The development of working memory functions occurs in a progressive manner across childhood. Initially children improve at basic perceptual and sensorimotor functions from age four through about age eight (Luciana & Nelson, 1998). During the period from later childhood through adolescence, young people experience physiological maturation of neural networks that enable the integration of task demands and memory items. Evidence suggests that adults are superior to grade-school-aged children on working memory tasks that tax executive functions such as spatial working memory tasks and planning tasks such as the Tower of London that involve implementation of strategies.

Changes in long-term memory are important for determining the amount of information that people encode and how likely they are to retrieve the relevant information at a necessary moment in order to facilitate reasoning. Typical memory experiments involve presenting a set of either words or objects to participants. When tested, the participant sees a set of the previous items mixed with new items. Successful performers are able to recognize the items that had been presented previously from those that are new. Classic research by Roger Shepard (1967) noted that adults are excellent at recognition memory for pictures, with participants recognizing a set of 600 items at over 90% accuracy after a 1-week delay. Several classic developmental studies of recognition memory indicate that children as young as five have similar recognition memory abilities (Brown & Scott, 1971; Daehler & Bukatko, 1977).

There may be more to this story; however, as other research suggests the strategies that children use for recognition memory are different than those employed by adults. This was realized when working memory research was first burgeoning in the 1970s. For example Dirks and Neisser (1977) presented children in first, third, and sixth grade sets with jumbled toys. They then removed or changed the location of some of the objects. The participants were asked to identify the changes that had been made to these crowded displays. The children showed step-wise increases in ability for processing these types of complex images, suggesting that there are recognition improvements that accompany aging and that by early adolescence, young people are better at identifying items when they are presented in simultaneous arrays that are more demanding of attentional resources. Another possibility suggested by these results is that strategies improve as children age and move into adolescence. A follow-up study conducted by Mandler and Robinson (1978) sheds additional light on this topic. The participants were at similar age levels to the Dirks and Neisser study (first, third, or fifth graders) and in this study they viewed sets of items presented simultaneously for later recognition. Some of the items were placed into organized scenes. For example, a clock, window, ladder, chair, crib, and staircase were presented together in a spatial arrangement that approximated a room scene. Other stimuli were presented in a disorganized scene, in which the items were placed in a random and nonsensible spatial arrangement. Participants of all ages were relatively poor at recognizing the items from the disorganized scenes, consistent with the position that recognition memory does not develop substantially as we age. Conversely, the older children were much better at recognition for the organized scenes than the younger children. This suggests that background knowledge in the form of schemas consisting of organized items develops as children move into adolescence and that this ability helps to drive better recognition performance.

Recall of information is typically more difficult than recognition, as the participant must retrieve the information actively in recall tests. These tasks are frequently presented in one of two formats, free recall, or cued recall. In a free recall task participants are presented with a set of words or pictures. After some period of delay, the participants are free to report as many of the items as they can remember from the original set. By contrast, cued recall experiments involve presenting words or pictures in pairs (such as book-basket). After the item set is presented, a delay is typically introduced followed by a test phase, in which the cue items are presented to the participants and they are asked to produce the associated word. For instance, if a participant was presented with "book-_____," he or she should report "basket." Free and cued recall studies tend to produce data that fit a serial position curve. This curve tends to show some basic characteristics (Fig. 6.8). First, the early items in the memory set tend to be reported at a high level because they have fewer items to interfere with them and may just appear to be more memorable, as they were the initial items in the set. This tendency for the items presented early in the set to be recalled at a higher level is known as a *primacy effect*. Second, the middle of the memory set shows a standard chance of recall lower than that of the early items. Third, the final items to be remembered tend to be recalled at a higher rate than the middle items, possibly attaining a probability of recall similar to the items at the beginning of the set. This is commonly known as a *recency effect*. These final items may be able to be produced from working memory provided that they are remembered after a delay is introduced before the memory test. The recency effect may also occur since the final items presented can benefit from a lack of interference, as no items were presented after them to disrupt the encoding process (Fig. 6.8).

Memory research with children and adolescents reveals that both groups show strong recency memory. There are, however, relatively strong improvements in recall that are associated with age (Cole, Frankel, & Sharp, 1971; Ornstein, Naus, & Liberty, 1975). By age 14, most teens will show strong primacy and recency effects, and they will also exhibit a higher probability of recalling items from the middle of the set than children. Meanwhile, there will usually be significant differences between children around age nine relative to children just beginning school, who are around age six. The six-year-olds tend to show a reduction in the primacy effect, while the 9-year-olds' patterns of recall probability are similar to those of 14-year-olds, but with a reduced chance of recall. What appears to differentiate the adolescent performance from that of the younger children is strategy use, or the use of mnemonics, which are memory aids that help us to organize material for

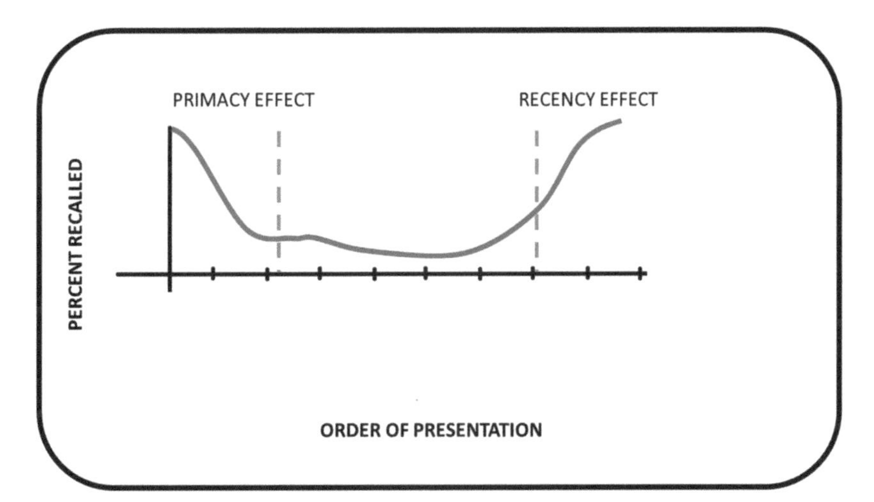

FIGURE 6.8 A serial position curve can be plotted for the probability of recall for a set of items based on their order of presentation. This idealized serial position curve approximates the major effects commonly observed in long-term memory experiments. Initially, the first items on the list are highly recalled. This is known as a primacy effect. The middle items are less well recalled. The final items may be better recalled due in part to a participant maintaining them in working memory following the presentation of the list. This is known as a recency effect.

later recall (Bjorklund & Green, 1991). The technique of using mental imagery by forming mental pictures of each of the items is one such mnemonic. Another mnemonic involves linking mental images with one another. Both techniques can enhance later recall. This effortful encoding strategy can prove useful in many reasoning situations that we will encounter in the later chapters of this book.

Memory and Inferences in Adolescence

Making inferences is one of the most important reasoning functions for which we rely upon memory. Relatively early on in the study of cognitive psychology, memory researchers isolated several key findings in memory and inference as children move into the adolescent years. In this section we will review some of the classic findings from this period of investigation into memory and inference. We take an active role in memory retrieval, often filling in information that we were not explicitly presented with. John Bransford and Jefferey Franks (1971) demonstrated this phenomenon in adults. They presented a series of statements about a particular situation. For example "The ants in the kitchen ate the jelly" and "The ants ate the sweet jelly" were both presented to participants. Later, during a recognition test, participants confidently inferred that they had heard the sentence "The sweet jelly was on the table." Note that this was actually a new statement that had not been previously presented; however, the idea that the jelly had been on a kitchen table is consistent with the schema that would be evoked for such a situation. Notice that the erroneous recognition of the sentence endorsing that "The jelly was on the table" has been presented and can be interpreted as an advantage for memory in many cases. While we should not make such inferences when testifying under oath, in most cases in life, drawing sensible inferences helps us to flesh out a situation in our minds. These inferences will often be correct much of the time, and they enable us to move beyond the information presented in verbatim to infer a likely new inference. These types of inferences are valuable in life, as they help us to gain a greater understanding of what likely occurred in a situation even if these were not directly presented to us in a verbatim format. When we make inductive inferences, we are moving beyond the information presented in order to do so.

In a related study, Paris and Lindauer (1976) evaluated children at the ages 7, 9, and 11 to determine whether they actively make inferences that move beyond the information provided. Participants were presented with statements of the following sort: "The workman dug a hole in the ground" and "The workman dug a hole in the ground with a shovel". Notice that in the first statement the tool being used (a shovel) was not actually presented, but could be inferred by the listener. The 11-year-olds were able to achieve higher sentence recall levels when a cue word (e.g., shovel) had been presented. These cue words helped the 11-year-olds to recall either presentation of the sentence (with or without shovel). In other words, as children age they become more effective at inferring a likely missing element of an idea and then this enables them to achieve higher recall levels when they need such information. By contrast, the younger groups were only able to achieve higher recall when the cue word had been presented explicitly, but not when it had been merely implied. In a second study, Paris and Lindauer asked participants to act out each sentence. The actions helped to make explicit that a shovel would be a very likely tool implied by the sentence. When acting out these statements, the younger children then benefitted from the presentation of the cue word during a later recall attempt. Notice that as young people enter adolescence they are able to use a more efficient method in order to make use of a memory inference. This suggests that they have become better at extracting an overall meaning or gist representation for each sentence. This is precisely the skill that is frequently needed for making inductive inferences or finding relevant prior solutions when confronted with a vexing problem to solve. The accrual of gist representations over the lifespan leads to a strong set of inferential capabilities later in life.

Another critical element needed in order to make effective inferences based on cues is a strong base of knowledge. This was nicely illustrated in a study by Brown, Smiley, Day, Townsend, and Lawton (1977). In this case, children at younger ages (second graders) were compared with a seventh grade sample. The task involved presenting a cover story about a fictional culture called the Targa people. In some instances the Targa were stated to be desert dwellers, while other participants were informed that the tribe was an Inuit group. The background knowledge manipulation was carried out in order to determine its effect on direct inferences that the young people in the experiment would make. Both recall and recognition tests were utilized, along with postexperimental interviews of the participants. In all test formats, the background information influenced the inferences that the young people made. For example, when presented with the literal item "The weather was bad", the sensible inference "the weather was hot" or "The weather was cold" tended to be mis-remembered as having been stated. The environment that had been indicated for the Targa people determined the specific inference about heat or cold. The rate of these inferred embellishments upon the information was higher for the seventh grade participants (79%) over the

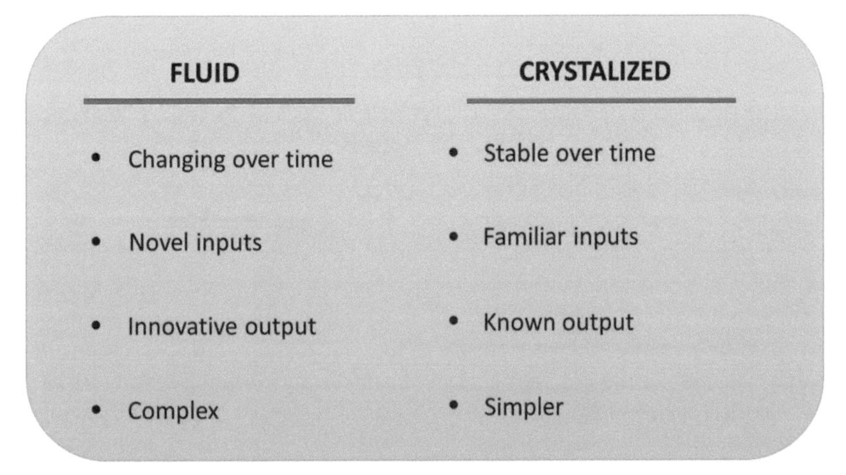

FIGURE 6.9 The characteristics of fluid versus crystalized intelligence. We discuss fluid reasoning in many sections of this book. Crystalized intelligence is made of the knowledge base that we have stored in semantic memory. This knowledge accrues over our lifetime and can facilitate fast and efficient inferences that are accurate much of the time.

second grade participants (59%). This study indicates that as young people enter adolescence, they tend to make greater use of background knowledge to actively inform and extend upon the facts that they are provided with. This will ultimately make them quicker and more accurate thinkers.

As we move through life, we continually make new inferences. While, the newly inferred information may not always be correct, if we are well informed with a wealth of knowledge, our inferences will be accurate much of the time. This enhanced knowledge base continues to accrue over our lifetimes and eventually enables us to make rapid, accurate inferences much of the time. The use of knowledge toward inferring new information is referred to as crystalized intelligence, as it is robust across the lifespan (Fig. 6.9).

BRAIN CHANGES ASSOCIATED WITH AGING ADULTS

The Aging Brain

There are a variety of physical changes to the body and the brain that occur with aging. Some of the most obvious signs of aging are changes to the skin, which can include age spots, wrinkles, and reductions in the subcutaneous fat layer beneath the skin. There are other notable physical changes with aging that include a reduction in overall muscle mass, graying and thinning hair, and an increase in body fat. There are reductions in bone density and joint flexibility. People frequently experience hearing and vision loss as they age, and these factors can impact how they respond to the environment around them (See Box 6.1). These types of structural changes also affect the brain.

There are strong cardiovascular effects that accompany aging. The overall integrity of the cardiovascular system decreases on average by about 1% per year in adulthood (Bortz, 2005). The cardiovascular system is comprised of the heart, the arteries, and the veins. The arteries are crucial for supplying oxygenated blood to the rest of the body, while the veins return the blood to the heart. The heart muscles are particularly prone to weakening over the course of life. Decreases in the strength of the heart muscles along with changes within the arteries can make the cardiovascular system function less efficiently over the lifespan (Nikitin et al., 2006). Andrew Betik and Russell Hepple examined changes in the maximal rate of oxygen utilization, or VO2 max in people as they age. VO2 max is an index of the capacity someone has for physical activity based on the efficiency of their cardiovascular system. These investigators estimated that reductions in muscle oxygen delivery can be attributed to reduced efficiency of the heart along with weakened distribution of the cardiovascular system in late middle age. In older adults, the ability of the muscles to absorb oxygen begins to decrease, and this further weakens the output of the system. This may be due to changes at the cellular level (mitochondrial dysfunction) in very old adults and can contribute to larger reductions in cardiovascular efficiency. On a positive note, regular aerobic exercise can reduce these losses of efficiency. Like many aspects of aging, there are large individual differences and lifestyle effects that can have major impacts in determining the degree of biological change that a person experiences.

The brain is a major recipient of resources from the bloodstream and brain health relies upon the efficiency of the cardiovascular system for delivery of both oxygen and glucose, which are critical in support of the nervous system. This makes cardiovascular health an important determinant of our reasoning abilities and

BOX 6.1

OVERCOMING SENSORY LOSS IN AGING...IT CAN TAKE TEAMWORK

Older adults frequently experience losses in hearing and vision as they age. A lifetime of exposure to noisy environments can lead to a variety of hearing impairments. These can include the loss of ability to perceive certain frequencies, tinnitus (hearing a chronic ringing sound), and a greatly reduced ability to hear overall. Likewise, visual impairments can include the effects of cataracts, reductions in elasticity of the eye muscles, astigmatism, and blurry vision. These challenges can be especially difficult in society, as they limit an older person's ability to notice nonverbal cues, to identify someone that they know, or to understand what is being said in a conversation. All too often we may jump to the conclusion that an older adult has a cognitive impairment when they do not appear to be aware of who is speaking to them, fail to answer a question, or appear to struggle to follow a conversation. These disabilities can lead to the perception that an older adult is "just not with it," or that they are experiencing memory problems.

Growing up, I was fortunate to have a close family that lived locally. My grandmother, Stella, and my great aunt, Loretta, lived close by Fig. 6.10. Stella and Loretta never missed a holiday gathering and the two of them lived into their nineties. Like many older adults, Stella suffered from cataracts and other visual impairments and was severely visually impaired for the last decade of her life. Loretta suffered hearing losses and had great difficulty following a conversation as she aged. We all became accustomed to speaking very loudly around Loretta, but honestly it could be difficult to tell if she had heard you, even if you felt you were near shouting. You have probably had an experience like this if you have ever tried to hear someone speak over a crackling poor phone connection, or tried to follow a conversation across a crowded noisy room. These disabilities could have been devastating to Stella and Loretta, but they used teamwork as a compensatory strategy.

Stella and Loretta were inseparable. They both became widows at relatively young ages and neither remarried. They lived next door to one another for the majority of their lives and frequently helped one another out, but still managed to live independent lives much of the time. When the family got together and conversations started up, Stella and Loretta would always sit together. In fact, Stella would sit on the side of Loretta next to her better ear. When someone started talking, Loretta would often identify the speaker for Stella, who had difficulty seeing. Stella would then kick into action and relay some of the discussion to Loretta, who often had difficulty tracking what was being said due to her hearing loss. I can recall many times where the conversation was much more effective when you could speak to the pair of them together.

Stella and Loretta's teamwork helped them to overcome major challenges presented by aging. They were very fortunate to have been such close sisters and to have lived very long and productive lives nearby one another. This example could be considered a form of practical intelligence, in which two people can compensate for one another's deficits and attain a symbiosis that allows them to function at a higher level of social engagement. The enhanced capability was helpful to both women and to the family members who could better communicate because of their shared efforts.

FIGURE 6.10 Loretta (left) and Stella (right). Vision and hearing loss are common in older adults. There are a variety of compensatory strategies that people can take to minimize the difficulties that age-related challenges cause in their daily lives.

cognition in general. One of the methods used to assess the health of a particular region of the brain is to conduct a single-photon emission computed tomography (SPECT) scan, which involves injecting a radioisotope into a person's bloodstream and then using a specialized scanner to image the uptake of the chemical within the brain. SPECT scanning enables researchers to determine the degree of blood flow to different areas of the brain.

Similar methods can be achieved by magnetically "tagging" blood molecules as they enter the brain using magnetic resonance imaging (MRI). In these cases, the health of the tissue can be determined by the overall blood flow to the brain, or regional blood flow within different areas of the brain. SPECT imaging, in particular, is commonly used to determine whether there are losses of brain function in aging due to dementia.

Some of the early studies of the aging brain suggested a very negative outlook. Initially, researchers had projected that large losses of neurons would occur over the course of the lifespan. This has been referred to as the *neuronal fallout model* (Triarhou, 1998) and is based on the general belief that it was not possible to regenerate any new neurons within the brain, a process now known as *neurogenesis*. This process was thought to be particularly acute in the PFC and the hippocampus. In the 1990s, new methods of estimating neuronal loss due to aging were developed. These methods showed a more positive outlook illustrating that the effects of aging in the PFC and hippocampus were likely overestimated (Burke & Barnes, 2006). There have been some difficulties with projecting these estimations at a population level, as some study samples have included individuals who are actively developing dementia, which falls into the category of disordered brain function, rather than reflecting the normal biology of aging (Yankner, Lu, & Loerch, 2008). The inclusion of individuals with dementia could bias a sample toward appearing more negative in outcome than one would otherwise find.

A prominent neuroimaging marker of cognitive aging has been found in the PFC. When younger and older individuals are compared using fMRI, several studies have indicated that the older adults show a bilateral pattern of activation involving both the left and right hemispheres in older adults, while the younger adults indicate a pattern that is more unilateral (Cabeza, 2002). Rypma and D'Esposito (2000) reported a relationship between PFC activation and speed during a working memory task. In this study, faster performing younger participants showed less dorsolateral PFC activation during working memory retrieval than younger individuals who were slower at the task. Meanwhile, older adults showed the opposite pattern, with faster performers showing greater activation in this region, while slower older adults showed less PFC activation. These results indicate that there are age-related changes in the dorsolateral PFC that are linked to cognitive efficiency in normal aging during working memory load conditions.

There are also white matter, or connectivity, changes in the brain that accompany aging in addition to cortical changes that occur. The density of white matter bundles can be measured using a type of MRI scan called diffusion tensor imaging (DTI). DTI research has indicated that white matter density tends to reduce over the course

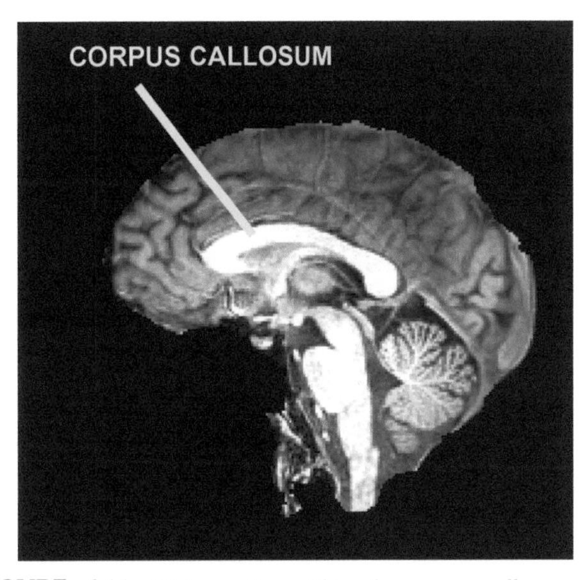

FIGURE 6.11 White matter within the corpus callosum can decrease in thickness as we age. This is particularly the case within the anterior part of the corpus callosum that connects to two frontal lobes, which are important for higher cognition and reasoning.

of the lifespan. The greatest reductions in white matter are observed within areas of the PFC and its associated band of white matter in the anterior corpus callosum, which connects the two hemispheres (Small, Chawla, Buonocore, Rapp, & Barnes, 2004) (Fig. 6.11). These reductions in connectivity among brain regions may be related to the patterns of cortical activation change that accompany aging and are likely linked to the differences in cognition that accompany healthy aging (Hedden & Gabrieli, 2004).

CHANGES IN COGNITION WITH AGING

Changes in Processing Speed

We have discussed some of the physical changes that accompany aging. In general, many systems in the body become less efficient and function at a lower level than they had earlier in life. This is true of the brain as well, and notably appears to influence the cortex and connectivity of the PFC, a region that is extensively linked to thinking and reasoning. How do these differences in biological efficiency relate to performance in cognitive tasks and in reasoning in older adults?

One of the most consistent changes in cognitive performance over the lifespan is a decrease in overall speed. Speed can be measured by a simple reaction time task. Such tasks include the trail-making test, on which participants connect a series of numbered dots in sequence while they are timed (Fig. 6.12). Similarly, simple response time can be measured by the speed at which a person completes a choice reaction time task, during

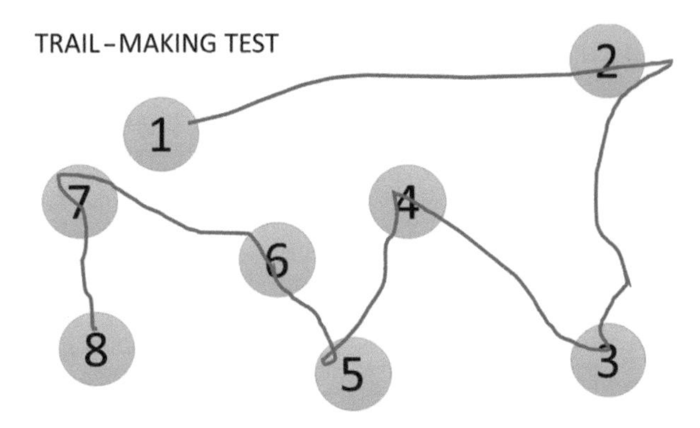

FIGURE 6.12 A trail-making task requires participants to connect numbered dots in a sequence as fast as they can. The time to complete the sequence can determine the processing speed of an individual.

which participants are asked to press a button to a colored shape or a letter, as quickly as they can. Younger adults tend to perform these types of tasks at a faster rate than older adults (Der & Deary, 2006).

A theory of cognitive aging maintains that people undergo a process of general slowing. Timothy Salthouse at the University of Virginia has been one of the major researchers examining this phenomenon. According to Salthouse (1996), there is a general slowing of cognitive processes as people move through the adult years. There are two challenges presented by age-related slowing. The first challenge is a timing problem that negatively impacts information processing. Salthouse termed this the *limited time mechanism*. Essentially, the claim is that as people age they can maintain information for only a limited time. This is especially problematic in the case of reasoning tasks that may involve numerous cognitive operations or steps that have to take place. The limited time mechanism would ensure that the early phase of task completion would dominate the amount of time available for processing, and this would mean that later cognitive processing would be reduced leading to errors.

This mechanism would limit the cognitive abilities of older adults particularly when they are faced with time limits. It would also be relevant when reasoning problems involve several steps that need to be carried out in a sequence. The second mechanism proposed by Salthouse is the *simultaneity mechanism*. This mechanism restricts a person's ability to use multiple sources or types of information at the same time due to age-related slowing. Essentially, the outputs of earlier cognitive processing on a complex task may be lost by the time that later processing can be accomplished. The simultaneity mechanism could operate when a reasoning problem involves multiple cognitive steps. These steps may include encoding information, elaborating upon that information, searching memory, rehearsing information in working memory, integrating information, and abstracting information. If each of these operations takes longer for older adults, it is more likely that they will be prone to errors. Effectively, a failure of the simultaneity mechanism in older adults could result in an inability to keep the outputs of multiple cognitive processes active for a sufficient amount of time to assemble an answer to a problem. In the later chapters of this book we will discuss numerous reasoning tasks that require either several steps to carry out, or in some cases deliberative thinking that is necessary to carry out a fully developed decision. Any such cases could be subject to reduced performance based on these mechanisms.

Changes in Attention

Attention is another cognitive domain that influences reasoning performance. Attention was an early focus of the field of cognitive psychology. Early cognitive researchers noted some key differences in attentional focus (Treisman & Gelade, 1980). When conditions allow a single feature to distinguish a stimulus from those that surround it then this produces a "pop-out" effect. Imagine that you are looking for your father who has white hair amid a group of younger people. You will likely find him rapidly. Such a search has been termed to rely on *parallel processing*, as it subjectively feels as if you can devote little focused effort to accomplish this search. By contrast, imagine that you are looking for your father who said he will be wearing a green coat. This time he is at a large convention of senior citizens, many of whom are wearing hats, and some will be wearing green coats. This situation will take a much more demanding search. Now white hair will not be sufficient to accomplish the search and there will be no similar "pop-out" effect. This time you will have to find the man with white hair *and* a green coat on amid a sea of other people, some of whom have white hair and some of whom are wearing green coats. This situation demands what is known as a conjunction search, as the conjunction of two features needs to be integrated in order to identify the matching item. This type of search is often referred to as requiring *serial processing*, as the search requires greater effort and controlled processing. The terms parallel processing and serial processing have somewhat fallen out of favor, as this distinction describes the subjective sense of the search process rather than a literal mechanism of attention (Thornton & Gilden, 2007). In both searches, the eyes must move around in order to locate the target item. Fig. 6.13 depicts two types of searches. In panel A the "pop-out" effect occurs as one feature alone (color) distinguishes the red circle from the surrounding green circles. Meanwhile, in panel B a conjunction search is required in order to isolate the target oddball item (the red oval) from other ovals that are green and other circles that are red.

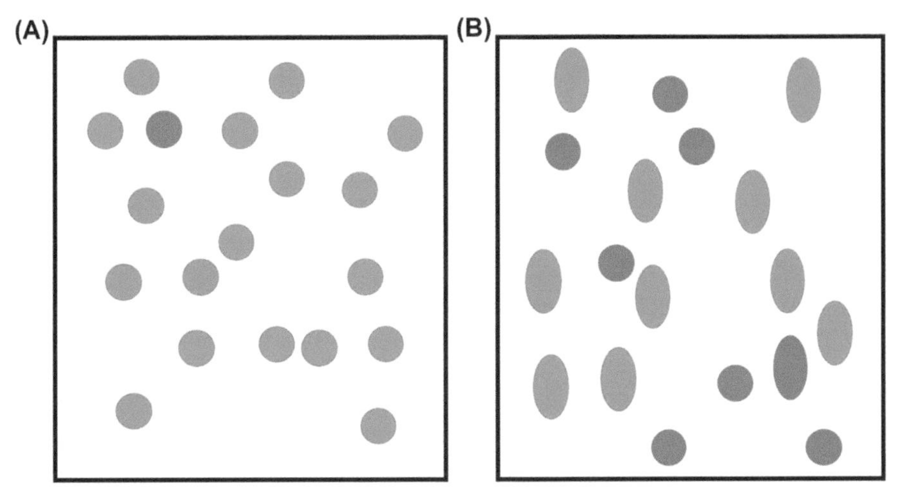

FIGURE 6.13 Try to find the oddball item in these two panels. (A) A "pop-out" effect occurs when the target item is different on the basis of one feature. Finding this target requires little focused attention. (B) A conjunction search must be carried out when two features need to be integrated in order to differentiate a target from surrounding items.

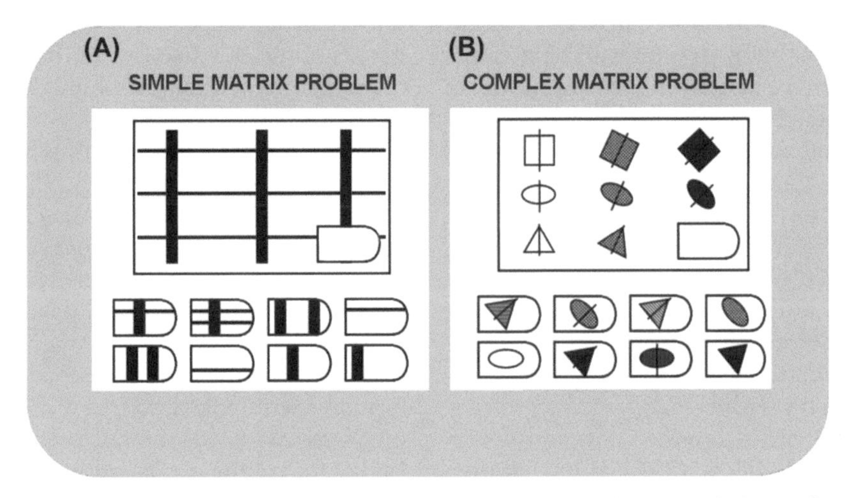

FIGURE 6.14 (A) A simple pattern match matrix reasoning problem can likely be carried out using little visual attention. (B) A more complicated matrix problem will require someone to carry out a more focused search that places more demand on attention.

For the purposes of aging research these attention search processes differentiate further in older adults. Both older and younger adults are highly efficient at carrying out the simple parallel processing search, such as the "pop-out" search depicted in Fig. 6.13A (Madden, Whiting, & Huettel, 2005). When older adults are asked to perform a conjunction search (Fig. 6.13B), they are often much less efficient and take longer to locate the oddball item that can only be differentiated on the basis of two features (Madden, Pierce, & Allen, 1996). In regard to reasoning skills, we have spent quite a bit of time in previous chapters discussing Raven's progressive matrices task, a test of visuospatial reasoning (Fig. 6.14). Note that in simple Raven's matrix problems a simple visual search is all that is required (Fig. 6.14A). Meanwhile, when a difficult Raven's matrix problem is encountered, it frequently requires the integration of multiple relational elements that become evident as you move your eyes across the rows and down the columns within the matrix. Adults can likely solve the simpler Raven's matrices problems effectively across the lifespan. Older adults show some degree of lost efficiency when solving more difficult Raven's matrices problems (Fig. 6.14B) (Salthouse, 1994). Note that the demands upon attention in visual search do not fully explain the performance on this type of integrative reasoning task, but the speed and efficiency with which one can process the visual elements of the matrices may be a gateway skill that facilitates performance in younger adults. Limitations in attention may limit the abilities of older adults in fluid reasoning tasks.

Changes in Working Memory

Earlier in the chapter we noted the importance of working memory during reasoning when we discussed the development of working memory ability that accompanies the shift from childhood to adolescence. As we discussed in that section, people will experience challenges in reasoning if they are unable to effectively encode relevant information. Similarly, retrieval of information from working memory is also a critical process ensuring that we are able to access knowledge at the appropriate time. Older adults generally show reductions in performance, compared to younger people, in tasks that involve manipulation or integration of the items that are being held in working memory (Glisky, 2007). There is some debate about the reasons for the decline in working memory efficiency that is associated with aging. As we described in an earlier section, the work of Rypma and D'Esposito (2000) indicated a difference of activation within the PFC that is linked to the speed of performance on a working memory task. This would suggest that the neural basis for efficiency in working memory may be impacted by structural and functional changes that occur within the PFC during the normal aging process. Other researchers have indicated that there may be a particular deficit with inhibitory control, another PFC linked function (Hasher, Zacks, & May, 1999). There is also neuroscience evidence from fMRI and EEG that supports a suppression deficit in working memory (Gazzaley, Cooney, Rissman, & D'esposito, 2005).

With regard to reasoning ability, there is a strong relationship between working memory and integrative reasoning on a task such as the Raven's progressive matrices. This relationship is specifically linked to performance time, which is often slower in older adults. In a relevant study, Adam Chuderski (2013) at Jagiellonian University in Poland compared the relationship between working memory and relational reasoning under normal conditions and under time pressure. Chuderski asked participants to encode items into working memory to estimate their capacity. He also asked participants to solve two types of relational reasoning problems, the Raven's advanced progressive matrices (Raven, 1938, 1960) and a four-term visual analogy test in the format "A is to B, as C is to D." Participants had to fill in an appropriate D term in order to complete a relational sequence that was shared between the A and B terms of the analogy. The A, B, and C terms in the analogies were shown as simple patterns of five latent rules (e.g., symmetry, rotation, change in size, color, thickness, number of objects). Participants in a time-pressured group were allowed 20 min to solve the Raven's matrices task and 15 min to complete the analogies. Chuderski also included a group under moderate time pressure, who had 40 min to complete Raven's matrices and 30 min to complete the analogies. There was also a group that did not experience time pressure having a full hour to work on each reasoning task. Results revealed that the participants under the greatest time pressure were severely restricted in reasoning performance. The results of a confirmatory factor analysis revealed that under high time-pressure conditions fluid reasoning ability, as measured by the Raven's matrices and analogies, was statistically indistinguishable from working memory capacity. In other words, relational reasoning is directly limited by working memory ability under those conditions to the point where the working memory and reasoning tasks are effectively limited by the same mental operations. Meanwhile, when time pressure was not applied, working memory capacity explained only 38% of variance in relational reasoning. While this study was conducted on younger adults, the results indicate that working memory capacity and speed are both important for relational reasoning abilities. The younger adults under time-pressured conditions were operating at a reduced reasoning capacity that was limited by working memory resources. This finding suggests that older adults may be slower in reasoning, but also exhibit declines in Raven's matrices performance that are linked to speed and working memory.

In a technological study of working memory in older and younger adults, Fiona McNab et al. (2015) evaluated individuals using a smartphone app. The task was similar to the electronic game *Simon Says*, in which a set of visual items was presented on a grid on the phone screen. On the screen, a 4-by-4 grid was presented with a set of red circles in different locations. Younger and older adult participants were asked to encode these circles and their locations within the grid (Fig. 6.15). In some conditions an interference set was then delivered in the form of several irrelevant yellow circles appearing at different locations on the grid. This app is available for smartphones if you would like to try it (www.greatbrainexperiment.com). Lastly, participants were asked to recall and touch the locations of the original red circles that they had encoded after a brief delay period. The investigators compared people's ability to ignore the irrelevant distractor items (yellow circles) when items are encoded into working memory and when items were being maintained during the delay between encoding and response. In older adults, working memory performance was reduced for all conditions. In individuals over age 50, performance was most disrupted when distracting items interfered during the delay period. This study suggests that distraction is a major source of working memory disruptions. It also indicates that more focused attention may be necessary for older adults to overcome disruptions during delay periods that ensue after they have encoded information. This study again indicates a working memory decline with aging that is more pronounced in the face of distraction by interfering items.

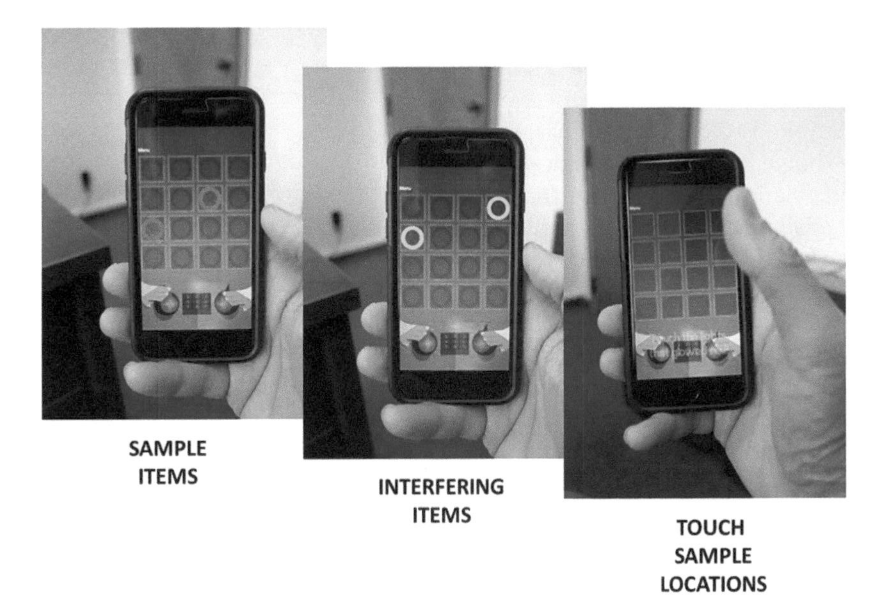

FIGURE 6.15 The smartphone app developed by researchers at the Wellcome Trust Centre for Neuroimaging at University College, London. McNab et al. (2015) evaluated working memory performance with and without distraction in older and younger adults. You can try this app if you are interested in this experiment (www.greatbrainexperiment.com).

The Role of Inhibitory Control in Reasoning as People Age

The inhibitory control hypothesis appears to explain some of the reductions in performance on working memory tasks that occur in older adults. A reduction in inhibitory control is also relevant for relational reasoning performance. There is some support for reasoning deficits being linked to the ability to control interference from patient studies. My colleagues and I published two prior studies evaluating relational reasoning in individuals with bilateral PFC damage (Krawczyk et al., 2008; Morrison et al., 2004). In these studies we tested participants on a series of verbal and picture analogy tasks, along with a visual relational task. The deficits experienced by PFC damaged patients in these studies were at least partly attributable to inhibitory control deficits. When participants saw semantic or perceptual distractors, they performed more poorly. We will discuss these studies and their implications further in Chapter 7. These types of PFC impairments also lead to deficits on other fluid reasoning tasks such as matrix reasoning tasks and transitive inference tasks (Waltz et al., 1999).

Viskontas, Morrison, Holyoak, Hummel, and Knowlton (2004) examined the role of interference in relational reasoning in older adults. In this study, the authors asked older and younger adults to solve variations of Robert Sternberg's (1977) *People Pieces* analogy task (Fig. 6.16). People Pieces presents two individuals side by side that can be evaluated on the basis of relational correspondences (size, gender, height, and clothing color). The task requires individuals to map the relational elements shared by these two people to another set of people presented in a second analogical pair. Viskontas et al. varied the relation numbers that existed in the first pair of people. On some complex problems multiple relations needed to be considered together in order to solve the analogy. In addition, the experimenters varied inhibitory control load on the participants. This was accomplished by presenting pairs of people who varied on a number of relational dimensions, but

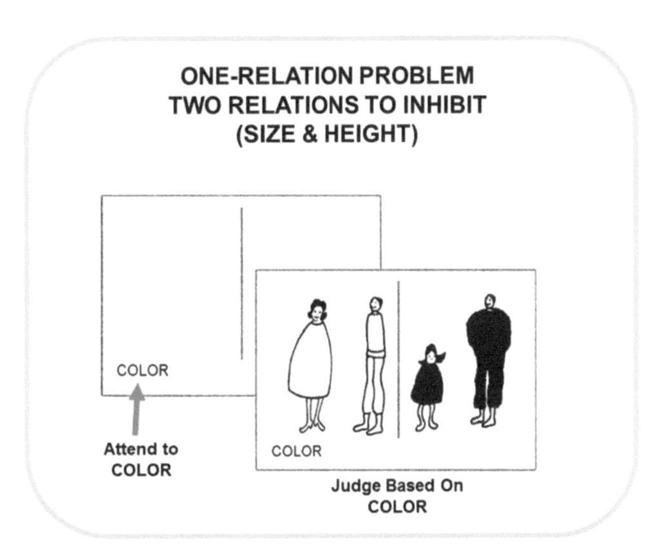

FIGURE 6.16 Robert Sternberg's (1977) *People Pieces* analogy task. Two individuals are presented (on the left side of the center line) that can be evaluated on the basis of relational correspondences (size, gender, height, and clothing color). The task requires individuals to map the relational elements shared by these two people to another set of people presented in a second analogical pair (on the right side of the line). In this example participants were cued to attend to the dimension "color" and answer the analogy based on this attribute only, while ignoring the others.

participants received a verbal cue indicating which specific dimensions should be used to answer the analogy. In some cases, irrelevant differences existed between the sample people pieces that had to be ignored by participants. Younger adults showed high performance at all but the highest level of relational complexity that had included the most dimensions of relational change. By contrast, older adults (aged 66–86 years) performed more poorly even at a medium level of relational complexity. Furthermore, the performance of the older adults was especially compromised when irrelevant relational dimensions were included in the problems. This result suggests that at least to some degree an inhibitory control deficit may underlie relational reasoning difficulties experienced by older adults.

PRACTICAL REASONING IN THE OLDER YEARS

Reasoning and Healthy Aging

We have discussed cognitive decline quite a bit in the previous sections of the chapter. What can we do about age-related declines in areas including working memory and reasoning? Studies have begun to evaluate what makes someone a more successful ager who can maintain their cognitive abilities later in life. Some of the emerging evidence on cognitive aging is positive (See Box 6.2).

Emily Rogalski at Northwestern University leads a study of highly robust aging adults. This is known

BOX 6.2

IT'S A DOG'S LIFE...AGING EFFECTS IN REASONING IN BORDER COLLIES

The research community has studied cognitive changes in humans to a large degree, what about reasoning over the lifespan of a dog? Lisa Wallis et al. (2016) at the Clever Dog Lab, a part of the Messerli Research Institute at the University of Veterinary Medicine in Vienna, Austria, investigated the effects of aging on pet border collies. They examined both memory and reasoning abilities in these dogs as they change over the lifespan.

How does one test a dog's ability to reason? The dogs were initially trained to use a touchscreen by poking it with their noses and receive a food reward. The dogs could then be tested on a variety of images. One of the key tests used in this study was an inferential reasoning by exclusion test. In reasoning by exclusion, an individual infers that they are choosing the correct answer on the basis of excluding a known incorrect answer. In an example with people, imagine that a mechanic is trying to isolate the cause of a car that won't start. The likely possibilities for this symptom are a fouled spark plug, a bad plug wire, or a weak ignition coil. After checking that the spark plugs and plug wiring look to be functional, the mechanic infers that the ignition coil is the bad part. This is reasoning by exclusion, as the ignition coil remains unknown, but it can be assumed to be the faulty part on the basis of excluding other causes. People make these types of inferences in other important areas including medical diagnoses and detective work. This is a task that has been used previously in developmental studies with children (Aust, Range, Steurer, & Huber, 2008) and also to evaluate the reasoning abilities of nonhuman primates (Call, 2006).

Before starting the inferential reasoning task, the dogs were trained to nose-poke different pictures of objects that appeared on the touchscreen. Some of the times the dogs earned a food reward when they touched certain objects (e.g., a picture of a basket). Other times they received no reward for touching a particular object (e.g., a picture of a telephone). After a period of time, the dogs became familiar with the objects and learned to choose those that would get them a reward when they were paired on the screen. Once this skill was mastered, the researchers provided new pairs of the items to assess reasoning by exclusion. On reasoning test trials, the dogs were presented with a pair of objects consisting of one new object that has unknown reward value (e.g., a picture of a book) and a picture of a previous object that is known to be unrewarded (e.g., the picture of a telephone). The dogs should choose the new objects in this case (the picture of a book). They do not know the reward value of this new item, but they know that the telephone does not lead to a food reward. The second key trial type presented the dogs with one of the new objects (e.g., the book) paired with another object picture never seen before (e.g., a picture of a clock). In this case, the dogs should again select the picture of the book, as it leads to a food reward. Dogs can use other tactics, such as avoiding the negative items (by not choosing the telephone picture) and always choosing the new item (which would lead to choices of the new clock picture). Only by consistently choosing in favor of the book did dogs exhibit reasoning by exclusion competency.

Dogs aged 5 months up to 13 years (very old for a dog) took part in the study (Fig. 6.17). The results indicated an age effect in which the older dogs were better at reasoning by exclusion than the younger dogs. Somewhat unexpectedly, the oldest dogs were still better than the younger

BOX 4.4 *(cont'd)*

FIGURE 6.17 Dogs aged 5 months up to 13 years took part in the study by Wallis et al. (2016) poking objects on a touchscreen with their noses. *Credit Lisa Wallis/Vetmeduni Vienna photo.*

dogs. Most previous canine cognition studies have been carried out in beagles that have been raised in laboratory environments. Wallis et al. speculated that pet dogs may lead more enriched lives and have greater mental stimulation. It is possible that the extra stimulation led to the high performance in the older dogs. In humans, cognitive flexibility tends to decrease as we get older. Wallis et al. offered the possibility that the younger dogs may be using more flexible strategies to solve the reasoning by exclusion task. This may account for the finding that the older dogs appeared to outperform the younger ones. Dogs are a fascinating species, and this study represents a creative effort to probe the mental lives of these friendly and clever animals (Fig. 6.18).

FIGURE 6.18 Pet dogs are intelligent and affectionate companions. It is possible that the life of a pet dog may be particularly enriched, and this may lead to healthy aging in the later years of a dog's life.

as the Northwestern University SuperAging Project. Rogalski et al. evaluate people over the age of 80 who have unusually high performance on memory tests. The goal is to characterize the features of these individuals' biology and lifestyle that may make them exceptionally resilient to many cognitive and neurological symptoms of aging and to understand why these individuals are able to respond this way. The definition of a SuperAger according to Rogalski is to perform at or above average normative values for individuals in their 50s and 60s on a test called the Rey Auditory Verbal Learning Test (RAVLT). SuperAgers also must fall within one standard deviation of the average range for their age and education for a set of other normative values for measures of executive function and semantic retrieval tests.

Rogalski et al. (Harrison, et al., 2012) have been carrying out structural imaging on these participants in order to assess brain differences that mark these exceptionally robust people. The research group has published an intriguing report comparing the SuperAgers to a group of cognitively average older adults within the same age range as the SuperAger group and to a younger group of middle-age controls averaging around 58 years old. Results of a cortical thickness analysis revealed that the older adult control group exhibited significant atrophy in several brain areas including the PFC, parietal cortex, and occipital lobes in comparison to the middle-aged control individuals. Meanwhile, the SuperAger group showed no statistically significant atrophy compared to the middle-aged control group, who were approximately 30 years younger and at an age at which cortical atrophy has not yet begun in many people. Furthermore, the left anterior cingulate, which is associated with cognitive control was significantly thicker in SuperAgers than in the middle-aged control group.

In addition to structural MRI findings, Rogalski et al. have examined the brains of SuperAgers after they have passed away. Post-mortem autopsy results of the brains of SuperAgers have shown a striking lack of plaques and tangles, which are neuropathologies that tend to accumulate over the lifetime and have been associated with Alzheimer's disease. Microstructural

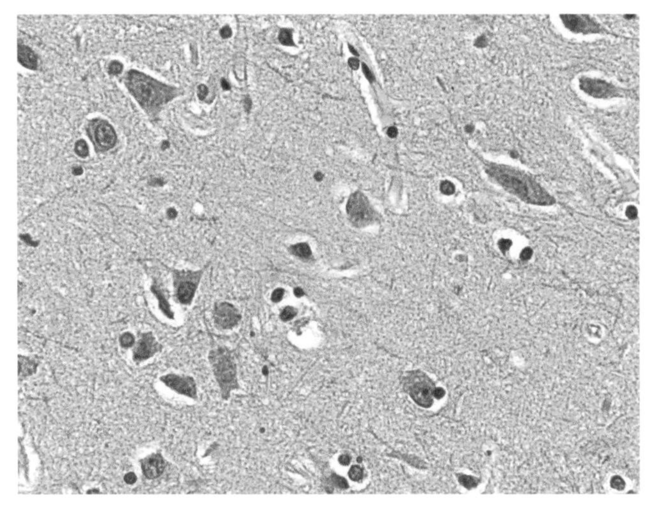

FIGURE 6.19 Spindle neurons (or von Economo neurons) are large and unusually shaped cells. The occurrence of spindle neurons is limited to larger mammals that are capable of complex behaviors. *By Nephron Wikimedia Commons.*

examination of the brains of these individuals has also revealed that the SuperAgers have a significantly higher density of spindle neurons in the anterior cingulate cortex, the same region that was found to be thicker in the structural images of these individuals. This high spindle neuron count was compared with that of cognitively average elderly controls (Rogalski et al., 2013). The spindle neurons are also called von Economo neurons after the eminent neuroanatomist Constantin von Economo who discovered them in 1929 (von Economo & Koskinas, 1929). These neurons are large and have a spindle-shaped cell body and thick dendrites. Other neurons tend to have many more dendrites, are smaller, and have different shaped cell bodies (Fig. 6.19). Spindle neurons have been found only in certain cortical regions including the anterior cingulate cortex, anterior insula, and frontopolar PFC. These neurons have been found in only a few species of mammals, and these include great apes, macaque monkeys, elephants, and some cetaceans (Raghanti et al., 2015). This suggests that the spindle neurons may be important for supporting higher cognitive functions in these complex mammals, many of which are species we discussed in Chapter 4 on comparative reasoning. The unique cingulate characteristics present in the SuperAgers may support their unusually efficient memory performance for their age group (Rogalski et al., 2013).

These results in SuperAgers give us reason for a positive outlook as we age. They demonstrate that high memory performance is possible throughout the lifespan. Additional studies will need to be conducted with larger groups of participants in order to better identify lifestyle factors and health habits that promote highly successful aging and brain health through the lifespan. Work is going on to carry out this important mission.

Results from healthy aging training studies are appearing regularly in the literature. My colleagues at UT Dallas conducted one of these studies entitled the Synapse Project (Park et al., 2014). In this study, the investigators enrolled older adults in a series of intervention training programs. In the programs older adults learned new skills, engaged working memory, utilized episodic memory, and carried out reasoning. They conducted three interventions over a 3-month period: learning to quilt, learning digital photography, or performing both tasks. Episodic memory was enhanced in these active conditions relative to lower demand conditions, in which participants met with a social group or performed simpler cognitive tasks. These results indicate that engagement in cognitively demanding and unfamiliar activities can lead to improvements in memory for older adults. Additional research is underway across the field to better understand the complex relationships between aging, learning, reasoning, and brain function. With better knowledge we may be able to better preserve brain functions and cognition throughout the lifespan.

SUMMARY

In this chapter we began by discussing the changes in the brain, cognition, and reasoning in adolescents. Some of the major changes between children and adolescents include maturation of the PFC and association cortex regions, enhanced working memory, improved strategies for encoding and making inferences based on schemas in long-term memory, and better relational reasoning capability. The teenage years are complex biologically with hormonal changes and alterations in brain areas associated with reward and risk. Teenagers may tend to take greater risks in decision making based on impulsivity and on a general lack of experience that limits their ability to use gist-based representations. Adolescent brain development continues with greater cortical maturation and interconnectivity among the networks of the brain that support task-based cognitive functions.

Toward the end of life people experience a variety of brain changes and cognitive declines. These include deficits in hearing, vision, joint flexibility, and less efficient cardiovascular health. The brain shows signs of reduced cortical thickness along with reduction in white matter density. These reductions in brain efficiency may be especially prominent in the PFC, which is linked to key cognitive skills necessary for reasoning. Working memory and long-term memory functions tend to decline in older adults along with controlled attention and integration of information. There is often a general slowing in cognition and performance speed on a range of tasks. The ability to inhibit distracting information may be especially impacted.

On a positive note, there are reports of extremely healthy agers, who are able to maintain high memory performance and cortical thickness well into their eighties. There are ongoing studies in the field to better understand how we can improve cognitive and brain health in later life by enhancing new learning and making other lifestyle adjustments.

END-OF-CHAPTER QUESTIONS

1. Adolescents undergo biological changes. Do you think that these changes are responsible for differences in reasoning relative to children, or might experience be the larger factor?
2. The PFC undergoes maturation during the period of adolescence. What reasoning functions are specifically linked to this area and which are not?
3. Memory improvements in the teenage years may be due to brain changes (maturation) or strategic changes (such as building schemas). Which of these factors may be more important for reasoning and why?
4. Risky decision making by teenagers may be related to biology or how people think about the world. Which explanation seems to be more likely in the majority of cases you can think of in which teens take risks?
5. Older adults tend to show declines in a variety of areas. Which declines appear to be most reversible by training or lifestyle factors?
6. Changes in the cognitive abilities of older adults include problems in working memory, speed of processing, and attention. Are these declines enough to explain reductions in reasoning performance on a task such as matrix reasoning?
7. Is the interference control challenge in older adults a problem for any people in your life? Do you think cognitive training could improve this challenge?
8. Some older adults remain highly active in their older years and maintain cognitive and brain health. What advice would you give an older adult to try to maintain brain and cognitive function?

References

Anderson, S. W., Bechara, A., Damasio, H., Tranel, D., & Damasio, A. R. (1994). Impairment of social and moral behavior related to early damage in human prefrontal cortex. *Nature Neuroscience*, *2*, 1032–1037.

Aust, U., Range, F., Steurer, M., & Huber, L. (2008). Inferential reasoning by exclusion in pigeons, dogs, and humans. *Animal Cognition*, *11*, 587–597.

Bjorklund, D. F. (1989). *Children's thinking: Developmental function and individual differences*. Pacific Grove, CA: Brooks/Cole.

Bjorklund, D. F., & Green, B. (1991). The adaptive nature of cognitive immaturity. *American Psychologist*, *47*, 46–54.

Bortz, W. M. (2005). Biological basis of determinants of health. *American Journal of Public Health*, *95*, 389–392. http://dx.doi.org/10.2105/AJPH.2003.033324.

Bransford, J. D., & Franks, J. J. (1971). The abstraction of linguistic ideas. *Cognitive Psychology*, *2*, 331–350.

Brown, A. L., & Scott, M. S. (1971). Recognition memory for pictures in preschool children. *Journal of Experimental Child Psychology*, *11*, 401–412.

Brown, A. L., Smiley, S. S., Day, J. D., Townsend, M. A., & Lawton, S. C. (1977). Intrusion of a thematic idea in children's comprehension and retention of stories. *Child Development*, 1454–1466.

Bunge, S. A., Helskog, E. H., & Wendelken, C. (2009). Left, but not right, rostrolateral prefrontal cortex meets a stringent test of the relational integration hypothesis. *NeuroImage*, *46*, 338–342.

Bunge, S. A., Wendelken, C., Badre, D., & Wagner, A. D. (2005). Analogical reasoning and prefrontal cortex: Evidence for separable retrieval and integration mechanisms. *Cerebral Cortex*, *15*, 239–249.

Burke, S. N., & Barnes, C. A. (2006). Neural plasticity in the ageing brain. *Nature Reviews Neuroscience*, *7*, 30–40.

Cabeza, R. (2002). Hemispheric asymmetry reduction in older adults: The HAROLD model. *Psychology and Aging*, *17*, 85.

Call, J. (2006). Inferences by exclusion in the great apes: The effect of age and species. *Animal Cognition*, *9*, 393–403.

Chuderski, A. (2013). When are fluid intelligence and working memory isomorphic and when are they not? *Intelligence*, *41*, 244–262. http://dx.doi.org/10.1016/j.intell.2013.04.003.

Cole, M., Frankel, F., & Sharp, D. (1971). Development of free recall learning in children. *Developmental Psychology*, *4*, 109.

Daehler, M. W., & Bukatko, D. (1977). Recognition memory for pictures in very young children: Evidence from attentional preferences using a continuous presentation procedure. *Child Development*, 693–696.

Der, G., & Deary, I. J. (2006). Age and sex differences in reaction time in adulthood: Results from the United Kingdom health and lifestyle survey. *Psychology and Aging*, *21*, 62.

Dirks, J., & Neisser, U. (1977). Memory for objects in real scenes: The development of recognition and recall. *Journal of Experimental Child Psychology*, *23*, 315–328.

von Economo, C., & Koskinas, G. N. (1929). *The cytoarchitectonics of the human cerebral cortex*. London: Oxford University Press.

Elkind, D., & Bowen, R. (1979). Imaginary audience behavior in children and adolescents. *Developmental Psychology*, *15*, 38–44.

Ferrer, E., O'Hare, E. D., & Bunge, S. A. (2009). Fluid reasoning and the developing brain. *Frontiers in Neuroscience*, *3*, 3.

Flavell, J. H., Beach, D. R., & Chinsky, J. M. (1966). Spontaneous verbal rehearsal in a memory task as a function of age. *Child Development*, 283–299.

Fuster, J. M. (2005). The cortical substrate of general intelligence. *Cortex*, *41*, 228–229.

Gazzaley, A., Cooney, J. W., Rissman, J., & D'esposito, M. (2005). Top-down suppression deficit underlies working memory impairment in normal aging. *Nature Neuroscience*, *8*, 1298–1300.

Glisky, E. L. (2007). Changes in cognitive function in human aging. In D. R. Riddle (Ed.), *Brain aging: Models, methods, and mechanisms*. Boca Raton, FL: CRC Press/Taylor & Francis.

Gogtay, N., Giedd, J. N., Lusk, L., Hayashi, K. M., Greenstein, D., Vaituzis, A. C., et al. (2004). Dynamic mapping of human cortical development during childhood through early adulthood. *Proceedings of the National Academy of Sciences*, *101*, 8174–8179.

Green, A. E., Fugelsang, J. A., Kraemer, D. J., Shamosh, N. A., & Dunbar, K. N. (2006). Frontopolar cortex mediates abstract integration in analogy. *Brain Research*, *1096*, 125–137.

Harrison, T. M., Weintraub, S., Mesulam, M.-M., & Rogalski, E. (2012). Superior memory and higher cortical volumes in unusually successful cognitive aging. *Journal of the International Neuropsychological Society: JINS*, *18*, 1081–1085. http://dx.doi.org/10.1017/S1355617712000847.

Hasher, L., Zacks, R. T., & May, C. P. (1999). Inhibitory control, circadian arousal, and age. In D. Gopher, & A. Koriat (Eds.), *Attention and performance XVII* (p. 653). Cambridge, MA: MIT Press.

Hedden, T., & Gabrieli, J. D. (2004). Insights into the ageing mind: A view from cognitive neuroscience. *Nature Reviews Neuroscience, 5*, 87–96.
Kahneman, D., & Tversky, A. (1979). Prospect theory: An analysis of decision under risk. *Econometrica, 47*, 263–291.
Krawczyk, D. C., Morrison, R. G., Viskontas, I., Holyoak, K. J., Chow, T. W., Mendez, M. F., et al. (2008). Distraction during relational reasoning: The role of prefrontal cortex in interference control. *Neuropsychologia, 46*, 2020–2032.
Laird, A. R., Eickhoff, S. B., Rottschy, C., Bzdok, D., Ray, K. L., & Fox, P. T. (2013). Networks of task co-activations. *NeuroImage, 80*, 505–514.
Luciana, M., & Nelson, C. A. (1998). The functional emergence of prefrontally-guided working memory systems in four-to eight-year-old children. *Neuropsychologia, 36*, 273–293.
de Macks, Z. A. O., Moor, B. G., Overgaauw, S., Güroğlu, B., Dahl, R. E., & Crone, E. A. (2011). Testosterone levels correspond with increased ventral striatum activation in response to monetary rewards in adolescents. *Developmental Cognitive Neuroscience, 1*, 506–516.
Madden, D. J., Pierce, T. W., & Allen, P. A. (1996). Adult age differences in the use of distractor homogeneity during visual search. *Psychology and Aging, 11*, 454.
Madden, D. J., Whiting, W. L., & Huettel, S. A. (2005). Age-related changes in neural activity during visual perception and attention. *Cognitive Neuroscience of Aging: Linking Cognitive and Cerebral Aging*, 157–185.
Mandler, J. M., & Robinson, C. A. (1978). Developmental changes in picture recognition. *Journal of Experimental Child Psychology, 26*, 122–136.
McNab, F., Zeidman, P., Rutledge, R. B., Smittenaar, P., Brown, H. R., Adams, R. A., et al. (2015). Age-related changes in working memory and the ability to ignore distraction. *Proceedings of the National Academy of Sciences, 112*, 6515–6518. http://dx.doi.org/10.1073/pnas.1504162112.
Morrison, R. G., Krawczyk, D. C., Holyoak, K. J., Hummel, J. E., Chow, T. W., Miller, B. L., et al. (2004). A neurocomputational model of analogical reasoning and its breakdown in frontotemporal lobar degeneration. *Journal of Cognitive Neuroscience, 16*, 260–271.
Nikitin, N. P., Loh, P. H., de Silva, R., Ghosh, J., Khaleva, O. Y., Goode, K., et al. (2006). Prognostic value of systolic mitral annular velocity measured with Doppler tissue imaging in patients with chronic heart failure caused by left ventricular systolic dysfunction. *Heart, 92*, 775–779.
Ornstein, P. A., Naus, M. J., & Liberty, C. (1975). Rehearsal and organizational processes in children's memory. *Child Development*, 818–830.
Paris, S. G., & Lindauer, B. K. (1976). The role of inference in children's comprehension and memory for sentences. *Cognitive Psychology, 8*, 217–227.
Park, D. C., Lodi-Smith, J., Drew, L., Haber, S., Hebrank, A., Bischof, G. N., et al. (2014). The impact of sustained engagement on cognitive function in older adults: The synapse project. *Psychological Science, 25*, 103–112. http://dx.doi.org/10.1177/0956797613499592.
Piaget, J. (1971). *Biology and knowledge: An essay on the relations between organic regulations and cognitive processes*.
Raghanti, M. A., Spurlock, L. B., Treichler, R. F., Weigel, E., Stimmelmayr, R., Butti, C., et al. (2015). An analysis of von Economo neurons in the cerebral cortex of cetaceans, artiodactyls, and perissodactyls. *Brain Structure and Function, 220*, 2303. http://dx.doi.org/10.1007/s00429-014-0792-y.
Raven, J. C. (1938). *Progressive matrices: A perceptual test of intelligence, 1938, sets A, B, C, D, and E*. London: H. K. Lewis.
Raven, J. C. (1960). *Standard progressive matrices: Sets A, B, C, D, & E*. London: H.K. Lewis & Co.
Reyna, V. F., & Brainerd, C. J. (2011). Dual processes in decision making and developmental neuroscience: A fuzzy-trace model. *Developmental Review, 31*, 180–206.
Reyna, V. F., & Farley, F. (2006). Risk and rationality in adolescent decision making: Implications for theory, practice, and public policy. *Psychological Science in the Public Interest, 7*, 1–44.
Reyna, V. F., Wilhelms, E. A., McCormick, M. J., & Weldon, R. B. (2015). Development of risky decision making: Fuzzy-trace theory and neurobiological perspectives. *Child Development Perspectives, 9*, 122–127.
Rogalski, E. J., Gefen, T., Shi, J., Samimi, M., Bigio, E., Weintraub, S., et al. (2013). Youthful memory capacity in old brains: Anatomic and genetic clues from the Northwestern SuperAging project. *Journal of Cognitive Neuroscience, 25*, 29–36.
Rypma, B., & D'Esposito, M. (2000). Isolating the neural mechanisms of age-related changes in human working memory. *Nature Neuroscience, 3*, 509–515.
Salthouse, T. A. (1994). The nature of the influence of speed on adult age differences in cognition. *Developmental Psychology, 30*(2), 240.
Salthouse, T. A. (1996). The processing-speed theory of adult age differences in cognition. *Psychological Review, 103*, 403–428.
Shepard, R. N. (1967). Recognition memory for words, sentences and pictures. *Journal of Verbal Learning and Verbal Behavior, 6*, 156–163.
Small, S. A., Chawla, M. K., Buonocore, M., Rapp, P. R., & Barnes, C. A. (2004). Imaging correlates of brain function in monkeys and rats isolates a hippocampal subregion differentially vulnerable to aging. *Proceedings of the National Academy of Sciences of the United States of America, 101*, 7181–7186.
Sternberg, R. J. (1977). *Intelligence, information processing and analogical reasoning: The componential analysis of human abilities*. Hillsdale, NJ: Erlbaum.
Thornton, T. L., & Gilden, D. L. (2007). Parallel and serial processes in visual search. *Psychological Review, 114*(1), 71.
Treisman, A. M., & Gelade, G. (1980). A feature-integration theory of attention. *Cognitive Psychology, 12*, 97–136.
Triarhou, L. C. (1998). Rate of neuronal fallout in a transsynaptic cerebellar model. *Brain Research Bulletin, 47*, 219–222.
Tversky, A., & Kahneman, D. (1985). The framing of decisions and the psychology of choice. In *Environmental impact assessment, technology assessment, and risk analysis* (pp. 107–129). Springer Berlin Heidelberg.
Vendetti, M. S., Matlen, B. J., Richland, L. E., & Bunge, S. A. (2015). Analogical reasoning in the classroom: Insights from cognitive science. *Mind, Brain, and Education, 9*, 100–106.
Viskontas, I. V., Morrison, R. G., Holyoak, K. J., Hummel, J. E., & Knowlton, B. J. (2004). Relational integration, inhibition and analogical reasoning in older adults. *Psychology and Aging, 19*, 581–591.
Wallis, L. J., Virányi, Z., Müller, C. A., Serisier, S., Huber, L., & Range, F. (2016). Aging effects on discrimination learning, logical reasoning and memory in pet dogs. *Age, 38*, 6. http://dx.doi.org/10.1007/s11357-015-9866-x.
Waltz, J. A., Knowlton, B. J., Holyoak, K. J., Boone, K. B., Mishkin, F. S., de Menezes Santos, M., et al. (1999). A system for relational reasoning in human prefrontal cortex. *Psychological Science, 10*, 119–125.
Wechsler, D., Kaplan, E., Fein, D., Kramer, J., Morris, R., Delis, D., et al. (2003). *Wechsler intelligence scale for children: Fourth edition (WISC-IV)*. [Assessment instrument]. San Antonio, TX: Pearson.
Wendelken, C., Ferrer, E., Whitaker, K. J., & Bunge, S. A. (2016). Fronto-parietal network reconfiguration supports the development of reasoning ability. *Cerebral Cortex, 26*, 2178–2190.
Yankner, B. A., Lu, T., & Loerch, P. (2008). The aging brain. *Annual Review of Pathology: Mechanisms of Disease, 3*, 41–66.

Further Reading

Betik, A. C., & Hepple, R. T. (2008). Determinants of VO2 max decline with aging: An integrated perspective. *Applied Physiology, Nutrition, and Metabolism, 33*, 130–140. http://dx.doi.org/10.1139/H07-174.

CHAPTER

7

Disorders of Reasoning

KEY THEMES

- Disorders of reasoning can involve numerous brain systems and several cognitive functions.
- Exogenous attention encompasses our involuntary orienting responses toward stimuli in the environment. Deficits of exogenous attention relevant to reasoning include attentional neglect, which is a failure to orient toward a particular location.
- Endogenous attention allows us to voluntarily focus. This type of attention is mediated by the frontal lobes. Deficits of endogenous attention can impact visuospatial reasoning.
- Working memory is commonly involved in reasoning about complex information when we must hold different types of information in mind to solve a problem.
- Long-term memory deficits can lead to inabilities to visualize future situations.
- Relational reasoning can be impaired when the frontal lobes are damaged. This condition can also lead to failures of inhibitory control, which also impact reasoning performance.
- Deficits in theory of mind lead to difficulties in reasoning with other people.
- The medial frontal cortex is important for helping us to avoid making risky financial decisions.
- Moral reasoning develops over the course of childhood and depends upon intact prefrontal cortex (PFC) for advanced reasoning to develop.

Reasoning
http://dx.doi.org/10.1016/B978-0-12-809285-9.00007-7

OUTLINE

REASONING DISORDERS

Introduction

In this chapter we turn to a topic that touches upon all of the aspects of reasoning that this book has covered so far, what happens when someone no longer exercises the ability to reason as effectively as they once had? A related question is what can be done about it? The answers to these questions are challenging, as the basis for the disorder must first be isolated before an intervention can be attempted to remediate reasoning. Isolating the problem can be complex, as different disruptions within the brain can compromise reasoning and different deficits in cognition can be involved. To better understand disordered reasoning, we must grapple with identifying the cause of the disorder. We will then need to understand what other cognitive abilities may be impacted by the disorder. Lastly, to address possible remedies, we will need to understand the capabilities of the individual affected into the future.

If someone has experienced a brain injury that is impacting his perceptual ability, which in turn reduces his ability to reason about space in the environment, there is a good chance that these skills can be recovered through targeted rehabilitation. Alternatively, if someone has been diagnosed with late stage dementia, there is almost surely a reduced chance for recovery of the lost cognitive ability, as the loss of neural networks in dementias is permanent and progressive. In this case it would be appropriate to consider alternative options, such as new training in strategies that may help to compensate for the loss in reasoning function.

This chapter will cover a range of challenges, circumstances, and examples relating to what happens when disorders of reasoning are experienced. We will begin with discussing the cognitive abilities that are necessary for intact reasoning and how disruption of these abilities can negatively impact performance. We will then discuss the impact of neurological injuries and how they can affect reasoning. This will be followed by a discussion of psychiatric conditions and how these can lead to disorders of thinking. We will then conclude by a further examination of specific types of reasoning and how they can be impacted by disorders.

This chapter is complementary to Chapter 3 on the neuroscience of reasoning. In that chapter we primarily reviewed the types of neurological recordings that can be obtained and how the results of such methods as electroencephalograms (EEG) and functional magnetic resonance imaging (fMRI) can tell us about the reasoning abilities of people solving problems within imaging environments. To make sense of these measures of brain activity, most investigators will attempt to understand how they are correlated with behavior. This means that while someone engages in reasoning of one form or another, activity from the device or recording method is gathered. This process can yield valuable information, but correlational measures do not allow us to draw clear cause-and-effect relationships. Determining causes is often strengthened by situations in which we can actively intervene on the reasoning process and change it. Transcranial magnetic stimulation offered some examples of cases in which relational reasoning, or discovering abstract correspondences among shapes, can be sped up by facilitatory magnetic stimulation. Examining the effect of disorders on reasoning offers an even stronger method for causal inference. If a brain area has been damaged or the neurotransmitter system depleted, then reliable effects can be observed. The observation of correlated behavioral changes and disrupted neural circuitry allows us to gain more confidence that the particular brain region or neurotransmitter in question is more strongly involved in governing a particular process, rather than just being active alongside that process.

An example can be illustrated with relational reasoning and the frontal lobes. We discussed the Raven's progressive matrices task in Chapter 3 (Fig. 7.1). This task is typically considered to measure fluid reasoning ability. Greater fMRI activation of the parietal cortex and prefrontal cortex (PFC) has been observed as relational information loads increase in Raven's matrices problems (Prabhakaran, Smith, Desmond, Glover, & Gabrieli, 1997). This coupling of increasing relational complexity with increased fMRI-based activity suggests that there is an association present. This co-occurrence may suggest that parietal cortex and PFC accomplish the processing of relational information. This is probably too strong an inference in some cases based on the methodology. There may be a direct association present, in which the demand to process relational information in the Raven's matrix problem leads directly to increased activity within these brain areas in order to accomplish the task. Alternatively one can imagine a situation in which neural networks within the parietal cortex actually process the visual and spatial inputs from the problem in question, while the PFC neurons activate in order to signal the presence of a task that requires attention. If this were the case, then the parietal cortex may be much more directly involved in the processing of relations, while the PFC has more of an indirect role. The true answer to this question remains elusive, based mostly on the limitations of current methodologies to actually read neural codes from thousands of cells and make sense of the information that they process at any given time. Also, we can rarely derive clear answers from one method alone when several cognitive processes are needed to solve a problem and are accompanied simultaneously by several active brain areas.

We can turn to a neuropsychological approach, in which we study an individual or multiple individuals who have damage restricted to a particular area of interest to better understand relational processing and how it is accomplished within the brain. Such individuals enable us to test whether the person can perform a Raven's progressive matrices task accurately. If they are able to accomplish the task, it would suggest that the damaged area is not essential for relational reasoning. Such a situation is possible when examining individuals who suffer from frontotemporal lobar degeneration (FTLD), which is a form of dementia. In FTLD both the frontal and temporal lobes become progressively atrophied over time due to neuronal loss. This neuronal loss can selectively occur and be isolated to the temporal

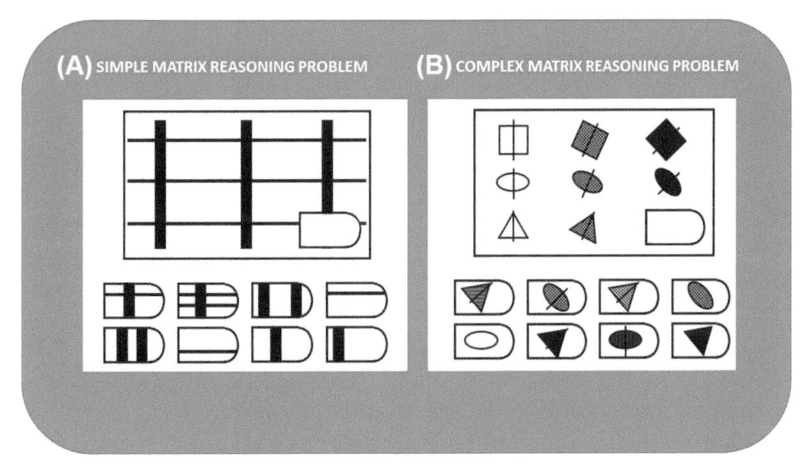

FIGURE 7.1 A sample image of some problems similar to those presented in the Raven's progressive matrices task. (A) A simple problem that involves visual matching. (B) A more challenging problem that requires the thinker to integrate multiple relations toward forming a solution.

lobes or frontal lobes in some patients. My colleagues and I had the opportunity to test frontal-variant FTLD patients, who suffered from diffuse and widespread damage to their frontal lobes, with little disruption to other areas of their brains. This rather unique population enabled us to test the hypothesis that the PFC was essential to processing relational information. Other individuals who have temporal-variant FTLD have cortical degeneration that is largely limited exclusively to the temporal lobes. Comparing frontal-variant to temporal-variant FTLD patients on reasoning performance allows for a stronger test of the PFC hypothesis. Through this research, we discovered that frontal-variant patients showed much more pronounced difficulties on relational reasoning tasks compared to temporal-variant FTLD patients. People could rarely reason effectively about relational information when damage to the PFC was present. Conversely, damage to the temporal lobes did not strongly affect relational reasoning ability. Such a situation provides researchers with greater leverage to infer a causal connection between the PFC and relational reasoning, rather than merely a correlational association. In this situation we must still dig deeper with further research in order to reveal the specific cognitive operations that the PFC is responsible for, but we can be more confident that neurons within this brain area are central to accomplishing the task.

COGNITIVE FACTORS IN REASONING DISORDERS

The concept that a single brain region is fully responsible for a reasoning task is no longer widely accepted within the scientific community. This stems from a broad convergence of findings indicating that neural networks participate in numerous functions. Reasoning tasks are also complex and typically require several mental operations to occur together, again making it likely that several areas of the brain will be important for the task to be carried out. In this section of the chapter we will discuss the cognitive factors that are important for reasoning. We need to appreciate the fact that disruption of one or more subprocesses that contribute to reasoning can in effect compromise overall reasoning performance in order to better understand the links between brain systems and reasoning disorders.

Disorders of Attention

Attention is central to many of the cognitive tasks that people engage in. This is the case for both involuntary tasks, such as detecting a change in the environment, and for voluntary tasks, such as choosing to pay attention to this chapter and ignore music that may be playing in the background. If someone has a reduced capacity for attention they will likely struggle with numerous types of tasks including multiple types of reasoning tasks. Before we discuss disorders of attention and how they can impair reasoning, it is first necessary to consider some critical distinctions among types of attention and how attention can be allocated in order to accomplish tasks effectively.

One of the first distinctions that we should make is that there is a difference between voluntary and involuntary attention. These categories can be differentiated at both qualitative and cognitive levels. The two categories are also supported by different brain systems. These two types of attention can work together, but impairment in either one of them will have differential effects on our cognition and reasoning.

The first system is known as the *exogenous* attention system. The exogenous system controls our experience of involuntary attention. For example, while I write this chapter I am sitting in a coffee shop with some ambient noise consisting of some folksy guitar music and people talking at the counter and nearby tables. Moments ago a rather loud refrigeration device kicked on and I became aware of this new sound. I am listening to the steady hum of a refrigerator as I write. I can tune this sound out relatively easily, but when its sound changes in volume or pitch I cannot help but notice the change. The change in sound grabs my attention forcing me to briefly attend to its source. Exogenous attention also applies when someone yells out, drops a tray of food, or unexpectedly lets out a loud laugh that cuts through the surrounding ambient noise. The ability to covertly monitor our environmental surroundings without demanding a lot of our active mental resources is extremely helpful and has high survival value. If a moving vehicle, or large animal, is unexpectedly bearing down on us, this involuntary ability to monitor the environment enables us to quickly orient toward it and execute an appropriate evasive action. We rely on our exogenous attention to keep us safe. The exogenous attention system is supported by areas of the parietal, temporal, and occipital lobes (refer to Fig. 7.2). Many of these same regions are also active in processing the visual features of objects and their locations in space. The parietal lobes, along with areas within the temporal lobes, help us to track moving objects. The exogenous attention system is also referred to as being the posterior attention system due to its neural support regions being stacked toward the back of the brain (Posner, Petersen, Fox, & Raichle, 1988).

Our second attention system is termed the *endogenous* attention system and controls our capacity to voluntarily direct our attention toward environmental stimuli. This system is the one that most of us call to mind when we hear the words "pay attention." The exogenous attention

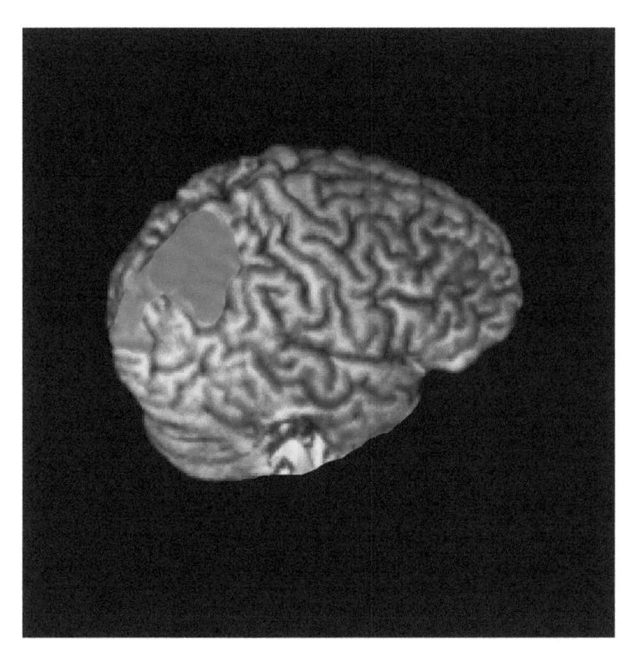

FIGURE 7.2 The exogenous attention system is comprised of several posterior brain areas. These include the parietal cortex and the occipital cortex.

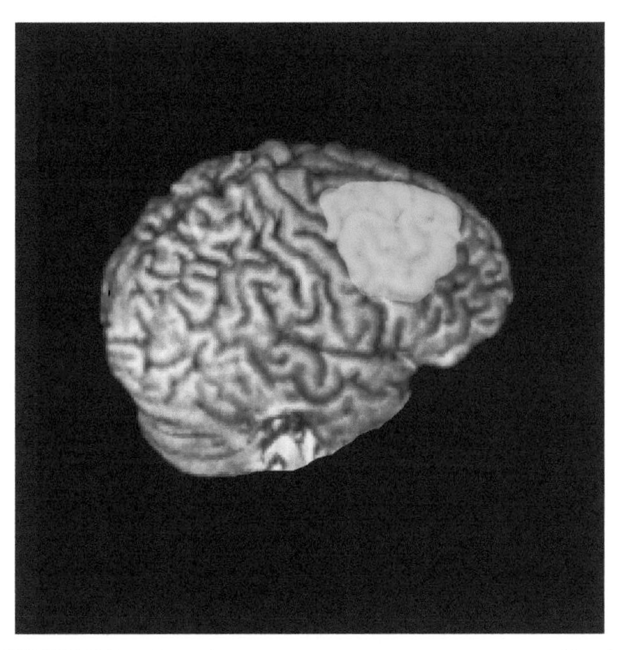

FIGURE 7.3 The endogenous attention system is supported by the frontal lobes.

system is able to be consciously directed to take in certain sights, sounds, smells, and other inputs in our environment. You use your exogenous attention system when you are riding on a subway in an unfamiliar city and are relying upon the driver to announce the current stop over the loudspeaker. In my experience drivers tend to mumble the stops rather quickly and far too softly, frequently through a crackling audio system. In such cases, you can find yourself actively straining with the effort of trying to attend to the loudspeaker announcement. Your directed attention also allows you to amplify certain sounds and tune out conversations that are going on around you, along with the mechanical noises made by the train. This system can also enhance visual information. For instance, if you are meeting a new acquaintance at a restaurant and he has said he will be wearing a blue shirt, you will probably find yourself overly focusing on everyone's shirt color, gender, and whether they appear to be looking for someone, as they enter the establishment. This voluntary sense of attention is primarily associated with the frontal lobes (Fig. 7.3). The endogenous system has also been referred to as the anterior attention system because of this anatomical focus (Posner et al., 1988).

Another striking feature of both the exogenous and endogenous attention systems is that they are limited in capacity. This feature becomes clear when you consider your ability to monitor sights and sounds within a complicated city environment, such as during a street festival. Such events may include music, yelling, laughing, animated conversations, as well as numerous sights and smells. We may fail to notice things going on around us when we are faced with these types of conditions. Similarly, there is a limit on how much voluntary attention we can apply toward a particular situation. In my subway audio system example, we are simply unable to separate the signal from the noise, and we will miss things when we are faced with high amounts of ambient sounds and a low degree of volume from the speaker. The capacity limitation on attention can reduce one's reasoning ability even in young and healthy individuals. Disorders that reduce attention further can lead to reductions in reasoning ability even under quieter and less complicated situations. Any distraction in the form of unregulated sensory inputs can lead a thinker to become distracted.

The exogenous, or posterior attention system, is reduced by damage to the parietal, temporal, or occipital lobes. This can occur in cases including strokes and traumatic brain injuries. A stroke occurs when the wall of a blood vessel ruptures within the brain. A stroke causes damage by two mechanisms. First, the regions that normally gain their nutrition, oxygen, and support from the ruptured vessel will lose their support due to the blood supply being eliminated. Second, blood will spill into the cortical area surrounding the site of the stroke and blood is toxic to neurons. This will further damage the areas proximal to the rupturing of the vessel. Trauma to the head through blunt force, or exposure to pressure (such as that from a blast wave) can also damage areas of the brain by bruising the cortex and possibly also stretching and shearing the axonal tracts of the brain. In either case, areas toward the posterior of the brain can be impacted and reduce one's exogenous attention system capacity.

Damage to the parietal cortex may result in difficulties locating objects in space. A particularly memorable example comes from neurological patient V. K. whose experience was described by Mel Goodale et al. (Jakobson, Archibald, Carey, & Goodale, 1991). V. K. experienced a stroke affecting the parietal cortex. She lost her ability to grasp items and determine where things were located in space. After the stroke, V. K. was also unable to appropriately orient her hand and adjust her grip appropriately in order to pick up objects, such as a small block of wood. Similarly, damage to the parietal cortex can also cause a disorder called neglect. Neglect is a form of attention deficit in which someone is no longer able to monitor certain areas of space. For instance, an individual with neglect will be unable to describe what is on one side of space. Neglect is typically caused by a stroke within one hemisphere and therefore results in one side of space being perceived and attended to intact, while the other hemisphere is unable to be attended. This condition is called unilateral neglect or hemispatial neglect.

Attentional neglect not only affects perception, but also the ability to visualize space associated with the affected areas. In other words, neglect causes a deficit in the mental representations associated with attention. This is known as *representational neglect* and was memorably illustrated by Italian researchers Bisiach and Luzzatti (1978), who described the mental imagery visualization behavior of two patients who had sustained strokes. The patients were asked to visualize and describe the famous city center Piazza Del Duomo in Milan (Fig. 7.4). Both patients had been to this location many times and knew it well. When asked to describe the Piazza, or square from the vantage point of the steps of the Duomo looking out, the patients described buildings that would have been on the left side of space (and north side of the Piazza) neglecting the buildings on the right side the square including the very famous Galleria entrance (Fig. 7.4). When asked to visualize the Piazza from the vantage point of looking at the Duomo (or cathedral) (Fig. 7.4), the same patients again neglected the right side of space, which was now populated in their mental images by all of the sites that they had just mentioned on the north side of the Piazza. From the viewpoint looking at the Duomo, the patients now listed off all of the other sites on the south side. Remarkable behavior after simply visualizing a switch of perspective! This study indicates how strongly linked attention is to our internal mental representations of space. These same processes are critical in visuospatial reasoning.

Spatial deficits in exogenous attention can affect visuospatial reasoning ability. One of the well-studied reasoning tasks is the Raven's matrices (Raven, 1938, 1960). We discussed this task extensively in Chapter 3. When people solve Raven's matrices, they rely heavily on visual and spatial processing. Freedman and Dexter (1991) reported on reasoning ability in individuals who had right hemisphere visual neglect, which had been acquired from either a stroke or the surgical removal of a brain tumor. These patients exhibited difficulties on the Raven's matrices task

FIGURE 7.4 Images of the two viewpoints in the Piazza del Duomo in Milan, Italy. Patients with damage to the left hemisphere exhibited representational neglect. When asked to imagine viewing the square from two different perspectives, the patients consistently named only landmarks that would have appeared on their left side of space.

relative to unimpaired controls, missing approximately 33% more problems than the control participants. Additionally, in this study individuals with cortical damage sustained from dementia showed similar attentional neglect symptoms. The dementia participants matched the performance of the right-hemisphere neglect patients showing reasoning deficits occurring at a nearly identical level. Perceiving intact representations of visuospatial reasoning problems is a necessary first step toward solving the problems. The ability to attend to the representations must also be possible or patients will show pronounced deficits in reasoning ability.

Problems with endogenous, or voluntary, attention can also impair reasoning ability. Tim Shallice (1982) had proposed that the PFC supports a supervisory attention system. This system was proposed to enable people to exert volitional control over task performance. Effective supervisory control enables someone to voluntarily disengage from one task and subsequently engage in another task. Shallice conceptualized failures of supervisory attention as a major determinant of whether someone is able to remain focused on a task and see it to completion. Patients who have injuries to the PFC from either strokes or brain injuries can be thought of as suffering from a disorder of the supervisory attention system.

French researchers P. H. Robert et al. (1997) studied the effects of supervisory attention on a sequence reasoning task comparing healthy individuals to those with a diagnosis of schizophrenia and difficulties exerting supervisory control. Robert et al. developed a verbal sequence reasoning task that required participants to rearrange sets of six words into complete sentences. The words were presented on a set of cards that was always given to the participant in a jumbled order. The first two words provided shared a strong semantic association. In "valid conditions" the participants used these first two words to arrange the cards into a sensible completion to the sentence, while the "invalid conditions" required the participant to first ignore the association of the first two words in order to successfully rearrange the cards into a sentence not containing those words in the order in which they had been presented.

Participants with schizophrenia performed lower than the healthy participants in both valid and invalid conditions. These participants with schizophrenia formed fewer correct sentences in invalid conditions than healthy individuals exhibiting a reduced ability to disengage (or inhibit) the strong semantic association in order to reengage attention toward forming a sentence that lacked this association. Robert et al. (1997) interpreted these results to indicate a verbal reasoning deficit in schizophrenia linked to a failure of the supervisory attention system.

Working Memory Disruptions and Reasoning

Working memory is another core cognitive ability that has a major impact on reasoning. The term working memory originated with Pribram, Ahumada, Hartog, and Roos (1964), who proposed the term to describe random access memory (RAM) in a computer system in the context of artificial intelligence. A classic example of working memory is the once common situation of having to dial an unfamiliar phone number. In the time before smartphones, one would have to physically look up the number in a hard copy telephone book and encode it into a form of transient memory that could be sustained by rehearsal, but would otherwise be lost quickly. Such a number could then be quickly dropped from memory simply by moving the focus of attention to the phone call itself (Fig. 7.5). This example highlights some important features of working memory. It is flexible, it can be maintained as long as the information continues to be rehearsed, it is quite limited in storage capacity, and can be easily overwritten by stopping the rehearsal and focusing attention elsewhere.

The concept of working memory derives from two sources, cognitive psychology and electrophysiology. In the 1960s, cognitive psychologists focused on identifying the differences between short-term memory and

FIGURE 7.5 A classic example of a working memory task involves looking up a telephone number in a printed telephone book and holding it in mind through rehearsal for a brief period of time. The number can promptly be discarded from the working memory buffer after it has been dialed into the phone. *From Wikimedia Commons.*

long-term memory. They conceptualized short-term memory to be dramatically shorter in duration. The contents of short-term memory were also believed to be more susceptible to disruption by interference from new or extraneous items. While long-term memory is thought to be permanent, possibly lasting over the course of the lifespan, short-term memory lasts only 20–30 s unless it is continually rehearsed. Neurophysiological evidence for short-term information storage was found in the patterns of action potential firing rates in frontal lobe neurons in nonhuman primates (Fuster & Alexander, 1971). These elevated neuronal firing responses were recorded while items were being stored over a delay period between an encoding phase and a subsequent memory test. Soon evidence continued to accrue from both cognitive psychology and electrophysiology that a distinct short-term form of memory existed separate from our longer-term knowledge store.

As research continued through the 1970s and 1980s, researchers began to conduct experiments varying the format of information being held over the short-term yielding a better understanding of this form of memory. Many of these studies used a *dual-task* method, in which two sources of information needed to be held at the same time. This was very difficult for people to accomplish if the two information streams to be held in working memory were both presented in the same format (e.g., verbal or spatial). When two verbal or two spatial sets are loaded into memory together, they tend to cause interference and lead to maintenance failures. Interestingly, two streams of information can be maintained relatively easily if they vary in their format. Verbal and spatial information can be maintained side-by-side with little loss of information from either stream. Visual and verbal information can also share working memory resources without much difficulty. Alan Baddeley (1986) advanced a theory maintaining that working memory consists of multiple components that performed different operations. In accordance with the dual-task interference results, Baddeley proposed that working memory includes both a *verbal buffer* and a separate *visuospatial buffer*. Each buffer could be loaded according to the format of the stimuli without interfering with the operations of one another. Coordination of these buffers was proposed to be accomplished by a *central executive*, a limited capacity routing system that could move incoming information into and out of the buffers. This central executive is roughly similar to endogenous attention, as its operations depended upon the direction of voluntary attention. Furthermore, the central executive was proposed to facilitate switching between contexts (Fig. 7.6). This construct in the Baddeley working memory model corresponds approximately to the supervisory attention system that had been proposed by Shallice (1982).

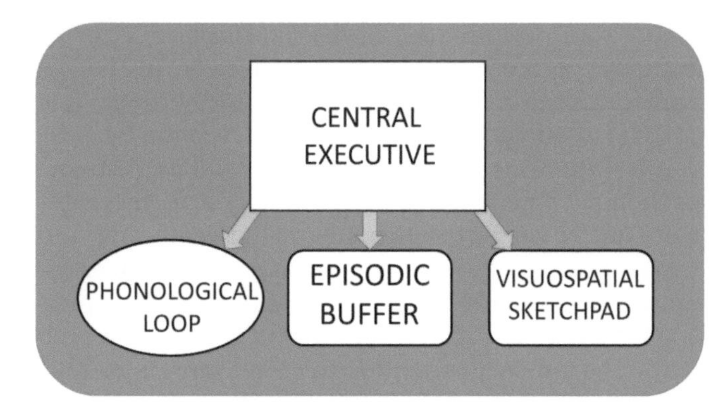

FIGURE 7.6 Alan Baddeley proposed four components in working memory. The verbal buffer maintains auditory and linguistic information. The visuospatial buffer is loaded by visual images and spatial sequences. These buffers are coordinated by a central executive that moves resources between the buffers and maintains or moves the focus of attention. An additional buffer was added in 2001 called the episodic buffer for temporary consideration of episodes of our lives.

An illustrative example of a central executive task is the verbal trails test (Baddeley, Chincotta, & Adlam, 2001). In a verbal trails test, a person has to maintain two sequences of information and must also alternate between them. For example, the first sequence of information may be the days of the week and the second sequence the months of the year. To complete this task effectively one would have to say aloud "January, Sunday, February, Monday, March, Tuesday, etc." with one word being said every second. The task becomes very demanding due to the need to maintain an alternating sequence back and forth at this brisk pace. This movement of focus from one type of information to another is one of the major features of central executive working memory operation. Most people can accomplish little else when they are engaged in performing the verbal trails task, as it loads heavily on the endogenous attention resources of the central executive. Reasoning may be thought to rely heavily upon working memory operations. Central executive operations such as switching, coordinating information, and manipulating information may be especially important for reasoning. Mathematical reasoning relies upon the ability to hold and manipulate numerical information in working memory, while verbal analogies or syllogistic reasoning (a form of deduction) relies more heavily on verbal maintenance processes.

A representative study investigating the relationship between working memory and reasoning was carried out by Süß, Oberauer, Wittmann, Wilhelm, and Schulze (2002) in Manheim, Germany. In this study, the investigators tested people on a relatively large battery of working memory tasks including both maintenance and manipulation tasks, such as performing a backward digit span. In this test, participants have to maintain a series of numbers, and these have to be repeated back in reverse ordering after the series has been presented. Such a task requires both maintenance of the incoming information being presented and the need to reorder the digits. Reordering will strongly involve central executive working memory resources. Other working memory tasks in the battery probed content-specific domains, such as verbal maintenance or arithmetic ability. The Daneman and Carpenter (1980) verbal working memory span test is an example of a task that imposes joint requirements on the individual. In this task, participants must judge the validity of sentences in a sequence and remember the final word of each sentence. Verbal maintenance is involved, but there is also a strong central executive demand to switch between evaluations of sentence material with the need to maintain individual words presented at the end of each sentence. Computation span is a similar task requiring solving subtraction and addition problems and then maintaining each outcome in order for a later test (Babcock & Salthouse, 1990). To fully tax central executive resources in working memory, Su¨ß et al. (2002) also included a dual-task condition requiring the coordination of two tasks being performed in parallel. In this study, a verbal and a math span tasks were used for this purpose requiring participants to maintain numbers while having to focus on solving arithmetic problems or maintain words while having to focus on categorizing other words.

Süß et al. (2002) also tested participants on a battery of reasoning tasks categorized into three domains. A *numerical reasoning domain* included tests such as solving mathematical text problems, equations, using frequency tables, and filling in numbers that followed specific sequences. A *figural reasoning* domain consisted of the following:

1. Analogies presented in the format "A is to B as C is to D" with figural patterns that needed to be completed.
2. A series of abstract drawings that had to be completed according to rules was included.
3. Classification of geometrical patterns based on prior examples.
4. Assembly of a geometric picture by rearranging parts.
5. A test in which two-dimensional images needed to be matched to their three-dimensional match provided that they were folded.

A *verbal reasoning domain* was also included. Tasks within this domain included the following:

1. Verbal word analogies.
2. Classifying sentences as facts or opinions.
3. Solving syllogisms that contained arbitrary premises (see Chapter 9 for more on this domain).
4. Solving syllogisms that contained real-world premises.
5. A test of word knowledge in which participants had to select oddball words that did not fit in a set.

Findings from the Süß et al. (2002) study revealed that working memory was highly related to reasoning. The reasoning domains had the strongest associations to working memory tasks overall amid other constructs that were tested including creativity and memory. The researchers had selected working memory and reasoning tasks that were unlikely to overlap significantly in terms of cognitive processes that they involve, so the strong associations were all the more impressive. Additionally, the working memory construct was globally related to reasoning ability, rather than being related to a specific subcomponent process or processes. In other words, verbal, numerical, and spatial domains were all strongly related to reasoning, as were dual-task studies. In relation to the Baddeley (1986) model of working memory, this set of findings indicates that the central executive, visuospatial buffer, and verbal buffer are all strongly related to reasoning ability and the specific format of the information present in the reasoning tasks does not strongly impact the association. Süß et al. suggested that reasoning is unlikely to depend upon any particular operation in working memory based on this pattern of associations.

Working memory deficits may occur in a variety of neurological and psychiatric conditions and can impair reasoning ability due to their strong linkage. Shelley Channon (1997) reported a clear example of this situation. In this study, Channon tested individuals with Parkinson disease on a working memory updating task and variations of deductive reasoning tasks. These participants were compared to healthy control participants. In the working memory task, individuals had to report the final six items of a series of randomly ordered consonants. Strings of items could consist of as many as 12 items requiring rapid updating within working memory. The results indicated that the group with Parkinson disease showed deficits in working memory updating relative to the control group.

The individuals with Parkinson's disease also showed reductions in deductive reasoning on a task that required hypothesis formation and testing out rules about variations in letter stimuli (refer to Fig. 7.7). Each trial consisted of two letters presented. The letters could vary in numbers of relational changes needed to understand their relationship on dimensions including size, color, letter, and sequence. Over the course of three trials, participants were shown two letters varying in dimensions and then given feedback about which was the correct letter. After three trials, participants would have had sufficient information to deduce the correct answer. The three trial sets were presented both in a serial fashion which would require maintaining prior information and in a parallel presentation, which reduced the working memory burden of the participant and allowed them to test out several types of possible deductive rules in order to understand the problem.

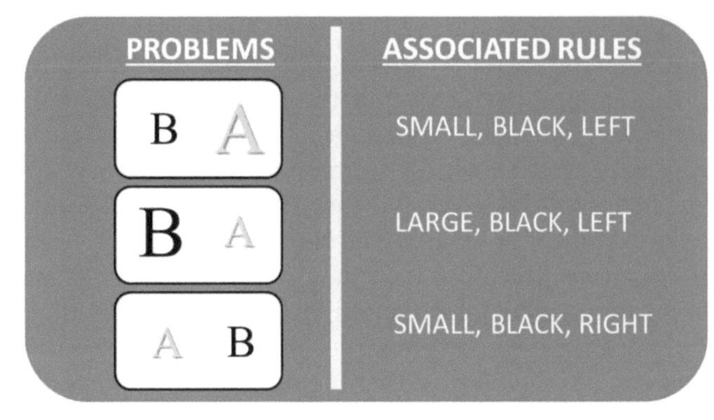

FIGURE 7.7 A deductive reasoning task appearing in the study by Channon (1997). Two letters were presented on each trial. The letters could vary in number of relational changes needed to understand their relationship on dimensions including size, color, letter, and sequence.

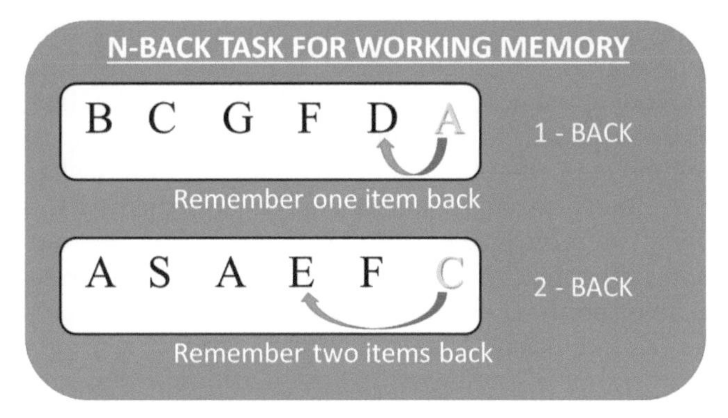

FIGURE 7.8 The n-back task is a demanding working memory task. In a 1-back task participants state whether each new letter is the same or different from the letter one position back. In a 2-back task this judgment of sameness is applied to a letter two back in the stream. This task becomes more complex in terms of maintenance and updating as the number of letters back increases.

In both problem formats individuals with Parkinson's disease showed deficits in reasoning compared to controls, with greater errors on the problems that had higher number of dimensions of variation possible. These results suggest that while reasoning and working memory remain separate constructs, they are interrelated with impairments in one affecting or co-occurring with impairments in the other.

Christina Fales et al. (2003) reported additional evidence of the links between disorders of working memory and reasoning. This study was an evaluation of working memory and relational reasoning in individuals diagnosed with Klinefelter syndrome, which is a condition involving a chromosomal deficit. Klinefelter syndrome can lead to mild cognitive impairments (MCIs) and verbal difficulties. The individuals with Klinefelter syndrome in this study performed more poorly on a working memory task known as the *n-back task*. The items in an n-back task can vary from an one-back condition to as many as three-back. In an one-back condition a series of letters are presented with a requirement that participants respond to each new letter by naming the previous letter "one back" (refer to Fig. 7.8). Likewise, in a two-back trial participants would have to respond to each letter in a sequence by repeating the item that was two items back. The working memory processing load continues to increase in a three-back condition, in which participants now must maintain a sequence of three items in a series while updating by dropping off the first item and retaining the latest item. The n-back task is challenging to interpret from a pure working memory operations perspective, as it involves numerous verbal maintenance, updating, and inhibitory control aspects of working memory. It remains one of the best examples of a task that can consistently tax working memory in an incrementally increasing fashion.

The individuals with Klinefelter syndrome also showed deficits on a transitive inference relational reasoning task requiring them to track the names of individuals and their relative heights, such as "Abe is taller than Bill, Bill is taller than Carl." These premises allow one to conclude that "Abe is taller than Carl" by making a relational comparison from each of these individuals to Bill. Participants were presented with cards having the names of the individuals printed on them and a correct response involved the positioning of the cards in the order "Abe, Bill, Carl" according to their respective heights. The individuals relationships in height were either presented in order as described in the example, but were also described out of order on other trials. These "out of order" presentations become more difficult from a relational reasoning perspective. For example, if problem stated that "Bob is taller than Jim" and "John is taller than Bob", this problem becomes more difficult, as one cannot use a simple "chaining strategy" that orders the individuals according to the sequence presented. Participants with Klinefelter syndrome performed at a lower level than controls on both ordered and nonordered problems. Interestingly, Fales et al. (2003) also tested the participant groups on Raven's matrices, another relational reasoning task with no differences being observed between the groups. In this study, Fales et al. presented evidence that working memory for letters and relational problems that are presented in a verbal format showed a similar impairment. Meanwhile visuospatial reasoning, as carried out in the Raven's matrices task was preserved. This study indicates that in some cases working memory and reasoning can be jointly dependent upon a particular format of information. These results are also consistent with the division between visuospatial and verbal working memory that Baddeley (1986) had proposed, in which the format of information will be cycled through one of two separate buffers.

The Effects of Long-Term Memory Loss on Reasoning Ability

A second important aspect of reasoning involves relying upon prior knowledge that can be called up from long-term memory. While working memory is brief in duration and can be disrupted by interference from incoming information, long-term memories tend to be highly resilient to degradation. One of the most rapidly available and successful ways to solve problems is to implement previous successful solutions. Such solutions are stored in long-term memory. These memories can be modified by the influences of further experience. The ability to use long-term memories to apply to new problems depends primarily upon the availability of the memory. Some long-term memories are more robust than others, particularly those that have been retrieved more often and over a variety of contexts. Long-term memories are made more, or less, accessible based on the number of cues that they are associated with. This may be the feature of long-term memories that most impacts their ability to be used to solve problems. If a novel problem presents itself, it is necessary to be able to access a prior successful solution or strategy that would be relevant.

An example of long-term memory being helpful for problem solving can be found when using computers. When we become temporarily familiar with using a particular computer program, we often learn shortcuts in order to move text, insert images, and save work. If we don't use these tricks for several months, we can find ourselves in a bit of a jam. My own experience with this challenge is all too memorable. It is very frustrating when you know how you would like to achieve an outcome in computing, but the means to do so is no longer available or obvious. In these cases, I typically find myself having to look up older work and try to reconstruct what I had done. This situation often has a bright side as we are able to more readily find and implement a previously successful solution if we can uncover enough cues to bring back the memories. This strategy often leads to more rapid solutions than having to invent a new method from scratch.

Amnesia is a condition in which people have lost the ability to access long-term memories. Juskenaite et al. (2014) investigated the effects of amnesia on mental imagery for solutions to problems from both the past and the future. Participants in this study suffered from a condition known as transient global amnesia. As the name implies, this is a neurological syndrome that onsets suddenly and is characterized by dense episodic memory impairments lasting up to 24h. This disorder provides an interesting isolated window into the effects of long-term memory loss on reasoning, as executive functions remain largely intact in individuals experiencing this form of amnesia. In the study, Juskenaite et al. asked participants to project themselves into their own future and past by reporting two different personal events for each of four different time periods: the previous 5years (except for the last calendar year), the previous year, the next year, and the next 5years (except for the next calendar year). Individuals with transient global amnesia reported fewer events for their future and past than unimpaired controls and were often unable to provide any description for these projected or remembered events. This result indicates that not only were the participants with transient global amnesia unable to remember long-term memories, they also struggled to project likely future events.

In a second study, the experimenters introduced an interesting modification by prompting the participants with a common everyday scenario that they were asked to provide an imagined set of details about. For example, participants were presented with the statements "you will be invited to celebrate a retirement" and "you will go to visit a friend in Paris." After participants read these statements, they tried to imagine the event occurring at a specific time and place in their daily lives. As in their first experiment, Juskenaite et al. (2014) asked the participants to describe these events within varying time frames (from 1year to 5years) into the future and past. In these cases, with details to work with, the participants with transient global amnesia were much more successful at both describing details from their past and in producing simulated details about how autobiographical scenarios would likely play out in the near future.

The results of the study by Juskenaite et al. (2014) suggest that long-term memory impairments influence reasoning about the future based on events in the past. If there is a selective long-term memory impairment, there can be value in prompting an amnesic person by naming a common scenario from daily life. When prompted to go to the hospital, for example, the participants were able to reason about the future by indicating that they would need to take health-related documents and other belongings in preparation for a hospital stay. Likewise, these prompts enhanced details about the past as well. One of the critical aspects of long-term memory that makes it effective for reasoning about the future is the ability to retrieve details. While transient global amnesic patients cannot effectively self-generate ideas, they appear to benefit from memory cues.

SOCIAL DEFICITS AND REASONING

Theory of Mind Deficits

Theory of mind refers to the awareness of the fact that other individuals have intentions. This ability enables much of our social reasoning and serves as a gateway skill toward reading others emotions (we discuss this concept further in Chapter 12). The level of awareness of intentions can vary from people who are a bit socially clueless or oblivious to those who are highly intuitive and can easily predict others reactions and relate to them based on what they are likely thinking. Theory of mind is also referred to as mentalization, which captures the idea that we can simulate what another individual may be thinking. The term theory of mind research originally emerged from chimpanzee research when investigators asked the question of whether chimpanzees possess the ability to attribute mental states to other individuals. If you remember some of the reasoning skills we discussed in Chapter 3 regarding Sarah the chimpanzee that was trained in symbolic language by David Premack's research group in the 1970s and 1980s, you may be tempted to conclude that chimpanzees do have theory of mind. Recall that one of the chimps, Gussie was able to withhold giving clues about where food was if she was presented with the "bad trainer," an individual who had a history of failing to share food with Gussie when she clued him in on the location of hidden food. This behavior suggests that Gussie was able to notice that this particular trainer was becoming aware of the hidden food, and she was then able to predict that he would likely keep it for himself (Premack & Premack, 1983).

Theory of mind has also been heavily investigated in children. Young children tend to have difficulties with predicting what other people will be likely to do. They tend to be egocentric, meaning they will assume that everyone is fully aware of the same information that they are, as we discussed in Chapter 5. As we age, people tend to acquire more insight into the thoughts and feelings of other individuals and are then able to make more accurate predictions about what is likely to occur in another person's mind during a particular situation.

Clear cases of individuals who have deficits in theory of mind are those on the autism spectrum. Such individuals are often very good at reasoning in tasks such as the Raven's matrices task. Visuospatial reasoning can even be considered a strength in many such individuals, especially those who would be considered to be on the mild end of the autism spectrum. My colleagues and I tested adolescents and young adults who were diagnosed with mild autism. These participants exhibited normal language acquisition and higher intelligence, but struggled with reading the emotions of others and participating in conversations due to inabilities to read the intentions and reactions of other people. These individuals were asked to perform a variety of social mentalization tasks, such as the "mind in the eyes" task which requires participants to infer a mental state of an individual based only on a photo of their eyes. Additionally, we tested for social knowledge, including the ability to infer the apparent intentions of shapes that act out social situations in animated videos, and the ability to understand the descriptions of actual social situations. We developed a virtual-reality (VR)-based social training intervention designed to allow individuals on the autism spectrum to practice social scenarios that occur in life. These included negotiating a lease for an apartment, meeting a new friend, and dealing with social pressure. The training was led by a clinician and every avatar participating in the VR social interactions was controlled by a real person, either the clinician leading the intervention, or a confederate who helped act out the scenario in the VR environment. In this way, we could enable participants to practice these social reasoning or problem-solving scenarios. Sure enough, when individuals practiced they did become better at these tasks. More impressive was the finding that the VR social training also improved their abilities on several of the other more standardized tests of reasoning including comprehension of scene analogies requiring understanding several relational elements together (Didehbani, Kandalaft, Allen, Krawczyk, & Chapman, 2016). We will discuss the potential for this and other VR-based tools in Chapter 13 that discusses the role of technology in reasoning.

NEUROLOGY OF REASONING DEFICITS

Cortical Change in Dementia and Aging

The brain begins to show the effects of aging along with the rest of the body. Over time our muscles become less powerful, our tendons less elastic, and our skin more wrinkled. Likewise our senses tend to weaken. We may need glasses as we get into our forties or fifties. By our sixties or seventies bifocal lenses may become a necessity. We may also find that some frequencies are no longer as easy to hear. These sensory changes begin to shed light on how our nervous system ages. The fact that corrective lenses are needed is a sign that some level of ability for the eyes to focus is lost. The need for bifocal lenses is a clearer sign that the eyes can no longer switch between long-distance and short-range vision and furthermore that our visual system is no longer optimized for either of these processes. This trend continues when we consider the fact that the sensory organs connect to the brain. What types of changes can we expect to occur within our cortex and its associated connective white matter as we age?

In the course of normal healthy aging the brain undergoes a series of biological processes. These changes are associated with differences in functioning that we can appreciate at a psychological level. Our cortex expands over the course of childhood, as we have discussed in Chapter 5. This period of cortical expansion is then followed by a pruning or thinning over the course of adolescence as the dominant neural pathways that support daily life activities strengthen, while other, less relevant connections are gradually eliminated. We experience further cortical thinning in older adulthood. These changes can result in the commonly observed "cognitive slowing" associated with the aging brain. This slowing of cognition applies to simple response time tasks that require detection of a target such as a letter "X" among a string of other consonants. This may be attributable in part to motor slowing, or a reduction in strength and effectiveness of the motor neurons associated with making a button press response. Of more interest to the topic of reasoning is the cognitive slowing that leads us to take longer to make complex decisions, solve novel math puzzles, or plan complex trips, parties, and meetings. A highly relevant question in an era of increasing lifespans is: at what point is slowing or forgetfulness no longer a sign of normal healthy aging, but rather a marker of disease that we should consider to be a warning sign necessitating a visit to a neurologist?

Many researchers working in the field of cognitive aging have become interested in the borderline cases between healthy cognitive aging and those that mark clinical syndromes and diseases. One such gray area has become known as MCI. MCI is primarily marked by reductions in memory performance, but a proper diagnosis of MCI will require a visit to a neuropsychologist or neurologist, who will conduct a series of standardized cognitive tests in order to assess function overall. If there are significant differences indicating that the individual being tested is markedly lower than age-matched control averages, then a diagnosis of MCI is applied. A person who is experiencing MCI is not necessarily destined to develop dementia and inevitable cognitive decline, but he will typically be asked to visit a neurologist more often and a clinical impression of his functional level will likely be observed at regular intervals. There is no clear consensus on whether individuals who have been determined to exhibit MCI will clearly develop dementia, but the category of impairment has led the pharmaceutical industry to focus on developing drugs to combat the cognitive decline. It is currently extremely difficult to treat damage to the brain inflicted by dementia using drug treatments. Many view the presence of MCI as a window of opportunity to intervene in an older person's life before they go on to potentially develop more serious forms of dementia. Regrettably, once one has been diagnosed with one of the dementias, there is not a cure. Once brain tissue has degenerated it cannot be regrown and dementias are progressive. This means that cognitive ability tends to decline more over time. If new drugs can be developed to delay the neuron loss and cortical degeneration associated with dementia, then the point at which an MCI diagnosis is made would be an opportune time to intervene pharmacologically.

In some cases an individual who has been categorized as having MCI may go on to develop dementia. Dementias are degenerative illnesses that cause the loss of neurons and degeneration of cortical tissue. The most well-known form of dementia is Alzheimer's disease (AD) (Longridge, 1939). This condition is perhaps most associated with memory loss. Initially, AD patients may appear forgetful by missing appointments, neglecting to buy groceries, and failing to pay bills. These are common events in the lives of any older adults and certainly could occur in healthy individuals regardless of age. People with MCI may also experience these forms of everyday memory loss. As AD advances, patients will often experience progressively more severe disruptions of memory to the point at which they are taken to see a neurologist. Often a dramatic moment can occur, such as when an older adult takes a walk in her neighborhood and forgets the way back home, or when an older adult causes an automobile accident on a lonely road. Such events are difficult to ignore and typically will result in a referral to see a neurologist or neuropsychologist.

The pathology of AD centers around cell losses in different areas of the brain. This process eventually can lead to severe disruptions of entire areas of the cortex. These disruptions frequently result in cognitive consequences in the form of reduced attention, loss of working memory, and inability to recall long-term memories. All of these factors can negatively impact reasoning ability. AD is progressive, meaning that the lost tissue is permanently eliminated and further insult to the brain will continue to occur. The course of AD also varies, with some individuals losing temporal lobe functions initially and others experiencing losses of PFC function. These variations in biological change lead to individual differences in patient cognition, speed of progression, and daily life impact. As we have discussed in the previous sections, reasoning ability depends upon effective attention, working memory, and long-term memory. As such, a person with AD will likely show a range of deficits in reasoning over the course of the progressive disease.

There are other forms of dementia as well, all of which share the common component of loss of neurons and disruptions of brain tissue. Ultimately all of the dementias will lead to alterations in the person's behavior. Types of dementia were initially associated with specific markers of neuropathology. For example, AD has been associated with the presence of fatty deposits within the brain commonly called plaques. Additionally AD neuropathology also includes disrupted axons commonly referred to as tangles (Johnson & Blum, 1970). Other forms of dementia have their own associated signs of neuropathology. Pick's disease is a form of dementia that has characteristic aggregates of proteins found within the cells (Jacoby, 1937). Dementia with Lewy bodies represents another case of a neurodegenerative condition with distinctive protein aggregations (Bethlem & Den Hartog Jager, 1960). This clinical feature is often associated with Parkinson's disease, which is another condition in which the nervous system progressively degenerates. These forms of dementia often begin by disrupting a particular area of the brain, or multiple areas, and eventually disrupt large areas of the brain.

DISORDERS OF RELATIONAL REASONING

Impairments of Visuospatial Relational Reasoning

Relational reasoning is considered to be a central intelligence measurement. Studies evaluating reasoning ability often use the Raven's progressive matrices test to evaluate relational reasoning ability. The Raven's matrices test has previously been shown to exhibit the strongest intercorrelation level with a wide variety of other tasks including verbal analogies, mathematical problem solving, evaluating similarities, and visuospatial thinking (Marshalek, Lohman, & Snow, 1983). While most people can reliably perform simple Raven's matrices problems, difficult Raven's problems are diagnostic of higher intelligence, as few people are commonly able to solve the most complex problems in the set. The difficulty of a Raven's matrices problem is determined by its complexity. The complexity is built into the problems by requiring people to discover multiple novel rules about how the figures change across the rows and down the columns of the matrix. Difficult Raven's problems also require people to integrate across multiple relations in the matrices (refer to Fig. 7.1 of this chapter). This becomes a demanding task that places considerable burden upon the processes of attention and working memory in order to solve the complex problems. Raven's matrices scores tend to decline as people age, but even a single training session in how to solve the problems was effective in improving Raven's scores, suggesting that older adults are still quite capable on this task despite declines in processing speed (Denney & Heidrich, 1990).

We can observe disorders in reasoning ability in cases in which there is a cognitive disorder linked to aging, such as dementia. This is often evident from the reports of loved ones and caregivers who notice that an individual who has developed dementia can no longer effectively manage her finances, daily affairs, or effectively schedule her life. These may be domain-specific problems, meaning that someone may experience challenges when dealing with money but can still get along rather well in social situations. It is also possible that the cortical decline that someone is experiencing may represent a domain-general deficit, as people with dementia often develop difficulties in more than one area of their daily lives. Such a deficit may be able to be detected using an intelligence task such as the Raven's matrices.

In 1999, James Waltz et al. tested the relational reasoning abilities of individuals with FTLD compared to healthy individuals. The investigators sought to test whether the functions of the frontal lobes were necessary for fluid reasoning as measured by the Raven's matrices task. More specifically, they tested the hypothesis that the difficulty of the Raven's matrices could be isolated to the number of relations that had to be simultaneously considered in order to complete the full pattern within each matrix. In this study, Waltz et al. also tested these same participants on a deductive reasoning task testing transitive inference ability, again with a relational load variation. The transitive inference task was the same one reported by Fales et al. (2003) in their test of reasoning in individuals with Klinefelter syndrome. In this task the participants were asked to arrange cards bearing the names of people from tallest to shortest based on some propositional statements. These would enable the participant to infer the appropriate height of the people on the basis of two statements.

The participants in this task included two variants of FTLD. Individuals with a condition termed *frontal-variant* FTLD had sustained a frontal lobe onset of cortical degeneration compared to those who had been assessed to have *temporal-variant* FTLD, who had primarily lost cortical function within the temporal lobes (Fig. 7.9). As FTLD progresses, most patients will exhibit greater deficits as the loss of cortex becomes more pronounced. In many such cases, both the temporal and frontal lobes become compromised. Prior to that occurrence, there are some cases in which patients can be evaluated at a time where they have diffuse damage to one single lobe bilaterally. Waltz et al. evaluated such individuals on Raven's matrices and transitive inference tasks.

The results of the Waltz et al. (1999) study demonstrated clear impairments in relational reasoning in frontal-variant FTLD patients restricted to only the cases in which integrating multiple relations were required. In the transitive inference task all groups performed very well on the simpler one-relation problems, while frontal-variant FTLD patients dropped dramatically to around 10% accuracy on the more difficult two-relation problems. The Raven's matrices task examples are displayed in Fig. 7.10. On this task participants with frontal-variant FTLD performed similarly to the other participant groups on the simpler one-relation matrices problems, but then dropped to near 15% accuracy on problems requiring the integration of two relations. Note that individuals with temporal-variant FTLD

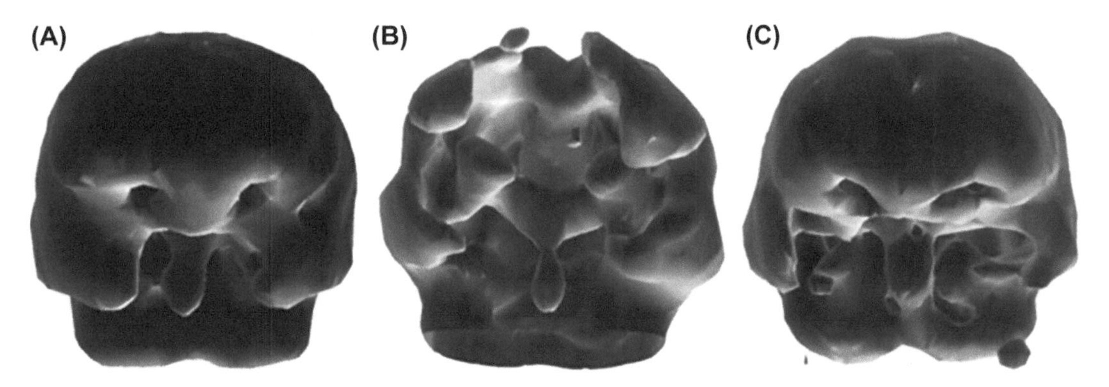

FIGURE 7.9 The instances of frontotemporal lobar degeneration as described in the study by Waltz et al. (1999). Neuroimaging scans of brain structure can reveal the brain atrophy patterns of patients. (A) A normal brain. (B) Frontal-variant patients had disruptions of the bilateral frontal lobes. (C) Temporal-variant patients showed cortical atrophy in the anterior temporal lobes. Neuroimaging scans of brain structure can reveal the brain atrophy patterns of patients. *From Waltz, J. A., et al. (March 1999). A system for relational reasoning in human prefrontal cortex. American Psychological Society, 10(2), Figure 1.*

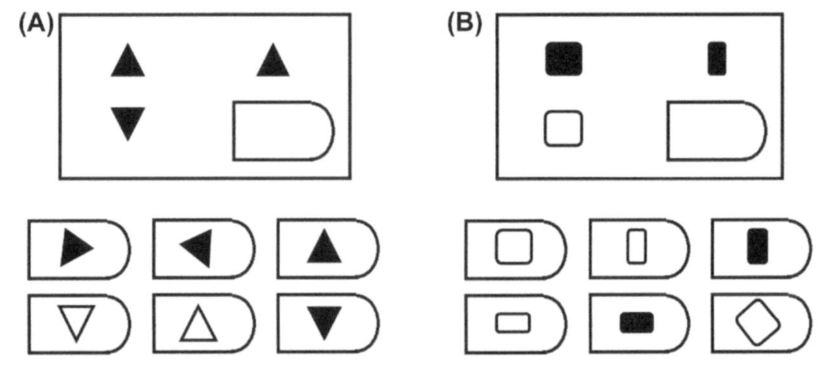

FIGURE 7.10 (A) Example of the simple (one-relation) stimuli adapted from the Raven's matrices as reported by Waltz et al. (1999). (B) Example of the two-relational stimuli that proved to be more difficult for the frontal-variant FTLD participants.

were essentially unimpaired in their ability to integrate across multiple relation problems in these tasks. The results of the investigation by Waltz et al. suggested that the frontal lobes are critical for completing the mental operations necessary to integrate multiple relations into a single answer. This integration process may also rely upon working memory abilities, inhibitory control, as well as attention maintenance. These possible component processes would be tested in follow-up studies.

The frontal lobe function of integrating multiple relations appears to be essential for reasoning ability. Another important aspect that may impact reasoning performance is that frontal-variant FTLD patients are often impulsive in their decision making and storytelling, and have difficulty in regulating their focus of attention. Impulse control may become more necessary when multiple relations are present in problems and therefore reasoning may be impacted alongside other cognitive phenomena.

In the early 2000s my colleagues and I set out to further specify the functional deficits present in FTLD and how these are relevant for reasoning. In one of our investigations we followed up on the Waltz et al. (1999) proposal that the PFC acts as a relational integration area. One of the tasks we developed was a simple linear reasoning task, which involved evaluating pictures of three people lined up left to right. There were only two relational changes possible among the people, greatly simplifying this task compared to the Raven's matrices and even the simpler variant of the Raven's matrices that Waltz et al. had created. In the linear relational reasoning task, we explained to our participants that the color of the clothing may change from the leftmost person to the middle person and again to the rightmost person. If that occurred, the color of the clothing became a progressively darker shade across the three pictures of the person (Fig. 7.11). The size of the person pictured could also vary moving from small to medium to large across the series. We included one-relation problems in which size or clothing color varied, but not both. We also included two-relation problems in which both size and clothing color always varied together. To evaluate these problems, participants were instructed to state "true" if all of the changes (size, clothing color, or both) completed their full sequence (moving from small to medium to large, for example). If the first person depicted in the series was small and the second person medium sized,

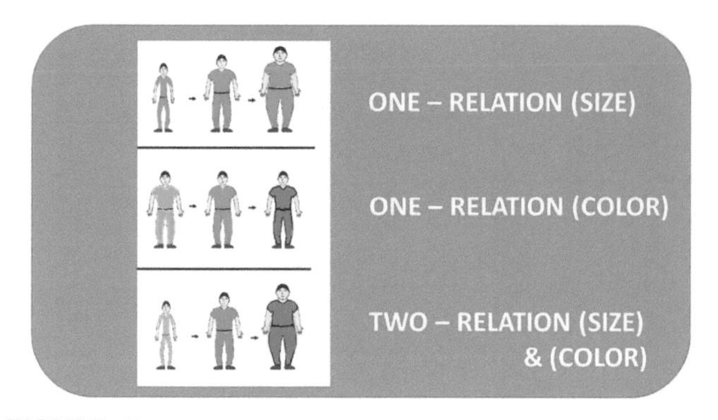

FIGURE 7.11 Examples of the stimuli presented in the linear relational reasoning task by Krawczyk et al. (2008). In one-relation problems, either color or size could change. In two-relation problems participants needed to integrate both color and size information to solve the problems.

but then the third person remained medium size, then participants were instructed to answer "false" in those cases. One-relation problems required that only the single relation be evaluated, while two-relation problems required participants to evaluate both relational changes to ensure that they both completed their sequence in order for the problem to qualify as "true." If either relation did not complete its sequence in the rightmost person, then the entire problem was "false." We were most interested in how frontal-variant FTLD patients would perform on these two-relation problems.

My colleagues and I followed the methodology of Waltz et al. (1999) evaluating temporal-variant FTLD patients, frontal-variant FTLD patients, and healthy age-matched control individuals on the relational people task we had developed (Fig. 7.9). As we had predicted, the task was very easy for the control participants and the temporal-variant FTLD patients. Not only did these participant groups perform near ceiling levels getting almost all of the problems correct, but they also solved the problems quickly (Krawczyk et al., 2008). By contrast, it was painstaking to test the frontal-variant FTLD patients on this task. These participants showed a pronounced impairment on the two-relation problems, but they also had difficulty with the one-relation cases. We had anticipated that such patients would be able to solve the one-relation problems successfully and generally speaking they were able to. A difficulty presented itself because we had randomized the problems, such that sometimes a one-relation problem followed a two-relation problem. The frontal-variant FTLD participants became *perseverative* on this task. Perseveration occurs when a person begins to behave in a certain way and then proceeds to apply that same behavior repeatedly on future trials even when they ought to stop and change response. In the linear relational reasoning task frontal-variant FTLD patients often provided the same answer repeatedly on trial after trial inappropriately. For example, a patient may have answered two "true" problems in a row and then continued to apply the "answer true" rule as the task progressed. This perseverative pattern reduced accuracy on the one-relation problems more than we had anticipated.

The linear pattern relational reasoning task revealed that relational integration is critical for reasoning and is impaired in this population, but it also suggested that a deficit in inhibitory control was also in part responsible for the impairments of this group. While performance was very poor for frontal-variant patients on the task, these participants appeared unable to even comprehend the instruction for two-relation problems. I can recall testing one individual who clearly understood the one-relation problems. This patient verbalized "small, then medium, then-large, true," or "light, then dark, then darker, true." On two-relation problems he would fixate on the first relation he noticed and proceed in the same manner answering on the basis of size or color. Eventually, I began to reiterate the instructions to see if it was possible for this participant to direct his attention to both relations. For example, on a two-relation problem in which size was true but clothing color was false, he would state "small, medium, large, true." When I pointed out that he ought to also consider whether clothing color changed, he would then state "light, darker, no change, false." I then would ask if he could consider both color and size together, at which point he would revert back to making a size-only judgment. Soon this testing session became exasperating for both myself and the participant.

This anecdotal example clarifies how essential intact PFC is for being able to integrate the outcomes from the two separate relations to produce one answer that takes into account the full complexity of the problem. Performance was made worse by interference from previous answers. On top of this challenge, we had added theme-based clothing to the people in some of the problems. For example, in some cases the person had on a fishing vest and hat, held a fish and a pole, or had on baseball equipment. We wished to see if frontal-variant patients could ignore these distracting elements. While healthy control participants and temporal-variant patients easily ignored these extraneous features, frontal-variant patients often paid close attention to the clothing, as well as the hair color and facial features of the cartoon images. This added to the difficulty that frontal-variant patients experienced on this task. Again, temporal-variant patients and healthy control participants easily ignored these extraneous features. This study indicated that relational reasoning impairments were profound in cases of frontal lobe damage. Additionally, the frontal lobe-mediated processes of filtering distracting elements also impacted performance. This suggests that relational integration is a core cognitive function necessary for relational reasoning, while voluntary and involuntary attention play a key supporting role in reasoning.

Challenges in Understanding and Applying Analogies

A special case of relational reasoning can be found in processing analogies. Some analogies are spatial and non-semantic, but most analogies involve some degree of understanding semantic information. Solving these semantically rich analogies is likely to require both relational integration and semantic knowledge. We will discuss analogical reasoning more in depth in Chapter 10.

My colleagues and I had an opportunity to test the degree to which FTLD impairs these multiple processes. Some of this work was published alongside the relational people task that we covered in the prior section (Krawczyk et al., 2008). In our first analogical reasoning study, we evaluated whether patients with FTLD could solve simple verbal analogies of the form "A is to B, as C is to D." For example, a participant would see the phrase "bear is to cave as bird is to _____," and they were asked to fill in the missing item. We also provided participants with two options based on a task developed earlier by Sternberg and Nigro (1980). In addition to the correct answer "nest," we also included an incorrect answer such as "wing." In this task we tested relational reasoning ability by asking participants to integrate across two relations. We also tested for semantic inhibitory control by varying the association strength between the third term (bird) in this example and the two answer choices. In some cases the problem could be made easier for the participant by having very highly associated third and fourth terms in the analogy. In other cases the problem could be made more difficult by providing an incorrect (nonrelational) item that was highly semantically associated to the third term in the analogy. An example of this is a problem such as "play is to game, as give is to ___" with "party" or "take" being offered as answer choices. In this case "give" is highly related to the incorrect answer "take," so if a participant is led by semantically similar items, he or she will have more difficulty blocking this "give-take" association in order to answer the problem correctly with "party."

In addition, we tested the ability of FTLD patients to solve scene analogy problems. This task emerged from the developmental psychology literature (Markman & Gentner, 1993). Simple situations were presented in these scene analogy problems (Fig. 7.12). In this case the first picture depicts a dog breaking away from a leash held by a man so that it can chase a cat. In relational terms, this picture shows a "restraint" relation. The man is the *agent* of a restraining relation and the dog is the *patient*. In this case, the dog has engaged as the agent of a chasing relation and

FIGURE 7.12 Scene analogy problems require relational reasoning and the ability to screen out distracting information. The situation depicted in the top picture shares relational and semantic similarity with the lower image. An analogical relational match can be made between the man in the top image and the tree in the bottom image. Meanwhile a possible semantic match exists between the man in the top picture and the man in the bottom picture.

the cat is the patient. In the second image showing the same dog breaking his leash in order to chase the man, there are some clear relational similarities. In the second scene, the agent of the restraining relation is the tree, while the patient of the restraining relation is again the dog. In this case, the dog has initiated a chasing relation as the agent, while the man is now the patient of the chasing relation. We asked our participants to select the item in the second picture that best corresponded with the man in the first picture. On analogical grounds, this would be the tree, as it is the agent of the restraining relation in the second image, just as the man was the agent of the restraining relation in the first image. The only problem with that answer is that the man is not a very compelling match to the tree based on his semantic properties. Indeed the man in the second picture shares a much higher degree of similarity semantically to the man in the first picture and even looks a lot like him perceptually in terms of his stature and how he is positioned. Young children will tend to make this man-to-man match, favoring perceptual and semantic similarity over relational similarity. As children age, they do much better at favoring the analogical match on such problems (Markman & Gentner, 1993).

Robert Morrison et al. used these verbal and scene-based analogies to test the analogical reasoning ability of individuals with FTLD. As in the prior studies we have reviewed (Krawczyk et al., 2008; Waltz et al., 1999), frontal-variant patients were compared to temporal-variant patients, and to healthy age-matched control participants (Fig. 7.9). Morrison et al. observed differences in both patient groups compared to controls. While control participants completed about 95% of the verbal analogies correctly, the temporal lobe participants performed at approximately 70% correct and the frontal patients were at about 80% correct. In this case it appears both relational reasoning and frontal lobe filtering functions are relevant, as is semantic knowledge. The primary clinical feature of temporal lobe FTLD patients is that they have a profound loss of memory and recognition for many semantic objects. This is referred to as semantic dementia as we have discussed earlier in this chapter. Interestingly, the problems could be divided according to the semantic relatedness between the third and fourth terms. In cases in which the correct analogical answer was more semantically related to the third term, semantic information would help the individual to choose the correct answer. Alternatively in cases where the incorrect answer was more related to the third term (such as "give-take" in the prior example) semantic information must be filtered in order to ignore that association strength and select the relationally correct answer. Frontal-variant FTLD patients showed worse performance on these latter types of problems that required semantic filtering. The frontal-variant patients were better at those problems that did not require semantic filtering. Meanwhile, the temporal-variant patients did not perform differently on the basis of the semantic relatedness of the terms. This suggests that their performance deficits relative to control participants were more related to a loss of semantic knowledge overall.

Results of the scene analogies told a similar story to the results from the verbal analogies problems. In these problems, both patient groups selected more perceptual matches than the healthy control participants. While these data were not conclusive, as it cannot be precisely known why participants selected the items they did, performance on the task was consistent with a semantic knowledge loss for the temporal-variant patients and a relational and semantic filtering loss for the frontal-variant patients (Morrison et al., 2004).

The role of inhibitory control in frontal-variant FTLD emerged as another important cognitive factor influencing reasoning, as indicated by the study by Morrison et al. (2004). We followed up on this verbal analogy study using a related picture analogy task. The picture format enabled us to further test the role of inhibitory control in reasoning. By using picture stimuli in place of words, we offered participants the possibility of focusing on other details within the images. This could be helpful in some cases, but could add further distraction in other cases. In the picture analogy task, as in the verbal analogy task, we included incorrect distracting elements. Fig. 7.13 depicts one of the problems from this study (Krawczyk et al., 2008). In the example, a picture of a radio is presented alongside a picture of an ear. This would likely inspire one to infer that the relation was something along the lines of "device that presents sensory input and the sensory organ that receives it." The next item in the analogy "television" would be similarly paired with the picture of the eye in order to complete

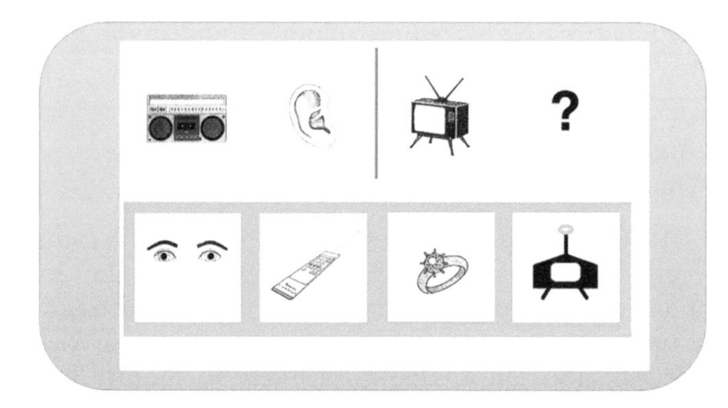

FIGURE 7.13 An analogy problem presented in picture form (Krawczyk, et al., 2008). Radio is to ear, as television is to? The correct relational answer is the picture of the eyes, but competitors exist in the form of a semantically close item to the television (remote control) and a perceptually similar item (the spaceship).

a relational match to the previous radio-ear pairing in order to form an analogy. Initially participants viewed the three terms in the analogy such as "radio-ear; television-?." They were asked to infer something that might go with television in the same way that the picture of the ear went with the radio. We then displayed a set of four new images below the problem with one correct answer (toolbox) presented along with three incorrect items. Among the incorrect choices, we presented a semantic memory distractor in the form of a remote control, an item which is highly associated with television. We also presented a perceptual distractor, a spaceship with a large window and an antenna, in order to provide an item that looked similar to the television. Lastly, an unrelated item was presented as well.

Results of the picture analogy task confirmed that participants who had frontal-variant FTLD were particularly impaired at solving analogies. The healthy control participants were excellent at the analogies and completed over 90% of them correctly. Temporal-variant FTLD patients also performed the task well completing analogies in around 75% of cases. The frontal-variant patients struggled with the task and completed analogies in only about half of the cases. The errors made by the frontal-variant patients were also striking. In about 35% of problems these patients chose the semantic distractors. This pattern of errors suggests that these patients rapidly abandoned the instruction to make a relational match and proceeded to choose the item that was most strongly associated with the third term in the analogy. The majority of other error responses were perceptual distractors for these participants. Such results suggest two possible interpretations. Either the frontal-variant patients have difficulty managing the relational comparisons in analogies and this impairs their reasoning or they are unable to block the strong associations between items such as hammer and nail and an inhibitory control deficit disrupts their ability to reason.

To further understand the role of the frontal lobes in analogical reasoning, we retested the patients on the picture analogy task after administering several other intervening tasks. In the second round of the picture analogy task we presented the same problems, but omitted the semantic and perceptual distractor items, so that all of the incorrect answer choices were irrelevant items. In this case, the performance of the frontal-variant patients was remarkably improved by about 35%. It appeared as if the distracting items had indeed derailed the patients from completing the correct answer. This result in combination with those presented by Morrison et al. (2004) provided a strong indication that the frontal lobes are important for filtering out irrelevant associations in analogical reasoning tasks. The inability to screen out distracting items is a key skill needed to allow people to maintain a focus on making the more complex relational mappings necessary to reason by analogy.

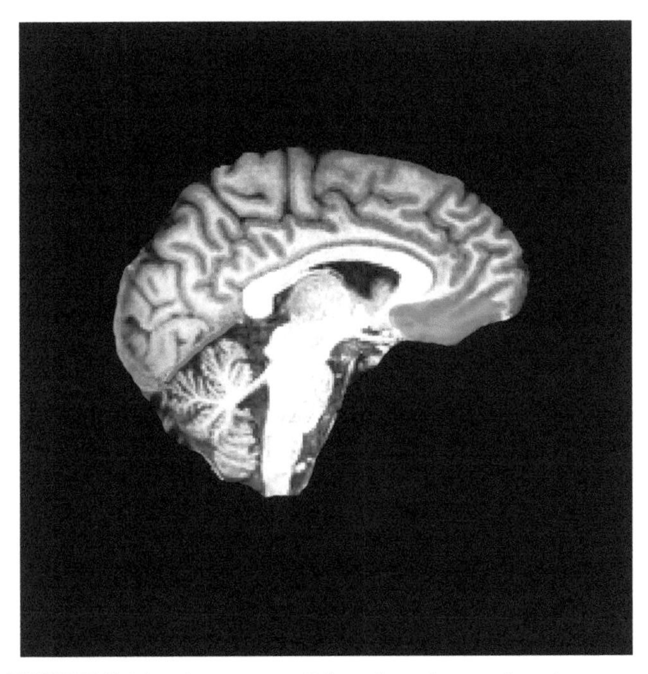

FIGURE 7.14 The ventromedial prefrontal cortex is an important area of the frontal lobe that is involved in linking situations to their emotional consequences.

DISORDERS OF DECISION MAKING

The Effects of Brain Damage on Decision Making

Some early decision research integrating the brain with behavior was carried out at the University of Iowa by neurologist Antonio Damasio et al. This research group saw a wide range of patients who had acquired brain injuries, especially those who had sustained damage to one or more areas of the PFC. Damasio (1994) wrote a book detailing the characteristics of several of these patients and was particularly intrigued by frontal lobe damaged patients, who appeared to show little to no cognitive deficit in laboratory environments when they performed memory or intelligence tasks. The situation outside of the clinic was very different. These same individuals often exhibited careless and risky decision making that ruined their finances, disrupted their relationships with family and friends, and made their living situations much worse than they had been prior to the injuries.

One particularly interesting case of a patient (referred to by his initials E. V. R.) is representative of this type of clinical cases. E. V. R. had little difficulty with standard neuropsychological tests of intelligence, attention, and memory. By contrast, his situation in life had deteriorated on the basis of poor decision making. E. V. R. had sustained damage to the lower middle portion of his PFC, an area known as the ventromedial prefrontal cortex (VMPFC) (Fig. 7.14). The VMPFC is a region that is strongly interconnected with the amygdala and is involved in relaying emotional information to the other regions of the PFC. Damasio et al. developed a decision task that they could run within the clinic to better capture the deficit that E. V. R. presented with in his daily life decisions. Surgeons removed a bilateral orbitofrontal tumor from E. V. R's brain. Comparisons of his behavior before and after the surgery were profound. Prior to neurosurgery E. V. R. had been a successful accountant and husband, had supported his siblings, and had earned promotions in his accounting work. After the surgery, he was repeatedly fired from jobs for tardiness and lack of organization. He experienced two divorces. He also began to exhibit unusual behavior including collecting broken appliances, old newspapers, and food packaging. E. V. R appeared to have acquired a selective decision-making impairment. Soon after his acquired brain injury, reckless business practices forced him to declare bankruptcy. He also took overly long amounts of time to make decisions of relatively minor importance. The task that Damasio et al. developed involved participants making card selections with financial consequences and has since become known as the Iowa gambling task.

Bechara, Damasio, Tranel, and Damasio (1997) reported initial results on the Iowa gambling task, which has subsequently been used in other investigations numerous times in the past two decades. Bechara et al. reported differences between individuals who had sustained medial orbitofrontal damage relative to healthy controls. The Iowa gambling task simulates real-world financial decisions through decisions made during a card game. The participant's goal is to achieve the most money possible based on making a series of card selections. Participants are initially given a "loan" of play money and instructed to maximize profits as they selected from four decks of cards with feedback after each card about whether the person won or lost money. Two of the card decks were risky, paying out large rewards for win cards, but penalizing the participants even more for loss cards. The two risky decks are therefore financially disadvantageous over the course of the task. The other decks were more conservative, paying out smaller rewards, but incurring smaller losses as well, making these decks financially advantageous in the long run. Participants made a total of 100 card decisions making it impractical to calculate the win and loss frequencies, so they had to rely on developing a gut feeling or automatic decisions about which decks were most advantageous to choose from. While healthy control participants learned that the risky decks were not good choices in the long run, the medial orbitofrontal damaged patients chose more often from the two risky decks leading to poor financial outcomes in this group overall. This result demonstrates the effectiveness of the Iowa gambling task in showing the devastating effects of medial PFC damage on financial decisions.

To further understand the medial PFC contributions to decision making, Bechara et al. recorded galvanic skin response (GSR) data in patients during the Iowa gambling task. GSR measures subtle differences in perspiration on the skin that accompany changes in emotional arousal. Participants in this study completed the gambling task, but were also asked to report their knowledge about the decks after every 10 selections. Most people selected cards from the risky decks for the first several trials before they experienced a loss card. Bechara et al. called this phase of the task the *prepunishment phase.* During these selections, participants did not show decision-related arousal in their GSR data. Healthy participants began to show anticipatory GSR activity after they experienced losses from the risky decks. Patients did not show these same arousal markers. After the next 10 selections, people entered a *pre-hunch phase* during which they could not state what the appropriate strategy was for card choices. At the midpoint of the gambling task healthy people realized that the risky decks were worse to select from. This period of the task was termed the *hunch period*, as people had a sense of which decks to avoid, but could not yet produce a clear reason why these decks should be avoided. Neurologically healthy individuals continued to show anticipatory GSR activity prior to making a choice from either of the risky decks. By contrast, patients with PFC damage could not describe any hunches about the task and continued to show arousal before selecting from the risky card decks in the form of anticipatory GSR activity. As people progressed through the gambling task the healthy control individuals caught on to the rules and could state that the two risky decks should be avoided, because they contained such large loss cards. Such people reached the *conceptual period* of the task, in which they had figured out the best decision strategy to use. Not all neurologically healthy individuals reached this conceptual period, but all continued to choose from the conservative decks operating on at least a hunch about how the gambling task worked. Only about 50% of the PFC-damaged patients reached the conceptual knowledge period and were able to state which decks were risky and which were conservative. Despite reaching this period, many of the patients persisted in making risky choices anyway. These individuals also continued to show GSR arousal responses prior to risky choices.

The Iowa gambling task tells us several important features about decision making. Decisions often require balancing tradeoffs between advantages and disadvantages. The conservative card decks represent better choices that will yield positive results overall. Such choices are analogous to making healthy lifestyle decisions. While it may be satisfying to stay up all night playing electronic games, we would be better served studying for exams and getting a quality night of sleep most of the time. Avoiding the risky card decks is representative of many situations where people choose to avoid risky conduct in the form of irresponsible drinking, recklessly speeding in a motor vehicle, and choosing to forgo risky sexual behavior with relative strangers. The gambling task studies also indicate that the PFC is an important brain area involved in governing our decision behavior. A core function of the PFC appears to be appropriately integrating emotional guidance in the form of arousal to risky situations with conceptual knowledge of how a decision outcome is likely to occur. Such knowledge could include our prior beliefs about alternatives, attributes, and heuristics that have helped to guide us in the past. These studies suggest that emotion plays a key role in helping to guide us away from situations that will be either not profitable or possibly dangerous.

DISRUPTIONS OF MORAL REASONING

Stages of Moral Development

Moral reasoning develops as children age. As we discussed in Chapter 5, moral development can be measured by a scale developed by Kohlberg (1958) and Kohlberg, Modgil, and Modgil (1986). Kohlberg's stages of moral thinking are based on an interview technique. In the interview children are asked about moral dilemmas. The first level of moral development that Kohlberg described is called preconventional reasoning, in which the moral world is viewed as being strongly delineated with actions that are reliably linked to punishments or rewards. Disruptions within the brain can impact moral development leading children to stall at this first preconventional stage of development.

Anderson, Bechara, Damasio, Tranel, and Damasio (1994) at the University of Iowa described the behavior of two frontal patients who showed a profound disruption of moral reasoning ability. The two patients were both under 16 months at the time of their acquired impairments. One patient suffered a devastating frontal lobe brain injury due to a car accident. The other experienced a tumor removal at 3 months of age. In both cases the orbital and medial sectors of the PFC were extensively damaged. These patients maintained normal intellectual functioning in many respects, similar to the adult patient E. V. R. described by Damasio (1994). Like E. V. R. these two frontal patients also showed deficits on the Iowa gambling task, showed low arousal levels toward risky decisions, and made poor financial and life decisions. At the time of the report, both patients were in their early 20s. They had both led irresponsible lives including numerous run-ins with the law, challenges at school, disciplinary issues, impulsivity, and poor dietary and social decisions. This behavioral profile is not especially unique for individuals who have experienced PFC damage. What is unique about these very early onset frontal lobe patients is that their moral development never moved beyond the preconventional stage, as classified by Kohlberg (which is characteristic of children around age 10). These patients showed problems with moral reasoning by providing inappropriate responses to social situations. Their responses to moral dilemmas were highly egocentric and primarily were geared toward the avoidance of punishment, rather than any sense of duty to society or sense of basic right and wrong. By contrast, Anderson et al. described a comparison to adult-onset patients with PFC damage. These patients can answer conceptual moral dilemma and social perspective taking tests appropriately, despite the fact that their behavior is often impulsive and inappropriate. This study suggests that the orbital and medial PFC is critical for the normal development of moral cognition. If an injury to the region occurs in very young children, they not only behave inappropriately, but they also lack the knowledge of how they ought to behave even when they have reached adulthood.

SUMMARY

Reasoning is highly dependent upon intact brain systems across the lifespan. There are a variety of levels of impairment that people experience based on what region of the brain is damaged and why. Reasoning involves the intake of new information, identifying and representing the context of the situation, manipulating information, recalling previous solutions, and projecting into the future to imagine the effectiveness of a solution.

Attention and memory are important for the representation of the current premises and context in a reasoning task. Attending to the relevant information and voluntarily focusing attention are both critical. Once information is attended to, it must be maintained in working memory and possibly transformed depending upon the nature of the task. The ability to recall relevant information from long-term memory at the appropriate time can also determine how we reason. Intact PFC and parietal cortex are essential for effective attention and working memory performance. Strokes and traumatic brain injuries can impair these areas. Dementias can lead to decline in attention, working memory, inhibitory control, and retrieval of long-term memories.

Social and emotional processing are important for reasoning about other people's behavior, making appropriate decisions based on risk, and moral reasoning. Theory of mind is one of the critical skills that enable us to effectively take the perspectives of others and negotiate effectively to solve problems in groups. This skill can also be important when working with other individuals. Frontal lobe impairments in the medial and orbitofrontal regions can lead to impaired decision making and excessive risk taking. Intact frontal regions are also important for advancing our moral development in addition to carrying out appropriate social behavior.

END-OF-CHAPTER THOUGHT QUESTIONS

1. Which cognitive function or brain system seems most central to carrying out reasoning?
2. Exogenous attention encompasses our involuntary orienting responses toward stimuli in the environment. Can you think of a situation in daily life in which this form of attention impacts our reasoning performance?
3. Working memory is commonly involved in reasoning. Is this the most important cognitive function involved in our reasoning ability? Why or why not?
4. Long-term memory deficits follow from many of the dementias. What type of reasoning might be most impacted by a long-term memory problem?
5. Frontal lobe damage can lead to challenges with relational complexity and inhibitory control. Which of these factors has a greater effect on reasoning?
6. Theory of mind helps us when reasoning with other people. Can you name some situations in daily life reasoning that require this skill?
7. The medial frontal cortex is important for helping us to avoid making risky financial decisions. What other factors are important for decision making?
8. Moral reasoning can be intact in individuals who acquired a frontal lobe injury later in life when tested in a lab environment. Why do such patients have difficulties translating moral knowledge into their actions in daily life?

References

Anderson, S. W., Bechara, A., Damasio, H., Tranel, D., & Damasio, A. R. (1994). Impairment of social and moral behavior related to early damage in human prefrontal cortex. *Nature Neuroscience, 2*, 1032–1037.

Babcock, R. L., & Salthouse, T. A. (1990). Effects of increased processing demands on age differences in working memory. *Psychology and Aging, 5*, 421–428 FS; CS.

Baddeley, A. (1986). *Working memory*. New York: Oxford University Press.

Baddeley, A., Chincotta, D., & Adlam, A. (2001). Working memory and the control of action: Evidence from task switching. *Journal of Experimental Psychology: General, 130*, 641–657.

Bechara, A., Damasio, H., Tranel, D., & Damasio, A. R. (1997). Deciding advantageously before knowing the advantageous strategy. *Science, 275*, 1293–1295.

Bethlem, J., & Den Hartog Jager, W. A. (1960). The incidence and characteristics of Lewy bodies in idiopathic paralysis agitans (Parkinson's disease). *Journal of Neurology Neurosurgery and Psychiatry, 23*, 74–80.

Bisiach, E., & Luzzatti, C. (1978). Unilateral neglect of representational space. *Cortex, 14*, 129–133.

Channon, S. (1997). Impairment in deductive reasoning and working memory in Parkinson's disease. *Behavioural Neurology, 10*, 1–8.

Damasio, A. R. (1994). *Descartes' error: Emotion, reason, and the human brain*. New York: G.P. Putnam.

Daneman, M., & Carpenter, P. A. (1980). Individual differences in working memory and reading. *Journal of Verbal Learning and Verbal Behavior, 19*, 450–466.

Denney, N. W., & Heidrich, S. M. (1990). Training effects on the Raven's progressive matrices in young, middle-aged, and elderly adults. *Psychology and Aging, 5*, 144–145.

Didehbani, N., Kandalaft, M., Allen, T., Krawczyk, D. C., & Chapman, S. B. (2016). Virtual reality social cognition training for children with high functioning autism. *Computers in Human Behavior, 62*, 703–711.

Fales, C. L., Knowlton, B. J., Holyoak, K. J., Geschwind, D. H., Swerdloff, R. S., & Gonzalo, I. G. (2003). Working memory and relational reasoning in Klinefelter syndrome. *Journal of the International Neuropsychology Society, 9*, 839–847.

Freedman, L., & Dexter, L. E. (1991). Visuospatial ability in cortical dementia. *Journal of Clinical Experimental Neuropsychology, 13*, 677–690.

Fuster, J. M., & Alexander, G. E. (1971). Neuron activity related to short-term memory. *Science, 173*, 652–654.

Jacoby, N. M. (1937). Pick's disease. *Proceedings of the Royal Society of Medicine, 31*, 115–117.

Jakobson, L. S., Archibald, Y. M., Carey, D. P., & Goodale, M. A. (1991). A kinematic analysis of reaching and grasping movements in a patient recovering from optic ataxia. *Neuropsychologia, 29*, 803–805 807–809.

Johnson, A. B., & Blum, N. R. (1970). Nucleoside phosphatase activities associated with the tangles and plaques of Alzheimer's disease: A histochemical study of natural and experimental neurofibrillary tangles. *Journal of Neuropathology and Experimental Neurology, 29*, 463–478.

Juskenaite, A., Quinette, P., Desgranges, B., de La Sayette, V., Viader, F., & Eustache, F. (2014). Mental simulation of future scenarios in transient global amnesia. *Neuropsychologia, 63*, 1–9. http://dx.doi.org/10.1016/j.neuropsychologia.2014.08.002.

Kohlberg, L. (1958). *The development of modes of moral thinking and choice in the years 10 to 16*. University of Chicago.

Kohlberg, L., Modgil, C., & Modgil, S. (1986). *Lawrence Kohlberg, consensus and controversy*. Philadelphia; London: Falmer Press.

Krawczyk, D. C., Morrison, R. G., Viskontas, I. V., Holyoak, K. J., Chow, T. W., Mendez, M., et al. (2008). Distraction during relational reasoning: The role of prefrontal cortex in interference control. *Neuropsychologia, 46*, 2020–2032.

Longridge, R. G. (1939). Alzheimer's disease. *Proceedings of the Royal Society of Medicine, 32*, 222–224.

Markman, A. B., & Gentner, D. (1993). Structural alignment during similarity comparisons. *Cognitive Psychology, 25*, 431–467.

Marshalek, B., Lohman, D. F., & Snow, R. E. (1983). The complexity continuum in the radex and hierarchical models of intelligence. *Intelligence, 7*, 107–128.

Morrison, R. G., Krawczyk, D. C., Holyoak, K. J., Hummel, J. E., Chow, T. W., Miller, B. L., et al. (2004). A neurocomputational model of analogical reasoning and its breakdown in frontotemporal lobar degeneration. *Journal of Cognitive Neuroscience, 16*, 260–271.

Posner, M., Petersen, S., Fox, P., & Raichle, M. (1988). Localization of cognitive operations in the human brain. *Science, 240*, 1627–1631. Retrieved from http://www.jstor.org/stable/1701013.

Prabhakaran, V., Smith, J. A., Desmond, J. E., Glover, G. H., & Gabrieli, J. D. (1997). Neural substrates of fluid reasoning: An fMRI study of neocortical activation during performance of the Raven's progressive matrices test. *Cognitive Psychology, 33*, 43–63.

Premack, D., & Premack, A. J. (1983). *The mind of an Ape* (1st ed.). New York: Norton.

Pribram, K. H., Ahumada, A., Hartog, J., & Roos, L. (1964). A progress report on the neurological processes disturbed by frontal lesions in primates. In J. M. Warren, & K. Akert (Eds.), *The frontal granular cortex and behavior* (pp. 28–55). New York: McGraw-Hill Book Company.

Raven, J. C. (1938). *Progressive matrices: A perceptual test of intelligence, 1938, sets A, B, C, D, and E*. London: H. K. Lewis.

Raven, J. C. (1960). *Standard progressive matrices: Sets A, B, C, D, & E*. London: H.K. Lewis & Co.

Robert, P. H., Migneco, V., Chaix, I., Berthet, L., Kazes, M., Danion, J. M., et al. (1997). Use of a sequencing task designed to stress the supervisory system in schizophrenic subjects. *Psychological Medicine, 27*, 1287–1294.

Shallice, T. (1982). Specific impairments of planning. *Philosophical Transactions of the Royal Society, London, Series B, 298*, 199–209.

Sternberg, R. J., & Nigro, G. (1980). Developmental patterns in the solution of verbal analogies. *Child Development, 51*, 27–38.

Süß, H.-M., Oberauer, K., Wittmann, W. W., Wilhelm, O., & Schulze, R. (2002). Working-memory capacity explains reasoning ability—and a little bit more. *Intelligence, 30*, 261–288.

Waltz, J. A., Knowlton, B. J., Holyoak, K. J., Boone, K. B., Mishkin, F. S., de Menezes Santos, M., et al. (1999). A system for relational reasoning in human prefrontal cortex. *Psychological Science, 10*, 119–125.

CHAPTER

8

Reasoning About Contingencies, Correlations, and Causes

KEY THEMES

- Causal reasoning has enabled some remarkable advances in medicine and health care. These have contributed to lengthening the average human lifespan dramatically over the past century.
- The ability to determine cause-and-effect relationships in complex conditions often requires us to equally weigh evidence that supports a possible causal theory and evidence that refutes it.
- When presented with a two-choice task where one option is better than the other, many animal species learn to choose the more probable option exclusively. This behavior is known as maximizing.
- When presented with a two-choice task in which both options are sometimes reinforced, humans tend to frequency match by choosing each option roughly the same proportion of the time that it is reinforced.
- Humans will find patterns in a series of outcome data, and this is dependent upon the left prefrontal cortex.
- People tend to experience an illusion of control in which they believe that they have more ability to influence outcomes than they actually do. This is especially true in cases with high reinforcement rates.
- People experience an illusion of willful action in which our thoughts can be misperceived as causing certain actions in the environment. This tendency may explain some supernatural beliefs.
- Both the spatial and temporal behavior of objects colliding will either establish or eliminate the likelihood that people will infer a cause-and-effect relationship.

Reasoning
http://dx.doi.org/10.1016/B978-0-12-809285-9.00008-9

OUTLINE

REASONING ABOUT CONTINGENCIES, CORRELATIONS, AND CAUSES

Introduction

In this chapter we will examine how people reason about situations involving the interactions of multiple elements. This type of thinking is called upon when we encounter situations involving two or more things, and it can vary in different ways. Sometimes events or situations can have elements that are correlated, which simply vary together, but have no direct influence on one another. In such cases a third variable may be influencing them jointly. Another situation involves contingency, in which one thing is dependent upon another, such as when a condition is set allowing another situation to occur. Lastly, objects can have a cause-and-effect relationship, which is a strong connection in which one thing causes another to happen.

Reasoning about the interrelations among people, objects, or parts makes up a core part of how we come to view the world around us. Contiguous, or correlated, events become connected in our minds. This topic has been discussed for centuries dating back to the philosophy of David Hume and others. There are many instances in our daily lives when two events occur and may appear to be connected in space or time. Such a relationship is up to us to infer. When we see a cue ball ricochet around a pool table and strike the eight ball, we infer that the eight ball's movement and trajectory have been caused by the action of the cue ball (Fig. 8.1). If a person is controlling the pool cue that preceded the movement of the cue ball, then we will attribute the cause of the cue ball's movement to the person. Going back another step further, we can assume that motor commands from the pool player's brain led to planned actions enabling the movement of the cue that, in turn, led to both the movements of both the cue ball and the eight ball around the pool table. We make many of these causal inferences effortlessly in cases in which movements and timing can be easily perceived. When the eight ball falls into a corner pocket, we readily infer that the "clunk" sound that follows after the disappearance of the ball was caused by it hitting the bottom of the pocket. We may have greater

FIGURE 8.1 When we observe the movement of objects colliding, we easily infer that one object can cause another to move. Timing and proximity influence these judgments about causes and effects. The game of pool or billiards provides an opportunity to witness numerous objects colliding and causing one another to move. We readily infer that a capable player was ultimately the cause of the balls moving to desired locations. Making sense of this game is based on the inferences that we make. *By Kucharek,* https://commons.wikimedia.org/w/index.php?curid=1022671.

difficulty assessing the relationships between events when we miss more of the information that accompanies a situation.

Causal events and mere correlations are commonly inferred in many daily life events involving situations that co-occur in time or space. These types of inferences regularly occur effortlessly when we witness simple co-occurrences. They also occur when events become complex. For example, we often have a difficult time inferring the cause of a delayed flight at an airport.

I was recently planning to catch a flight out of Reagan National airport in Washington, DC, which is a very busy location. It had rained all day around the city. My terminal was jam-packed with people and I began getting notifications that my flight was going to be delayed. I also began to receive notifications that the gate for my flight had changed. I was mostly operating in the dark in this situation and feeling a bit helpless. The circumstances suggested that my flight had been impacted by events in the airport, which I presumed were caused by the inclement weather, but this was not necessarily true. Beyond this general guess, I possessed little information about my situation. I began to grow increasingly uneasy as I saw the growing number of weary travelers sitting in nearly every seat and milling around the various gate areas. Under such conditions, we tend to hypothesize about a possible cause or set of causes, and we then proceed to evaluate these possibilities as new events come to light. My gate change coincided with the presence of rain that day. This led me to formulate a ballpark

FIGURE 8.2 Complex situations such as routing airline travel at major airports involve numerous interrelated situations. It can be exceedingly difficult to establish cause and effect when trying to determine why a flight is delayed or a gate has changed. Without understanding the entire situation at the airport, we can find ourselves at a loss to explain our circumstances.

hypothesis that poor weather was responsible for my flight delay, but why the gate change? To answer this question, I had to attempt to trace the influence of events surrounding other flights leaving the airport. Many possibilities existed. Perhaps the plane that was supposed to take off from my flight's original gate was still there, as it was not currently safe to take off. Perhaps, wind and rain earlier in the day had delayed a whole series of flights and these planes were currently causing congestion on the runways. Another possibility is that the pilot who was slotted to work my flight could be coming in from another city and was not able to gain clearance for landing as was scheduled.

Daily events such as airport delays can lead us to feel powerless and emotional. I couldn't help but feel somewhat irritable. We may begin to think to ourselves "why can't they move on with getting the flights out?" "why can't they just board these planes and clear out the terminal (Fig. 8.2)"? Like many people in that type of situation, I was at a loss to be angry with any one individual or group. This inability to assign blame was related to the fact that I could not assess the actual cause of the delay due to the discrepancies between my theories about what might be occurring at the airport and the limited information that I had to go on a the time. My flight eventually boarded and I was immensely grateful to have made it home quite late that evening.

Throughout history, complex information has led people to generate equally complex theories about what to do in challenging situations. Weather provides a classic example of this. We are relatively powerless to change anything about the weather. It simply arrives and we have to deal with it. Meteorologists are still not precisely capable of predicting the weather in any given place until it is nearly upon us due to the numerous complex influences such as wind speed, direction, and barometric pressure. This indeterminacy likely led to supernatural beliefs in earlier times. There would have been little means to predict what was coming next in an era before radar, satellite imagery, and the field of meteorology. People were simply left with vague information to go on and therefore they generated complex theories involving the animacy of nature and human-like deities that were thought to control weather outcomes as retaliation or reward for the actions of people on Earth. Numerous cultures had inferred the existence of gods that influenced the weather.

It is not an accident that people tend to infer that some type of agent is causing events to happen behind the scenes. This tendency is in part due to an analogy that we set up comparing our own thoughts to those of others. The feeling of causing something extends outward and even mostly random events such as lightning strikes and tornado paths become effects in our minds. Without knowledge of the mechanisms that cause events like these, we will often infer the existence of an elaborate, but incorrect, cause for an event.

In the following sections we will review how people reason about events in the world that are either independent, correlated, or causal. The chapter begins with a discussion of the importance of establishing the degree of relatedness between different events. This is perhaps most obvious in a domain such as medicine. When medical personnel are presented with a set of symptoms, it is of paramount importance that they establish the cause for the symptom. Only then can a treatment be recommended and implemented. Likewise, in law we must clearly delineate the cause of death for a victim and whether or not someone was at fault when someone has suffered harm. Lives are on the line in death penalty cases and the ability to establish whether someone caused harm to other individuals is critical to the outcomes.

We will focus the remainder of the sections of the chapter on the ways that people reason about the influences of objects or events upon one another. These represent people's attempts to explain contingency data and establish possible causes and effects in the world. As a species capable of complex thinking, humans tend to generate hypotheses. We then set about testing these predictions by examining the relatedness between what we think is a cause and what we think is an effect. There are times in which we are correct and other times where we can be completely wrong. The challenge for people making causal inferences is that we often do not have enough data to know that we are incorrect. We will also examine cases in which we ascribe causes to factors that are out of our control and others that are subject to our own willed action. In all cases we can be misled by the incoming data. We can be tricked by illusions, we can fail to perceive information that is present, and we can also erroneously infer information that is not present.

We will conclude the chapter with a further discussion of real-world reasoning about contingencies in medicine and financial investing. Here we will examine how people link causes to effects under the complex and incomplete conditions that we experience in daily life. Throughout the chapter we will discuss situations in which there is brain-based evidence about how we link information in causal ways, or correlational ways.

ESTABLISHING CAUSE AS A BUILDING BLOCK FOR KNOWLEDGE

The Importance of Establishing Cause

Perhaps one of the most important moments in the history of science occurred in the middle of the 1800s when Louis Pasteur isolated the cause of infectious diseases using a microscope. For centuries prior to the discovery of what would become known as "germs," people had been unable to isolate the causes of most diseases. They were left to deal with the symptoms of illnesses, but had no strongly effective treatments, as they could not understand the mechanisms that led to the occurrence of a disease.

Fig. 8.3 shows a graph of life expectancy from the year 1543 through 2011 in the United States and in Great Britain (Roser, 2017). Notice that there is a general pattern to the lifespan data from the 1500s through the early 1800s, with people averaging somewhere between 30 and 40 years of age at death. What should we make of the data in this graph? What causal factors influenced the average lifespan in the past relative to modern times? These data were of course influenced by the very high death rates caused by infant mortality and the high likelihood of people dying from diseases in childhood. Note that many women during this period died in childbirth as well, so there are some rather sobering events that served to reduce the average lifespan during the earlier periods in human history. Notice also that the pattern fluctuated. There were some very bad years when the mean age dropped down into the 20s. These were likely years when serious communicable diseases wreaked

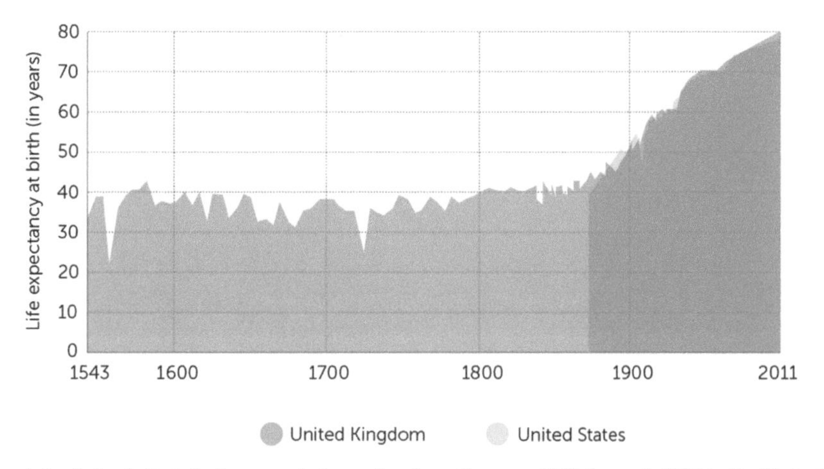

FIGURE 8.3 Average age at death is plotted during a period ranging from the year 1543 through 2011 in the United States and in Great Britain. Notice the patterns that exist until the mid-1800s and the remarkable increases in lifespan that lead up to modern times. We can apply causal reasoning to try to understand the patterns present in lifespan data. *Adapted from www.ourworldindata.org. Reproduced by Ashton Miller.*

havoc on human populations in many regions of England. Notice also that the pattern shows a more stable pattern that emerged by 1800 with the average lifespan hovering around age 40. By today's standards we would not jump for joy at this relatively minor improvement. The pattern takes on a strikingly different trend as the 1800s came to a close and we moved into the 1900s. At this point there was an almost logarithmic increase in lifespan in England and the United States with the population achieving an average lifespan around 50 by 1900. The population average lifespan would then continue to increase, rising about 10 years on average each decade. One might ask what is the cause of this pattern? Why did we see this remarkable lengthening in lifespan? For a likely answer we must look to the causal reasoning involved in medical diagnosis during the period beginning in the mid-1800s.

Ignorance about the cause of disease led to devastating effects on people's health. Not only were physicians unable to develop cures for diseases, they could barely develop effective treatments. This ineffectiveness in medical practice led to few treatments and this in turn led to premature deaths. Lack of medical knowledge may have also led to needless suffering due to ineffective attempts at treatment such as bleeding an ill person or putting someone with a virus out in cold weather to try to get fresh air. These actions were not due to cruelty on anyone's part, but rather due to simple ignorance that led to inaccurate theories and then ineffective attempts to treat patients. Added to this challenging situation was the fact that physicians did not understand how disease was transmitted. This lack of causal knowledge enhanced the potential for a devastating illness wiping out whole towns and villages and becoming an epidemic.

The period of initial stability in which England saw a stable population that lived to about age 40 on average may have been due to some advances in medicine. By this point in history, British doctors were trained formally and though they would have had little knowledge of disease and physiology relative to what we have in the modern era, the information that they did have was widely distributed by this principled training. Science was also advancing rapidly and single lens microscopes were moving toward providing the evidence needed to understand bacteria and infections. Despite the advances, physicians during this period used practices that were highly counterproductive because they did not understand the cause of infections. They did not understand the importance of basic cleanliness and sterilization in controlling disease. This led to the widespread use of germ-ridden instruments in treatments and surgeries. The leading theories about the causes of disease dated back centuries in many cases and were poorly informed by scientific evidence. In the early AD 200s the physiologist Galen had postulated that health stemmed from four "humors" (black bile, yellow bile, blood, and phlegm). He suggested that illness was caused by an imbalance among these humors (Stearns, 2011). Other theories included *miasma theory*, which essentially stated that bad air was the cause of disease. Note that miasma theory is not entirely off base, as there are certainly cases in which pathogens can become airborne. It was, however, quite far from accurate as a true cause of disease. There was also the germ of a correct concept when theorists postulated that *spontaneous putrefaction* was responsible for disease by disrupting the balance of the four humors (Stearns, 2011). Notice that putrefaction of meat and other foods can lead to illness if these substances are ingested. Despite the association, the mechanism of action was again incorrect and would not lead to appropriate treatment of infection.

The key to understanding infectious disease largely came about due to advances in the use of the microscope. It was the French chemist Louis Pasteur who "opened

FIGURE 8.4 The work of French chemist Louis Pasteur (1882–95) in the mid-1800s led to an understanding of the role that microorganisms play in spreading disease. This breakthrough in causal reasoning enabled people to develop effective medical procedures and treatments. Ultimately this knowledge would contribute to greatly extended average lifespans. *From Nadar [Public domain], via Wikimedia Commons.*

the world of the infinitely small" (Gaynes, 2011, p. 143). Pasteur had been working to understand the process of fermentation at the University of Lille and the Ecole Normale in Paris (Fig. 8.4). Up until the middle of the 1800s fermentation (such as the process that produces alcohol) was commonly believed to be caused by a simple chemical transformation. Pasteur's work indicated that fermentation was actually caused by the actions of microorganisms. He realized that microorganisms could differ in their characteristics including shape, size, nutritional needs, and endurance. His publications in 1860 revolutionized the state-of-the-art in chemistry and biology. His discoveries opened up a gateway toward the possibility of understanding the nature of infection and its role in causing diseases. These breakthroughs in the study of fermentation would later lead Pasteur to investigate the role of heat in killing bacteria (Gaynes, 2011). Pasteur's contributions to microbiology led not only to the critical methods of sterilization, but they would eventually contribute to the development of antibiotic drugs, which played a major role in further extending the average lifespan in developed nations (Fig. 8.5).

Causal reasoning is the primary cognitive process that led to the breakthroughs enabling our much longer lifespans through improvements in healthcare. After development of the germ theory of disease, medicine saw an incredible burgeoning of scientific experimentation in the 1900s. This new era of experimentation

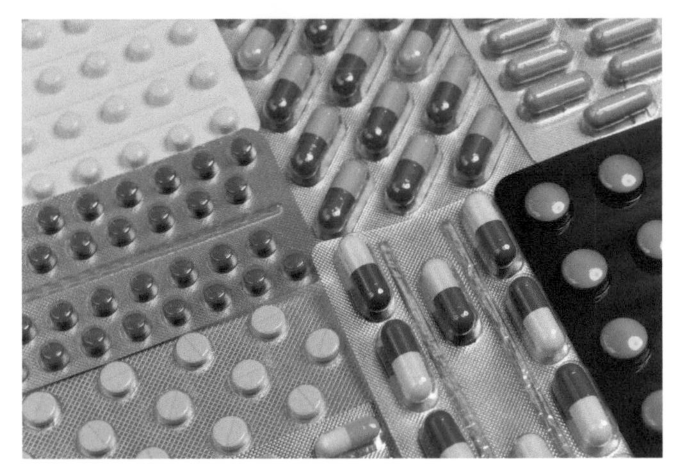

FIGURE 8.5 The development of antibiotic drugs followed from a combination of new information about microbiology and the causal connections between bacteria and disease. *From Shutterstock.*

continued to yield new advances in human understanding and enabled a new arsenal of tools that could be used to fight disease. This led to many of the modern day procedures that are used to control the transmission of disease. This may be viewed in retrospect as a case of additional data collection leading to the refinement and improvement of theories of disease. Once the relevant data were available, the theories could be revised to a point at which the cause of a disease became a more manageable problem to solve. For effective causal reasoning, we must have a hypothesis to test, data to refute or support that hypothesis, and the ability to update our theories in order to accommodate the implications of the new data. This cyclical process of inquiry enabled people to gain an understanding of the mechanisms of disease. Knowledge of the appropriate mechanisms of action for a phenomenon enables people to intervene in the process to disable the cause of undesirable events.

CORRELATION OR CAUSATION?

The Challenge of Establishing Causal Links

There are many events or situations in the environment that are correlated with one another. Some subset of those situations meet the conditions of having a causal relationship. In order to have a causal influence, an event must not have occurred without a second precursor event that caused it. There are many examples of relational correspondences between analogous situations. Relations that have causal power are particularly important when people draw analogies between situations in order to make accurate predictions. If there is a causal element in one situation and we see this operate again in another, we can make accurate predictions based on the cause. Conversely, if we misread a causal influence

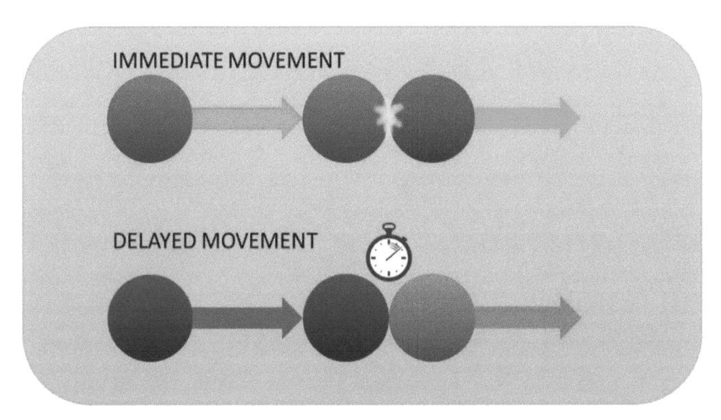

FIGURE 8.6 The timing of movements between two seemingly related events can determine whether we infer one to cause the other. People are much less likely to infer a causal connection between Event A and Event B if an overly long delay occurs between Event A and Event B.

that is not actually present, it can severely undermine our ability to take appropriate actions.

There are several factors that influence whether someone comes to view an event as having *causal power*, which is the ability to produce an effect. The timing between cause and effect is one of the most important features used to establish causal relationships. When two events (Event A and Event B) occur at precisely the same time, we will rarely see Event A as having caused Event B or vice versa. If Event A occurs and many minutes later (or even seconds later in the case of physical causation) Event B occurs, we are also unlikely to infer that Event A caused Event B. For such a delay to occur within a causally linked pair, we would have to have good reasons to hypothesize the existence of a very slow acting mechanism that allows Event A to cause Event B with a time lag (Fig. 8.6). We readily infer cause only when Event A is rapidly followed by Event B. The occurrence of these conditions can even mislead us to infer a causal relationship when Event A does not have causal power over Event B. You can imagine that some unseen force may jointly cause both Event A and Event B, but that Event A occurs a split second before Event B. Such a situation may lead us to erroneously infer that Event A has caused Event B. For example, imagine that you are looking out a window and see a paper bag move toward a crow that is standing on the lawn (Event A). The crow flies off (Event B) and you infer that the paper bag must have spooked the bird causing it to fly away. In actuality, the same gust of strong wind that caused the bag to move had also made the crow uncomfortable causing it to fly away. In this instance the wind acted as an unseen force causing both Event A and Event B.

Spatial proximity is another important feature that can distinguish events with causal power over those that do not have a cause/effect relationship. In the familiar case of collisions, two objects must appear to touch and then rapidly move apart. When objects do not touch, but both move in a temporal order suggesting a causal connection, we would have to infer the action of an unseen force. An object can have an unseen force such as magnetism enabling it to cause another object to move despite a lack of physical contact. Conversely, if objects touch and remain in contact for too long it also disrupts our tendency to draw a causal inference. Again, some sort of unseen transference of a substance or property (such as a fuel) could have occurred and the fuel may provide a mechanism for one object to have caused another to move. When these mechanisms are not apparent, we are less likely to infer the existence of a cause in that situation.

Lastly, we are capable of causing situations to occur ourselves. This can be very obvious, such as when we physically cause an object to move. Less obvious are unseen causes that may exist between our thoughts and actions. These types of connections can become quite difficult for people to understand, as we do not fully understand our own thoughts or the mechanisms of causal inference at the level of the brain. This lack of information can lead us to experience illusions involving cause and our agency over actions. We will examine these fascinating cases further in the sections that follow.

THE CHALLENGES OF ESTABLISHING CAUSE

Illusory Conjunctions

People have a remarkable ability to look for the presence of patterns in the information that we gather. This curiosity to seek an understanding of the mechanisms of action is an intimate part of our intellect. To a large degree this drive emerges from the desire to understand causes. What caused the tree to fall on my car? What caused this unusual rash on my arm? What caused this stock price to drop leading me to lose money? These are the types of causal questions that we routinely seek to answer, and this action sometimes leads us to remarkable discoveries that we would not have otherwise known about. At the best of times this type of motivated reasoning in search of a cause can allow us uncover surprising, unintuitive, and remarkable mechanisms for things. This understanding can allow us to stop a problem from happening, develop a solution, or achieve justice.

There is a clear downside to our zealous causal reasoning. There are times when we lead ourselves to find patterns that are not actually present. This happens routinely, as the environment around us is full of variables that interact and obscure the influences of one another. Some of the complex systems we seek to understand involve too many interactive variables to isolate the actual causal links. For example, technology, the economy, and weather are complex systems that contain many interactive mechanisms. This complexity stifles our attempts to predict the economy or weather with sufficient precision. Many of us routinely experience technology interface problems because we do not possess enough information about the software and hardware that drives our devices. Perhaps the major challenge in grappling with causal relationships within complex mechanisms is that we often do not have enough information to understand a root cause. The key information that we do not know may not be apparent to us. People often forge onward attempting to make sense of information even when a clear pattern is not present in the data that we have available to consider. This can lead people astray in their reasoning.

The search for a cause goes back to child development as we discussed in Chapter 5. In an early developmental study, Inhelder and Piaget (1958) offered children and adolescents ranging from age 10 to 15 years the opportunity to make sense of the possible relationship between hair color and eye color. The children observed a series of cards depicting images of persons who had either blue or brown eyes and either blonde or brown hair color. Once the children had the full card set they were free to sort them, group them, and evaluate the information as they saw fit. The younger children in the study showed a lack of systematic investigation. They tended to gravitate toward sorting the cards based on the first pattern that they noticed, rather than fully evaluating all of the four possible combinations of blue eyed and brown eyed people. For example if a younger child noticed a relationship between brown hair and brown eyes, she would tend to base her estimate of how related these were based on that positive evidence, underweighing the other combination of a brown eyed person with blonde hair. The children's confidence in their judgments was not shaken by the presence of instances that they had failed to account for. Notice also the tendency for the child to base an estimated relationship between variables on positive cases, rather than seeking to find an exception to the rule that she has adopted. This is likely one of the roots of the challenges that people have in testing a hypothesis that they believe to be correct. A similar tendency operates when people fail to change strategy in making decisions at a card table, even after they have lost a considerable amount of money (Fig. 8.7).

FIGURE 8.7 Seizing upon perceived patterns in outcomes can lead us to avoid noticing other contradictory information that refutes the pattern. This tendency to overemphasize a preferred hypothesis can lead us into trouble in card games such as Black Jack. *By Poliander80 (http://creativecommons.org/licenses/by-sa/4.0), via Wikimedia Commons.*

Frequency Matching and Maximizing in Two-Choice Decision Tasks

Some remarkable results occur when people evaluate cause in a two-choice situation. In these experiments, people have great difficulty sorting between correlations and causes within the data. These results come from relatively simple two-choice psychology lab tasks. John Yellott (1969) reported a fascinating experiment capturing people's tendencies to attribute causal information. Yellott presented participants with two lights positioned side-by-side. The participants were asked to predict whether the light on the right or the left would illuminate before each experimental trial. After the prediction, the participant observed one of the lights flash. The location of the illuminated light was random in the majority of the experiments, but one of the lights illuminated more frequently, turning on 80% of the time. Interestingly, participants predicted that this more probable light would illuminate at approximately the 80% level. For example, if the light on the left illuminated at random 80% of the time and the light on the right illuminated the remaining 20% of the time, then the participant tended to predict that the left light would illuminate 8 out of 10 times and the right light 2 out of 10 times. The experiment changed without the subjects' knowledge in the final set of trials. Here, Yellott changed things up, with the light illuminating at whichever location the participant predicted it would. For this part of the experiment the light really did behave according to the participant's predictions. Yellott stopped the experiment after the participants had completed 50 predictions on which they would have been correct with the light bulb location illuminating in direct response to their predicted location. Yellott asked the participants what they thought was causing the light at this point. A large majority of the participants claimed that they had determined the pattern that caused the lights to illuminate and they still continued to prefer choosing the left light that had illuminated 80% of the time during the earlier phase of the experiment. In other words, the participants failed to

FIGURE 8.8 Simple prediction tasks in which people try to predict outcomes can become very complex. People frequently infer that they understand the sequence of two light bulbs even when the outcomes vary randomly. This result demonstrates a tendency for people to overreach in their theories about cause and effect.

realize that they had no control over the location of the light in the earlier trials and had complete control during the final portion of the trials. This was because they had become so convinced that they really could figure out the cause of the light during the whole experiment!

When people are asked what caused the light pattern they will tend to provide a lengthy and complicated description of how they learned an elaborate sequence that predicted the behavior of the light bulbs. For example, they may state something like "I finally got it, when I started predicting the bulb on the right three times in a row, then the bulb on the left twice, followed by an alternation of right and left for the next four trials." Yellott's (1969) study demonstrates that people search for causal sequences in random information and can become led to believe that they understand a sequence even when it is actually random or variable (Fig. 8.8).

The tendency for people to try to find causal influences and clear patterns within random data leads to a suboptimal strategy. Let's imagine that Phil is offered a choice of two lights that will illuminate in a sequence of trials. The left light illuminates 80% of the time. Phil is likely to predict that the left light will activate about 80% of the time and the right light about 20% of the time. If we think about this from a probability perspective the probability of the left light illuminating can be represented by P, while the probability for the right light is $(1-P)$. When we compute the probability of correctly predicting the location of the illuminated light, we arrive at $(.8\times.8+.2\times.2)=68\%$ success. This tendency to select each light to the degree that each option is successful is known as frequency matching, as the person selects each option in a percentage that matches its successful occurrence (Estes, 1961). The obvious problem with frequency matching strategy is that Phil could be successful 80% of the time if he adopted the much simpler strategy of always predicting the more probable option (the left light) and accumulate success every time that this bulb illuminates. This more profitable strategy is known as maximizing, as the decision maker is going to achieve the maximal amount of success when taking that strategy (Wolford, Miller, & Gazzaniga, 2000). Remarkably, animal studies of these types of sequence prediction tasks reveal that other species will adopt a maximizing strategy when they are faced with a two-choice problem (typically linking outcomes to food rewards). Other species including rats and pigeons will perform such experiments provided they are well compensated for their efforts (Hinson & Staddon, 1983). As we discussed in Chapter 4 many animals are keenly sensitive to the amount of effort that they put forth in relation to the number of calories that they obtain.

Why do animal species with less complex brains outperform people on simple binary choice tasks? In other words, why do people fail to adopt a maximizing strategy when many other species will in this type of task? Instead people persist in often predicting the less-probable option in the hopes of attaining successful predictions on every trial within the task. Psychologists Wolford et al. (2000) proposed that people carry out a frequency matching strategy because they formulate and test causal hypotheses about the behavior of the two lights. People, unlike animals, tend to believe that they can figure out the best strategy and that they can discern the cause of the light sequence that they are presented with. This can lead people to infer a wide variety of unusual and complicated rules. It leads people to conduct experimentation with the outcomes, and to overthinking that leaves the person to be worse off than if they had taken a simpler path by always predicting the choice that led to success more often. While we may be less successful at these simple probability decision tasks, the fact that we conceptualize the task to be more complex than it is certainly helps us in other reasoning contexts. It is as if our tendency toward complex pattern judgment gets in the way of noticing the simpler and more optimal strategy.

The Neuroscience of Frequency Matching and Maximizing

What about the brain basis of matching and maximizing behaviors? There is relatively strong evidence that the striatal dopaminergic system of the basal ganglia participates in reward seeking behavior and can generate error signals when reward-seeking decisions are inaccurate (Schultz, Dayan, & Montague, 1997). Such a signal would be of tremendous value to an organism seeking to make the best predictions, as error signals provide feedback so that future decisions can be adjusted. It is possible that this basal ganglia system underlies the

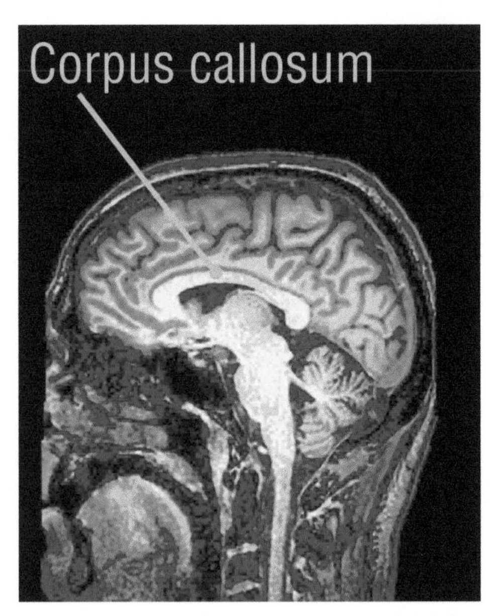

FIGURE 8.9 The corpus callosum is a densely packed band of white matter, the axons of the brain that provide a connection between the two hemispheres of the brain. Corpus callosotomy, or split-brain surgeries, involved surgically cutting this band of white matter in order to prevent seizures from spreading from one side of the body to the other.

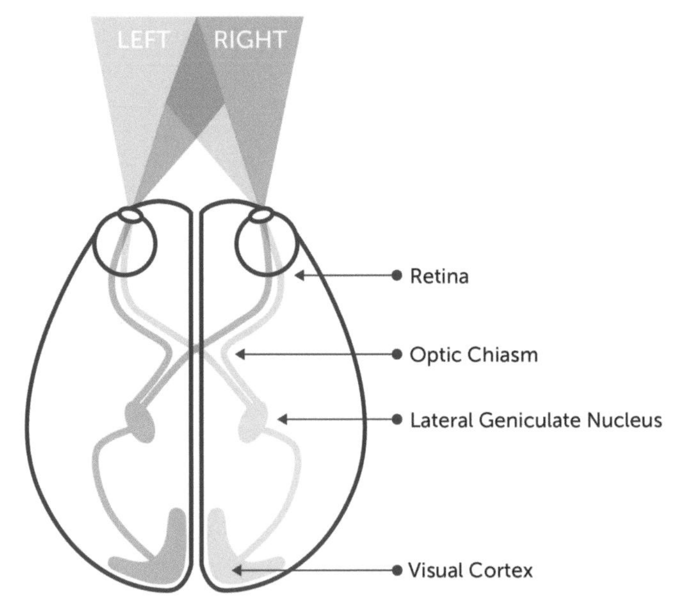

FIGURE 8.10 Information can be presented preferentially to one hemisphere over the other in cases in which people have had a corpus callosotomy surgery. The setup for the Wolford et al. (2000) study showed a binary choice task to each hemisphere independently by isolating each choice pair to one side of a computer screen. *Reproduced by Ashton Miller; Adapted from Wolford, G., Miller, M. B., & Gazzaniga, M. (2000). The left hemisphere's role in hypothesis formation.* Journal of Neuroscience, 20*(RC64), 1–4.*

predictions that nonhuman species make on two-choice tasks, since maximizing behavior is common in nonhuman species and is also the optimal strategy. Performing these tasks by engaging in a pure reinforcement learning process should eventually lead toward a maximizing strategy. Wolford et al. (2000) postulated that in humans, the left frontal lobe is especially involved in searching for clues toward establishing causality in patterns. They noted that the left frontal lobe tends to be more active in language-based tasks and may have a privileged role in generating symbolic representations.

Wolford et al. (2000) tested their hypothesis about the left frontal lobe as a region important for underlying the search for causal information in random pattern sequences in a patient study. A split-brain operation was performed in the past in order to control epileptic seizure activity from spreading from one hemisphere to a bilateral seizure. The corpus callosum is a dense band of white matter fibers that connect the two hemispheres (Fig. 8.9). A split-brain surgery involved cutting the corpus callosum in order to control the spread of neural activity. This type of surgery made it possible to present information predominantly to one hemisphere over the other by asking the patient to maintain eye fixation on the middle of a screen and then presenting information visually on one side or the other. For example, when information is presented to the right side of space it will enter the brain via the left hemisphere. This is due to the crossing fibers that connect the visual system to the hemispheres. In a non-split-brain participant the information would quickly transfer between hemispheres due to the connections provided by the corpus callosum. When the corpus callosum has been severed, the information will remain "locked" or isolated into the hemisphere opposite to the visual field to which the information was presented. Wolford et al. suspected that neural networks within the left hemisphere preferentially control people's causal search behavior. If that is true, then split-brain patients should show different prediction strategies when they predominantly rely upon the left hemisphere over the right hemisphere in an outcome prediction task.

Wolford et al. presented split-brain participants with a red and a green square displayed one above the other on each side of a computer screen. Participants predicted that either the green or the red square would be correct on each of a series of trials. The pair of squares on the right side of the screen was assigned an 80% versus 20% split on which would be correct on any given trial. The left hemisphere had a 70%–30% split on red-green correctness (Fig. 8.10). Remarkably, when participants made predictions about the squares on the left side of the screen (using predominantly the right hemisphere), they employed a maximizing strategy by nearly always predicting the square that lit up 80% of the time. This suggests that the right hemisphere does not carry out the matching behavior that is characteristic of someone who is actively testing causal hypotheses about the options. For these squares processed by the right hemisphere, the participant was correct roughly the full 80% of the time. Meanwhile, when the stimuli were presented to the left hemisphere from the right side of the screen, these same participants carried out a matching strategy by selecting the more probable square about 70% of the time and predicting the less probable square the remaining 30% of the time. In other words, Wolford et al. (2000) demonstrated maximizing behavior when the split-brain participants predominantly processed the prediction task using the right hemisphere and matching behavior in these same participants when they processed the task primarily with their left hemisphere. This result supports the idea that searching for a particular pattern that will allow one to fully predict the task is dominated by the left hemisphere over the right.

To further evaluate the cortical basis for probability matching and the search for causal explanations, Wolford et al. (2000) reported the results of a second study evaluating the role of the frontal lobes in prediction behavior. In modern neuroscience, we rarely view information processing as involving an entire cerebral hemisphere. To narrow down the anatomical focus for making complex predictions, Wolford, Miller, and Gazzaniga tested the hypothesis that the left prefrontal cortex (PFC) was likely responsible for the probability matching behavior. They cleverly tested this hypothesis by recruiting individuals who had a stroke that had selectively damaged one side of the PFC. Five of the participants had right sided PFC damage and one participant had left PFC damage. The experimenters presented the same split-screen prediction task (Fig. 8.10) with participants guessing which square was most likely to illuminate on each trial. In this case, the patients with damage to the right hemisphere all exhibited a frequency matching strategy. This indicates that they were searching for causes for the squares illuminating and testing hypotheses, all the while making suboptimal decisions and being incorrect more often than if they simply chose the more probable square on every trial. The patient who had left hemisphere PFC damage exhibited the maximization strategy, further supporting the hypothesis that the left PFC is important for seeking to explain patterns in incoming data. This study serves as a further indication that the PFC is important for formulating and testing causal hypotheses in the face of complex probabilistic data. Interestingly, there appears to be a hemispheric asymmetry in which the left, but not right PFC, drives this hypothesis testing behavior.

The context of these prediction tasks also likely influences human behavior. When people enter a laboratory environment, they may infer that their behavior will be observed and scrutinized by others. They may anticipate that experiments involve "cover stories" that the experimenters have manufactured in order to obscure the actual purpose of the study. Thus, people may believe that they can determine the correct way to perform a selection task when faced with varying amounts of correct feedback. In other words, people may assume that in a lab environment they are faced with an artificial system that the experimenters have constructed to determine whether participants can learn to perform the task at a fully correct level on all trials. For participants, learning such a task may mean figuring out how to receive positive feedback on all possible trials. People may hold themselves to this unrealistic standard of achievement in labs and therefore fail to notice that a maximizing strategy, in which they get mostly positive feedback, is actually the correct strategy. Human participants, unlike animals, may be resistant to accept the idea that they have limited control over the outcomes. Many participants may feel as if they can cause a positive outcome every time and therefore there may be a particular sequence that the experimenters developed that can be discovered and adopted. In effect people may believe they can outsmart the experimental task. Such thinking likely leads to the tendency for people to develop increasingly complex hypotheses about the order in which the correct choices need to be predicted in order to always get the prediction correct. These hypotheses can then be tested in the form of applying them to new trials, only to realize that they are never quite correct and need to be continually modified.

ILLUSIONS OF CONTROL

Inferring Control Over Outcomes

When people set to work testing hypotheses, they may preferentially tend toward looking for evidence that supports their predictions. There may be a tendency to look for positive feedback when people are asked to predict outcomes. In other words, if success is being achieved most of the time, then people often misinterpret this occurrence as an indication that they are on the correct path toward solving or mastering the task. People tend to undervalue negative outcomes, or trials, during which the current hypothesis is not supported. Unfortunately, this same tendency makes it difficult for people to properly test scientific theories as well. People tend to assume that there has been a glitch or a rare occurrence when their current theory, or hypothesis, is not supported. Meanwhile studies that do show support for a preferred hypothesis may be overly emphasized. People's tendency to overweigh positive outcomes in prediction tasks can lead them to falsely believe that they control the outcomes and have found ways to cause or control outcomes that are actually either random or mostly removed from their control.

An early study evaluating people's ability to reason about cause demonstrated people's tendency to infer illusory correlations between situations. Jan Smedslund (1963)

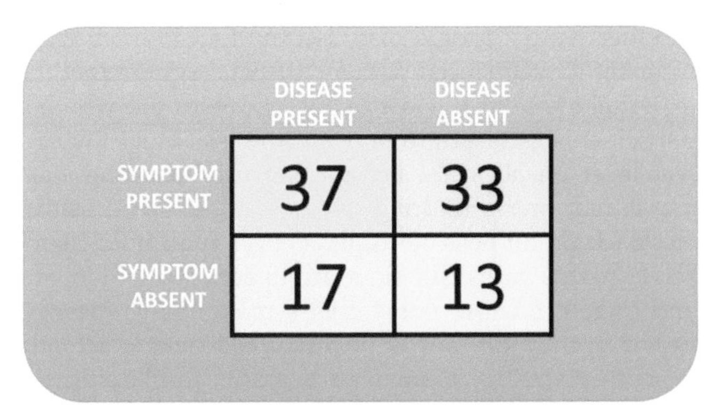

FIGURE 8.11 Jan Smedslund (1963) investigated the connection between symptoms and diseases in a study of nurses.

conducted an experiment inspired by the connection between symptoms and diseases in the medical world. Smedslund asked nurses to evaluate potential relationships between medical patients' symptoms and possible diagnoses. The nurses were provided with a deck of 100 cards bearing a series of letters that were meant to represent different symptoms that patients express. They were offered a second deck of 100 cards with a different series of letters representing possible diagnoses for these patients. Fig. 8.11 shows a table with the four possible contingency relationships that existed in the Smedslund experiment (symptom present and disease present, symptom present and disease absent, symptom absent and disease present, and symptom absent and disease absent). This type of table can be produced for the evaluation of any type of contingency relationship. As shown in Fig. 8.11, the cell in the upper left contains 37 cards for which both the symptom and the disease were present. The upper right cell indicates that there are 33 cases in which the symptom was present in the absence of the disease. In the lower portion of Fig. 8.11 there are 17 cards indicating the number of patients who did not express symptoms, but did have the disease, and 13 cards indicating patients with neither the symptom nor the disease. There are a few key points to notice about the setup of Smedslund's experiment. The highest number of cards indicated both the symptom and the disease being present. Also notice that there is an overall high prevalence of the symptoms with a full 70 of the 100 cards indicating people who are expressing these symptoms.

Results from Jan Smedslund's nursing study indicated that people tended to overestimate the correlation between symptoms and disease. He asked the nursing sample to estimate the relationships between the symptom and the disease finding that the nurses tended to endorse a higher correlation than what actually existed in the cases that had been provided. The nurses may have been overly swayed by the high prevalence of the symptoms overall and the fact that the largest number of cases (over one-third of the sample) consisted of symptom-present and disease-present cases. This pattern of results suggested that people tend to look for the positive relationships rather than calculate the actual proportions of four possible cases (Fig. 8.11). The study may serve to indicate that people tend to pay too little attention to the instances in which the symptoms occur without the disease (also a high number in this study). This experiment may illustrate a case of neglecting the possibility that the disease does not actually cause the symptoms. It is a difficult situation to properly evaluate because health is a highly complex domain. Few diseases cause a clear-cut symptom that is always present in all people with the disease. Instead healthcare professionals are often presented with people whose symptoms are masked by other factors such as allergies or amounts of sleep. Nurses also evaluate people who express symptoms mimicking a particular disease that are attributable to another cause entirely. Those involved in healthcare must learn to operate based on incomplete information in those types of cases. They must also engage in generating plausible theories that have to be tested by evidence during the workflow of seeing patients. One of the striking features of the study by Smedslund (1963) was the finding that a large majority of nurses appeared to be unaware of their overestimation of the correlation between symptom and disease. They had inferred a causal link between disease and symptom that was not supported by the frequency data that they had been presented with. Later work by Vallée-Tourangeau et al. (1998) indicated that Smedslund's original results may be complicated by the relatively high base-rate for the disease overall and that people are better capable of making judgments that come closer to the real frequencies when positive instances of a disease (or symptoms) are lower.

Jenkins and Ward (1965) conducted a landmark illusion of control study evaluating the effects of numbers of positive instances that people experience. In this study, participants entered a lab and were asked to press buttons in order to achieve particular outcomes. Participants were offered two buttons to press and two light bulbs that may or may not illuminate. This experimental setup resembles that used by Yellott (1969) who had investigated people's tendency to frequency match on two-choice prediction tasks. The light bulbs in the Jenkins and Ward study were labeled "score" and "no score" with the participant trying to make the "score" light illuminate as often as possible on the basis of pressing one of these two buttons (Fig. 8.12). Jenkins and Ward were interested in the degree to which people claimed that they could control the sequence of each light illuminating. In other words, they asked people how strongly the buttons were causally linked to the outcomes. The theory behind this experiment is that the actual amount of

FIGURE 8.12 Jenkins and Ward (1965) offered participants a prediction task in which they were trying to illuminate the "score" bulb. If this bulb illuminated frequently, people misattributed a high degree of control to themselves.

control is determined by the probabilities of success for each of the two buttons that can be pressed on each trial. These probabilities can be represented by the terms *P* and *Q* with the actual level of causal influence being represented by the probability of "scoring" when pressing Button 1 (*P*) minus the probability of "scoring" by pressing Button 2 (*Q*). So, if Button 1 illuminates the score light on 80% of the trials (probability *P*) and Button 2 leads to a score on 25% of the trials (probability *Q*) then the actual control would equal (*P* – Q), which in this instance amounts to (80% – 25%) = 55% control. Participants in the experiments by Jenkins and Ward had an opportunity to participate in 60 trials with the instruction that they would be asked to estimate their degree of control over the outcomes. The instructions provided to participants are shown below:

> After each problem you are to indicate your judgment of control by putting an "X" some place on the scale: at 100 if complete control has been achieved, at 0 if no control has been achieved, and somewhere between these extremes if some but not complete control has been achieved over the outcomes. Complete control means that you can produce the score light or the no score light (alternatively, the circle or the square) on any trial by your choice of responses. No control means that you have found no way to make response choices so as to influence the outcomes. Intermediate degrees of control mean that your choice of responses influences which outcome appears even though it does not completely determine the outcome. *Jenkins and Ward (1965, p. 6)*

When Jenkins and Ward allowed participants to try their hand at attempting to illuminate the score light using the two buttons, the participants behaved rather differently than would be predicted based purely on probabilities. The participants were also asked to estimate how much control they had over the outcomes of the two bulbs. The overall result from several studies indicated that people paid the most attention to the percentage of times that the most probable button resulted in the score light illuminating. For example, if one of the buttons yielded an 80% success rate for the score light after it had been pushed, participants rated their control of the outcomes to be high. Remarkably, participants estimated their control over the outcomes to be similar even when the buttons had no effect on the outcomes. In these cases the score light illuminated 80% of the time regardless of which button had been pressed. This result suggests that people tend to attribute causal influence over successful outcomes to themselves even when they do not actually possess a clear level of control. This illusory sense of control may stem from the same type of mental process that leads people to continually try pressing a button or making a prediction in favor of a less probable outcome in the hopes that they may cause the desired outcome every time. It also suggests that people may tend to take too much credit for positive outcomes and are not likely to realistically evaluate the probability of an outcome that is less common. People may overthink the situation emphasizing positive outcomes and thereby miss the fact that they are off the mark in terms of tracking actual probabilities as can be seen in Box 8.1.

There are additional factors that influence people's beliefs about their ability to cause outcomes. Lauren Alloy and Lyn Abramson (1979) conducted an interesting study evaluating the effects of emotion on people's estimates of control over contingency data. In this case Alloy and Abramson manipulated the mental state of participants by looking at differences between participants reporting depressive symptoms over those that did not. Depressive symptoms were evaluated using a self-report measure called the Beck Depression Inventory. This instrument allows an assessment of recent experience with depressive symptoms, rather than an acute in-the-moment assessment of depressive state. In this study, the experimenters also linked the outcomes to financial incentives.

Results indicated that both financial outcomes and depressive states influenced the assessment of control over outcomes by participants. First, all participants tended to report that they had a high degree of control over outcomes when they were paid 25 cents for each positive outcome. This result was obtained when both depressed and non-depressed participants had to press a button to see if they could make a rewarding outcome occur. Second, both participant groups tended to claim to have low levels of control when a successful outcome meant no loss of money, while a failed outcome meant that they lost 25 cents. Interesting differences were observed for the participants who had depressive symptoms, who (accurately) claimed to have little control over outcomes that were *not* linked to their pressing of a button for rewarding positive outcomes (with successful outcomes paying them 25 cents). In summary, non-depressed people tend to show an illusion of control under conditions in which they can gain money

BOX 8.1

TRYING TO CONTROL OUTCOMES IN DAILY LIFE

People tend to overestimate their level of control over outcomes in studies in which they receive a lot of positive feedback. They also tend to estimate having much less control over situations in which little positive feedback is offered. What do these results suggest about our daily lives?

For a period of my life I lived in Berkeley, California, and did a lot of walking around the campus. There was a relatively high volume of car traffic in the area, and you had to be fairly careful crossing the streets on foot. Like most cities, the streets had crosswalks with traffic lights. In order to illuminate the protected walk signal, pedestrians had to have already pressed the "walk" button prior to a green light occurring on the traffic signal. If you failed to press the button, you got no "walk" signal on that cycle of the traffic signal light. Pressing the button to call the "walk" signal did not influence the timing of the traffic lights. Pressing the button simply meant that you would get the "walk" signal when the green traffic light cycled through. Nonetheless, I can recall on many occasions someone, usually in a rush, pressing the call button for the walk sign numerous times and even edging out into the street directly after pressing it, as if the button would immediately cause the car traffic to receive a red light so that the pedestrians could cross (Fig. 8.13). It would be nice for pedestrians (and rather difficult for vehicle drivers) if crosswalks worked that way. In the end, the "walk" signal appears no matter how many times we may have pushed the button. It is a positive outcome and may reinforce our button-jabbing behavior even if one press is all it actually takes.

Similarly, have you ever been waiting for an elevator to return to the ground floor and seen someone repeatedly press the call button to go up or down? I am often struck by people's tendency to press the button over and over, even if it is already clearly lit. I have never heard of an elevator that prioritizes the next stop as being the floor that has registered the most call button presses!

These circumstances emulate the conditions of causal reasoning experiments. My own personal theory is that people tend to feel a bit helpless when they find themselves waiting for a mechanical device to allow them to proceed. The presence of a button to call the elevator, or the crosswalk signal, offers a refuge for the impatient. While this button may have already been pressed and there is little else it can do to speed one on his or her way, it offers a psychological tool that gives people an opportunity to exert an attempt at control. By repeatedly pressing the button, it may feel like you're doing all you can to get the elevator to your floor. I have even found myself engaging in this relatively silly behavior. I was recently running late for a meeting and found myself repeatedly poking at the call button for the elevator, which was obviously already on its way down to my floor. Somehow it just felt right to keep on pressing. When under duress, repeated pressing seemed like I was doing everything in my power to make to that meeting, even though my rational brain knew better!

FIGURE 8.13 We sometimes find ourselves trying to exert control over crossing signals and elevators by simply pressing a button repeatedly.

for successes, even when they actually have no control over those successes. When people potentially can lose money for failed outcomes, they appear to become much more tuned into whether they possess causal control over the outcomes. Under these conditions people more appropriately deny having much control over outcomes. Fascinatingly, these experiments suggest that a tendency toward depression makes people more realistic as well. A belief that one controls aspects of one's life tends to be linked to positive mental health outcomes. The participants who are more realistic about perceived control but tend toward depression may be sadder but wiser.

There may be some positive mental health benefits to believing that we have control over outcomes as demonstrated in this experiment. Alloy and Tabachnik (1984) suggested that the role of depressive symptoms in the generation of illusory control may be attributable to an expectation that one will be able to control outcomes. The participants who are non-depressed tend to believe that they will be able to find a way to control outcomes and will proceed to generate and test hypotheses about these possibilities extensively. Meanwhile, the depressed individuals may not believe that they can control outcomes and therefore they do not continue to try to find patterns. The higher the reinforcement (and the higher the valence of reinforcement, in the case of monetary consequences), the more control the participants attribute to themselves, provided that they expect to be able to control the outcomes.

Helena Matute revisited the illusion of control in 1996. Matute conducted a version of the study in which a tone was presented to the participant as an outcome for choosing to press (or not press) a single button. Participants in an avoidance condition were asked to try to avoid annoying computerized sounds by either pressing the button or choosing not to press it. In a second condition, participants pressed or did not press the button on half of the trials in order to determine the degree of control that they had over the outcomes. In both conditions, the participants received positive outcomes on 75% of the trials (by avoiding the noise presentation). Surprisingly, the instruction type influenced the degree of control over outcomes reported by the participants. The noise-avoidance participants claimed to have a high degree of control (which was an illusion, as they did not control the outcomes at all). Critically, this group had tended to press the button on the majority of trials. Meanwhile the participants who were instructed to press the button on only half of the trials experienced a much larger proportion of instances in which they did not press the button and did receive a positive outcome anyway. This group, unlike the noise-avoidance participants, experienced more negative outcomes that successfully deterred them from over-crediting themselves for the occurrence of positive outcomes. It may be that depression and punishment tend to dampen enthusiasm over pressing buttons and attributing credit to oneself when judging contingencies.

Hypotheses, Guessing, and the Brain

More recent efforts have sought to determine the role of the brain in guessing and hypothesis testing situations. In an early neuroimaging study of contingency judgment, Elliott, Rees, and Dolan (1999) at University College, London, asked participants to make predictions and evaluate hypotheses while undergoing functional magnetic resonance imaging (fMRI) scanning. In this study, participants were asked to guess which one of two playing cards was "correct" on each trial. This was essentially a guessing task and participants had no actual control over outcomes, as it had been predetermined by the experimenters how often they would be informed that they were correct, regardless of which card they selected. Participants were informed after each guess whether or not they had chosen correctly. A graphic depicting a green bar was displayed for the participants, and it indicated a cumulative monetary value obtained during the task. This bar was adjusted up or down depending on whether the card was shown to be correct or not. Participants reported that they had developed their own decision strategies based on the colors, suits, or numbers of the cards, just as people tend to hypothesize about possible sequences that will yield positive outcomes in illusion of control tasks. Neuroimaging results indicated that different brain areas were involved in processing the wins and losses associated with the card selections. The most consistent activation related to winning monetary rewards occurred within the thalamus, striatum, and subgenual cingulate gyrus. Meanwhile, the hippocampus was most associated with losing money for incorrect guesses. These same brain regions showed sustained activity when participants hit upon a streak of either successive wins or losses (which had been planned by the experimenters). There were also some brain areas that were associated with the general process of making card guesses. These included the orbitofrontal cortex, the caudate nucleus, and insular cortex. All of these regions were active in all cases and showed sustained activity during streaks of either outcome. The authors suggested that the areas associated with streaks may underlie the emotional highs that gamblers associate with winning a number of times. Such a situation occurs in many casinos in the slot machine areas (Fig. 8.14). Activation in

FIGURE 8.14 Gambling devices such as slot machines can sometimes pay out small amounts of money repeatedly. This can make a gambler feel that they are "on a roll" and encourage further betting due to an enhanced sense of control over outcomes. *From Shutterstock.*

the orbitofrontal cortex, caudate nucleus, and insular cortex was associated with task periods that participants reported as being the most exciting, such as when they felt that they had figured out the correct pattern that would yield the most gain in the experiment.

PERCEIVING CAUSAL INFLUENCE

The Launching Effect

In his landmark book entitled *The Perception of Causality*, Belgian psychologist Albert Michotte discussed the way that we perceive events to be causal. Michotte's famous example was to show people images of two balls, similar to those that could be observed interacting on a billiards table. Michotte claimed that we perceive causal influence automatically and with little effort, provided that the characteristics of a situation are consistent with our expectations about cause. Michotte demonstrated cases in which one of the balls approached the other one, they touched very briefly, then the second ball moved off. In these cases people tend to report that the first ball caused the second ball to move. Michotte called such instances "launching effects" claiming that these events could be perceived with little effort. The viewer would infer that the first ball (*the launcher*) confers movement upon the second ball (*the target*). The causal inference in this task can only be obtained under certain critical conditions. First, the effect will only be perceived if the contact between the launcher and the target is very brief. When the launcher is seen to linger next to the target for multiple seconds, then the target is perceived to move away based on some other force not applied by the launcher. This effect is also disrupted if the target begins to move off before the launcher has arrived. Secondly, the speed at which the launcher and the target move must be within a close range. For example, the causal inference will not occur if the launcher arrives at too high a rate of speed relative to the subsequent movement of the target. The causal connection will also be broken for most people if the target moves off at too high a rate of speed relative to the initial speed of the launcher. Third, the launcher and the target must move in the same direction for the causal inference to hold. If, for example, the launcher moves from right to left and the target then moves from top to bottom, this would not be seen as a plausible result and no cause would be perceived.

Fugelsang, Roser, Corballis, Gazzaniga, and Dunbar (2005) investigated the brain basis for such causal inferences using a billiard ball style task similar to that used by Michotte in his original work (1963). In this study, participants were presented with visual displays of different types of ball collisions (Fig. 8.15). In causal trials, participants viewed Ball A move toward Ball B and then stop as the two touched very briefly, before Ball B moved away in the same direction that Ball A had been moving just before. These are the conditions that Michotte referred to as eliciting a launching effect, in which people perceive a causal influence from Ball A to Ball B. The participants also viewed two other ball collisions that occurred under conditions that are not typically seen as causal by participants. In a noncausal temporal delay condition, Ball A moved toward Ball B and stopped. The two balls then touched for 170 ms, a relatively long time with no movement. After the delay, Ball B moved off in the same direction that Ball A had been moving earlier. This type of event is not perceived to be causal, as it fails to replicate action that can be observed in actual collisions in which one ball imparts a causal influence on the other. A second noncausal trial condition involved a spatial delay. In this case, Ball A moved toward Ball B and stopped prior to actually touching Ball B. There remained over 1 cm of distance between the two balls, when Ball B moved off; again in the same direction that Ball A had moved. Fugelsang et al. were interested in the brain regions associated with making these causal and noncausal inferences under conditions that had been specified by earlier research.

FIGURE 8.15 Ball collisions make excellent stimuli for causal reasoning investigations. The speed and direction of the two balls will determine whether people tend to infer that one has caused the other to move. Additionally, timing is also important in influencing the perception of causality in collision experiments.

The results indicated that causal influences involved several unique brain areas. When the causal contiguous trials were analyzed for activation that subtracted out the effects of the noncausal stimuli, the investigators reported activation within the right middle frontal gyrus and the right inferior parietal lobule. This indicated that primarily right hemispheric regions were involved in the perception of causal structure from perceptual displays. These results indicated that making a causal inference leads to increases in activation in prefrontal and parietal lobe regions that are often associated with higher-order executive function and attention resources. These additional right hemisphere areas activate in addition to visual processing regions that are involved in the perception of simple motion, which involve the occipital and ventral temporal cortex.

A follow-up study conducted by Benjamin Straube and Anjan Chatterjee (2010) reexamined the brain basis for causal inferences associated with ball collision tasks. The authors noted that there are individual differences in perception that underlie the causal inference. They set up displays depicting a red ball colliding with a blue ball. The blue ball moved first, rolling horizontally for a duration of 1 s. It then stopped once it contacted the red ball. Next, the red ball moved in one of two possible ways. First, it varied somewhat in the direction that it moved proceeding away from the blue ball at different angles. Second, the red ball varied in terms of its delay between being contacted by the blue ball and beginning its own movement. The investigators held the velocity and the length of the ball's trajectories constant across both of the conditions. Straube and Chatterjee reported interesting results driven by individual differences in the perception of cause and effect. The investigators used a statistical technique called logistical regression in order to derive an individually sensitive set of characteristics for the precise conditions during which different people perceived the blue ball to cause the red ball to move. This analysis included the stimulus parameters of angle, time delay, and their interaction to reveal large individual differences. Furthermore, they demonstrated that there were no consistent effects on people's reported causal inferences of the interaction between the ball approach angle and the time delay prior to movement of the red ball.

The neuroimaging results of Straube and Chatterjee extend and complement those reported by Fugelsang et al. (2005). These authors indicated that increases in the participants' sensitivity to the angle of movement of the red ball were associated with increases in activation within the right postcentral gyrus. These active clusters extended across the superior and inferior parietal cortex as well. Participants' increasing sensitivity to the time delay for the red ball to move was correlated with increases in activation within the left putamen, a region of the basal ganglia. The associations between the brain and inferences were based on correlations of inter-individual differences in activation in response to the spatial and temporal properties of the displays. The results also reflect inter-individual differences in

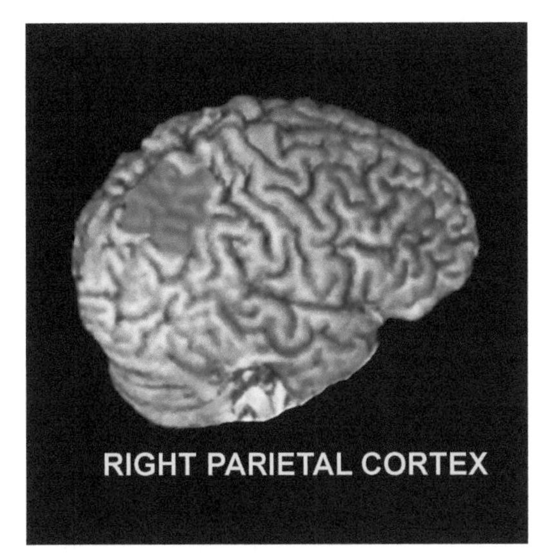

FIGURE 8.16 The right parietal cortex appears to be an area that is important for evaluating the spatial aspects of ball movements for possible causal connections between balls that we observe colliding under varying conditions.

the specific qualities, or settings of spatial and temporal cues, that participants had used to form their judgments about the causal influence of the blue ball on the red ball. These results indicate right parietal activation as having a strong relationship to causal judgments related to space (Fig. 8.16). This is consistent with the prior report by Fugelsang et al. (2005) indicating a right hemisphere parietal involvement in causal judgments. Interestingly, the putamen was sensitive to timing, and this result was novel at the time of its reporting. The results reported by Straube and Chatterjee indicate that spatial and temporal aspects of these perceptual causal judgments are independent of one another. While both spatial angles and timing are important for conveying causal influence, these dimensions appear to be separate at the level of the brain, with different regions tracking different aspects of information that relate to inferring cause.

Illusions of Conscious Willed Action

The subjective sense of control may also arise in situations in which our thoughts appear to link to actions that follow. The illusion of control can extend from predicting the likelihood of different events to other situations in which it appears that our thinking is linked to some particular action. This type of link may be the reason that people infer that the mind has mystical properties and that extrasensory perception may be possible, as it can appear that we sometimes influence physical events with only our thoughts. This phenomenon may be why people sometimes claim that supernatural or magical events have taken place, which simply has no physical explanation.

Psychologist Daniel Wegner has spent considerable time investigating the sense of conscious will in action. He was particularly interested in cases in which the cause of an event was not under someone's control, yet they reported having the experience that it was. These types of illusions can provide interesting clues about how we reason about cause in our daily life as we try to make sense of complex occurrences that we experience.

Wegner and Wheatley (1999) reported the results of a task called the "I Spy" test to investigate the illusion of conscious willful action. They designed the test based on inspiration from the *Ouija Board*, a popular board game that is played in groups, in which people report losing track of the cause for the actions of the board (refer to Box 8.2). Participants in the "I Spy" test arrived at an experimental testing room at about the same time as a

BOX 8.2

THE OUIJA BOARD: ANSWERING QUESTIONS ABOUT THE PAST, PRESENT, AND FUTURE

A very common scene from horror films involves a group of people sitting down at a table in a candlelit room to perform a séance. The séance involves a leader of the ceremony, called a medium, who is billed as an expert at contacting long deceased people from the spirit world. Often times a movie scene depicting a séance features a table that shakes, the candles being blown out, and someone either scribbles or speaks a profound message from the great beyond once the spirit being contacted has entered the room. Where do such ideas come from?

In the 1800s, the life expectancy was much lower than it is today (Fig. 8.3). Deaths by disease, child birth, and warfare were far more common than they are currently. Also, people were perhaps more apt to believe in supernatural causes, including the presence of ghosts, witches, and magic. People in past centuries had access to far less information about the mechanisms of action of disease, natural disasters, and weather. At the core of this thinking is the notion that events in the world can be caused by unseen forces. These forces were thought to be linked to the spirits of deceased individuals and that such individuals can be communicated with if one knows the appropriate procedures and has access to a medium to lead the ceremony. Holding a séance to commune with the dead was not uncommon in the United States during the 1800s. Even First Lady Mary Todd Lincoln held séances intended to contact family members (McRobbie, 2013). It was relatively common for a séance to involve a group of people experiencing a shaking table, or surprising movements within the room, even though no single person appeared to be deliberately causing the furniture to move. The table might shake or slide due to the combined minimal efforts of numerous people in the group.

In 1891 a new product was unveiled claiming to be "Ouija, the Wonderful Talking Board." The new product was billed as being a magical device that answered questions "about the past, present, and future with marvelous accuracy" and that it was "a link between the known and unknown, the material and immaterial" (McRobbie, 2013). Businessman Charles Kennard and a small group of investors launched the Kennard Novelty Company in 1890 in order to market this entertainment device to the masses. In some sense a Ouija board offered people an opportunity to communicate with the spirit world in a somewhat lighthearted and efficient way by asking questions and appearing to receive answers from the spirit realm.

The Ouija "talking board" is essentially a combination of a supernatural device and a board game all in one. It consists of a flat piece of wood or cardboard that contains the letters of the alphabet displayed above the numbers zero through nine. It also features the words "yes" and "no" at the corners of the board and the word "goodbye" displayed in the lower part of the board (Fig. 8.17). The action associated with the board is achieved by a small device called a "planchette," which is moved around the board. When people engage the Ouija board, multiple individuals lightly touch the planchette and ask a question of the mysterious board. The planchette then eerily moves toward the different numbers, letters, or words answering people's questions. In a well-performed Ouija session, no single person feels the experience of having purposefully moved the planchette. This makes the Ouija board seem somewhat mystical, as the collective efforts of several people are responsible for the movement indicating letters or numbers.

The Ouija board was originally sold as a combination game and mystical device. How does the Ouija board appear to have moved to letters and numbers that actually spell out coherent answers to the questions? The answer may lie in the phenomenon of priming. You will find discussions of priming throughout this book, as it relates to many aspects of reasoning and decision making. Priming involves exposing people to some information, often for a short duration. Once exposed, people will be "primed"

FIGURE 8.17 The Ouija board is a game that was developed in the late 1800s. People place their hands on the small "planchette" marker and it appears to move of its own will to spell out the answers to questions using numbers, letters, and words. The collective nature of the exercise can blur the perception of cause and effect and make it appear that nobody is directly causing the board to "answer" these questions. Such conditions sometimes encourage people to draw inferences that supernatural influences have intervened.

or more likely to retrieve information related to what they have just seen. For instance, I recently had a somewhat mystical experience when I found myself thinking about a neighbor while in the checkout line at the grocery store. Almost supernaturally, that neighbor appeared and said hello to me. It seemed to be a remarkable coincidence. I believe priming may be the root cause of this occurrence. There is a good chance I saw my neighbor rapidly, but with too little exposure to register consciously. Even subliminal presentations of visual images can influence our later processing, as we will discuss in Chapter 11 on decision making. The unconscious presentation of a neighbor can prime thoughts about that person and then consciously seeing that neighbor appears mystical. Priming can occur through hearing words as well. When multiple people take part in the Ouija board question session, all it would take is some discussion or even a picture in the room to potentially prime people toward moving the planchette in a unified way to spell out a sensible answer to a question.

Incidentally, the name "Ouija" is supposedly a term for "good luck" (McRobbie, 2013). The story behind the name is that Helen Peters, supposedly a talented medium in the late 1800s and the sister-in-law of one of the Kennard Novelty Company investors, claimed that the board itself had generated that name when she asked it what it should be called. Interestingly, she was wearing a locket at the time of the Ouija board's naming event that bore the name "Ouija." Could this have been a case of priming, in which the locket's inscription inspired a collective effort to spell out the name using the planchette?

Ouija boards have been a fun and somewhat mysterious pastime for over 125 years. They offer the players a unique situation involving sensible answers to questions that appear to arise without direct cause by any one individual. The game in a sense plays upon peoples' illusions of causality, where there is not a conscious cause-and-effect mapping between the thinking of the players and the board's mysterious outputs.

confederate, a person who was actually an experimenter posing as a second participant. Both the participant and the confederate were led into a lab room and seated at a table containing a small board that had been fixed to the top of a computer mouse. A monitor was present in the room and depicted a set of 50 random small objects such as plastic toy cars, dinosaurs, and animals. The mouse pointer was also evident on the screen. The participant and confederate lightly touched the edges of the board fixed to the computer mouse and were instructed to make circular movements stopping about every 30 s. Participants were instructed that the study was being conducted in order to examine people's feelings of intention about actions and how these feelings come and go (Wegner & Wheatley, 1999). The participant and confederate were asked to rate the amount of intentionality that they experienced prior to each time the mouse-board made a stop. They did so on a 0–100 scale with zero representing the statement "I allowed the stop to happen" and 100 indicating, "I intended to make the stop."

The participants listened to music that was presented through headphones when they were supposed to make the board stop. Words were also presented through the headphones as primes, or cues, that were associated with objects depicted on the computer screen (e.g., cars, dinosaurs, etc.). The confederate had a specific role on four types of stops that had to be made. In these "forced stops" the confederate deliberately steered the mouse-board over to one of the objects depicting the prime word that the participant heard on that trial. For example, if the participant heard the word "car" at a certain "forced stop," the confederate moved the mouse-board over to the image of the car on screen. The timing of the "forced stops" was also set up in advance so that the prime word was presented either 30s before the stop, 5s before the stop, 1s before the stop, or 1s after the stop. The variation in timing was included to examine the effects of timing and event order linking the prime word to the stop. In other stops, the participant heard prime words and was free to dominate the steering and stopping of the board.

The "I Spy" experiment was primarily designed to assess the degree of intentionality that a participant places on an event that was suggested to them in the form of a word, but they did not actually cause these events. Analysis of the "forced stop" conditions allowed Wegner and Wheatley (1999) to evaluate this situation by looking at the perceived level of intent to make the stop that the participants reported. Any reported intent to make the stop was illusory, as the experimenter had assumed control of the mouse-board during those events. The least powerful illusions of intent occurred in situations in which the prime word was presented 30s before the "forced stop," or 1s after the "forced stop." Participants reported a conscious willed stop at around 45% on the 0–100 scale. The most fascinating result occurred on the instances in which the prime word was presented 5 s or 1s before the "forced stop" (on that same item named). In these cases, participants reported a significantly increased level of intent to stop the mouse-board at rates of approximately 60% intention to stop the board. Notice that in all "forced stop" events, the confederate was in full control of the timing and location of the stop. This result suggests that the prime word had a strong influence on participants' perception of intent to cause an action (the stopping of the mouse-board) in terms of both timing and location of the stopping point.

Wegner and Wheatley described the experience of willful action as being a theory that we entertain within our conscious minds. Results like those demonstrated in the "I Spy" task indicate that people seek to discover links between their thoughts and the actions that occur within the environment. Much of the time our thinking is predictive of what we will do. There are clearly times, however, when our thoughts are not directly coupled with actions that we control in the environment. In such cases, we may still infer that we have played a role in making these things happen. In other words, we take credit for playing the role of an agent in influencing a physical situation if that situation fits well enough with what is on our minds. The experiment also represents a case of priming, in which the mere presentation of a word can suggest a concept to the person. The activation of that concept can then fit with an occurrence, such as the mouse-board cursor landing upon the object that links to the primed concept that had been presented via the headphones. If the activation of the concept occurs just before an action (the "forced stop" in the "I Spy" experiment), then people theorize a causal connection between themselves and the action, despite the fact that they actually had little, if any, degree of control over the timing and location of the "forced stop" event. Wegner and Wheatley (1999) compared this illusion to the situation in which a magician has secretly placed a rabbit inside his hat, while distracting the audience at a magic show. The magician then loudly pronounces magic words, taps the hat with a wand, and gloriously produces the rabbit from inside his hat. The timing of the events can trick our minds into producing an illusory feeling of cause that the words and wand actions have led to the rabbit's appearance. This illusory experience likely influences us in our daily lives as we attempt to infer cause and effect within our complex daily lives using our relatively limited degree of consciousness and focused attention.

The Ouija board (see Box 8.2) has been the subject of additional research on causal inference within psychological studies. A research group comprised of Hélén Gauchou, Sid Fels, and Ron Rensink at the University of British Columbia became interested in the interplay between subliminal, or implicit, influence and our conscious experience in situations similar to that illustrated by the use of a Ouija board.

Gauchou, Rensink, and Fels conducted two interesting studies reported in 2012. This group reported a study in which a confederate player was introduced into a Ouija board session with experimental participants. The researchers introduced a cover story indicating that the participant was going to use the Ouija board along with another participant (the confederate). They told participants that they would be blindfolded during the session at the Ouija board, during which they would answer a series of questions. The participants directly observed that the confederate was also blindfolded prior to having their own vision obscured. In reality, the confederate just pretended to participate in the Ouija board session answering a series of questions aloud to the participant. The confederate did not touch the planchette (the control device on the board) again during the test question period, so indeed the participants were acting alone in using the board. The participants answered a series of yes/no questions about world events (e.g., Is Buenos Aires the capital of Brazil?). People were at chance (around 50% correct) on these questions when they simply answered on their own without the act of using the Ouija board. Meanwhile, when using the Ouija board under the mistaken impression that they were working with a partner, their performance increased to around 65% correct! Note that they were acting entirely alone in both cases, but the illusion that someone else was helping to steer the board toward "yes" or "no" answers appeared to help them make better use of their own implicit knowledge. This knowledge then gave them an edge in estimating the answers to the questions about world events compared to when they simply tried to answer alone. This result suggests that we are open to suggestion and that even the appearance of participation by another individual can give us a greater sense of capability than we might otherwise believe that we have alone. This experiment also highlights the importance of our own perceptions in how we make use of implicit knowledge.

SUMMARY

As we close this chapter it is again worth considering how all of these fascinating phenomena lead us to make the causal inferences that we do. The illusion of control is perhaps one of the phenomena that allows us to forge ahead and investigate outcomes and their mechanisms at a detailed level. While the idea of operating under an illusion can seem unpalatable, this same sense allows us to generate theories and test hypotheses. We are capable of remarkable insight when we apply that drive to explain things, toward gathering more evidence, and isolating the mechanisms at work in different complex phenomena.

We began this chapter discussing some of the most important achievements in causal reasoning. The ability to test hypotheses in order to discover mechanisms proved to be the key to isolating germs or microorganisms. Knowledge of these pathogens enabled the revolution in medicine that we have observed over the past 150 years. The average human lifespan has quite literally been doubled in large part on the basis of understanding the causes for disease.

We then discussed the challenges that people face when confronted with illusory conjunctions between events that appear to cause one another. The mind can play tricks on us when we attempt to read patterns and make predictions. The left PFC may be particularly involved in creating theoretical inferences and making predictions about outcomes. Features of a situation including the timing of how feedback is delivered and the schedule of reinforcement can lead us to believe that we have more control than we do over outcomes, or that a particular event is causing another event, when it is actually not. The characteristics of physical events in space and time influence whether we perceive a causal connection. The parietal cortex is involved in perceiving causal information in space, while the basal ganglia are involved in evaluating the timing of events for establishing causation.

Lastly, we sometimes experience an illusion that our thoughts are controlling circumstances. This experience often results from implicit cognitive phenomena including priming and the perception that others may be helping us to perform a task, when they are not actually doing so.

END OF CHAPTER THOUGHT QUESTIONS

1. Causal reasoning has enabled some remarkable advances in medicine. What are some other achievements in human history that we can attribute to our causal reasoning abilities?
2. People often place too much emphasis on evidence that supports a possible causal theory and look too little for evidence that refutes it. Why do you think this happens?
3. What does maximizing behavior (in which a thinker always chooses the more probable of two choices) tell us about the brain of a species that tends to do this?
4. Do you think that people would move from a matching strategy toward a maximizing strategy in a two-choice task if they were to make hundreds or thousands of choices on this task?
5. Can you think of some situations in life in which people think that they see patterns in outcome data that are likely not there?
6. People tend to experience an illusion of control. They are more prone to this illusion when they can win money for successful outcomes. What might this behavior tell us about people with gambling problems?
7. People experience an illusion of willful action. Can you think of some cases in daily life in which this can lead to problems?
8. When two objects collide, we often immediately sense a causal relationship between them. Can you think of situations in which an invisible force imparts movement on an object and you do not see the interaction as being causal?

References

Alloy, L. B., & Abramson, L. Y. (1979). Judgment of contingency in depressed and nondepressed students: Sadder but wiser? *Journal of Experimental Psychology: General, 108*, 441–485.

Alloy, L. B., & Tabachnik, N. (1984). Assessment of covariation by humans and animals: The joint influence of prior expectations and current situational information. *Psychological Review, 91*, 112–149.

Elliott, R., Rees, G., & Dolan, R. J. (1999). Ventromedial prefrontal cortex mediates guessing. *Neuropsychologia, 37*, 403–411.

Estes, W. K. (1961). A descriptive approach to the dynamics of choice behavior. *Behavioral Science, 6*, 177–184.

Fugelsang, J. A., Roser, M. E., Corballis, P. M., Gazzaniga, M. S., & Dunbar, K. N. (2005). Brain mechanisms underlying perceptual causality. *Brain Research, Cognitive Brain Research, 24*, 41–47.

Gauchou, H. L., Rensink, R. A., & Fels, S. (2012). Expression of nonconscious knowledge via ideomotor actions. *Consciousness and Cognition, 21*, 976–982. http://dx.doi.org/10.1016/j.concog.2012.01.016.

Gaynes, R. (2011). *Germ theory: Medical pioneers in infectious diseases*. Washington, US: ASM Press. Retrieved from http://www.ebrary.com.libproxy.utdallas.edu.

Hinson, J. M., & Staddon, J. E. R. (1983). Matching, maximizing and hillclimbing. *Journal of the Experimental Analysis of Behavior, 40*, 321–331.

Inhelder, B., & Piaget, J. (1958). *The growth of logical thinking from childhood to adolescence*. New York, NY: Basic Books.

Jenkins, H. H., & Ward, W. C. (1965). Judgment of contingency between responses and outcomes. *Psychological Monographs, 79*.

Matute, H. (1996). Illusion of control: Detecting response-outcome independence in analytic but not in naturalistic conditions. *Psychological Science, 7*, 289–293.

McRobbie, L. R. (October 28, 2013). The strange and mysterious history of the ouija board. *Smithsonian Magazine*. Retrieved from http://www.smithsonianmag.com/history/the-strange-and-mysterious-history-of-the-ouija-board-5860627/.

Michotte, A. E. (1963). *The perception of causality (T. R. Miles and E. Miles, Trans.)*. London: Methuen (Original published in 1946).

Roser, M. (2017). *Life expectancy*. Retrieved from https://ourworldindata.org/life-expectancy/.

Schultz, W., Dayan, P., & Montague, P. R. (1997). A neural substrate of prediction and reward. *Science, 275*, 1593–1599.

Smedslund, J. (1963). The concept of correlation in adults. *Scandinavian Journal of Psychology, 4*, 165–173.

Stearns, J. K. (2011). *Infectious ideas: Contagion in premodern Islamic and Christian thought in the Western Mediterranean*. Baltimore, MD: Johns Hopkins University Press.

Straube, B., & Chatterjee, A. (2010). Space and time in perceptual causality. *Frontiers in Human Neuroscience, 4*(28). https://doi.org/10.3389/fnhum.2010.00028.

Vallée-Tourangeau, F., Hollingsworth, L., & Murphy, R. A. (1998). "Attentional bias" in correlation judgments: Smedslund (1963) revisited. *Scandinavian Journal of Psychology, 2*, 221–233.

Wegner, D., & Wheatley, T. (1999). Apparent mental causation: Sources of the experience will. *American Psychologist, 54*, 480–492.

Wolford, G., Miller, M. B., & Gazzaniga, M. (2000). The left hemisphere's role in hypothesis formation. *Journal of Neuroscience, 20*, RC64.

Yellott, J. I., Jr. (1969). Probability learning with noncontingent success. *Journal of Mathematical Psychology, 6*, 541–575.

CHAPTER

9

Deduction and Induction

KEY THEMES

- In deductive reasoning people are provided with premises and they assess the validity of a conclusion. This conclusion is always guaranteed to be valid, unlike inductive reasoning problems.
- There are four key cases within deductive reasoning: affirming the antecedent (valid), denying the antecedent (invalid), affirming the consequent (invalid), and denying the consequent (valid).
- The context of a deductive reasoning problem influences how successful people will be. Familiar contexts are more likely to evoke valid deductive reasoning.
- The Wason card selection task measures deductive reasoning. People tend to select cards that represent affirming the antecedent (a valid approach) and affirming the consequent (an invalid approach). The correct answer involves selecting the card that represents denying the consequent (another valid approach).
- When the Wason card selection task is presented as a permission or obligation situation about human behavior, people are much more likely to carry out both valid approaches in the task.
- Evolutionary psychology has explained deductive reasoning success on permission and obligation problems as relating to the adaptive tendency of people to check for cheating behavior.

Reasoning
http://dx.doi.org/10.1016/B978-0-12-809285-9.00009-0

- Inductive inferences do not guarantee valid reasoning, but many situations in life require that we draw these types of inferences. While they may not always be correct, they may better represent real-world behavior.
- Knowledge influences inductive reasoning with similarity judgments affecting whether people will infer new information about a new instance based on information about a prior instance.
- Inductive reasoning varies based on culture, which emphasizes that people's background knowledge is critical for determining the types of inferences that they will make.

OUTLINE

DEDUCTIVE AND INDUCTIVE REASONING

Introduction

Deductive and inductive reasoning have been considered to be core constructs in the study of reasoning. These forms date back to ancient times with clear roots in classical Greek philosophy. Aristotle emphasized the value of deductive reasoning as a source of powerful inferences in his writings. Deductive reasoning can reliably yield a valid conclusion based on the premises provided and the premises must also be valid for the approach to be successful. Inductive reasoning by contrast may yield a valid inference and is likely to move us beyond the current known information. Inductive reasoning comes with a price; however, in the form of a greater probability of an invalid inference if we inappropriately move beyond the information that we currently have. In other words, deduction is a safe bet moving us from currently known information to a relatively safe and valid conclusion, while inductive reasoning gives us a greater leap forward in terms of new knowledge, but comes with a greater opportunity for making erroneous conclusions.

Our focus in this chapter will be on the types of psychological approaches that have been applied toward understanding deduction and induction. Experimental psychology differs from philosophy, the other main discipline associated with deduction and induction. While philosophy has focused on the certainty and the validity of these types of reasoning, the psychological approach has focused primarily on gaining experimental evidence. This means that the psychology of reasoning deals with the many complexities that arise when we attempt to use the deductive approach, the inductive approach, or a combination of the two. The outcomes of experiments have also led to theories of reasoning that then face the added challenge of ensuring compatibility with neuroscience and biological evidence, which we have already reviewed in detail in Chapters 3 and 7.

DEFINING DEDUCTION

Examples of Deductive Logic

When we think of deductive logic, we are typically asked to consider premises and then draw a conclusion that follows from those premises. The premises are often provided in the form "if P, then Q" statements. For example we may be provided with the following premises:

If an animal is a dog, then it has a tail

This animal is a dog…

In this example the premises set up the conclusion "…then this animal has a tail." This example provides us with a *valid* inference, which is correctly drawn based on the premises. This is also a straightforward example. In philosophical terms, the initial phrase, "If an animal is a dog…" is referred to as the *antecedent*. The second part of the premise, "…then it has a tail" is called the *consequent*. The antecedent is restated exactly as it initially appeared in the premises (this animal is a dog) in this straightforward example. In other words, the antecedent has been *affirmed* in this case. This is an example of a valid inference referred to as *affirming the antecedent*.

Let's now consider a second important case within deductive reasoning that is not valid. Given the same premises as we considered before, but with one argument negated, we have the following premises:

If an animal is a dog, then it has a tail

This animal is not a dog…

In this case one may be tempted to conclude "then it does not have a tail." Unfortunately, this proves to be an *invalid* conclusion. The logic only flows in one direction from "if P…" to "…then Q." One cannot reverse this flow and move in the other direction to make a conclusion. Assuming you knew absolutely nothing about dogs, you would be at a loss to conclude anything else regarding the status of tails in the example. We only have the premise that if it is a dog, then it has a tail. If this is not a dog that we are being asked to consider, then we have nothing to conclude about the property of having a tail. The term for the negation of the "P term" in this instance is *denying*. The broader class of this example is referred to as *denying the antecedent*.

In addition to the antecedent, we can also consider situations that focus on the *consequent* (...then it has a tail). Consider the following example:

If an animal is a dog, then it has a tail

This animal does not have a tail...

The example above allows us to validly conclude that indeed this is not a dog, as all dogs must possess tails according to the rules. No tail, no chance of being a dog. In this case we again have a "not" statement, or a denial of information. This type of example is referred to as *denying the consequent*.

We must also consider one last situation in which we evaluate information about the consequent. Representing this case is the following example:

If an animal is a dog, then it has a tail

This animal does have a tail...

In this last example we may be tempted to conclude "...then this animal is a dog." Be careful here. Again, we must consider the direction that the information flows. In this instance we receive affirming information stating that a tail exists. In other words this is a case of *affirming the consequent* (this animal does have a tail). The problem here is that we cannot state anything definitive about the antecedent based on the information provided. The argument does not flow in reverse. The presence of a tail does not guarantee dog status. This is clearly true in life, where many animals have tails, but it is also important to keep in mind for any deductive problem, as categories are not mutually exclusive. In other words the property of "having a tail" does not apply only to dogs (Fig. 9.1).

To summarize, we have considered four different situations that involve the "if P, then Q" structure of the premises. Two of these types, *affirming the antecedent* and *denying the consequent*, provide valid conclusions. The other two cases, *denying the antecedent* and *affirming the consequent*, yield invalid conclusions. The situation can be clarified by using Venn diagrams, which are helpful for depicting situations involving the types of class inclusions that we have been considering. Fig. 9.2 provides a Venn diagram of the dogs and tails scenario. In the large class of "animals," the subclass "animals with tails" is fully contained within the large circle. We can also consider subcategories represented by even smaller circles. The "dogs" subcategory falls completely within the "animals with tails" class. This is helpful, as we refer back to the earlier examples. The valid situations in which we consider the statement

FIGURE 9.1 "If this animal is a dog, then it has a tail." The direction of the statement is key to determining whether you are making a valid deductive inference.

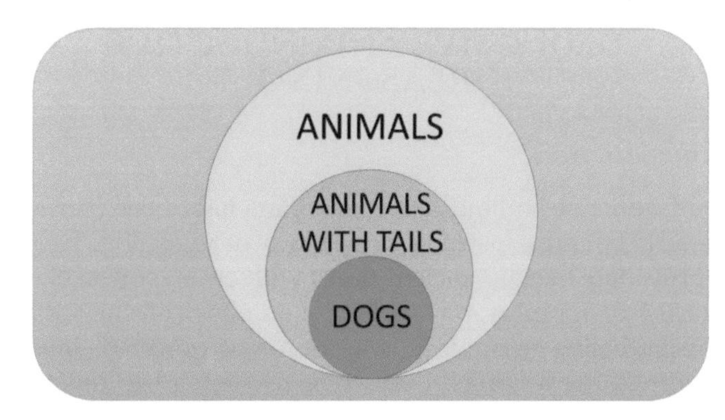

FIGURE 9.2 A Venn diagram that helps us to evaluate the deductive arguments in the example "If an animal is a dog, then it has a tail." The circles represent categories with all members of that category falling within the circle. Any space outside the circle contains all nonmembers. The "animals" category has both members who have tails and those that do not. The "animals with tails" category contains a subcategory "dogs" with dogs falling entirely within the "animals with tails" circle. Thus, there are no exceptions in the form of dogs who have no tails. This is a pictoral representation of the original situation "If an animal is a dog, then it has a tail."

"if an animal is a dog, then it has a tail, this animal is a dog…therefore it has a tail" (affirming the antecedent) and the statement "if an animal is a dog, then it has a tail, this animal does not have a tail…therefore it is not a dog" (denying the consequent), are seen in the Venn diagram (Fig. 9.2). The Venn diagram supports the validity of these statements. In the case of affirming the antecedent, the dog membership status fits entirely within animals that have tails. There are no exceptional dogs to consider that would make the rule invalid by not having a tail. Likewise in the situation representing denying the consequent, "if there is not a tail" there is not any possibility of falling within the dog membership status area depicted in the Venn diagram.

Fig. 9.2 also helps us to verify our thinking on the invalid statements as well. When we consider the invalid inference "If an animal is a dog, then it has a tail, this animal has a tail…therefore it is a dog" (affirming the consequent), we can see why this inference is invalid. In the "animals with tails" circle there is clearly additional room in the category for animals that are not dogs. Similarly in the other invalid case "If an animal is a dog, then it has a tail, this animal is not a dog…therefore it does not have a tail" (denying the antecedent), we observe in Fig. 9.2 that the instances of animals that fall outside the dog membership circle clearly include other animals that happen to have tails. It should be noted that Venn diagrams can be used for any category representation and may be especially handy when we are asked to evaluate unfamiliar instances. For example, if you have little experience with marine biology, it may be very helpful to use Venn diagrams to categorize the types of species that exist and to determine which genus and class they belong.

Another handy rule of thumb for assessing the validity of deductive arguments is shown in Fig. 9.3. In this case we have a quad chart showing the two cases of P and Q (antecedent and consequent) across the top along with the two cases of negation (denying) presented along the right side. The valid cases are highlighted in green (affirming the antecedent and denying the consequent) and the invalid cases are highlighted in red (denying the antecedent and affirming the consequent). A sound understanding of these four situations will be essential for evaluating the results of the experiments that we will discuss next in this chapter.

In addition to the terms we have used thus far, you may have encountered the Latin descriptions of these same four cases of valid and invalid deduction. *Modus Ponens*, literally translated as "the mood that affirms," refers to the situation in which we affirm the antecedent. Such a case occurs when we are presented with the argument "if P then Q, P…" and we go on to infer "then Q." *Modus Tollens* or "the mood that denies" refers to the case that we have previously referred to as denying the consequent. Such is the case with the argument

Valid and *Invalid* Arguments for Conditional Logic

Affirming the Antecedent (*modus ponens*)	Denying the Consequent (*modus tollens*)
1. $P \rightarrow Q$ 2. P 3. Q	1. $P \rightarrow Q$ 2. $\sim Q$ 3. $\sim P$
Affirming the Consequent (INVALID)	**Denying the Antecedent (INVALID)**
1. $P \rightarrow Q$ 2. Q 3. P	1. $P \rightarrow Q$ 2. $\sim P$ 3. $\sim Q$

FIGURE 9.3 A quad chart can be helpful to understand deductive arguments and their validity. Across the top the categories antecedent (the P term in "if P, then Q") and consequent (the Q term, in "if P, then Q") are presented. Along the left side we can "affirm" either the antecedent or consequent, or we can "deny" either term. The valid conclusions are highlighted in green, while the invalid cases are presented in red.

"if P, then Q, not Q"; therefore, we can validly conclude "not P," as Q must logically follow P. There are not equivalent terms for the invalid cases (affirming the consequent and denying the antecedent); however, these cases have more simply been referred to collectively as *logical fallacies*.

Before we move on it is also important to note that the validity of an inference does not necessarily guarantee that it will be true in the world. Validity only allows us to conclude that if the premises are accurate, we have deduced a sound conclusion. If the premises are not true in the world, we can still have a valid situation by deductive logic, even though this would prove relatively unhelpful in most of our real-world reasoning.

Challenges in Deductive Reasoning

The study of deduction has long existed in the realm of pure philosophy. Experimental psychologists began to evaluate deductive reasoning by first gathering behavioral data on this topic in the early 1900s. Researchers began to focus on the factors that influenced people's ability to reason deductively. In this early period of psychological research, a pioneering reasoning researcher named Minna Cheves Wilkins (1928) published her work on syllogistic reasoning. Wilkins presented deduction problems in a standardized format as follows:

Some A are B, all C are A, therefore…

Fig. 9.4 provides examples of the materials used in this study. Participants in this early work were college

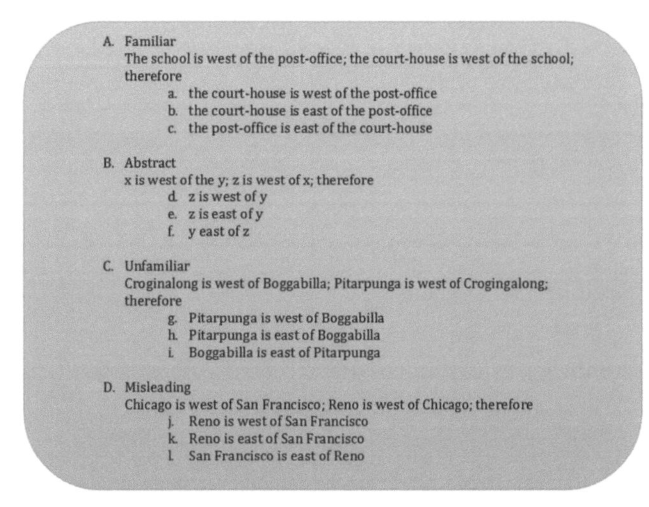
A. Familiar
The school is west of the post-office; the court-house is west of the school; therefore
a. the court-house is west of the post-office
b. the court-house is east of the post-office
c. the post-office is east of the court-house

B. Abstract
x is west of the y; z is west of x; therefore
d. z is west of y
e. z is east of y
f. y east of z

C. Unfamiliar
Croginalong is west of Boggabilla; Pitarpunga is west of Crogingalong; therefore
g. Pitarpunga is west of Boggabilla
h. Pitarpunga is east of Boggabilla
i. Boggabilla is east of Pitarpunga

D. Misleading
Chicago is west of San Francisco; Reno is west of Chicago; therefore
j. Reno is west of San Francisco
k. Reno is east of San Francisco
l. San Francisco is east of Reno

FIGURE 9.4 Deductive reasoning materials. *From Wilkins, M. C., (1928). The effect of changed material on ability to do formal syllogistic reasoning.* Archives of Psychology, 102, *1–83.*

students who were then asked to select among three possible conclusions that could be drawn from the given premises. Wilkins presented syllogisms in several formats including premises that were generally familiar as in part A of Fig. 9.4, which were true statements about familiar geographical locations. Other premises were presented in an abstract format providing relational statements about letters. Fig. 9.4C shows examples of the unfamiliar category in which fictional locations were substituted into the problem. Lastly, deliberately misleading materials were presented in which factually incorrect statements had to be evaluated logically.

In essence, Wilkins tested the effects of biases about the world that are derived from prior knowledge and how these biases influence reasoning. Results indicated that people were most accurate with condition A, the familiar statements. Wilkins suggested that these problems may have been solved in a different manner than the other three condition types, as the most efficient way to tackle these problems would have perhaps been to simply note that the spatial relations, or factual statements, did indeed fit with prior world knowledge. Meanwhile, participants showed lower percentages of correctness with the unfamiliar and misleading problems. The misleading category presented the most difficulty for participants, as it specifically placed logical validity at odds with what is actually true out in the world.

Other early psychological research groups further explored the situations in which people tended to endorse invalid inferences. Woodworth and Sells (1935) reported that people were inaccurate in assessing the validity of problems when they were asked to verify the validity of a conclusion. Janis and Frick (1943) found additional evidence of logical errors in deductive problems and emphasized that people's prior beliefs are one of the major reasons that they fail to reason accurately about validity. This phenomenon became known as the *belief-bias effect*. For example, if you consider the following statement:

If a car engine is out of fuel, then it sputters

The engine sputters,

Then it is out of fuel.

This particular deductive reasoning argument is not valid, as it represents a case of affirming the consequent, a logical error (refer to Fig. 9.2). The difficulty in this case is that the conclusion that the car is out of gas is highly intuitive. This conclusion simply fits with most people's prior beliefs about car engines. There is a very good chance that a sputtering engine is due to a fuel shortage, so the conclusion appears likely. In everyday life we would probably check the fuel gauge immediately if we were confronted with a sputtering engine. The challenge here is that a lack of fuel is a common reason for a sputtering engine, but there are other possible causes, some related to fuel, others due to possible mechanical failures. While an engine out of fuel must sputter according to the premises, we cannot reason in reverse that a sputtering engine must be out of fuel. Other examples can be constructed in experimental conditions to show this effect. The belief-bias effect is similar to some of the observations noted earlier by Wilkins (1928) who had asked participants to evaluate misleading syllogisms that contained information explicitly in conflict with facts about the world. Thus a major finding in deductive reasoning is that people will abandon their assessment of logical validity when an expedient answer is readily available. This may simply indicate that our powers of deductive reasoning are not the only tool available to us. It takes mental effort to go through the motions of assessing deductive problems and many instances in life demand only correlation or association. When a quick and obvious answer pops out at us, many people will endorse it without further analysis. This is very similar to the sort of fast thinking that is commonly associated with the use of heuristics described later in Chapter 11.

Another common challenge to our deductive reasoning occurs with the presentation of negative framing. Processing double-negatives is often difficult in speech, but this can be especially problematic when negative terms are presented in deduction problems. One can observe this effect when negative information is embedded within one of the straightforward syllogisms presented by Wilkins (1928). Consider the following example:

The school is not west of the post-office; the court-house is not west of the school; therefore

a. the court-house is not west of the post-office

b. the court-house is not east of the post-office

c. the post-office is not west of the court-house

The answer to this brain-twister is: "The court-house is not west of the post-office." Placement of the word "not" in front of "west" or "east" radically changes the ease with which we can process this problem. Spatially, the buildings must be arranged in the following order moving from east-to-west: (1) court-house, (2) school, (3) post-office. The negative information "*not* west" prevents us from processing the locations in a straightforward way. In attempting to solve this syllogism you may also find yourself using mental imagery to place the buildings in their proper arrangement so that no building is inappropriately "not west" of another. It is a further challenge that we rarely use negative phrases to describe a spatial arrangement, since it is both easier to state that it is "east" of another building and others will more readily understand our meaning.

The perils of negative wording have a long history in the deductive logic field with major contributions appearing in the early 1960s by Peter Wason et al. Wason was able to reliably show that people take longer to produce conclusions in response to negatively phrased deductive logic premises. Wason and Jones (1963) noted that these effects continued even after participants practiced deduction tasks that included negative information. Additional complications were noted when nonsense syllables were substituted into the phrases in place of the words "is" and "not" in the task. This novel phrasing led participants to produce even longer response times to solve problems. These findings indicate that people will take extra time to grapple with assessing negative statements. This results in less efficient processing compared to cases in which they are asked to evaluate positively phrased information. The tendency that negatively phrased sentences are more difficult to process fits with prior work in psycholinguistics. The longer response times toward negative information is also a nice example of *confirmation bias*, one of the key decision heuristics discussed by Wason in 1960. Confirmation biases lead people to seek evidence that fits with their current hypothesis, rather than seek information that would disconfirm a hypothesis. We will discuss this phenomenon in greater detail in a later section of Chapter 11 that discusses decision making.

DEDUCTIVE REASONING IN THE LABORATORY

Laboratory Methods for Evaluating Deducting Logic

Laboratory tests for deductive reasoning began with the early work of Wilkins (1928) and Woodworth and Sells (1935). In these early experimental deduction tasks, participants were presented with a set of premises that fell into one of the four major deductive classes (affirming the antecedent, denying the consequent, affirming the consequent, or denying the antecedent). Participants in these experiments were typically asked to state whether a given conclusion was valid, or were asked to select the valid conclusion from among a set of conclusions that were presented. In some cases participants have been asked to state any conclusion that follows from the premises (Evans, Newstead, & Byrne, 1993), though such experiments risk yielding complicated data due to their open-endedness. Common challenges for participants in these experiments include cases in which the content is negated, modified, or made to be misleading to tax the robustness of people's deductive reasoning in response to these challenges.

The Wason Card Selection Task

Peter Wason at University College, London, developed a novel way to test deductive reasoning that departed from the traditional syllogism format. His task achieved remarkable results that remain robust these many decades later. Wason (1966) developed a card selection task, in which participants were presented with four different cards and a set of premises that needed to be tested by turning over some of the cards. Participants in the original version of the task evaluated a set of cards displaying the following: a vowel facing up, a consonant facing up, an even number facing up, and an odd number facing up. The participants were asked to evaluate whether the experimenter was lying if they uttered the statement "If a card has a vowel on one side then it has an even number on the other side." An example of this task is depicted in Fig. 9.5. Go ahead and look at this figure now and see if you can pick the *fewest number of cards* that will prove that rule false.

Participants in the original Wason card selection task were initially asked to decide which of the cards

FIGURE 9.5 An example of the "Wason card selection task" in which participants are asked to evaluate the statement "If a card has a vowel on one side then it must have an even number on the other side." Depicted are four cards face-up showing a vowel "E", a consonant "Z", an even number "6," and an odd number "7." The correct answers are to turn over the card with "E" as it must have an even number on the other side (affirming the antecedent) and "7" as it must not have a vowel on the other side (denying the consequent).

had to be turned over to evaluate that rule. Next, the participants were asked to physically turn over the cards and state which of them proved that the rule was false. This experiment originally appeared in a chapter on reasoning by Wason in which he did not formally present the actual data. Rather, Wason described the gist of the results and these have been replicated across many later experiments. Most people selected the card with the vowel ("E" card in Fig. 9.5). This is a straightforward way to test for the rule and it conforms to the *Modus Ponens* classification we have discussed earlier in the chapter, in which thinkers must check a situation that affirms the antecedent. A majority of these original participants also selected the even number ("6" in Fig. 9.5). No participants selected the card with the consonant ("Z" in Fig. 9.5), and few selected the card showing the odd number ("7" in Fig. 9.5).

If you have taken the time to try this experiment yourself as shown in Fig. 9.5, you know that the correct answer to the task is to select the cards showing E and 7. I have covered the Wason card selection task numerous times in both undergraduate and graduate level classes. Without exception, I have seen students stumble on this result and fail to grasp why the card showing 7 should be selected. In one of my more memorable graduate seminar classes, several of the students (who would rank among our finest young scholars of psychology) simply didn't accept the results, as they are so unintuitive. For this reason, I will go to some length to describe how the cards map onto the four key deductive cases that are depicted in Fig. 9.3. Please bear with me on this, as it is one of the most well-studied tasks in psychology and the results are initially difficult to grasp.

First, there is no reason to select either the card with the Z on it or that with the 6. Few people choose to flip over the Z card in this task, as consonants are relatively obvious as being irrelevant to evaluating the "if vowel, then even number" rule. Strikingly, it is intuitively appealing to people to check the card with the 6 facing up. It is so appealing that it can be hard to talk people out of checking the 6. If you think back to Fig. 9.3 you can see why checking the 6 card is unnecessary. This card represents the fallacy of affirming the consequent. Consider the rephrasing of the Wason card selection task as follows: "If there is a vowel, such as E, then there must be an even number, such as 6." We are correct to reason in the forward direction "if E, then 6," but we cannot apply the rule in reverse "If 6, then also E." It was never stated that "If there is an even number on one side, then there must be a vowel on the other side." So we need not check the card with the 6, as whatever is on the other side is irrelevant to the rule (it could have a picture of a cute puppy on the other side and it would not disprove the rule in question).

Secondly, we have to check the card with E facing up and the card displaying 7 to disprove the rule. Again, the card with E is highly intuitive to most people. Almost everyone will start by wanting to check on E, as it transparently tests the rule in the forward direction. Then we come to the 7 card…and the troubles begin. My class demonstration typically falls apart at this point, hands shoot into the air, students revolt, and chaos ensues. Few people will select this card in the first place, and few will believe you when you insist to them that it has to be checked. This case represents an absolute conundrum for lab-based deductive reasoning, an unfamiliar rule that has no consequences in the real world and seems utterly perplexing to many people at first glance.

To convince the skeptic, we can rephrase the rule again as follows: "If there is a vowel, such as E, then there must be an even number, not an odd number such as 7." This awkward phrasing reinforces the fact that you have to check the exception to the rule. If the back of the 7 card contains any of the vowels, you have discovered a problem with the rule. To return to my rather silly example, if the 7 card has a picture of a cute puppy, then you also have a problem with the rule. It simply has to have a vowel on the other side, or the rule goes up in smoke. In formal terms, you must check the situation in which there is a denial of the consequent and in the Wason card selection task the 7 card represents this critical valid case in deduction, *Modus Tollens*.

To summarize, the Wason card selection task has two correct answers in the form of card selections. One is highly intuitive and involves simply checking the case that represents affirming the antecedent. The other correct answer involves checking the case that involves denying the consequent and this is very unintuitive, because people do not tend to look for the exception to the rule that they have to test. In the original Wason card selection task (1966) participants were asked to turn over each card and describe which of the cards would prove that the premise statement was a lie. In this case, nearly all of the participants selected the vowel case as necessary and that it would prove that the premises were a lie. Again, the card representing the *Modus Ponens* case is highly intuitive. Many expressed that the even number was informative, though it did not disprove the rule (in fact it is neither informative nor capable of disproving the rule). Very few people correctly stated that the odd numbered card was capable of disproving the rule. Wason noted that participants had exhibited a "bias toward verification" in wanting to select the even numbered card. In this way, people preferred to see that something proved true, rather than to actively attempt to find an error in the logic. Follow-up work by Wason (1966) indicated that approximately 50% of people will only select the card that affirms the antecedent (the card displaying a vowel) and only about 16% of people accurately selected the denying the consequent card (the card displaying an odd number). Despite efforts to intervene with experimental procedures that emphasized testing falsity, Wason's participants failed to show appreciable improvements on the selection task.

The Role of Context on the Wason Card Selection Task

To this point we have dealt exclusively with the case in which people try to solve the Wason card selection task reasoning about abstract rules involving letters and numbers. Perhaps it is not surprising that people do not readily understand that they must check the logical validity of all of the possibilities by using a *Modus Tollens* procedure. We usually function just fine in life by checking for obvious associations in the world. Also, few situations in life are truly causal in nature. Rather many interactions in the world are only correlational and most will have some type of exceptions. We discussed these distinctions in much greater detail in Chapter 8. We rarely encounter situations that have absolute rules that only work one way. To better evaluate people's capacity to reason deductively in everyday life, it is necessary to find one of these rare "all-or-nothing" scenarios that can only work in one particular way and then see how people do on deductive reasoning.

A few studies in the 1970s indicated that real-world materials about postage stamp prices and car transit to different cities could improve deductive reasoning performance on a card selection task modeled after the original Wason card selection task (Johnson-Laird & Wason, 1972). This phenomenon became known as the *thematic materials effect* (Griggs & Cox, 1982). Wason and Shapiro (1971) showed a thematic materials effect by asking participants to evaluate the rule "Everytime

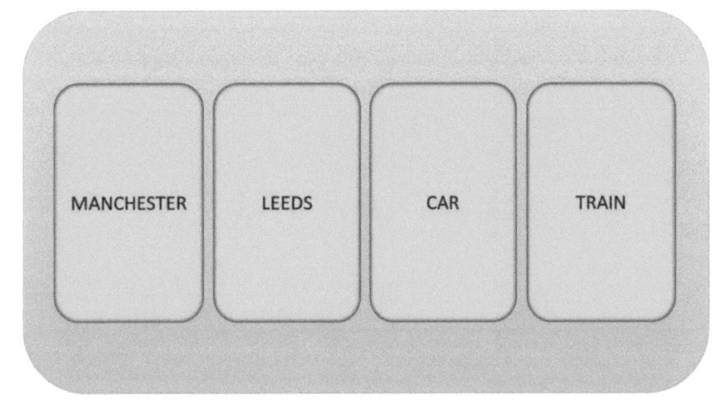

FIGURE 9.6 A Wason card selection task that elicits a thematic materials effect. Participants evaluate the rule "Every time I go to Manchester I travel by car." Cards included "Manchester" (the valid affirming the antecedent), "Leeds" (irrelevant card), "car" (irrelevant card), and the critical "train" (the valid denying the consequent card) (Wason & Shapiro, 1971).

I go to Manchester I travel by car." The cards in this task included "Manchester" (the valid affirming the antecedent), "Leeds" (irrelevant card), "car" (irrelevant card), and the critical "train" (the valid denying the consequent card) (refer to Fig. 9.6). Relative to the original abstract letter and number task on which only 13% of participants correctly chose both of the valid cards, a remarkably high proportion (67%) of the participants solved the thematic version correctly having selected both of the valid cards ("Manchester" and "train").

One of the very clear examples of an "all-or-nothing" situation in life is when we have rules to obey. Laws make excellent rules for deduction because a well-crafted law simply cannot be violated, and it is abundantly clear to most people when a law is being broken. Griggs and Cox (1982) developed a very powerful thematic version of the card selection task modeled after this premise. In their "drinking age problem" Griggs and Cox asked participants to imagine that they were a police officer charged with making sure that people are conforming to rules of conduct. The cards in this version of the selection task were said to represent people who are sitting at a table and presented on each of four cards is either a person's age or the type of drink that they are having. The rule to be checked was stated as follows: "if a person is drinking a beer, then the person must be over 19 years old" (note that this study took place in Florida at a time when the legal drinking age was 19 years old). Participants were asked to select the card or cards that would have to be turned over to determine whether anyone at the table was violating the rule. Refer to Fig. 9.7 for a diagram of the drinking age problem. Note that the cards available correspond to the standard selections present in other versions of the Wason card selection task (P, Q, not P, and not Q). In this version the cards face up read: "drinking a beer, drinking Coke, 16 years of age, and 22 years

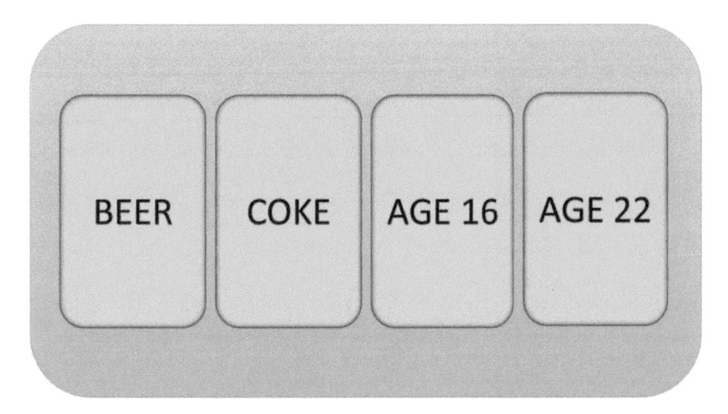

FIGURE 9.7 An example of the "Wason card selection task" in which participants are asked to evaluate the statement "If a person is drinking a beer, then the person must be over 19 years of age" (Griggs & Cox, 1982).

of age." Participants in this study were also tested on a more standard Wason card selection task with abstract materials and these versions were run twice each in a within-subjects design. The participants were asked to evaluate the rule "if a card has an A on one side, then it has a 3 on the other side" with the card faces reading A, B, 2, and 3. The inclusion of both the abstract card selection task and the thematic drinking age problem enabled the researchers to compare both for possible differences in accuracy.

The results of the Griggs and Cox drinking age problem were remarkable. Seventy-three percent of participants provided both of the valid, correct answers in the drinking age problem (drinking a beer and 16 years of age). Meanwhile, no participant selected the correct two valid cards together in the abstract problem using letters and numbers. Furthermore, 53% of participants selected the valid card (P, or affirming the antecedent) paired with the commonly selected invalid card (Q, or affirming the consequent). No participants selected that most common valid/invalid pairing in the drinking age problem. You probably also notice how intuitive the correct answers are in this version of the task. Checking the 16-year-old would be equivalent to a modern police officer checking the drink of a person who looks suspiciously underage for drinking alcohol. Notice however that the 16-year-old is logically equivalent to the highly unintuitive card with the number 7 in the standard version of the Wason card selection task. You likely also recognize how strange it would be to select the erroneous choice of the card representing the person who is 22 years of age, a card equivalent to the often mistaken choice of the even numbered card in the standard version of the task. Just as the selection of the even number in the abstract problem is not informative, checking the drink of someone clearly over the legal age is clearly irrelevant to checking the rule in question on the drinking age problem. In other words, if the problem is relevant to a circumstance in real life in which a concrete rule exists, people are familiar with that rule, and the rule has consequences, then people are able to show strong capacity for deductive reasoning.

The drinking age problem relates to what people should or should not do in their daily life conduct. Sometimes this type of theme is centered on permissions or obligations and their enforcement is termed *deontic reasoning*. Deontic logic relates to the evaluation of what we ought to do, may do, or ought *not* to do (Bucciarelli & Johnson-Laird, 2005). When people evaluate deontic statements about our behavior, these statements may be treated quite differently than pure and unapplied logical statements concerning arbitrary rules and abstract content, such as those studies applied to letters and numbers.

Another interesting modification to the Wason card selection task can be found in the realm of postal regulations. This set of experimental materials represents another example of deontic reasoning about regulations and permissions that occur in everyday life. Like the drinking age problem, the postage version of the card selection task can yield surprisingly different results than those that are commonly obtained with the original abstract form of the task. Research examining the postage version of the Wason card selection task has included participants evaluating a rule about the price of postage on the front of an envelope and the status of the back flap of the envelope. Again, the rules are both clear, as they relate to permissions and obligations. There is no gray area or exception to paying a particular amount of money for a good or service. These materials are also consistent with real life, much like the drinking age version of the problem. One of the best known examples of this postage task was reported by Johnson-Laird, Legrenzi, and Sonino Legrenzi (1972). Participants in this study acted as postal workers and evaluated whether the rule "If a letter is sealed, then it has a 50 lire stamp on it" was violated. With this themed set of materials, participants were correct an impressive 81% of the time in selecting the two correct cards (representing the choices of affirming the antecedent, and denying the consequent) (Fig. 9.8).

By the 1980s a series of Wason card selection tasks had been conducted and interesting results continued to mount. Errors on the task became a well-established phenomenon in the reasoning literature. Abstract selection tasks asking people to evaluate letter-number materials proved to be extremely difficult for people who would reliably fail to choose the card that would represent an exception to the rule (the valid card representing denying the consequent). Meanwhile, many participants would continue to choose the uninformative (invalid) selection that affirms the consequent. The relative success that individuals experienced when evaluating the cards related to real-world deontic reasoning situations suggested that the poor performance observed on the classic abstract Wason task involving the rule about vowels and

FIGURE 9.8 There is no ambiguity about the price of postage when we need to mail a letter. Goods and services have clear pricing requirements that must be met and people know this. When Wason card selection tasks have been embedded in the form of a postal purchasing requirement, people are much more accurate at selecting the correct cards (Johnson-Laird et al., 1972).

even numbers may be a reflection of how people process the information rather than a fundamental lack of competence at deductive reasoning. The details of how information processing occurs and why some situations evoke superior deductive reasoning performance to others became the focus of the next generation of deductive reasoning research aimed at understanding people's reasoning about the card selection task.

Theories of Syllogistic Reasoning

Theories of syllogistic reasoning had to evolve rapidly in the 20th century to accommodate the emerging results that were being obtained as experimental evidence increased. Early theories sought to explain why people often fail to correctly evaluate the validity of the consequent, choosing (invalidly) to affirm the consequent, rather than (validly) deny it. Woodworth and Sells (1935) proposed the *atmosphere effect*. Simply stated, the idea was that "all X are Y" also implied that "all Y are X" by an unspecified, but likely general phenomenon. This makes some intuitive sense because we are so rarely presented with inviolable rules of directionality when evaluating relationships. We live in a world full of exceptions and many things are correlated, but do not contain a causal relationship (as we noted in Chapter 8). These factors suggest that people may consider an "all X are Y" rule to also imply an "all Y are X" rule. Chapman and Chapman (1959) advanced another early theory suggesting that the common errors of deductive reasoning may be due to individuals committing conversion errors. A conversion error is made when one assumes that the rule "if X, then Y" also implies the reverse statement "if Y, then X." The Chapmans also proposed that errors in reasoning may be due to misinterpretations of the directionality of the rule and secondly, that participants likely engage in probabilistic reasoning, rather than pure deduction. Probabilistic reasoning occurs when someone assumes that something is likely based on probability. In the case of deductive reasoning, "if X, then Y" implies that the X and Y terms share some common property and therefore it may be seen to imply that the reverse case "if Y, then X" would also be likely to be valid. In explaining the *belief-bias effect*, Janis and Frick (1943) also maintained that the thinker is likely evaluating what he or she already knows about the world, rather than relying entirely upon only the logical terms provided.

FIGURE 9.9 A schema of an office will likely include a standard set of objects, such as a desk, chairs, a computer, papers, and books.

The rich history of research evaluating deduction using the Wason card selection task and the evolution of variations of this task led to more theoretical development. Cheng and Holyoak (1985) advanced an important hypothesis termed *Pragmatic Reasoning Schemas*. A schema is a unit of knowledge about the world as we have discussed in other chapters. For example, we have a schema for an office that applies to no one office in particular, but it contains information that is generally true about most offices and therefore would likely apply if you were to enter a new office. Most people's office schemas would include a desk, a chair, a computer, writing tools, paper, possibly some pictures, or personal items. The office would likely not contain a cat, a camping chair, or a bed. Schemas help us to establish what is likely in the environment around us, thereby freeing up our attention to devote to other incoming information (Fig. 9.9).

We also form schemas about behavior and commonly experienced situations in life. You probably have a well-developed schema for checking out of a grocery store. This schema would likely contain several common elements. For example, you are likely to be the person who unloads your cart or basket, an employee will most likely scan and bag the items, and you will be asked to scan your card or coupons prior to paying using a credit card machine or cash. You are likely to expect guidance from the cashier during the process. Some possibilities are not part of this schema. You are very unlikely to be given items free of charge, you are not expected to buy items for full price if they are on sale, and you are not obliged to pay for additional items that you do not want. Notice in this example that some of the aspects to this situation deal with obligations and permissions. In other words, they inform your deontic reasoning process. Patricia Cheng and Keith Holyoak (1985) proposed that we develop schemas for certain situations and among those schemas are some that contain the rules that will apply in daily life. These rules will help us get through our day quickly and efficiently, but may either enhance or impair our deontic reasoning depending on the situation in question and how familiar we are with it.

Some of the strongest evidence for the use of pragmatic reasoning schemas comes from Cheng and Holyoak's (1985) study of a variation of the postage stamp problem that had been introduced previously by Johnson-Laird et al. (1972). The researchers reasoned that if a card selection situation evoked a schema, people would be likely to use that schema-based knowledge to solve the card selection task. Thus, card selection errors could occur due to people failing to evoke the correct schema, or if someone retrieved a different schema that is inappropriate for the situation, then they may also make errors. A schema about permissions applies very nicely to the postage problem, or to the equivalent drinking age problem that we discussed earlier. In both cases, the critical step of inferring the presence of a situation that violates the rule would be important in maintaining enforcement of that rule. If people rely upon schemas in card selection tasks, then it stands to reason that performance should vary depending upon whether the participants have preexisting experience with a given permission situation. This led to a very creative cross-cultural study in which participants from Hong Kong were compared to those in Michigan. The Hong Kong participants were familiar with a postage rule stating that a sealed envelope must contain more postage than one with the flap merely tucked into the back. Such a rule had been put into effect in Hong Kong months prior to the study taking place, and this experiment took place during a time prior to email and other electronic correspondence, so people were likely aware of the postal rule due to the more frequent use of traditional paper-based mail. The postal version of the Wason card selection task required participants to evaluate the rule "If an envelope is sealed, then it must have a 20 cent stamp." Cards to be evaluated stated: "back of sealed envelope," "back of unsealed envelope," "10 cent stamp," and "20 cent stamp" (Fig. 9.10). The results were very clear. The postal rule was difficult for the Michigan participants with about 60% being correct on the task, which is still a rather high percentage relative to a standard abstract card selection task. Meanwhile, nearly 90% of the participants in Hong Kong got the postage problem correct.

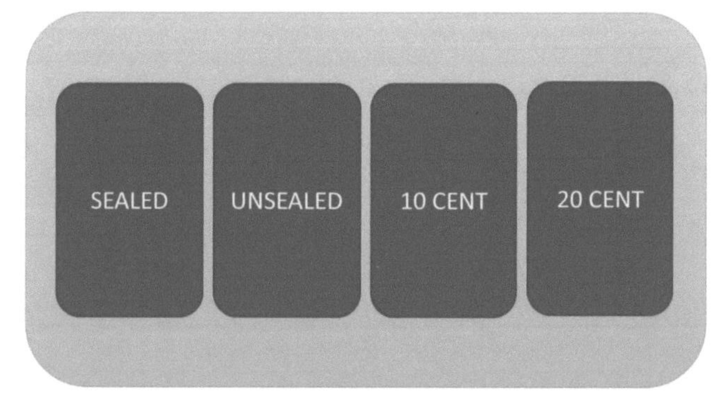

FIGURE 9.10 The Wason card selection task variants used by Cheng and Holyoak (1985). Shown are the rules and cards for the postage problem.

Cheng and Holyoak (1985) also included a problem about the people's history with the disease cholera and people being admitted or denied access into the Philippines in order to guard against any general individual differences in reasoning abilities influencing the results. Both participant groups were not familiar with this type of situation and performed similarly with around 60% correct. These results are similar to the Michigan participants' performance on the postage case. Additionally, Cheng and Holyoak provided a rationale for the rules and included a separate reasoning attempt for participants. In this case the instructions explained the background behind the postage rule and the cholera immigration rule. Knowledge of the background boosted both the Hong Kong participants' performance on the cholera problem to around 90% and the performance of the Michigan participants to 90% for both the postage problem and the cholera problem (refer to Fig. 9.11).

Pragmatic reasoning schemas remain one of the compelling theories that explains why participants are able to effectively handle evaluations of the violation case in a permission rule in some instances but not others. These results are consistent with the belief-bias effect that we discussed earlier with both theories claiming that consistency with background information is critical to how people perform on a deductive reasoning case.

Evolutionary Psychology and Deductive Reasoning

The last topic we will review in deductive reasoning comes from the field of evolutionary psychology.

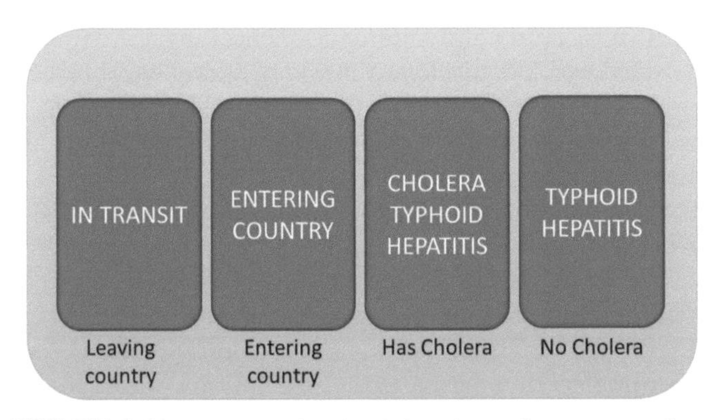

FIGURE 9.11 Cheng and Holyoak (1985) tested participants from Michigan and those from Hong Kong. The cholera problem asked people to act as an immigration officer and evaluate the rule "If a participant is entering the country, then they must be vaccinated for cholera." Neither participant group was especially successful on the cholera problem, which was unfamiliar to both participant groups.

Evolutionary psychology is a discipline that seeks to understand the roots of human behavior by explaining how behavior can benefit an individual and therefore would be selected for by evolution. The Evolutionary Theory of Natural Selection becomes relevant to deductive reasoning in the case of certain variations of the Wason card selection task that relate to detecting cheating among individuals.

Detecting cheating is important in evolution due to the concept of reciprocal altruism (Trivers, 1971). Altruism refers to a set of behaviors that do not outwardly benefit the individual. For example, if a bystander jumps into a rapidly moving flooding river to save a child, only to escape near-drowning himself, this individual is engaging in altruistic behavior. There is not a tangible benefit to this good samaritan, as he is not a relative of the child and he is in fact risking his own life in attempting the rescue. A key feature of truly altruistic actions is that they must be directed at benefitting nonrelatives. This is important, as relatives are valuable to the individual as viewed through the lens of evolutionary theory and therefore benefit the individual committing the beneficial act. The evolutionary biologist J. B. S. Haldane once jokingly stated that he "would willingly die for two brothers or eight cousins." This sentiment captures what is known as *kin selection*, the idea that people are more likely to act in ways that will benefit family members over strangers. Such behavior makes sense from an evolutionary perspective, as Haldane's "two brothers" would each share roughly 50% of his genetic inheritance. Therefore, the two brothers together would be equal to Haldane himself in terms of pure genetics. Likewise, an aunt or uncle should be 50% genetically similar to a parent, or 25% similar to Haldane, making the cousins roughly 12.5% similar to Haldane (requiring eight cousins to equal the genetic makeup of one Haldane). All of these calculations are made on the basis of *evolutionary fitness*, a perspective that considers the function of an organism, humans included, to be the advancement of the genes one passes on to future generations. The term *inclusive fitness* describes this relatedness factor. Our relatives share genes with us; therefore we share in their reproductive success by genetic association. Their genes win out by being passed on and we share some portion of those genes, so we also win by association. This is the essence of inclusive fitness and why we should generally be helpful to our immediate relatives, at least from a "good of the genes" mindset. This perspective was eloquently articulated and defended by British evolutionary biologist, Richard Dawkins in his book *The Selfish Gene* (1989).

Kin selection aside, are there good reasons to be altruistic toward strangers, such as the fictional child in the river scenario? If one is helpful to a stranger at a cost to oneself, this may constitute true altruism, a situation in which help is offered to another with no benefit to oneself. Such a behavior is considered to be illogical from an evolutionary fitness perspective, as the stranger is unlikely to share any appreciable genetic relatedness to oneself, when that person is compared to the genetic relatedness of one's relatives. There is however another possibility that makes altruism beneficial even when viewed from an evolutionary perspective. This brings us to the "reciprocal" part of reciprocal altruism. We should help strangers provided there is a good chance that they will help us in the future. Imagine yourself as a member of a hunter-gatherer tribe thousands of years ago. Imagine you just completed a successful hunt and are lugging a load of mammoth meat back to your tribe. You have an abundance of food and encounter another individual from a neighboring tribe who is hungry and may not survive if he does not get some nutrients soon. Should you give some food to the stranger? From an inclusive fitness standpoint, perhaps not, as this person likely shares little if any genetic inheritance with you. Of course the charitable thing to do would be to help the hungry stranger out because doing so would make you feel better about helping your fellow early hominid. It may actually be quite helpful to offer this person some food from a purely selfish viewpoint as well. This hungry stranger from a distant tribe may be starving now and therefore unlikely to live if food does not arrive soon. If you share food with him, he will survive and there may be a time in the future when he will return the favor. Everyone encounters challenges at some point and any chance of gaining future benefits may improve your inclusive fitness. Like a form of "rainy day" fund distributed among other people in your environment, it will help to have friends when you are the starving individual one day in the future. You decide to help this fellow out, and some months later after experiencing a harrowing pursuit by a sabre tooth tiger, you find yourself lost and injured. This poor soul whom you once helped, takes you back to his home, introduces you around the tribe, and they nurse you back to health. You live to hunt another day and increase the chances of passing your genes on to future generations. Everyone wins in that scenario.

People across cultures generally dislike cheaters who fail to return a favor. Imagine another situation in which the stranger whom you fed months ago leaves you without care in your wounded state. You almost die out in the elements only to eventually find help from some of your long-lost kin. You will be unlikely to help the stranger ever again. You may in fact now be in favor of launching a raid on the stranger's tribe to take the resources of this ungrateful and untrustworthy scoundrel. You have been cheated and now experience a viscerally negative reaction when thinking about this person. Once someone is revealed to be untrustworthy, this is a very difficult reputation to shed. This is the type of trait that affects our reasoning. The behavior of crowds around election seasons also demonstrates this tendency to reject political candidates suspected to be untrustworthy.

Let's now return to the Wason card selection task. Leda Cosmides and John Tooby are evolutionary psychologists who have championed the idea that we have developed a cognitive module honed for detecting cheaters. This position has been termed *Social Contract Theory*, as cheaters represent cases of breaking social contracts (Cosmides & Tooby, 1992). The ability to detect individuals who cheat would confer upon us the ability to maximize our inclusive fitness by having a well-developed sense of when to trust others. This may explain why people have a universal negative reaction to cheaters. Research on variations of the card selection task indicate that people are unusually good at selecting the unintuitive, but valid, inference of checking the negative case, as we have discussed in the drinking age problem by Griggs and Cox (1982). In essence the 16-year-old in the drinking age problem is a cheater attempting to skirt the drinking age law.

Stone, Cosmides, Tooby, Kroll, and Knight (2002) provided one of the most compelling brain-based cases for the ability to detect cheaters representing a special case in deductive reasoning and supporting social contract theory. In this experiment, evaluating patients with neurological damage tested the evolutionary position. In cases where the medial temporal lobes and orbitofrontal cortex are diffusely damaged, peoples' ability to respond in emotionally appropriate ways becomes impaired (Fig. 9.12). These abilities are relevant to situations involving cheating, and such individuals may struggle to appropriately detect the breaking of social contracts. Stone et al. found one such patient, referred to as patient R. M., who had sustained bilateral damage after a bicycle accident. The researchers tested R. M. on the original abstract version of the Wason card selection task dealing with the evaluation of letters and numbers. Patient R. M. also completed two contextual variations of the card selection task, a social contract problem, and a precaution problem. The social contract problems included evaluation of social rules such as "If you go canoeing on the lake, then you have to have a clean bunk house." The four cards in this task featured the terms "canoeing on the lake," "not canoeing on the lake," "cleaned the bunk house," and "did not clean the bunk house." This version of the card selection task represents a social contract situation. One in which you need to have performed a particular action in order to take advantage of a particular benefit, and people generally do quite well on these types of problems. Stone et al. found that uninjured

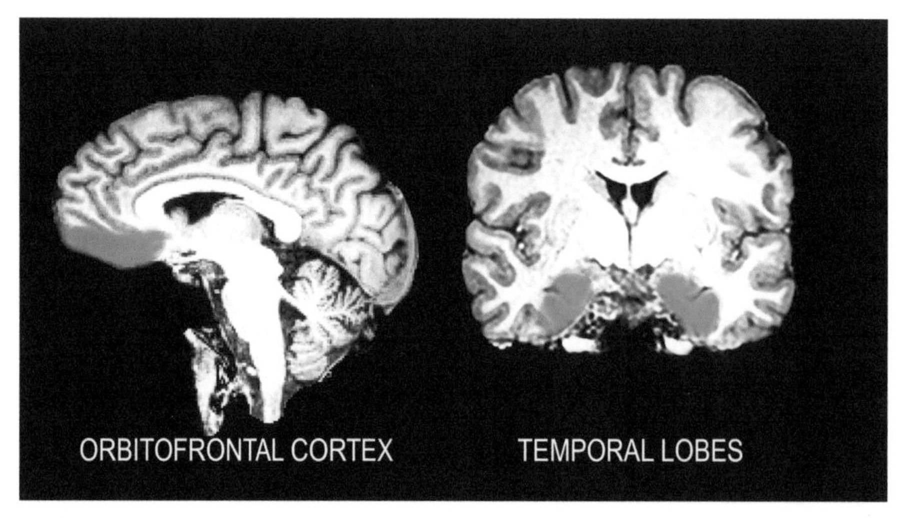

FIGURE 9.12 The orbitofrontal cortex and medial temporal lobes are important areas for processing emotions. These areas appear to be important when we evaluate situations involving cheating that occur in our daily lives.

control participants got approximately 70% of these problems correct. It is not too difficult to spot the potential cheater who did not clean the bunk house. We intuitively want to check on this person's status in terms of accepting the benefit of canoeing. People also tend to do well evaluating precaution problems, such as evaluating the rule "If you work with toxic chemicals, then you have to wear a safety mask." This context also encourages high deductive reasoning performance in most people, with healthy controls in this study reaching a performance level of 71% correct. Patient R. M., who exhibited a lack of social knowledge on other neuropsychological measures, was highly accurate on the precaution problems (70%) and inaccurate on the abstract letter-number problems (17%). Strikingly, R. M. was very poor on the social contract problems that required an evaluation of a cheater, with his average performance being around 40% on these problems. This performance level is over 30% lower than his performance on the precaution problems. The results suggest that R. M. suffers from the effects of a brain injury that has selectively reduced his ability to reason deductively about social contracts, as he was able to perform within the normal range on the precaution problem. This result supports the idea that we have evolved an ability to detect cheaters and that our limbic system circuitry including the bilateral amygdalae, orbitofrontal cortex, and temporal poles contribute to making these emotion-laden evaluations.

In daily life, R. M. would be unlikely to be able to evaluate the trustworthiness of other individuals. As such, he might inappropriately help untrustworthy people, who would seek to take advantage of him. This behavior fits with the profile of others who have limbic and frontal cortical damage, as such people commonly make errors with money and tend to fall victim to financial scams.

DEDUCTIVE REASONING IN EVERYDAY LIFE

Comparisons Between Reasoning in the Laboratory Versus Reasoning in the Real World

When considering people's deductive reasoning ability, we should perhaps be most concerned with what people actually do in everyday life. We began this section of the chapter by discussing the differences between formal logic that derive from philosophy, as well as from the psychology of deduction, which involves experimental evidence from people solving problems in laboratories. Two central findings are that people vary in their ability to perform deduction tasks and that the context of the task greatly affects performance. This second point gets us closer to everyday life reasoning. People will frequently add extra information to the given premises of a situation based on their backgrounds. They may also selectively ignore some details of the premises, if they are at odds with real-world experience and background knowledge.

Several experimental psychologists have drawn a distinction between reasoning in everyday life and lab-based reasoning. The early American cognitive psychologist Frederic Bartlett (1958) described the syllogistic reasoning ability often studied in the lab as an example of *closed-system thinking*. The implication is that a well-defined set of skills is used in deduction, and people do not move outside of those skills when solving deduction problems. This position captures the insight that lab-based deduction tasks tend to be solved using distinct steps and one does not move beyond these steps. Revlis (1975) provided a representative example of this closed-system process which is pictured in Fig. 9.13. In this case the reasoner begins with: (1) an encoding stage, in which the premises are read and in which the processes necessary to establish a database operate; (2) moves to a composite stage, in which logical operations are applied in order to produce a single representation of the information provided; (3) next there is a conclusion-encoding stage; and lastly (4) a comparison stage, in which the composite information is compared with the contents of the conclusion. This perspective predicts that numerous cognitive subprocesses must be applied in order to reason deductively, but also that these subprocesses are limited or "closed" as Bartlett described it. If one were to follow these stages, they should successfully generate valid conclusions across all instances of the task.

By contrast, Bartlett described everyday reasoning as *adventurous thinking*, capturing the fact that in daily life people will add extra information onto the premises, and that the process is less-bounded than that used in closed-thinking. Adventurous thinking may deviate from the stepwise process described by Revlis (Fig. 9.13) to incorporate wider domains of knowledge, the effects of emotion, and the availability of recent experiences in affecting reasoning. Other researchers have used the terms *well-defined* and *ill-defined* to illustrate a similar distinction. Glass, Holyoak, and Kiger (1979) described well-defined problems as providing certain premises in the form of information about aspects such as the rules or operations, as well as the goal state with all information being specified completely. Well-defined problems would include the deductive reasoning tasks described thus far in the chapter such as syllogistic reasoning and the Wason card selection task. Meanwhile, ill-defined problems are those in which some aspects are unspecified for the reasoner.

Many problems found in daily life have this ill-defined quality. For example, imagine that you are shopping for shoes. You may set up your own constraints on

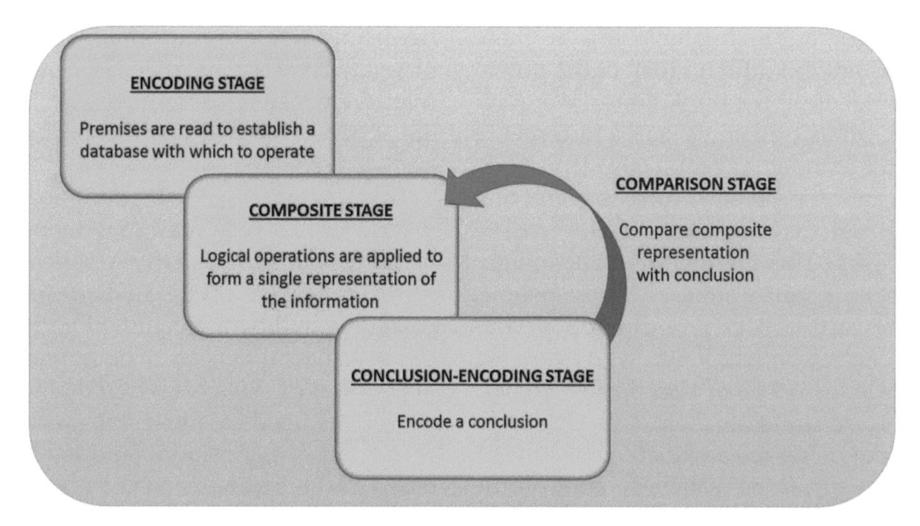

FIGURE 9.13 Revlis's (1975) stepwise process model for deductive reasoning. This is an example of a closed-system process. The thinker begins with: (1) an encoding stage, where the premises are read in order to establish a database to operate with; (2) a composite stage, sees logical operations applied in order to produce a single representation of the information; (3) a conclusion-encoding stage in which a conclusion is generated; and lastly (4) a comparison stage, in which the composite information is compared with the contents of the conclusion.

color or style. You go in seeking black shoes, but instead you buy a pair of very stylish brown boots on sale. You have not failed the logic test, as there was not a clear correct answer owing to the fact that the information in the "shoe shopping" problem was not fully specified. In essence, buying shoes is an ill-defined problem, one that becomes very difficult to evaluate in terms of correctness, as success is determined by the happiness of the buyer.

This chapter has focused primarily on well-defined, or closed-system, thinking with regard to deductive reasoning. This is in part due to the lack of literature examining the processes of deduction outside the lab. Many real-world studies instead fall into the domain of decision making. In particular, multi-attribute decision making is an area in which deductive reasoning likely applies, but we rarely label multi-attribute decision tasks as being deductive reasoning tasks. An example is choosing a college or university to attend. In some sense there is not an objectively correct or incorrect answer to this problem. Instead there are gradations of better and worse choices based a several attributes including size, location, tuition, financial aid, and quality of instruction. The aspiring college student will typically make a final decision on this task and provided it satisfies most of the premises (which the student generated for him or herself) the effort can be considered a success. Deductive reasoning in such a situation may occur when we consider the distance of a school from our home. Some schools are objectively closer to home and therefore can be clearly compared to the distance of other schools. The fact that there may be multiple objectively better schools and these need not all align to form a single correct choice makes college selection an ill-defined problem. The fact that all information is not given at the outset in the form of a premise makes this problem less able to be solved using a stepwise process model as can be used in a well-defined or closed-system problem. We will focus more on multi-attribute decision making and examine the college selection task further in Chapter 11.

A last point in this section is that ill-defined problems are difficult to study. The data gathered from a standard Wason card selection task yields objectively correct answers that can be easily quantified. By contrast, ill-defined problems may result in much more variable data that will be more difficult to analyze. There is a trade-off between creating a task that is well-defined and straightforward to understand as a researcher and creating a task that is truly representative of everyday reasoning.

Evidence From Cultural Differences in Deduction

There are some challenges to deductive reasoning performance that are clearly related to cultural background, and this reveals some important features about the cultural basis of reasoning. People often fail to consider all of the possible interpretations of a premise (Galotti, 1989). For example, the premise "If X, then Y" allows for the following possibilities: (1) X and Y are both true, (2) X and Y are both false, and (3) X is false and Y is true. Similarly, the premise "All X are Y" makes possible the interpretation that there are some Y that are not also X. This is clear from the Venn diagram depicted in Fig. 9.14. Another possible inference from "all X are Y," is the common interpretation that X and Y have complete

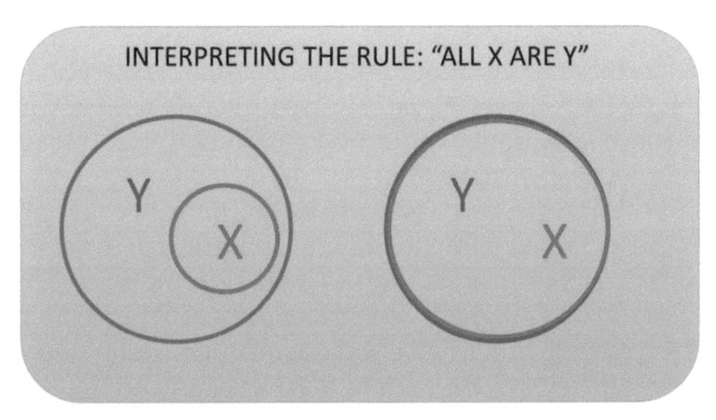

FIGURE 9.14 A Venn diagram representing two different ways to evaluate the statement: "all X are Y."

overlap, such that there would not be a Y that is not also an X. Johnson-Laird (1983) emphasized that deductive reasoning errors may result from people failing to identify all of the possibilities that follow from the premises. It is probably also fair to say that people rarely create Venn diagrams for problems (as in Fig. 9.14), as they are time consuming and often unnecessary in the context of everyday reasoning. The subset of thinkers schooled in philosophy of logic would be more likely to consider Venn diagrams when considering these classic deduction problems.

Studies evaluating reasoning in other cultures have frequently found that non-Western educated people perform worse on deductive reasoning tasks relative to those that have been trained in these problems in educational environments. In many deduction tasks, familiarity with the premises serves as a leading indicator of success. This position supports the idea that recognizing the pattern associated with certain premises may be a key to success in deduction, rather than engaging in a rule-based routine enacted by a cognitive module. This brings us back to the likely possibility that everyday reasoning tasks are typically ill-defined and therefore, we may not engage in deduction under such circumstances. This was elegantly demonstrated by Cheng and Holyoak (1985) in their study of the Wason card selection task in individuals varying in cultural background who were familiar or unfamiliar with the postal rule, which we had discussed earlier in the chapter.

INDUCTIVE REASONING

Induction: From Example to Generalization

To this point in the chapter we have emphasized the study of deduction, a situation in which a sure answer can be arrived at provided that the premises are true. Regardless of what is true in the real world, deduction assures us of making a valid inference. We now turn to the very different process of inductive reasoning, a situation in which the validity and truthfulness inferences are not guaranteed. Most of us use inductive reasoning extensively over the lifespan. Induction is involved when we make category judgments, and it is intimately linked to the similarity estimates that we develop in childhood. Our inductive reasoning expands as we age and gain more knowledge about the world. Inductive reasoning evolves to accommodate our growing abilities to assess similarity and apply our knowledge in order to make sound inferences. When we become experts in a particular domain, our knowledge and our associated inferences change yet again. In this section of the chapter we will review research about how we induce inferences in different types of situations, what factors drive inductive reasoning, and how these processes change with age and expertise.

Let's turn to some examples of inductive reasoning. To illustrate the classic distinction between inductive and deductive reasoning, we will start by revisiting the drinking age problem from earlier in the chapter (Griggs & Cox, 1982). In the drinking age problem you play the role of a law enforcement agent in charge of enforcing the rule "If you are drinking alcohol, you must be over the legal age limit." In a deductive case such as this one, you can accurately deduce that you must check the age of people who are drinking beer, and you must check the glass of anyone who looks too young to drink, as they are possibly rule-violators. This is all that is needed to perform the job and you should be highly accurate if you follow this deductive approach.

Now imagine a variation on this scenario. You are again in the law enforcement business and this time you hear that there is a new trend in which teenagers have begun to drink at certain bars during certain hours, but the location varies, so that they cannot easily be caught. You are charged with figuring out where and when the teenagers will show up to drink illegally. This will essentially require detective work. You will have to frequent bars across town at different hours and attempt to piece together how to stop the behavior. As you begin to check bars, you find that there are a lot of young people on the west side of town before 10 p.m. and that more young people show up on the east side of town after 10 p.m. You begin to explore this pattern and after a certain number of instances you decide that you are going to report the following "Young people move from the bars on the west to the east side of town through the course of the evening." You recommend that greater police presence be placed accordingly. This is inductive reasoning. You do not know the rule that the teenagers are following. You are using your mental prowess to figure out the pattern through experience and then try to induce a rule that is being followed.

Notice that there is no guarantee of accuracy in this case. You may have made mistakes in your analysis. You may find that you were not looking into the right type of bar on the right date. There may be a more subtle pattern that you failed to notice, such as the days of the week. You work Sunday through Thursday and this pattern only applies to teens that drink during the weekdays, but the majority who frequent bars actually do so on Friday and Saturday and these teens follow a different pattern moving from the south to the north side of town. There may in fact be no pattern at all, but you are determined to make sense of the situation anyway by imposing rules and predictability on your model of the situation. This process bears a strong resemblance to the causal reasoning scenarios we discussed in Chapter 8.

Philosophers of logic have sometimes assigned inductive reasoning a lower status, as it cannot guarantee validity and may very well lead to false inferences. The fact that deductive reasoning alone can produce valid conclusions places it above induction in one's reasoning toolkit. Notice however, that the majority of our reasoning in everyday life is probably inductive rather than deductive. We rarely have clearcut black and white rules to follow and evaluate. These features of the environment make deduction inappropriate in many instances. In a world full of exceptions to the rules and shades of gray, we must frequently engage in *probabilistic reasoning*. This is at the core of induction, taking category knowledge that we have gained through experience and attempting to apply it to new cases in order to infer *a likely* aspect of a situation, not a guaranteed aspect. Induction is also found in other forms of reasoning. When we use analogies to infer new information, we are often inducing a new rule based on a prior situation. If we take into account statistical information in order to draw inferences, we are acting probabilistically and using induction. Scientists also make use of induction when forming theories for later testing. A theory typically develops from a bed rock of replicated findings that begin to make the world appear as if it works in a certain way. A scientist induces a rule about a situation and then sets about testing hypotheses to see if indeed the rule holds. None of these situations would be possible without our use of inductive reasoning.

Children make heavy use of inductive reasoning. There is a common tendency for children to make speech errors around the ages of two or three. One such error is called *overgeneralization*. This can be observed when a small child has learned that the four-legged furry brown animal who lives in the house is called a dog. Similarly the child labels her toy bear and horse to also be "dogs." In this simple case, the child has overgeneralized the use of the word "dog" to too many four-legged, brown animals. This is an example of a child making an inductive inference. Much of our early learning has this same quality. We observe something about the world, take note of the characteristics, and then apply that same observation to other examples that approximate having those same properties. This is how we build our semantic knowledge. Children also make a similar speech error known as over-regularization. For instance, my son had learned that the past tense of "run" was "ran." He took this a step further by inferring that the past tense of "won" would be "wan." Similar errors occur when children apply the common "–ed" ending to words, yielding charming and funny errors such as: "standed, taked, hurted, drinked, and holded."

While incorrect, children's language errors make sense to us. We know what the child means and why he or she made the error. These represent instances of the child actively applying rules that they have either been told about or have induced themselves based on many instances. Since we can only explicitly teach language to children on occasion, much of what is learned is simply picked up by being spoken to and by listening to speech around us. As with all inductive inferences (see Box 9.1), we must remain open to the possibility that we have over-applied or misread a rule that we had induced in a different situation.

BOX 9.1

INDUCTIVE INFERENCE AND STEREOTYPING

Recently, there have been a series of news stories about police violence against African Americans resulting in the deaths of several individuals. This has led to public protests and demonstrations. The protests often focused upon stopping the police from unjustly using excessive force or poor judgment against members of the African American community. Such incidents and the associated protests have garnered considerable

Continued

BOX 9.1 *(cont'd)*

media attention. Each reported new incident further strains the often fragile relationships that exist between ethnic minority populations and police forces across the United States.

Initially, there was a controversial incident in Florida. There was an altercation between George Zimmerman and Trayvon Martin. Zimmerman reported a head injury inflicted by Martin and claimed that he fatally shot the teen in self-defense, which was legal due to Florida's "stand-your-ground" law in which using deadly force against another citizen is permissible in a situation in which one feels their life is under threat. Zimmerman was acquitted of second-degree murder and of manslaughter sparking community outrage. Many people have claimed that race was a factor in the killing, as Martin was an African American male. While Zimmerman was not a police officer, this incident has been categorized in repeated media stories as being related to other incidents in which an African American man was killed during a police stop, such as when Ezell Ford was shot by the Los Angeles Police Department during a police interaction in 2014, or when Alton Sterling was killed by the Baton Rouge Police Department in Louisiana. Soon, many in society and in the media were categorizing every incident in which an African American man was killed at the hands of police as being related only to race. Continuing media coverage was fueling this state of affairs. Protesters across the United States called for police accountability and reform of the criminal justice system. Meanwhile, police forces struggled to redeem their reputation and police chiefs across the country focused heavily on maintaining control of their officer's behavior, their policies, and how those policies are being carried out in US cities.

Let's now consider this situation from a psychology of reasoning standpoint. The incidents that have received the most media attention have begun with police officers being called to a scene to investigate a crime of some sort. The police in such situations are likely to be on edge. Officers may anticipate a conflict with a potentially armed citizen who may be in a desperate state-of-mind and may resort to violence rather than be apprehended. Race may factor into the situation. African American males are incarcerated at a rate nearly four times as high than Caucasian or Hispanic males. If the police stop takes place in a location populated by a majority African American population, then officers who patrol that area will likely make a greater proportion of arrests of African Americans. Together, these factors may lead to officers stereotyping African Americans as being criminals. No such inference is warranted. The majority of people of all ethnic backgrounds do not commit crimes, or become incarcerated. Nonetheless, if the experience of repeatedly arresting a higher proportion of African Americans occurs, an inductive inference may be made that there is a greater tendency for African American citizens to be involved in committing crimes. Such an inference, whether made explicitly or subconsciously may lead an officer (of any ethnic background) to enter into a situation with an African American individual and assume that there will be a greater chance of that individual having committed a crime and that the individual will resist arrest. This type of reasoning may be at the core of racial or ethnic bias. It may contribute to why police stops have escalated to actual killings.

One can ask the question: did the police act appropriately when someone is killed? A common gut reaction is to react with outrage toward the officers involved. On further examination, it may not be clear if an officer's behavior was warranted. In reality, police officers must sometimes use deadly force against an individual whom they judge to be physically threatening to themselves or others. Each different situation contains different individuals, different circumstances, and requires different actions. While it is extremely unfair of police officers to treat people differently based on their ethnic backgrounds, it is also extremely unfair of citizens to assume that all police officers have acted solely based on racial stereotyping and that all of them have used faulty and racist judgment in all cases.

This is clearly an emotional issue and one that involves inductive reasoning on both sides. Some police act more aggressively toward potential suspects who are African American due to inductive inferences about the population. Some African Americans assume that the police are going to treat them unfairly due to inductive inferences about the properties of the police and their attitudes. The American media frequently make inductive inferences that situations in which ethnic minorities are stopped by the police that race prevails as the impetus.

While it is extremely frustrating the fact remains that stereotyping is part of human interaction and that inductive inferences contribute to such judgments. Inductive inferences are prone to error. A remedy to some of these challenges may be for people to carefully consider the individuals and circumstances in each incident based on the best possible information, rather than glossing over specific details and making inductive inferences about how the situation played out based on previous situations that appear similar.

INDUCTIVE REASONING IN THE LABORATORY

Inductive Reasoning Tasks

We can reason about whether a conclusion is warranted based on a premise in inductive reasoning. This is much the same as our ability to reason deductively from premise to conclusion. Fig. 9.15 shows images related to an example of a premise "Blue whales feed using baleen plates" and a possible conclusion "Fin whales feed using baleen plates." The job of the thinker in this case is to evaluate whether that conclusion is warranted. Think for a moment about whether this conclusion is likely to be true. When I consider this question, a few observations come to mind. First off, experience with whales matters a great deal. If you are completely unfamiliar with fin whales you are probably in a poor position to evaluate the likelihood of this conclusion and somewhat left guessing. Alternatively, if you are a whale expert this will be trivial to evaluate based on your existing knowledge. If you are somewhere in the middle, having perhaps heard that there is an animal called a fin whale, and you are more familiar with the much more famous blue whale, you will have to decide how likely the conclusion is and balance this judgment with your background knowledge. Interestingly, a key factor in making an inductive inference about fin whales is your knowledge about blue whales. Many people are aware that the blue whale is the largest of the great whales and that this whale feeds on tiny oceanic animals called plankton using its baleen plates to filter the plankton out of the water. Knowing this, you must then establish whether the lesser known fin whale also feeds this way. You may also know that another well-known great whale, the sperm whale, feeds on the massive giant squid and uses teeth to feed. This sets up a thought challenge: is the fin whale a toothed whale that likely eats squid, or is it a baleen whale that eats plankton? Similarity is the key here. Does the fin whale look like the blue whale? Does the fin whale dive to great depths like the sperm whale? How big is a fin whale? These types of questions will provide a kind of evidence with which to determine the likelihood of the conclusion.

In answer to the question, Fig. 9.15 shows a fin whale of enormous size, only slightly smaller than the blue whale. It is extremely closely related to the blue whale genetically, and does in fact feed on plankton using baleen plates. If you knew about any of these traits, the conclusion was likely much easier than if you did not.

FACTORS THAT AFFECT INDUCTIVE REASONING

Similarity as a Key Property for Inductive Inferences

The similarity of one thing to another was established in psychological research as a key determinant of inductive inferences. Lance Rips (1975) published an influential study evaluating this factor in an experiment in which people judged the traits of animals on the basis of facts about other animals, much like the fin whale example we just considered. In the article, Rips explained that inductive reasoning situations require three factors: a set

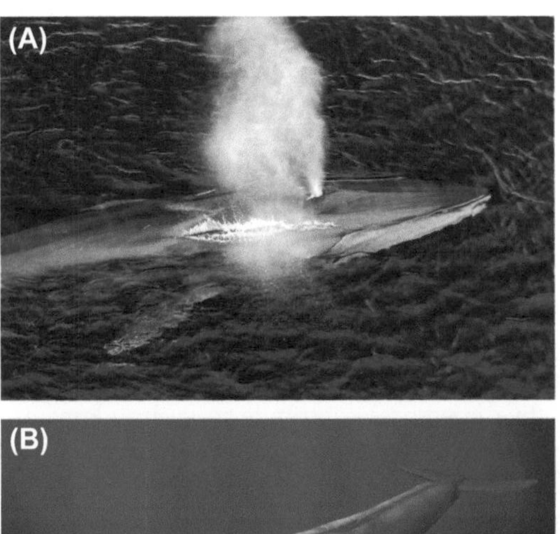

FIGURE 9.15 An example of inductive reasoning problem. (A) The fin whale. (B) The closely related blue whale. (C) The sperm whale.

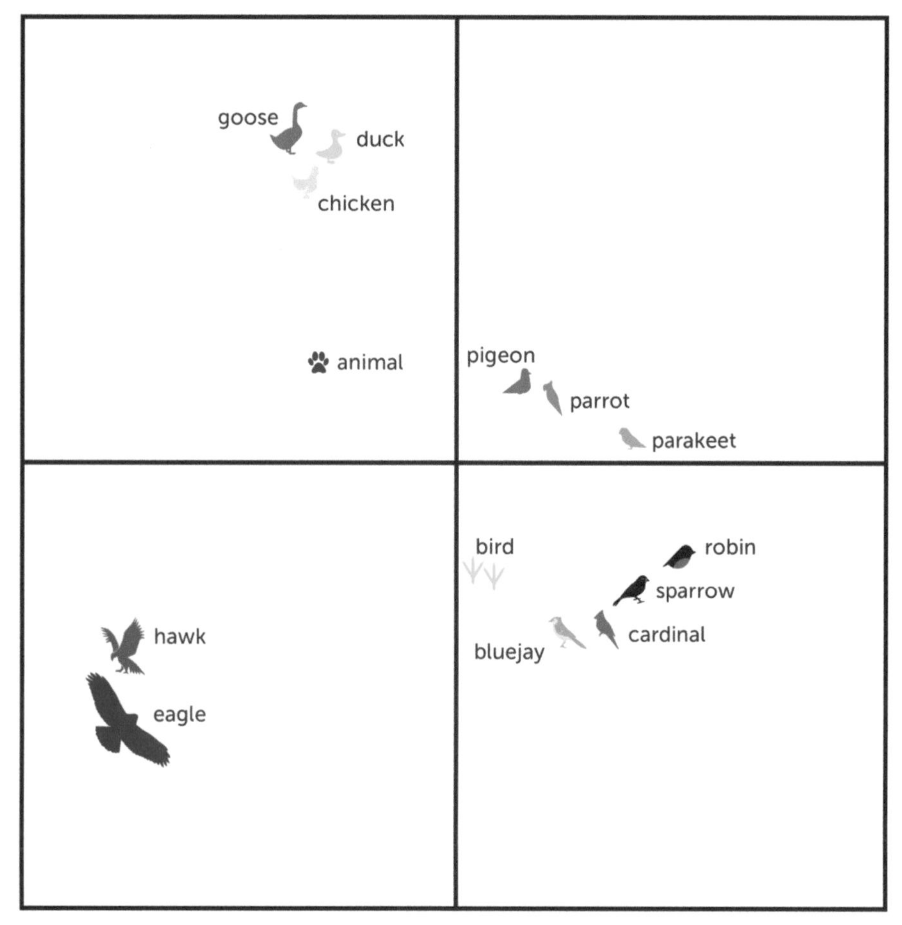

FIGURE 9.16 Rips (1975) plotted items such as birds to represent their similarity spatially. Inductive inferences are strongly guided by how similar two items are with a greater tendency to draw an inductive inference between similar items. *Image by Ashton Miller.*

of instances that can be used as a reference, a specific property that the instances could consistently possess, and a specification of the instances that possess the property. Given those factors, people can then make a conclusion about whether a new instance has that property. Participants in this study were presented with a scenario about the spread of contagious disease on an isolated island containing very few species. Among the species on this hypothetical island were the following: sparrows, robins, eagles, hawks, ducks, geese, ostriches, and bats. Note that the first seven of these are all birds. All of these birds, except the ostrich, are instances of relatively common birds. Rips included these six common bird species as critical items, ones that could be evaluated based on their similarity to one another. Meanwhile, ostriches and bats are somewhat similar to the others, but can be considered outliers due to their lack of overall similarity to more common birds. Participants in the experiment were told that all members of one of the species on the island had a contagious disease and were asked to provide an estimate of the percentage of members of each of the other species that are also likely to have the disease (Fig. 9.16).

Results of the Rips (1975) study provided strong evidence that the likelihood of an inductive inference is highly related to the similarity of the case in the conclusion to that in the premise. For example, in Rips' Experiment 1, participants were told that all eagles had the disease. In this case they were likely to conclude that a large majority of hawks also had the disease, while fewer geese and ducks had it. By contrast, when the participants were told that all ducks had the disease, they estimated that a very high percentage of geese also had the disease, while few hawks and eagles were estimated to have it. These elegant results indicate that when we see two entities as similar, we expect that they will behave in a similar fashion with regard to unknown properties that we are asked to evaluate.

It quickly becomes apparent that when people apply inductive reasoning to human behavior they run the risk of stereotyping. This can lead to offending people in large numbers (see Box 9.1). Rips (1975) provided an interesting remedy to this challenge in Experiment 3 of this classic paper. Rips informed the participants that "Scientists know that any of the other animals could also contract the disease." Adding this phrase to the instructions resulted in much more even distributions of estimates across the different species. In other words, the typicality (or how typical an example of a category any one item is) no longer had a strong effect on the estimated disease numbers once that statement was included. This finding suggests that reminding people that all possibilities are open can have immediate effects on how seriously we take an inference. If someone begins to stereotype, applying the behavior of few to that of many, it may help to remind that person that "Any individual can behave in the manner described."

Inductive Reasoning and Homogeneous Category Effects

Inductive inferences about people and their behavior may vary based on the shear diversity of human beings. Nisbett, Krantz, Jepson, and Kunda (1983) reported an experimental effect illustrating this point. In this paper, the authors argued that inductive inferences should be stronger when several key features are present in the linkage between a given instance and the population that it may be similar to. Specifically, people should reach more confident conclusions when they reason from a larger number of instances to one new instance than when they reason from a small number of instances toward a new instance. People should also be more confident when reasoning from a narrow range of well-defined instances toward a new instance than when reasoning from a wide-ranging, poorly defined set of instances to a new instance. People should also take into account any known statistics about the sample from which one is drawing a conclusion. This would include use of the base rate, or the prevalence of a particular characteristic of interest in the population from which it is drawn. The application to human behavior based on these assertions is clear, and reasoning that is guided by these factors would likely result in better informed inductive inferences that reflect representativeness in the world to a greater degree than if these factors are ignored. There are certainly times that people do ignore statistical information when reasoning (as will be further discussed in Chapter 11 covering judgment and decision making), but there are also times when people do account for these and other important factors that can lead to improved inductive reasoning.

In the study by Nisbett et al. (1983) people were given a scenario about a newly discovered island and properties about its features. Conclusions needed to be drawn about the characteristics of the native people, animals, and resources. Examples of the experimental stimuli are shown in Fig. 9.17. In their first experiment, Nisbett et al. presented participants with a statement about an animal, person, or thing and asked participants to estimate the percent of others like the item, or individual, described that would share that same characteristic. For instance consider the following statement:

Imagine that you are an explorer who has landed on a little known island in the Southeastern Pacific. You encounter several new animals, people, and objects. You observe the properties of your "samples" and you need to make guesses about how common these properties would be in other animals, people or objects of the same type.

Suppose you encounter a new bird, the shreeble. It is blue in color. What percent of all shreebles on the island do you expect to be blue?

(This and the subsequent questions were followed by " percent. Why did you guess this percent?")

Suppose the shreeble you encounter is found to nest in a eucalyptus tree, a type of tree which is fairly common on the island. What percent of all shreebles on the island do you expect to nest in eucalyptus trees?

Suppose you encounter a native, who is a member of a tribe he calls the Barratos. He is obese. What percent of the male Barratos do you expect to be obese?

Suppose the Barratos man is brown in color. What percent of male Barratos do you expect to be brown (as opposed to red, yellow, black or white)?

Suppose you encounter what the physicist on your expedition describes as an extremely rare element called floridium. Upon being heated to a very high temperature, it burns with a green flame. What percent of all samples of floridium found on the island do you expect to burn with a green flame?

Suppose the sample of floridium, when drawn into a filament, is found to conduct electricity. What percent of all samples of floridium found on the island do you expect to conduct electricity?

Suppose you encounter a new bird, the shreeble. You see three such birds. They are all blue in color. What percent of all shreebles on the island do you expect to be blue?

FIGURE 9.17 Nisbett et al. (1983) experimental stimuli.

> Suppose you encounter a new bird, the shreeble. You see three such birds. They are all blue in color. What percent of all shreebles on the island do you expect to be blue?

In this example, we do not know how the color of these three shreebles relates to that of all of the other shreeble birds on the island. Perhaps they are sexually dimorphic, with the male being colored differently than the female, as many birds are. Alternatively, perhaps they change color as they age. By contrast, some statements such as that regarding a metal called floridium (see Fig. 9.17) indicated a property related to chemistry (that the flame is green when the metal is burned). This is likely to be a homogeneous property across many instances, as chemical properties are stable. Lastly, the question about whether members of the Barratos Tribe will be obese (Fig. 9.17) represents a relatively nonhomogeneous characteristic. Additionally, the experimental materials specified the number of individuals that were present in the sample and this included samples including 1, 3, or 20 individuals.

Results indicated that categories that were perceived to be homogeneous were rated to be highly likely to share a characteristic (such as the flame color of floridium). Meanwhile those categories that were thought to be heterogeneous such as the weight of the Barratos people were judged to be less likely to universally share that characteristic. Interestingly, the number of cases specified in the sample also made a large difference in the study. When only one individual or instance was considered, there was a large difference in estimated numbers of cases to share that property (refer to Fig. 9.17). Near 95% of floridium samples were estimated to burn green on the basis of only one described sample. This percent did not change dramatically at the higher sample rates of 3 or 20 cases. Meanwhile the midrange homogeneous characteristic of shreeble bird color was estimated to be around 70% on the basis of 1 bird, but jumped to over 80% when 3 or 20 birds were described. Lastly, the heterogeneous quality of Barratos people's obesity rate was estimated at under 40% on the basis of one individual, jumping up to over 50% when three obese individuals were described and then over 70% when 20 individuals were described.

The study by Nisbett et al. (1983) also examined the effect of expertise on inductive reasoning. Participants were either familiar with drama or football. A scenario was described in which a coach or director evaluates new, young talented actors, or football players, and these individuals prove to be worse over time then when they were first evaluated. Participants selected a likely explanation for the discrepancy between the strong initial impression and the subsequent lower assessment. Those who were experienced with football preferred a statistical explanation for the lower performance in that domain. For example, participants familiar with football tended to select the explanation that "The brilliant performances at try-out are not typical of those boys' general abilities," "They probably just made some plays at the try-out that were much better than usual for them." These statements indicate a lack of statistical sampling as the reason for the initial incorrect impression. Conversely, the participants familiar with drama tended to select a nonstatistical explanation for the unusual first impression. The effect reversed for the participant groups when they considered the drama scenario with those participants familiar with drama, now preferring the statistically sound explanation for the change in talent evaluation. Clearly statistics matter, and we weigh them with greater importance, the more familiar we are with a particular domain.

The studies by Nisbett et al. (1983) enable us to place additional constraints on people's inductive reasoning performance. First, the tendency of individuals to infer a new property about a given case is influenced by the number of items from which a comparison case has been sampled. This can be observed commonly in psychological research in which a study with a very small sample size does not reveal statistically significant differences, while one with a large sample size is much more likely to show differences that are present in the population. Nisbett et al. further showed that the homogeneity of cases matters quite a bit. Domains such as chemistry produce large estimates of probability based on even one case, while heterogeneous domains resulted in markedly lower estimates of probability. People factor in the mechanisms for things having properties, and they account for the variability within those mechanisms.

Other Factors Influencing the Strength of Inductive Inferences

In addition to similarity, there are a variety of interesting factors that influence the strength of an argument and therefore the likelihood that an inductive conclusion will be judged to have followed from the premises. Several of these phenomena were outlined by Osherson, Smith, Wilkie, López, and Shafir (1990). To illustrate this, in one example Osherson et al. noted the following comparison between two different inductive scenarios:

A. *Grizzly bears love onions.*
Polar bears love onions.
Therefore, all bears love onions.

B. *Owls prey on small rodents.*
Therefore, rattlesnakes prey on small rodents.

Note that in the example above most people would judge the conclusion in B (that rattlesnakes prey on small rodents) is more probable than that in A (that all bears love onions). Despite this real-world judgment, the strength of A is higher than that of B, as A presents two examples that are similar and these examples from A are from a similar category. Meanwhile, in B, the example provided (owls) has little similarity to that described in the conclusion (rattlesnakes).

Typicality of the items in question will lead to stronger conclusions. As an example, consider the following:

A. *Robins have a higher potassium concentration in their blood than humans.*
All birds have a higher potassium concentration in their blood than humans.

B. *Penguins have a higher potassium concentration in their blood than humans.*
All birds have a higher potassium concentration in their blood than humans.

In this example, the strength of A is higher than that of B, as robins are thought to be more representative, or typical, of birds relative to penguins.

In this next example a phenomenon referred to as *premise diversity* is illustrated. Consider the following:

A. *Hippopotamuses have a higher sodium concentration in their blood than humans.*
Hamsters have a higher sodium concentration in their blood than humans.
All mammals have a higher sodium concentration in their blood than humans.

B. *Hippopotamuses have a higher sodium concentration in their blood than humans.*
Rhinoceroses have a higher sodium concentration in their blood than humans.
All mammals have a higher sodium concentration in their blood than humans.

Which do you think has the stronger conclusion? Here we should note that A presents two very different members of the mammal category, the hippopotamus and the hamster. These category members vary dramatically in size, as well as the types of animals they are related to, their diet, behavior, and other notable features. Meanwhile, example B presents us with premises concerning two very similar mammals the hippopotamus and the rhinoceros. These animals are not only similar in size, but they live in similar parts of the world, and have similar names. The likelihood that a conclusion based on the premises in A applies to *all* mammals appears relatively high, as two very different mammals share the property of the higher potassium concentration than humans. If very wide-ranging species share the property, then it appears plausible that all animals share this property. Meanwhile, the probability that a conclusion based on the two highly similar premise species (hippopotamus and rhinoceros) applies to all mammals seems much less likely. There is a high probability of relatedness between these large African mammals and therefore a strong probability that they share a relatively uncommon property that would not likely be true of all other possible mammals.

Another phenomenon specified by Osherson et al. (1990) is known as *conclusion specificity*. This example below illustrates this situation:

A. *Blue jays require Vitamin K for the liver to function.*
Falcons require Vitamin K for the liver to function.
All Birds require Vitamin K for the liver to function.

B. *Blue jays require Vitamin K for the liver to function.*
Falcons require Vitamin K for the liver to function.
All Animals require Vitamin K for the liver to function.

Here were are presented with identical premises in both A and B. What differs is the amount of other possible examples that the inductive inference applies to. In A, we need only conclude that blue jays and falcons are representative enough of the category "birds" to warrant a conclusion that "All birds require Vitamin K for their livers to function." Meanwhile, in B, we are asked to evaluate these same two bird categories, but make the inductive conclusion that "The Vitamin K liver requirement applies not only to all birds, but the much broader class of animals in general." You can see that A is more likely to yield a correct inference, as it runs much less risk of overextending the premise beyond a reasonable inference.

In addition to the types of category members included in the premises, it is also possible to alter the strength of an inductive inference on the number of items. In a phenomenon named premise monotonicity, Osherson et al. (1990) described the following example:

A. *Foxes use Vitamin K to produce clotting agents in their blood.*
Pigs use Vitamin K to produce clotting agents in their blood.
Wolves use Vitamin K to produce clotting agents in their blood.
Gorillas use Vitamin K to produce clotting agents in their blood.

B. *Pigs use Vitamin K to produce clotting agents in their blood.*
Wolves use Vitamin K to produce clotting agents in their blood.
Gorillas use Vitamin K to produce clotting agents in their blood.

In this example above, people tend to find A to be more compelling, as it contains an extra premise statement that supports the conclusion. A and B are identical in all other respects. Stated more simply, premise monotonicity maintains that a greater number of items in support of an inductive inference will make for a stronger conclusion.

The sheer number of items in the premise does not entirely account for inference strength. Consider the following example in which the addition of extra information weakens the conclusion:

A. *Crows secrete uric acid crystals.*
Peacocks secrete uric acid crystals.
All birds secrete uric acid crystals.

B. *Crows secrete uric acid crystals.*
Peacocks secrete uric acid crystals.
Rabbits secrete uric acid crystals.
All birds secrete uric acid crystals.

This example represents a mixed argument. In example A we have two typical members of the bird category (crows and peacocks) and the conclusion is applied to birds alone as a category. Meanwhile, in example B, the same information is presented in the premises, but the strength of the conclusion about birds is weakened by the addition of the information that rabbits also secrete uric acid crystals. In this instance, a mammal having the property casts doubt on whether this would apply to all birds, as the mixing of categories makes the inductive strength lower.

One last case of how the quality and quantity of content affects inductive inference is also interesting to consider. The example below illustrates a situation known as an inclusion fallacy:

A. *Robins have an ulnar artery.*
Birds have an ulnar artery.
B. *Robins have an ulnar artery.*
Ostriches have an ulnar artery.

In this case you may believe that A has more inferential strength. Robins are quite prototypical birds, and intuition tells us that they make good representatives of the category bird; therefore, making the inductive inference that all birds have an ulnar artery appears sound. Meanwhile, in example B we are faced with the comparison of robins to ostriches. This is not a favorable comparison, as the two bird exemplars vary so widely on several dimensions. Therefore, we do not see example B as a strong inductive inference. This is where we get into trouble. Notice that example A specifies that if robins have an ulnar artery, then all birds have one too and this would include the ostrich. In a sense, example A includes example B implicitly within it. This makes it impossible for A to be more likely than B, as there are any number of exceptional birds that may not have an ulnar artery that would make A false, while only the vascular system of the ostrich is needed to make B accurate. This is called a *conjunction fallacy* in the judgment and decision-making field. Simply stated, the probability that a conjunction of two or more features exist cannot be more probable than the existence of any of the single features alone. This is another unintuitive complication that can make inductive inference more challenging. It also provides another nice illustration of the importance that perceived similarity plays in how people go about evaluating inductive inferences.

Expertise and Cultural Background Influence Induction

We have emphasized many results that suggest that the way people think about the world affects their inductive inferences. Therefore, it would be reasonable to assume that people who have different background knowledge based on different experiences may vary in how they evaluate the strength of inductive inferences. An alternative possibility is that inductive inferences have a universal quality that is shared among most cultures and even those that vary widely across the world. Such a possibility would be similar to Paul Ekman's findings on the evaluation of emotional faces being nearly universal (Ekman & Friesen, 1971). Like many topics in science, the answer probably lies somewhere in between. There is evidence that inductive strength has some strong commonality across cultures, but that also varies depending upon the expertise of individuals.

In previous research on categorization, Rosch, Mervis, Gray, Johnson, and Boyes-Braem (1976) indicated that there are privileged levels within taxonomies. In other words, people have a psychological preference for a particular level at which to describe and think about a category. Rosch et al. (1976) specified the existence of a *superordinate category* level, which provides a very broad overview of a category. Examples include "animal" and "tool." Somewhat more specific is the *basic category* level, which would include examples of descriptions such as "dog" and "hammer." Even more specific is a *subordinate category* level, which features very specific examples such as "golden retriever" and "ball-peen hammer." We tend to describe things at the basic level most often, as we rarely need the specificity of the subordinate level and often times the superordinate level description is not specific enough. Many basic categories are similar to one another, while sharing little similarity with members of other categories (Rosch et al., 1976). Critically, all of these similarity judgments are relevant to inductive inference because we will tend to assume that if one category member has a particular property it is likely to be shared by other members of that category.

A key study demonstrating cultural similarity in inductive inference comes from Coley, Medin, and Atran (1997) who asked individuals from two different cultures to evaluate the strength of different inductive inferences. Western-educated students in the United States and the Itzaj, Maya natives from the rainforests of Guatemala were included in this study. The researchers noted that prior work indicated that the boundaries between what is considered to be the basic level vary among categories and that cultures differ as to where they set these boundaries for a particular category. The basic level has been termed the *privileged level*, as it is the level at which most people will think about a category. For people in the United States "fish" is considered to be at the basic level, while for members of traditional societies "trout" is considered to be the location of the basic level for animals. These differences are likely influenced by expertise and Coley et al. reasoned that the groups might differ regarding what level they would consider to be the privileged level for making inductive inferences.

The researchers asked people from both cultures to evaluate inferences about hypothetical diseases about animals described at different levels of specificity. The researchers found that the privileged level for induction appeared for both groups at a similar level, such as "shark" or "sparrow" for example, rather than "hammerhead shark" or "bird." This suggests that there is some level of universality in making inferences. The interesting cultural difference finding was that the basic level for Americans was above the privileged level for inductive inferences, while the privileged level for the Itzaj was aligned with their basic level.

In a related study, Lopez, Atran, Coley, Medin, and Smith (1997) also compared students in the United States to the Itzaj people finding further differences. This study further revealed cultural differences about inferences in cases in which the category differed between the two groups. For instance the Itzaj considered foxes to be more similar to cats than to dogs, while Americans viewed foxes as being more similar to dogs than cats. This difference of categorization resulted in strength differences about inductive inferences. The Itzaj would be more likely to claim that a fox had a particular trait that a cat was stated to have, than to claim that a fox would have a trait that a dog was stated to have. The reverse was the case for Americans, who judged inferences about foxes to be stronger if they were linked to dog rather than cat traits. This result again, supports the existence of common mental representations for inductive inference across culture, but suggests that differences in the knowledge base among cultures can result in different conclusions.

Experts within a particular culture can also vary from their fellow societal members on the basis of their background knowledge. This was demonstrated by Douglas Medin et al. in a study from 1997. The researchers asked three types of tree experts (landscapers, tree maintenance workers, and tree classification experts) to evaluate different inductive conclusions about trees. Consistent with expertise level, those who were experts at classification sorted different types of trees on the basis of their scientific categorization and endorsed inferences between tree types that were biologically similar. By contrast, maintenance and landscape workers tended to sort the trees on their functional qualities such as size and ability to provide shade and endorsed some inferences based on these properties rather than strictly biological differences. These results again support the idea that experience shapes our tendencies to sort things in the world and this has a large effect on what types of inductive inferences we are likely to find compelling (Fig. 9.18).

FIGURE 9.18 How should we consider the properties of a tree? The answer varies depending on one's experience.

SUMMARY

Deductive and inductive reasoning have long histories dating back to early philosophy. One of the primary distinctions between these two forms of reasoning is that deduction leads to a valid conclusion, while induction does not. Deduction is strongly influenced by the specific premises that are offered to an individual. People will frequently fail to follow directionality information when reasoning deductively. Schemas and other information may suggest to us that two things that vary asymmetrically (such as "all A's are B's) are actually symmetrical (inferring that this means all B's are also A's). People rarely notice that they need to check a deductive situation in which they must see if there is a rule violation. This tendency accompanies our general preference for confirming a hypothesis, rather than disconfirming one. Evolutionary theory suggests that the human brain is optimized to detect cheating behavior in others, as this type of behavior undermines the value of acting altruistically toward people who cheat. Evidence from patient studies suggest that this capacity involves the medial temporal lobes and prefrontal cortex.

Inductive reasoning is strongly guided by our experience. Over the course of our lives we begin to see some things as similar and others as dissimilar. Similarity among items is a major determinant of whether we will apply an inductive inference toward a class of items. In addition to raw similarity, knowledge about the mechanisms of something will also influence our tendency to infer a property in a new item on the basis of its existence in the population overall. Inductive inferences can lead to stereotyping when we apply a set of properties widely toward a class of individuals seen as similar. Inductive inferences vary across cultures further illustrating the idea that these inferences are based on our prior knowledge about the world.

END-OF-CHAPTER THOUGHT QUESTIONS

1. How much do you think we use deductive reasoning in our daily lives compared to the amount we use inductive reasoning?
2. Name the four key cases within deductive reasoning and state which two are valid and which two are not valid.
3. Familiar contexts are often more likely to evoke valid deductive reasoning. What are some modern day examples of times in your life when you use deduction?
4. What are some reasons that people struggle in study after study to check the odd numbered card in the Wason card selection task?
5. Are we better at reasoning about a permission or obligation situation? Can you think of examples in your daily life that invoke use of these schemas?
6. Evolutionary psychology has explained deductive reasoning success on permission and obligation problems as relating the desire to check for cheating behavior. Could schemas account for this same result?
7. Name some of the major factors that influence the strength of an inductive inference.
8. Inductive reasoning varies based on culture. Many countries have diverse populations. Can you think of any cultural differences that occur within your country leading to differences in induction?

References

Bartlett, F. (1958). *Thinking: An experimental and social study*. New York: Basic Books.

Bucciarelli, M., & Johnson-Laird, P. N. (2005). Naïve deontics: A theory of meaning, representation, and reasoning. *Cognitive Psychology, 50*, 159–193.

Chapman, L. J., & Chapman, J. P. (1959). Atmosphere effect re-examined. *Journal of Experimental Psychology, 58*, 220–226.

Cheng, P. W., & Holyoak, K. J. (1985). Pragmatic reasoning schemas. *Cognitive Psychology, 17*(4), 391–416.

Coley, J. D., Medin, D. L., & Atran, S. (1997). Does rank have its privilege? Inductive inferences within folkbiological taxonomies. *Cognition, 64*, 73–112.

Cosmides, L., & Tooby, J. (1992). Cognitive adaptations for social exchange. In J. H. Barkow, L. Cosmides, & J. Tooby (Eds.), *The adapted mind: Evolutionary psychology and the generation of culture* (pp. 163–228). New York: Oxford University Press.

Dawkins, R. (1989). *The selfish gene*. Oxford: Oxford University Press.

Ekman, P., & Friesen, W. V. (1971). Constants across cultures in the face and emotion. *Journal of Personality and Social Psychology, 17*, 124–129.

Evans, J. B. T., St., Newstead, S., & Byrne, R. M. J. (1993). *Human reasoning: The psychology of deduction*. East Sussex, UK: Lawrence Erlbaum Associates, Ltd.

Galotti, K. (1989). Approaches to studying formal and everyday reasoning. *Psychological Bulletin, 105*, 331–351.

Glass, A. L., Holyoak, K. J., & Kiger, J. I. (1979). Role of antonymy relations in semantic judgments. *Journal of Experimental Psychology: Human, Learning and Memory, 5*, 598–606.

Griggs, R. A., & Cox, J. R. (1982). The elusive thematic-materials effect in Wason's selection task. *British Journal of Psychology, 73*, 407–420.

Janis, I. L., & Frick, F. (1943). The relationship between attitudes toward conclusions and errors in judging logical validity of syllogisms. *Journal of Experimental Psychology, 33*, 73–77.

Johnson-Laird, P. N. (1983). *Mental models: Towards a cognitive science of language, inference, and consciousness*. Cambridge, MA: Harvard University Press.

Johnson-Laird, P. N., Legrenzi, P., & Sonino Legrenzi, M. (1972). Reasoning and a sense of reality. *British Journal of Psychology, 63*, 395–400.

Johnson-Laird, P. N., & Wason, P. C. (1972). *Psychology of reasoning*. London: Batsford. Cambridge, MA: Harvard University Press.

Lopez, A., Atran, S., Coley, J. D., Medin, D. L., & Smith, E. E. (1997). The tree of life: Universal and cultural features of folkbiological taxonomies and inductions. *Cognitive Psychology, 32*, 251–295.

Nisbett, R. E., Krantz, D. H., Jepson, C., & Kunda, Z. (1983). The use of statistical heuristics in everyday inductive reasoning. *Psychological Review, 90*, 339–363.

Osherson, D. N., Smith, E. E., Wilkie, O., López, A., & Shafir, E. (1990). Category based induction. *Psychological Review, 97*, 185–200.

Revlis, R. (1975). Syllogistic reasoning: Logical decisions from a complex data base. In R. Falmagne (Ed.), *Reasoning: Representation and process in children and adults* (pp. 93–135). Hillsdale, NJ: Erlbaum.

Rips, L. J. (1975). Inductive judgments about natural categories. *Journal of Verbal Learning and Verbal Behavior, 14*, 665–681.

Rosch, E., Mervis, C. B., Gray, W. D., Johnson, D. M., & Boyes-Braem, P. (1976). Basic objects in natural categories. *Cognitive psychology, 8*(3), 382–439.

Stone, V., Cosmides, L., Tooby, J., Kroll, N., & Knight, R. (2002). Selective impairment of reasoning about social exchange in a patient with bilateral limbic system damage. *Proceedings of the National Academy of Sciences, 99*, 11531–11536.

Trivers, R. (1971). The evolution of reciprocal altruism. *The Quarterly Review of Biology, 46*, 35–57.

Wason, P. C. (1960). On the failure to eliminate hypotheses in a conceptual task. *Quarterly Journal of Experimental Psychology, 12*, 129–140.

Wason, P. C. (1966). Reasoning. In B. Foss (Ed.), *New horizons in psychology* (pp. 135–151). Harmondsworth: Penguin Books.

Wason, P. C., & Jones, S. (1963). Negatives: Denotation and connotation. *British Journal of Psychology, 54*, 299–307.

Wason, P. C., & Shapiro, D. (1971). Natural and contrived experience in a reasoning problem. *Quarterly Journal of Experimental Psychology, 23*, 63–71.

Wilkins, M. C. (1928). The effect of changed material on ability to do formal syllogistic reasoning. *Archives of Psychology, 102*, 1–83.

Woodworth, R. S., & Sells, S. B. (1935). An atmosphere effect in formal syllogistic reasoning. *Journal of Experimental Psychology, 18*, 451–460.

Further Reading

Heit, E. (2000). Properties of inductive reasoning. *Psychonomic Bulletin & Review, 7*, 569–592.

Sloman, S. A. (1998). Categorical inference is not a tree: The myth of inheritance hierarchies. *Cognitive Psychology, 35*, 1–33.

CHAPTER

10

Analogical Reasoning

KEY THEMES

- Analogies enable a form of inductive reasoning that allows us to learn about a new situation based on prior knowledge about a similar situation.
- Analogical reasoning is based primarily on structural correspondences, which are abstract properties of situations that are revealed when we consider how objects, people, places, or things relate to one another.
- A source analog and a target analog are placed into correspondence to form an analogy. The target analog is the current unfamiliar situation we wish to learn about, while the source analog is a previously known familiar situation.
- For most analogical comparisons to be successful there are constraints that apply. Most analogies require one item in the source analog to correspond to only one other item in the target analog. This prevents the analogy from leading to mixed inferences from multiple situations that are less likely to be valid.
- Structural similarity refers to the relationships among elements within a situation matching to similar corresponding relationships in an analogous situation.
- Superficial similarity is based on perceptual features or properties that are shared between two situations that may be placed into correspondence in an analogy.
- Analogies have alignable properties that can be placed into correspondence. They also have non-alignable properties that do not allow correspondences. Many analogies with greater structural similarity have alignable properties.

Reasoning
http://dx.doi.org/10.1016/B978-0-12-809285-9.00010-7

- Source analogs can be difficult to retrieve. This is often the case when the two parts of an analogy come from different domains.
- Long-distant, or creative, analogies are those in which the two domains placed into correspondence are very different from one another semantically.
- Insight problem solving may yield creative analogies through coarsely coded information. The right hemisphere of the brain supports this type of reasoning.
- Analogies in daily life vary in terms of their purpose, the depth of the analogy, and the distance between domains. Expertise can also influence the formation of analogies as well.

OUTLINE

REASONING BY ANALOGY

Introduction

This chapter focuses on analogical reasoning, a type of reasoning in which we compare two situations and decide how they relate to one another. The goal of reasoning by analogy is to gain new insights about one or both situations being compared. Analogical inferences arise from the process of induction. As we discussed in Chapter 9, inductive reasoning can provide us with new information, but it does not assure us that any new conclusions are valid. As a result, analogical reasoning requires some degree of caution. At times analogies provide remarkably novel and creative insights, while at other times analogical inferences can be misleading and result in erroneous thinking.

Scientific discovery has always relied upon prior information to inform inferences about new ideas. In the early 1900s there were groundbreaking discoveries about the microstructure of life forms, the planet, and the universe. These new domains of knowledge had few clear reference points to prior information. This influx of new information resulted in highly interesting, creative, and colorful analogies being made comparing a new domain to other well-understood situations from the past. One such analogy has been made recently in order to explain the difficult concepts concerning the Higgs boson, a particle physics phenomenon publicized as being the "God particle." Researchers at the CERN Large Hadron Collider located near Geneva, Switzerland, observed a new particle that is consistent with the Higgs boson. Shortly after this discovery, Peter Higgs was awarded the 2013 Nobel Prize in physics for his contributions to this field. There has been a proposition in particle physics that something called a Higgs field exists with its associated boson. These phenomena have to do with particles and the mass that they possess at different times entering and leaving the field. The Higgs boson was a very difficult concept to communicate to the public, a common situation in particle physics, as nonexperts often have little background knowledge needed to understand modern discoveries in this area. The Higgs field was so difficult to describe that in 1993 the science minister of the United Kingdom proclaimed that he would award a fine bottle of champagne to anyone who could provide a convincing way to explain the Higgs field and the Higgs boson to the public.

The champagne for the best analogy was awarded to David Miller, a physicist at University College, London, for his analogy likening a Higgs field to a cocktail party and the boson to a rumor moving through the room ("Best explanation of the Higgs boson?," 2012). The Higgs field is made up of particles. Each particle in the analogy corresponds to each person at the party who collectively form a mass of people corresponding to the field itself. The Higgs field gives mass to particles that occupy space in the field, a difficult concept for most people to grasp. The analogy provides a clear and interesting way to describe the dynamics associated with the Higgs field. If an average person enters the party, they move easily through the room, and no significant change in the spatial arrangement of the party guests occurs. By contrast, if a famous person moves into the room they are likely to be mobbed by the other party guests, dramatically altering the mass of people occupying space around him or her at the party. Just as the surrounding elements in a Higgs field give mass to a particle, the party guests increase the spatial mass surrounding the celebrity who entered the room. Meanwhile, the boson is an alteration in the Higgs field. This is nicely and colorfully explained by Miller's analogy, in which the boson corresponds to a rumor being transmitted among the party guests. The news causes temporary alterations in the spatial arrangement of the party guests, as they lean in to hear about the rumor and lean out in order to spread it to others. In this way, the rumor moves about within the context of the cocktail party providing different spatial dynamics, much as the boson moves about manipulating the Higgs field. While, these correspondences between the particles in the Higgs field and the guests at the cocktail party are imperfect and do not fully represent the reality of the Higgs boson concept, the analogy turns a complicated and abstract phenomenon into one that is concrete and easily understood. What's more is that the analogy also makes the concept memorable and provides a nice example with which one can readily explain the difficult concept to others. Analogies make excellent teaching tools and ones that can be readily applied in classroom settings (see Box 10.1).

BOX 10.1

ATOMIC ANALOGIES: IS IT A DESSERT OR A SOLAR SYSTEM?

One of the classic analogical reasoning cases in science comes from the domain of particle physics. In the early 1900s little was known about the structure of the atom. Scientists were struggling to understand how electrons were organized. Recordings were much more limited than they are today, so scientists drew conclusions about the new field based on analogies to other well-known domains.

J. J. Thomson had proposed a model to explain the organization of the atom. Thomson proposed that the electrons, which carry a negative charge, were likely embedded within a much larger positively charged spherical field. This became known as the "plum pudding model" of the atom, in which electrons were imagined to act as stationary surrounding elements. The colorful analogy draws comparisons between this proposed organization of the atom to an English dessert that is equivalent to a modern day blueberry or raisin muffin. In the plum pudding model, the electrons correspond to raisins, while the field of positively charged mass corresponds to the bread surround. Both the electrons and the raisins are embedded in their respective surrounding matter.

An experiment was conducted in 1910 by Ernest Marsden, a young associate of Ernest Rutherford and Hans Geiger, two well-known physicists interested in atomic organization. Marsden fired highly dense α particles through a thin gold foil to see how much the particles would scatter after passing through the foil. If Thomson's plum pudding model was correct, there should be very little scattering, as it was imagined that the α particles would not face much resistance due to the comparative lack of mass that the plum pudding atom was imagined to have. The results were very surprising and suggested a different structure. While the majority of α particles passed through, others deflected at extreme angles and even reversed course entirely, as if they had encountered a very dense mass in the gold foil atoms. This result did not square with the plum pudding model of the atom, as that model lacked anything of sufficient mass to deflect the α particles in so strong a manner as the researchers had observed.

The unexpected results of Marsden's experiment led Ernest Rutherford to propose an alternative model of the atom. Rutherford's model was organized differently than Thomson's plum pudding model. Rutherford conceived of the electrons acting in a dynamic way, orbiting around the central nucleus of the atom. The Rutherford model evokes comparisons to the organization of the solar system, which was well-understood at the time that the model was conceived. Rutherford placed the nucleus of the atom in correspondence with the sun, as both are large central objects that have the greatest mass. The dense mass of the nucleus could explain the dramatic deflections observed by Marsden. The Rutherford model placed electrons in positions around the central nucleus. This placement calls to mind a correspondence with the planets orbiting around the sun (Fig. 10.1). The colorful use of analogical descriptions clarifies the differences between these two models. The analogies also make the models easily understandable to particle physics novices, such as most students who take introductory chemistry. The analogies make an ethereal and unobservable phenomenon concrete and tangible.

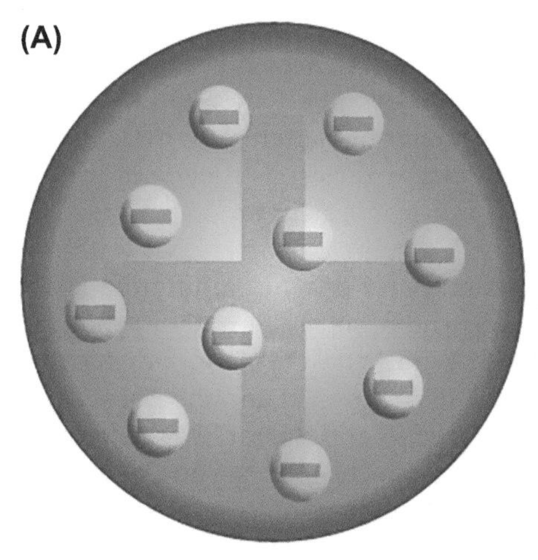

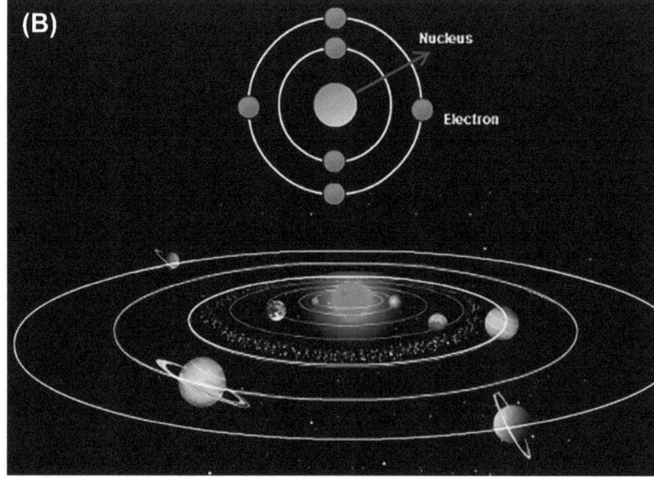

FIGURE 10.1 (A) The plum pudding model of the atom as proposed by J. J. Thomson drew an analogy between the organization of the atom and that of a dessert consisting of a surrounding field corresponding to cookie dough or bread, along with electrons distributed in positions around the field corresponding to raisins or blueberries positioned in various places around the dough. (B) The atom-to-solar system analogy as evoked by the alternative Rutherford model of the atom. In this case the central mass of the sun corresponds to the central mass of the nucleus. Here the tiny orbiting electrons correspond to the comparatively tiny planets orbiting the sun. Establishing these correspondences makes the organization of the atom clear to the casual student of introductory chemistry classes. *(A) Credit boundless.com. (B) Credit Slideshare.com.*

Analogies also play a major role in public life. Politicians are very likely to draw upon analogies when describing the implications of public policy actions. When President Barack Obama was advocating for reforming the US healthcare system in 2009, he used an analogy between healthcare costs and flood waters. Obama claimed that by failing to change the course of increasing healthcare premiums, the number of uninsured people would continue to swell to a point of eventually eroding the benefits provided. This analogy calls to mind an advancing water level swelling up to overrun the land. In using these words, President Obama communicated an emotional image of healthcare policy as a crisis that had to be dealt with before it was too late. By 2010, the Patient Protection and Affordable Care Act was passed into law and numerous opponents to the law drew their own analogies. Some of the analogies were intended to evoke emotional outrage toward the new policy. In one such case, the law was compared to the Challenger explosion of 1986 in which a space shuttle exploded during launch. This analogy calls to mind a catastrophe and was intended to evoke rapidly impending panic after the law, as the Challenger explosion occurred within the first 2 min after liftoff. One of the more colorful analogies came from Justice Antonin Scalia of the US Supreme Court who stated that "Everybody has to buy food sooner or later." He said, "Therefore, you can make people buy broccoli." This analogy implied that requiring Americans to buy health insurance would be good for them and they would have to do it eventually, but the comparison also implied that an inappropriate attempt to control people's freedom was also taking place.

An individual's political leanings and how they view the president often strongly determines whether someone will draw a positively or negatively valenced analogy about healthcare. Whether an analogy is good or bad depends on the effectiveness of the analogy to clearly drive home a point to the audience. The value of the analogy is also based on the contents of the two domains. In this chapter, we will grapple with the factors that make a good analogy. Good analogies tend to be viewed as clever, creative, and oftentimes humorous. Bad analogies, by contrast, may make us feel confused, or even groan at their inelegance. Worse yet, a bad analogy can have an effect on the audience that is opposite to what the person drawing the analogy intended.

In this chapter, we will discuss the psychology of analogies, as well as the latest evidence about how the brain forms analogies. You may recall from Chapter 4 that there is evidence for analogical reasoning ability in other species, notably the crow and other primates, but those analogical capabilities appear to be limited to simple comparisons between shapes or object properties. Human analogical reasoning can be much more complex. Semantic memory and people's estimations of similarity are closely linked to the analogies that they will generate, as we have already discussed with regard to inductive inferences. In Chapter 3 we discussed several of the brain areas that are important for enabling analogical reasoning to occur. This is an interesting form of reasoning that reaches its maximum level of complexity in humans and is tied to our language ability. We are a highly symbolic species and our ability to label objects, people, and situations facilitates our capability to draw interesting analogies. People's strong semantic memory capabilities allow us to store world knowledge and access it for the purpose of making interesting comparisons to educate, influence, or entertain others. Analogies are also a highly social phenomenon, as we will discuss in the sections on real-world analogy use. Frequently, the person drawing the analogy will do so for a specific purpose. He or she may wish to evoke fear, outrage, or even humor in those who are told about the comparison.

We will begin this chapter with a discussion of relational information. Seeing the world around us in terms of relations among objects enables us to form larger sets of systematically connected information that can be used to create broader and more complex analogies.

RELATIONAL CORRESPONDENCES: THE BUILDING BLOCKS OF ANALOGIES

Attending to Relational Matches in Situations

When children are engaged in learning a language, they will begin to describe comparisons between different situations. This occurs as children attempt to understand new situations through comparisons to other prior situations that they have experienced. As we have discussed in Chapter 5, children undergo a gradual shift from egocentric behavior, in which they are highly focused upon their own internal thoughts concerning concrete representations, toward a gradually broader and more abstract view of the world. This has been described as a "relational shift" by Dedre Gentner (1988). Gentner hypothesized that children initially view the world with an eye toward how individual objects are similar to one another. Later, they go on to view the world through the broader lens of relational similarity, in which they will compare situations involving multiple objects, or people, to other complex situations that consist of multiple elements. Analogies of this type rely upon relational correspondences, ones in which common relational elements form the basis for similarity.

Research on similarity has supported the idea that attention to relations is a key component of analogical reasoning and that children gradually come to view relational matches as being properties of interest. This developmental trend was nicely illustrated by

Richland, Morrison, and Holyoak (2006) who showed participants pairs of drawings depicting scenes that could be compared to one another (refer to Fig. 10.2). We discussed work on scene analogies previously in Chapters 5 and 7. Picture analogies were originally developed to test the emerging relational and analogical abilities in children as they aged. In Fig. 10.2, we are presented with two such scenes. The top image (Fig. 10.2A) shows a dog that is chasing a cat, who is in turn chasing a mouse. Notice that the cat has an arrow pointing to it. The participant in this task is instructed to pay attention to the item that is indicated by the arrow and try to find the best matching item in the lower picture. In Fig. 10.2B, we see a woman who is chasing a boy, who is in turn, chasing a girl. When the pictures are described this way, you can probably already anticipate that the boy will make the best match to the cat. Note that the woman could also be a match to the cat, as the cat chases a mouse just like the woman chases the boy. Alternatively, the girl could also correspond to the cat, as the cat is chased by the dog, and the girl is chased by the boy. Only the boy matches the cat in terms of playing both roles in a chasing relation. When we consider these people and animals to be players in a chasing relation, we could apply the term *agents* to the dog and the cat, as they each actively chase another animal, while we can apply the term *patients* to the cat and the mouse in a chasing relation, as they are both chased by another animal. Likewise, the boy is both an agent and a patient of a chasing relation, making him the best match to the cat, who also occupies both the agent and patient roles. Results showed a progressive increase in correct analogical matches in these problems. Children at the youngest age group, 3 to 4 years old, were slightly below 50% in relational matching. This percentage climbed to around 60% in children 5 to 6 years old. The percentage of relational matches rose to around 90% by ages 13 to 14 years. Why do children appear to miss these matches early on? What makes relational matching so much more obvious by the early teen years? Relational matching likely has a lot to do with how similarities are processed.

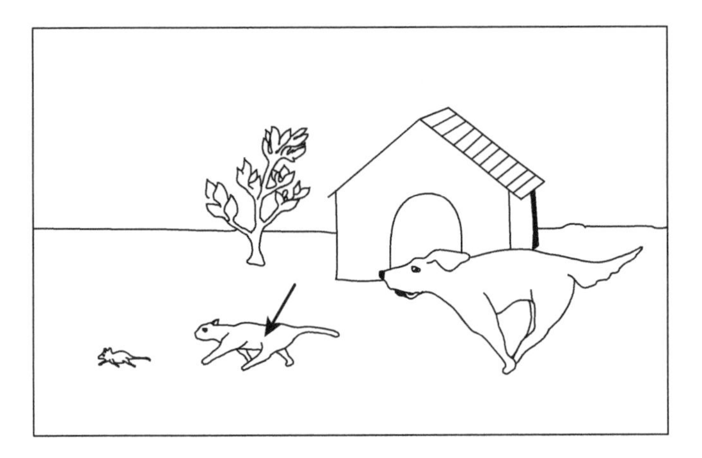

FIGURE 10.2 An example of scene analogies that were used in a developmental study by Richland et al. (2006). In the example, first refer to part A in which a cat plays a critical dual-role in the situation. The cat is chasing a mouse (as the agent of a chasing relation) and is also being chased by a dog (as the patient of a chasing relation). The cat is marked by an arrow to alert participants that it is the item for which they must find a match in the second scene. Participants were asked to select which item in part B matches the cat in scene A. As can be seen, the boy matches the cat, as he is both chasing a girl and being chased by a woman. One can test relational reasoning ability by asking children of varying ages to indicate a match for the critical item (the cat). Notice that any other match in scene B could be made on the basis of being an agent or a patient of a chasing relation, but only the boy satisfies both of these requirements. *Adapted from Richland, L. E., Morrison, R. G., & Holyoak, K. J. (2006). Children's development of analogical reasoning: Insights from scene analogy problems. Journal of Experimental Child Psychology, 94, 249–273.*

As children age, they progressively show the ability to match items relationally. Additionally, the types of relational matches used by children also become progressively more complex over the course of development. Gentner and Rattermann (1998) described the initial relational matches as being based on object features. For example, the roundness of a ball can be compared to that of a balloon or a globe. Later in childhood, additional properties may be compared. These properties may lead to "on" relations, for example, when children compare an apple that is on a table with a book that is on a desk. Relational matches become even more complex later on as children begin to notice higher-order similarities that include sets of relational information that can be matched. For example, early on in childhood, a four-year-old child may make a comparison between his rain boots and those of a friend, noting the similarities in shape and color. Later on, by age six, armed with the knowledge acquired at his grade school, the child makes comparisons between his own kindergarten class and his older sister's sixth grade class. He notes that the teachers are similar, but that the behavior is much different from the boys' and girls' sixth grade class than that of his own. He also notices that the teacher uses different tactics to control the behavior of the

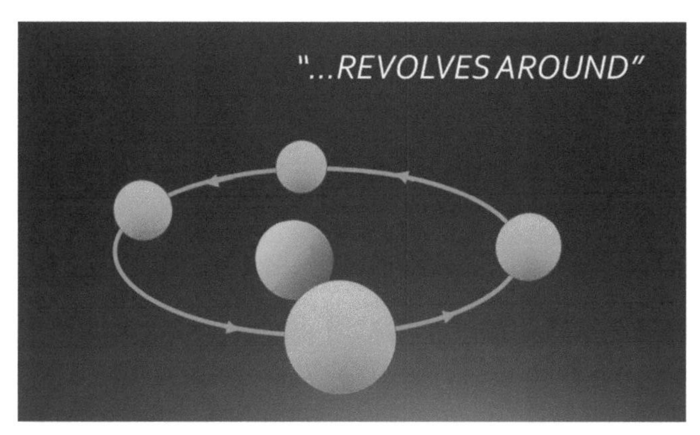

FIGURE 10.3 Relational terms include above, below, rotates, and revolves around. Matches based on these relational properties lead children to form analogies rather than make simpler feature-based matches.

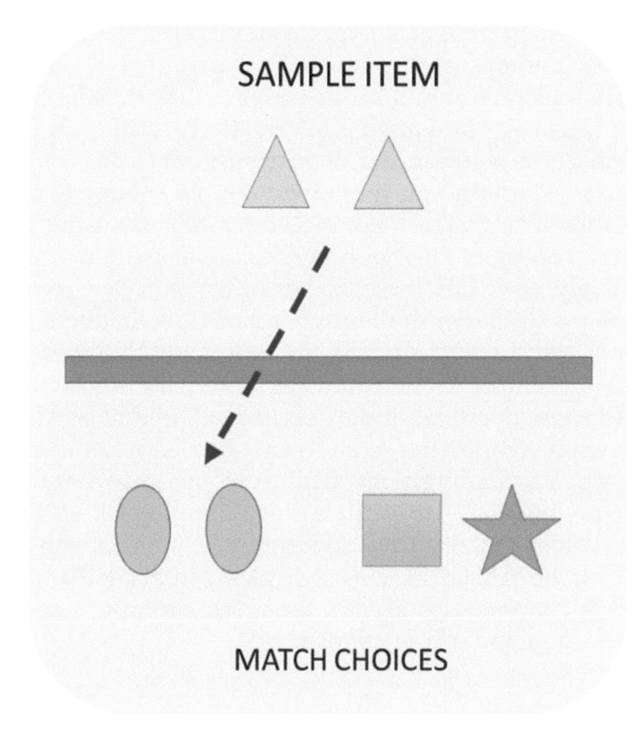

FIGURE 10.4 The relational match-to-sample task has proven to be difficult for most species to master. In this case, the sample (*triangle, triangle*) should be paired with the match item (*oval, oval*) on the basis of a sameness relation. Only primates and crows show a clear ability to solve these tasks. Nonhuman primates often require hundreds of trials of practice before they are capable of demonstrating success on this task.

class overall. In comparing the classes across grade levels, the boy is now thinking at a higher-order level, making comparisons across systems of relations. When limited to the simpler rain boot comparison, the boy's comparisons focused on concrete attributes such as shape, size, and color. Later on, his awareness shifted to understanding similarity between the teachers and also the relations that exist between the teachers and students. In making a class-to-class analogy, the boy has demonstrated a new ability to form complex analogies. This will enable him to make new inductive inferences about his sister's school day based on his own growing knowledge about the goings on in his own classroom.

Critically, Gentner (1988) has argued that the knowledge of relational information is most central to analogical matching, rather than simply the age of a child. Knowledge about relationships that exist in the world can enable children to make new analogical comparisons. Gentner has argued that the use of relational language is one of the keys to enabling this process to occur. For example, children who are exposed to terms such as "under," "above," "taller than," "quicker than," and "revolves around" are likely to begin to view similarities between objects, people, and relational systems, such as grade school classes as being in the form of higher-order comparisons (Fig. 10.3).

EVALUATING SIMILARITY

What Makes Objects Similar?

People, places, and things can be seen on a scale ranging from identical to extremely different. Some of the most basic forms of similarity involve making literal object matches. Such matches can be made relatively easily by many species. Matching an apple to an apple is a straightforward process that need not involve anything beyond perceptual feature comparisons. This is probably why many species, including primates, birds, and rodents can all succeed on match-to-sample tasks as we had described in detail earlier in Chapter 4. Symbolic coding such as language can facilitate simple perceptual matching. Once children begin to build their vocabularies, they may no longer need to go through a period of study in the process of making perceptual comparisons. They simply know that there is a fruit called an apple, that it is edible, and that other apples that exist are also likely to share these same properties.

Similarity judgments get much more complex simply by making them relational, rather than perceptual. In Chapter 4, we discussed the relational match-to-sample task. Fig. 10.4 shows an example of this task in which the participant is presented with a sample of two items (e.g., two triangles) and is asked to encode the relation between these items (a "sameness" relation in this case). The sample is taken away and now the participant must make a relational match by selecting among two new pairs of relational items. In Fig. 10.4, the correct answer is to select the two ovals on the basis of the sameness relation. The star and square are incorrect, as these items share a "difference" relation, despite their angular appearance, which could suggest that this is the correct answer based on having sharper angles than the two ovals. On perceptual grounds alone, perhaps one could make a case that the star-square picture would be the better match. People who are sensitive to relational information readily see the connection between the two triangles and two ovals. This is especially true for individuals old enough to use relational language. Once it is understood that similarity can occur at a higher-order relational level, we may actively seek out these higher-order, non-perceptual types of similarity in the environment. Evaluating similarity in this way unlocks an almost limitless potential to sample from situations and make inductive inferences about what may occur next, diagnose what may have gone wrong, and generally learn about the world. Fascinatingly, hooded crows are also capable of solving the relational match-to-sample task within just a few practice trials (Smirnova, Zorina, Obozova, & Wasserman, 2015). This ability may indicate the tendency for a species to view the world around them in terms of higher-order comparisons.

Similarity Between Situations

Moving beyond mere appearance matching among single entities unlocks the ability to make comparisons among whole situations. This is a powerful tool in our reasoning abilities, as noticing similar situations allows one to make fluid comparisons and infer the existence of potentially new instances. In this section we consider evidence regarding how we make these similarity judgments and how we can understand relational correspondences alongside featural correspondences.

In an influential paper, Goldstone, Medin, and Gentner (1991) discussed the role of relational information in similarity judgments. Goldstone and his colleagues asked readers to consider a situation in which they viewed a tiger and her cub. This situation can call to mind a mother robin and her nestling. The two situations share some similar properties including the fact that they both involve living things that have eyes, hearts, skin, etc. The situations also share relational information, such as the fact that one animal is the parent of the other and that one provides food for the other. It is somewhat unclear how this type of information is represented. One possibility is that relational information is represented in just the same way as object information. In the tiger and robin comparison this could be represented in terms of proximity. The tiger cub is near the larger tiger, just as the nestling is near the adult robin. This representation appears to miss a key aspect of the way that people make relational comparisons. A parenting relation or a caregiving relation is another (and possibly more satisfying) way to represent the relations among these pairs of animals. The mother tiger cares for the baby tiger, just as the mother robin cares for the nestling. This is a strikingly human inference, as the *caring for* relationship need not be observed perceptually. This relational property can simply be inferred based on the common associations known to exist between parent and infant interactions (at least in species that are known to care for their young).

How does overall perceived similarity operate when relational information is placed in opposition to feature-based (or semantic) similarity? Which type of information is more likely to be used to form an inference? My colleagues and I were interested in this question and designed an experiment that would evaluate people's preferences for making similarity matches. We designed an experiment to investigate what type of information people would use most for inferences made between two analogous situations. The two situations featured in the experimental materials involved multiple possible correspondences based on workplace interactions between employees, one of whom cheated another out of credit for a new product (Krawczyk, Holyoak, & Hummel, 2004).

In this study, we structured the materials so that there were very clear correspondences that competed with one another. The situation involved a lighting corporation, called Brightech, that was unveiling a new product. The company boss charged two employees with making a presentation about the progress. In this case one of the employees gave the presentation and took most of the credit for himself. In other words, he cheated another person out of credit for the work. The two employees involved in the cheating situation were described as having distinctive prior careers, such as being former astronauts, ex-professional wrestlers, or ex-navy SEALs. After participants read about the cheating situation, the story continued describing the boss's thoughts about a prior situation, in which he had witnessed a different pair of employees engage in the same kind of behavior. In that prior situation one employee cheated another out of credit for work that they had jointly conducted. As a remedy for the unfair behavior, the boss ensured that a new pairing occurred between a person who had been cheated and one who had cheated another. The boss was sure that the person who was cheated was aware of the other employee's cheating behavior. He offered the person who was cheated an opportunity to take credit from the former cheater. We asked participants to read about a new situation at the Offstar corporation that had similar occurrences involving cheating relations (Fig. 10.5). In all cases, there were rather transparent similarities in which two pairs of employees engage in joint work only to have one person take all of the credit. In this new story at Offstar, the employees had distinctive former professions that mirrored those that had been included in the prior story about Brightech. The new situation at Offstar involved a boss who was planning to pair up two of the employees to restore equity

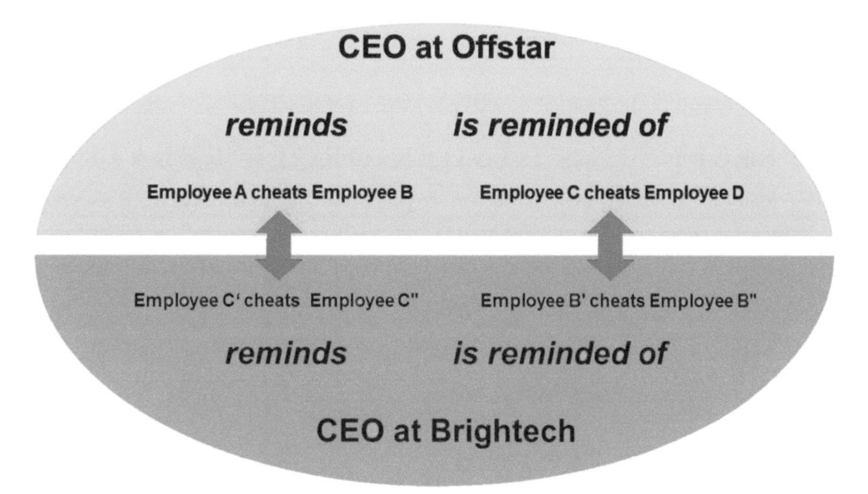

FIGURE 10.5 The analogical correspondence between two scenarios in which semantic elements and relational elements compete (Krawczyk et al., 2004).

between employees by allowing one to present the new idea that both had worked on. Participants in the experiment were asked to infer which employees the boss of Offstar would ask to do the new presentation. If participants used the analogy to Brightech and formed a relational inference, they should choose the employee who cheated in an older situation and one who was cheated in the more recent one. Such a relational inference would represent a higher-order relational mapping, as both cheating and reminding relations are needed to generate this inference. A compelling case could also be made for a simpler analogy matching up the semantic attributes of the individuals. If an ex-navy SEAL and an ex-professional wrestler were paired up in the Brightech story, it may make sense that individuals at Offstar with these same traits would also be paired up. Such a scenario is a relational correspondence, but one based on attributes rather than on the cheating relations that were also present.

Results from the Krawczyk et al. (2004) study indicated that people made extensive use of analogical information when listing potential inferences that could follow from that situation. The vast majority of participants inferred either the higher-order relational inference (based on the cheating relations), the attribute relational inference (based on prior professions), or both. About half of the inferences generated were either attribute or relational analogical inferences. The other half were sensible other inferences that were plausible for the company scenario, but were not analogical. Thus, people appear to be sensitive to relational structure when making inferences. Further, our participants were sensitive to both the higher-order relational structure of the analogies, as well as the distinctive attribute information that was provided in the situation (see Box 10.2).

Analogical Similarity

The ability to notice and use relational elements in order to form correspondences and establish similarity provides a foundation for reasoning on the basis of analogies. Both common attributes and relational correspondences can evoke analogical connections. Once these correspondences are noticed, it can lead to the selection and mapping of an analogy and potentially new inferences. Let's now discuss the mental processes that take place in order to form analogies and use them to expand upon one's knowledge.

The terminology used to describe the steps that go into analogical reasoning is depicted in Fig. 10.7. A previously understood domain of knowledge that serves as a candidate for forming an analogy is known as a *source analog*. The source typically will contain several elements (such as the sun and planets in the solar system source analog) along with some relational information that describes how the different elements interact. A relatively new domain that is under consideration is called the *target analog*. The target may be somewhat unknown or at least less-well understood than the source analog. In our earlier example, analogical pairing between the Higgs field and the cocktail party, the cocktail party represents a well-understood source analog, while the Higgs field is a less-well understood target analog. This analogy works to inform an individual about the properties of the Higgs field because the source analog is well-known. By imagining the target analog, the Higgs field, as a cocktail party, one can begin to infer new information about the Higgs field. The degree of similarity influences whether there will be an analogy set up.

The analogy is set up through a process of reminding. Typically, some characteristics of the target analogy will cue a memory of one or more potential source analogs. Some properties of the source and target analogs

BOX 10.2

REPRESENTING HIGHER-ORDER RELATIONS

There are a variety of ways to make statements about situations. If you see a helicopter flying over a city freeway you may state that "the helicopter is flying over the cars on the freeway." Alternatively, you may state that "the cars are driving beneath that helicopter." Both statements convey the same information. The use of the term "over" or the term "beneath" are just two different ways to state the same relations.

Let's now return to the atom-to-solar system analogy that we had discussed in Box 10.1. This analogy was made by Rutherford and the critical elements that are needed for the analogy to be meaningful are related to the mass of the objects, the movements of the objects, and the overall behavior of the system. An atom does not have a lot of semantic overlap with a solar system. Any inferences about life on the planets or the composition of a particular planet versus another would yield misleading and irrelevant inferences about atoms. Even something like the size of the planets must be ignored if the analogy is to be successful. For the analogy to be effective, it is essential that the size differential between the sun and the planets is emphasized. The relative spaces between the planets themselves, and between the planet and the sun must also be emphasized. Lastly, the movement of the planets orbiting the sun must be emphasized. From those key relations, we can then form a meaningful analogy between the solar system and the atom that clarifies Rutherford's model of the atom and how it differed from incorrect ideas of atomic structure that had existed previously.

We can use a relational representation called *predicate calculus* to represent these key relations. Predicate calculus comes from the domain of formal logic. Its' representations resemble mathematical proofs. The structure of predicate calculus removes most of the semantic details of a situation distilling it down to the core relational elements that need to be placed into correspondence. This tool has been widely used in modeling efforts in which computers are programmed to simulate analogical reasoning.

Simple attributes can be represented in this format as follows:

LARGE (sun)
SMALL (planets)
ROUND (sun)

Relational elements involving two entities can also be represented in this format. In this case a relational term is placed in capitals along with two role-filler items. An *agent* is positioned first, followed by a *patient*. The representations of the Rutherford model of the atom would look something like the following:

ATTRACTS (sun, planets)
REVOLVE (planets, sun)
LARGER (sun, planets)
DISTANT (planets, sun)

In these examples the agent and patient roles flip depending on which term is considered to be the active part of the relational correspondence. Such relational descriptions involve first-order relations, those that directly occur between the agent and the patient objects or items.

When a causal connection needs to be represented, it can be placed out front of the other relational elements as follows:

CAUSE [ATTRACTS (sun, planets), REVOLVES (planets, sun)]

When considered together the Rutherford model can be understood in terms of correspondences as indicated in Fig. 10.6. This more elaborated causal description involves what are known as higher-order relations. The causal relation between the *attracts* relation and the *revolves* relation represents a second-order relation, or a relation among relations. Such complex relational structures become characteristic of complex human reasoning and enable the mental insights that people are capable of.

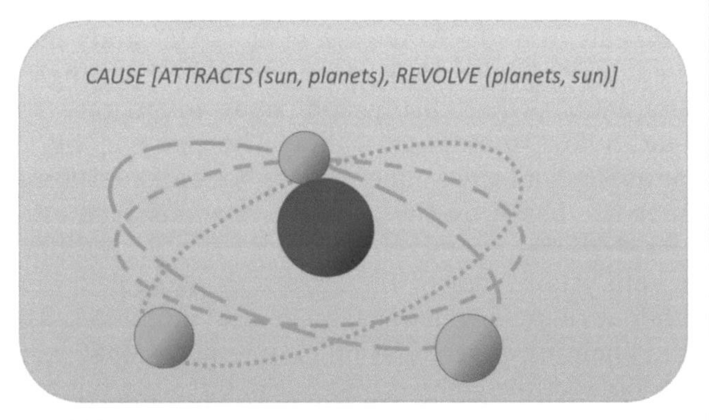

FIGURE 10.6 The Rutherford model of the atom can be described using predicate calculus.

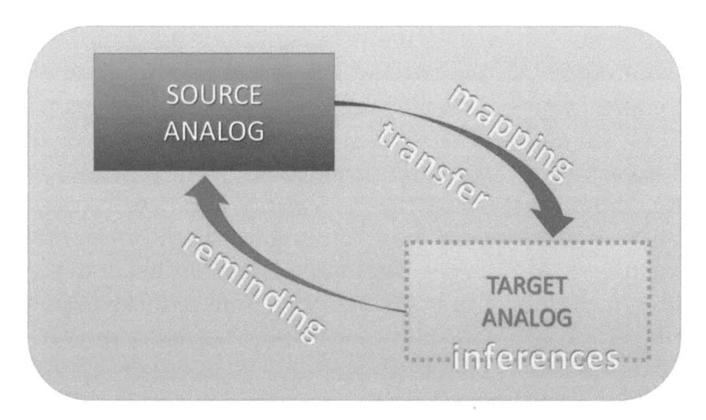

FIGURE 10.7 Several key components involved in analogical reasoning are depicted. The source is a well-known domain of knowledge represented by a large rectangle. The target is represented as an incomplete rectangle to indicate that less is known about this domain. Arrows indicate that the characteristics of the target lead to retrieval of a source analog and that mapping and transfer can occur next if the source provides an effective analog to the target. The target is then filled out by inferences that complete the scenario.

must be similar in order to establish the analogy. The degree of similarity between the source analog domain and that of the target analog will influence the ease with which the analogy is established. This process can be facilitated if the two analogs share features and semantic properties. In other words, attribute similarity will help to establish the analogy. If there is little attribute similarity, the target domain may not evoke a source analog. Alongside attribute similarities, the source and target analogs will need to share some degree of relational similarity. The degree of overlap of relational properties will determine how effective the analogy is. This will in turn determine how useful the analogy is for yielding accurate inferences.

Once a candidate source analog has been retrieved, it can be placed into correspondence with the target analog in a process called mapping. Mapping involves placing different items in each analog into correspondence. In the Rutherford atom-solar system analogy this involves placing the nucleus into correspondence with the sun and each of the planets into correspondence with each of the electrons. The mapping process will be complete to the extent that there are elements that share relational correspondences. The mapping process is critical to forming analogies, as misalignment or incomplete correspondence between the analogs can result in muddled analogies that are either unclear or even misleading.

After an analogy has been mapped, new inferences can be made. This process of making new inferences about the target domain on the basis of what is observed to occur in the source domain is called transfer. For an example of analogical transfer, let's consider the case of the Rutherford atom model. With the sun-nucleus and planets-electrons mappings completed, we can now consider other characteristics of the solar system and consider whether they may apply in the atom. We know that the planets rotate and that they revolve around the sun. With this in mind, we may decide that the electrons probably rotate and that they revolve around the nucleus. These are reasonable inferences. While electrons are not actually spinning orbs, they possess a property called angular momentum, which one can consider to be acting the same way "spin" does for a planet. It is also appropriate to infer that the electrons revolve around the nucleus based on some understanding of physics, though true revolution does not literally describe the dynamics of electrons. The analogy can break down if we go on to infer that the electrons vary in size and that their speed and distance from the electron mirrors that of the solar system. If we were to also infer that pieces of electrons likely break off to form miniature chunks of charge that revolve around the electron, as a moon revolves around a planet, we would have taken the analogy too far and are going to begin to generate false inferences.

When we use analogies and begin to consider an analog as a system of relational elements, we can begin to form new categories. This type of learning by analogy can ultimately lead to a schema, a set of general semantic knowledge that applies across a wide array of domains. We have discussed semantic memory and schemas already in Chapter 3 when we covered how semantic dementia affects reasoning. We also discussed schemas in Chapter 9 regarding cultural explanations of the Wason Card Selection Task. Schemas can include relational elements, and these need not be tied to one specific instance. Rather, one of the key properties of a schema is that it overcomes particular instances and comes to stand for a whole class of cases once it has become established.

PLACING ITEMS INTO CORRESPONDENCE

How Do Objects Correspond in an Analogy When People Make Inferences?

The correspondence between elements within the analogy have to be constrained in order to ensure that analogies will be effective in helping to form valid inferences that are also likely to be true in the world. Gentner (1983) emphasized that a one-to-one constraint needed to operate on mappings. In other words valid inferences could only be drawn when one item in a source situation matches to only one other item in a target situation.

The need for a one-to-one constraint makes sense if we imagine a situation in which certain items can be "cross-mapped." Cross-mapping occurs when single item in the target analog matches to two possible elements in the source analog. The possibility of cross-mapping is introduced when multiple potential source analogs are available to apply toward a single target analog. Public policy and wartime scenarios lend themselves especially well to analogies of this type. Given the complexity of large-scale human interactions and the prevalence of previous examples and conflicts, one can easily call to mind numerous other situations that might serve to inform decisions about a current problem and how to solve it.

In a memorable example of wartime analogical reasoning, Barbara Spellman and Keith Holyoak posed a question to participants about former president George H. W. Bush, "If Saddam is Hitler then who is George Bush?" In their study, Spellman and Holyoak (1992) asked participants to map people and countries from World War II to their potential counterparts in the Persian Gulf War. In 1990, former Iraqi leader Saddam Hussain had invaded neighboring Kuwait occupying territory within that land. This led to a coalition of forces countering the Iraqi invasion in early 1991, led prominently by the United States. Iraq's invasion of Kuwait could be compared to numerous other invasions through history. A clear comparison could be made between Iraq's Saddam Hussein and German leader Adolf Hitler, who escalated World War II by invading and occupying much of Europe. Hitler's invasion of Poland initiated a large-scale takeover of much of Western Europe that led to warfare on an unprecedented scale. In the target scenario, in which Saddam Hussein and Iraq invade and occupy Kuwait, one could wonder whether this action would lead to escalated warfare among several nations following the invasion. Spellman and Holyoak (1992) asked participants to provide potential matches from World War II for Persian Gulf War people and countries. Instructions were presented as follows:

> Many people have drawn an analogy between the Gulf situation and the situation in Europe prior to World War II. Regardless of whether or not you think this analogy is appropriate, we would like to know what you think the analogy really means. Suppose someone says, "Hussein is analogous to Hitler." For each of the people or countries listed below that are involved in the Gulf crisis, please write down the most natural match in the World War II situation (from the point of view of someone who thinks Hussein is analogous to Hitler). If you think there is no good match, write "none." *Spellman and Holyoak (1992, p. 916)*

Participants were asked to provide possible correspondences for Iraq, the United States, Kuwait, Saudi Arabia, and George Bush, who had been the US president at the time.

Results indicated that participants were sensitive to multiple possible matches between some of the countries. Two relatively uncomplicated mappings were observed for Iraq and Kuwait. A large majority of people mapped World War II-era Germany to Iraq in 1991. Poland and Austria were common mappings for Kuwait. More interesting mappings were reported for the United States, George H. W. Bush, and Saudi Arabia. Participants viewed the United States in 1991 as corresponding to either the World War II-era United States, or the World War II-era Great Britain. Both of these nations played prominent roles in opposing Germany in World War II. George H. W. Bush was viewed as either Franklin Delano Roosevelt, the US president during World War II, or Winston Churchill, who led Great Britain during World War II. This leaves open the possibility that the United States in 1991 could be cross-mapped to either of these nations. Such a cross-mapping could enable mixed inferences that would likely be erroneous. A one-to-one mapping constraint is needed to avoid mixing information from two simultaneous source analogs. Such a mixing of information could lead to incorrect inferences. For example, if the United States in 1991 was viewed as being both Great Britain and the United States in World War II, it may lead to the conclusion that the 1991 United States should remain neutral toward Iraq's invasion of Kuwait (as the United States had in the earlier portion of World War II) and further that the United States should begin a bombing campaign against Iraq (as Great Britain had in World War II). Such mixed inferences should be avoided, as the motivations and circumstances that drove wartime policies in World War II varied at different time points between these two nations. Key to understanding the possibility of mixed analogies are the mappings that participants made for George H. W. Bush and for Saudi Arabia. Most participants placed Bush into correspondence with Roosevelt if they also mapped the 1991 United States to the World War II-era United States. Such a mapping keeps everything straight. People who mapped Bush to Churchill also tended to map the 1991 United States to World War II-era Great Britain. The Saudi Arabia mapping clarified some of these mappings. Saudi Arabia was often mapped to the World War II-era Great Britain when the 1991 United States mapped to the World War II-era United States, meanwhile Saudi Arabia was more often mapped to France if the alternative World War II-era Great Britain mapped to the United States in 1991. Spellman and Holyoak noted that people were sensitive to the possibility that one element in an analogy could correspond to multiple other elements. They likened the actual operation of this two-to-one analogy to the famous Necker cube illusion in which two possible interpretations of the cube exist (tilting up or tilting down) but that each interpretation is incompatible with the other and therefore only one stable interpretation is viewed at a time (Fig. 10.8).

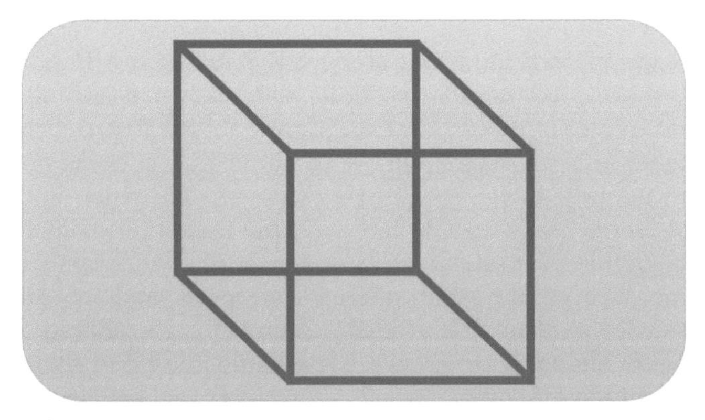

FIGURE 10.8 The Necker cube illusion can be perceived as a cube that tilts up or down; however, both of these possibilities are mutually exclusive. We do not see the cube as tilting in both directions at the same time. Similarly, the analogical mappings for the United States in 1991 during the Persian Gulf War consist of two possibilities: The United States or Great Britain during World War II. As in the Necker cube illusion, people likely consider only one possible source analog at a time, rather than consider both of them together. Keeping things straight in this way is important, as it allows people to avoid drawing invalid mixed inferences based on combining both source analogs.

Other research has also confirmed the importance of a one-to-one constraint on analogical mapping. Both perceptual features and semantic similarity between analogs are important in this process. When semantic similarity strongly competes with relational similarity the result is a potential *cross-mapping*. This type of mapping can potentially cause a misleading analogy to be formed. Semantic similarity should not drive a mapping unless relational correspondences also exist between the domains. My colleagues and I conducted a study in which we carefully controlled the semantic characteristics of the potential analogs by presenting scenarios on fictional planets in which alien species interacted (Krawczyk, Holyoak, & Hummel, 2005). We asked participants to predict what would likely occur on the basis of an analogy between the planets. In the absence of any background knowledge about these hypothetical planets, participants could choose to place aliens into correspondence on the basis of their semantic features, such as jumping, napping, or having unusual horns. The aliens could also be placed into correspondence based on their relational roles, such as being neighbors or competitors. We constructed the materials for this experiment to enable people to form a mixed inference drawing together predictions based on two alternate source analogs. Results indicated that people were very sensitive to both potential mappings, but when asked to explicitly generate inferences, participants kept each source analog separate. This result reinforces the idea that a one-to-one constraint is a core property that is enforced when people generate new inferences based on analogy. At the very least, the one-to-one constraint is prioritized and people tend to keep their source analogs separate.

SYSTEM MAPPINGS

Alignable and Non-Alignable Properties

Using relational mappings allows us to move beyond simple similarity comparisons. These mappings enable us to evaluate similarity of a much broader scale. When two situations include multiple correspondences among numerous elements that can be mapped together, this is referred to as a *system mapping*. The system mapping will often place several different objects, living things, or mechanical parts in correspondence and will frequently include correspondences between multiple relations. The analogy between World War II and the 1991 Persian Gulf War is a nice example of a system mapping. In that comparison not only can we compare different countries to one another, or different leaders, we can compare the entire war in the source to the entire war in the target. This large-scale system mapping between wars has the most potential for providing new insights from the source to the target. For example, World War II involved a vast postwar reconstruction period and included steps by the Allied powers to rebuild and resuscitate the countries that had been invaded and occupied by the Axis powers. We might infer that similar processes would follow the Persian Gulf War during which coalition forces would attempt to restore occupied land in Kuwait to a sustainable level resembling its preinvasion status. A system mapping on this scale may also enable us to rethink aspects of World War II in light of new information that we learn about the more recent Persian Gulf War. For example, Iraqi forces were very interested in securing oil drilling sites in Kuwait for exploitation by the new regime. Similarly, we may consider whether Germany had a similar interest in securing the natural resources of the countries that they occupied such as Poland and France. These opportunities for drawing new inferences and creating new schemas are made possible by the numerous elements that can be placed into correspondence in a system-mapped analogy.

One challenge that arises with a complex system-mapped analogy is that not all situations are similar and even among those situations that have high systematic levels of similarity, there will almost surely be differences that exist. These differences can serve to misalign a comparison and risk the possibility of inaccurate or misleading inferences to be generated from it. For instance, both German and Soviet forces carried out the invasion of Poland in World War II at the start of that war. Russia's leader Joseph Stalin debated the meaning of German aggression and initially concluded that it could help to serve Soviet interests in the future. Because of the complex set of relations in that situation, the analogy might cause one to look for an Iraqi ally who had contributed to the invasion of Kuwait. This could lead

FIGURE 10.9 System-level mappings can take place between analogs. In a case in which we compare an aircraft to a car, we see similarities (both have tires, both run on fuel, etc.), as well as differences (airplanes have wings, cars have turn signals). Differences are alignable when they exist on the same scale (e.g., comparing the number of seats in the plane and in the car). There are also non-alignable differnces that center upon a property that is not shared between the two domains (e.g., presence of wings). *From Shutterstock.com.*

the thinker to assume that there must have been an internal debate about the actions of Iraq and how the invasion would serve the needs of that hypothetical partner nation. This is in fact all wrong, as there was not a second nation assisting Iraq with the invasion of Kuwait. These key differences between situations matter quite a bit in determining the utility of an analogy. Differences need not take away the value of analogy as long as the thinker has a grasp of the facts in both situations. Again, background knowledge is a critical determinant of the value of any act of analogical reasoning.

The types of differences that exist between situations can be categorized according to whether they align with one another. *Alignable differences* are those that exist on a common dimension. For example, when considering the degree of similarity between a car and a large passenger airplane (Fig. 10.9) we can consider the presence of seats as something these vehicles have in common, but the plane has many more seats. The difference in seat numbers is considered to be an alignable difference as this variable differs on a common dimension between the two analogs (Gentner & Markman, 1997). We can also note that the airplane has wings and the car does not. This type of difference would be called a non-alignable difference, as there is no common dimension in this case. One analog has the property (wings) and the other does not. People notice the differences between alignable and non-alignable properties and treat these types of categories differently.

Paul Slovic and Douglas MacPhillamy (1974) reported on a classic case of alignable and non-alignable properties. Participants in this study evaluated the likely grade-point-averages (GPA) of hypothetical individuals who had taken tests. The information provided to participants indicated that student A had taken a unique test such as quantitative skills and student B had also taken a unique test, such as English ability. Participants were also presented with a common test, such as one evaluating their *Need for Achievement*, for which both students had scores. When we apply the terminology of alignable properties, the common *Need for Achievement* measure would be an alignable property and the difference in scores for students A and B is considered to be an alignable difference. Meanwhile, the other scores (*Quantitative ability* and *English ability*) that are unique to one student only are non-alignable differences. There is no way to compare the students on those particular dimensions. Results showed that participants overestimated the importance of the alignable differences in predicting student GPAs. In other words, the student who scored higher on the common *Need for Achievement* measure was estimated to be a better performer in overall GPA than the unique tests. This result was obtained even after the experimenters had warned people not to overvalue the common score and when people were given rapid feedback about the accuracy of their judgments. Slovic and MacPhillamy (1974) indicated that their results represented what they called a "common-dimension effect." Similar results have also been obtained in later studies, indicating that alignable differences occupy a privileged place in people's ability to evaluate similarity among multidimensional domains, which would include those that are able to be mapped at a relational system level.

The preference for making evaluations on the basis of alignable differences can be applied toward analogies. When we think about analogies between complex situations such as wars, cities, or ecosystems, we are likely to focus most on the common elements shared by both domains, rather than focusing on how these may be different. Perhaps this is why David Miller's analogy discussed earlier in the chapter likening a Higgs field to a cocktail party is considered to be so effective. We can relate the particles in the Higgs field with the people at a cocktail party focusing on their distribution in space and how they move in response to a change in the field (or a rumor moving through the room). We are focusing on alignable properties in this case. We can appreciate that there are differences in the way that particles and people behave, but they do not undermine the effectiveness of the linked commonalities that these diverse situations share. Note that we are not distracted from the relational comparisons by the fact that cocktail party guests drink alcohol and received invitations to the party location, while particles in a Higgs field do no such things. The analogy can be effective as we focus on the relevant alignable properties alone.

SEMANTIC MEMORY AND THE ROLE OF ASSOCIATIONS

Creative Analogies Based on Long-Distance Mappings

Analogies between situations are often viewed as creative if they involve very dissimilar domains. For instance a comparison between remote domains such as an office building and a beehive; this analogy has a humorous and clever quality to it. If an analogy is made between domains that are too similar to one another, it may appear to be uncreative. There comes a point where the similarity level can be so high that one can question whether a comparison even qualifies as being an analogy. Sometimes these can be more accurately characterized as being literal comparisons. Analogical reasoning researcher Adam Green at Georgetown University described the differences between analogies that are close in terms of semantic space such as the example: *salmon is to drift net as shrimp is to trawl net* and those that are far in terms of semantic space such as: *salmon is to drift net as sample characteristic is to prediction interval*. Green claimed that two key characteristics of creativity are novelty and usefulness. Creative analogies that are distant in semantic space are likely to be novel, but their true value may ultimately depend upon how useful they prove to be in fulfilling the requirements of a given situation (Green, 2017).

Creative analogies drawn between diverse situations have been termed "long distance" analogies by Kevin Dunbar and his colleagues. We have already discussed one of the studies by Green, Fugelsang, Kraemer, Shamosh, and Dunbar (2006) in Chapter 3. In that study, the participants were provided with different types of analogies to evaluate. Within-domain analogies involved items that were semantically close. For example, the analogy *nose is to scent, as tongue is to taste*, all of the items represent the category "parts of the face." By contrast, distant cross-domain analogies were also evaluated by participants. An analogy such as *nose is to scent, as antenna is to signal* is comprised of a facial feature and its function compared with an electronic device and its function. This type of analogy is distant in that the terms come from very diverse semantic categories. Participants in the experiment were asked to determine whether the analogy was true or false. A false analogy example is *nose is to scent, as eyelash is to mascara*. Green et al. (2006) used fMRI to evaluate the regional brain responses associated with distant over close analogies. The left frontopolar prefrontal cortex showed this type of response with greater activity associated with semantic distance. This finding supports the idea of the cross-domain, distant semantic analogies are fundamentally different from analogies that involve highly similar terms.

To further test the effect of the frontopolar cortex on long-distance, or creative, analogies Green et al. (2017) used a technique called transcranial direct current stimulation (tDCS) to influence activity within the brain. Box 10.3 describes the technique, which involves applying a current to the scalp. This current runs through the underlying brain tissue and can be targeted to pass through particular regions influencing the activity of neurons. In this study, Green and colleagues applied either tDCS or a sham control nonstimulation to the left frontal lobes of the participants. The tDCS was targeted at a left frontopolar prefrontal region that had been found to enhance creative responses which is close to the area that had previously been found to be activated in response to increases in semantic distance among analogies (Green et al., 2006). The tDCS was applied for 20 min and has been found to shift the excitability of the stimulated cortex for around an hour after stimulation. Participants in this study were asked to complete a verb-generation task during which they were presented with a noun and asked to say a verb that was related to that noun in a creative way. The tDCS increased the creativity or remoteness of the verbs generated.

Green et al. (2017) were also interested in the effect that this greater excitability of the frontopolar cortex would have on analogical distance. For this aspect of the experiment, participants were asked to complete a checklist of relational matches that could be paired with a related word pair. For instance, the words *kitten* and *cat* were presented and possible matches included close analogical mappings, such as *puppy* and *dog*, as well as distant mapping such as *spark* and *fire*. As with the word-generation task, creativity of the analogy matches selected also increased when people had been provided with the tDCS stimulation compared to when they had received sham stimulation. This outcome was determined by computing the semantic distance between the concepts. This innovative study provides further evidence that the distance of analogical mappings appears to be governed by different brain areas and increases in the activity of the left frontopolar region enhances people's ability to detect and generate creative analogies between distant concepts. This satisfies the distance requirement for creative analogies, but leaves the usefulness quality somewhat less clear. Future work may be necessary to determine what abilities enable people to formulate particularly useful analogies. This is the topic of the next section dealing with insight in creative problem solving by analogy.

Semantic Networks and Insight

Semantic retrieval is one of the most critical cognitive process if the creativity level of an analogy can be determined by the distance and usefulness of the comparison. The ability to make a large mental leap comparing two situations requires someone to retrieve a memory that shares relational structure to the current situation. This is

BOX 10.3

TRANSCRANIAL DIRECT CURRENT STIMULATION

For many years people have been interested in stimulating the brain by applying electricity from outside the head. Ever since neurons were understood to communicate electrically, researchers and healthcare providers have experimented with ways to apply electrical currents to brain circuits. Some of these efforts involved trying to improve health and brain function, while other efforts were carried out purely to experiment on the brain.

One of the methods gaining widespread use is transcranial direct current stimulation (tDCS) (Fig. 10.10). This method involves placing a cap on a person's head that is fitted with electrodes. The electrodes can apply a direct current to the scalp. The current passes through hair, skin, and eventually crosses the skull (Datta et al., 2009). The application of current to the head in this manner can induce cortical changes (Fregni & Pascual-Leone, 2007). The current intensity values typically vary between 1 and 3 mA, which are relatively low and therefore present limited risk to a person undergoing tDCS stimulation.

There are several factors that influence the effects of tDCS. First, the positioning of stimulation electrodes on the scalp affects which areas of the brain will receive stimulation. Because the current must travel through the hair, the skin, the skull, and meninges in order to reach the brain, the technique cannot be targeted perfectly toward one brain area, as these parts of the head can deflect, or block some of the current (Philip et al., 2017). Second, stimulation of the cranial nerves can also influence the effects of tDCS on different brain regions, both stimulating nontargeted areas and deflecting stimulation from intended target areas. Third, both the strength and direction are important for modulating neuronal activity once the electrical stimulation reaches the brain through the skull, or stimulation of the cranial nerves. The applied current can alter the voltage levels across the membranes of neurons with higher field strengths inducing larger changes. The actual effects of the current application on neurons is modulated by a part of the tDCS cap. These caps will have two electrodes called the anode and the cathode. Stimulating from the anode increases the likelihood of neuronal spiking, in effect stimulating the region to become more electrically active. Meanwhile, passing current via the cathode will decrease the likelihood of neurons firing, effectively inhibiting an area of the brain.

The technique is seeing widespread use as both a neural enhancement device and as a means to inhibit neural firing. Both methods may have therapeutic value and both are valuable tools that can be used in order to evaluate the involvement of different brain areas within cognitive and motor tasks.

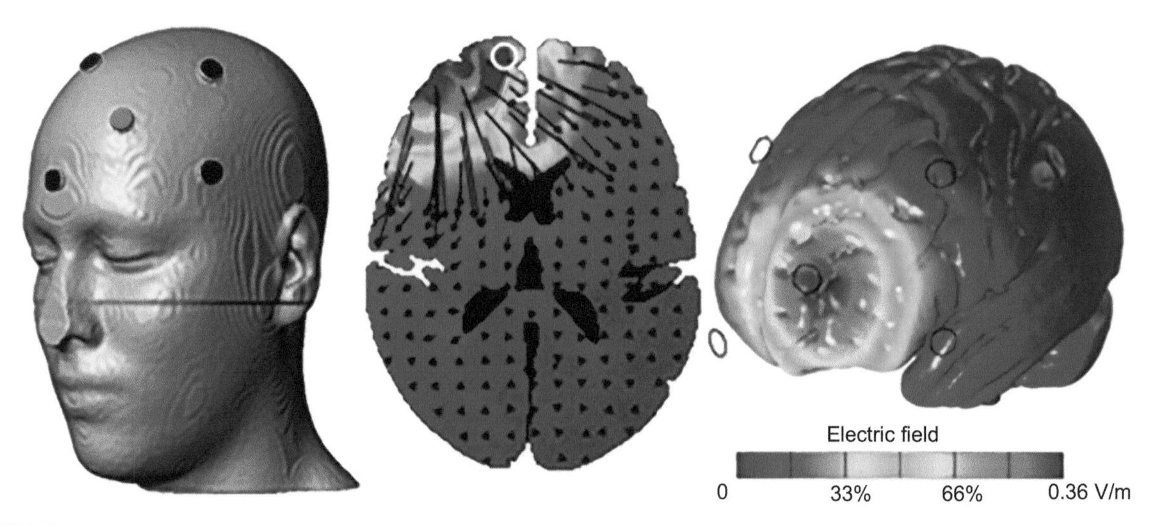

FIGURE 10.10 Transcranial direct current stimulation (tDCS) is a technique that involves placing a stimulation cap over someone's head and applying a current through the brain. This technique can stimulate different areas of the brain electrically and in some cases influence our thinking and perception (Green et al., 2017). *Credit Adam Green.*

particularly challenging since the distance requirement ensures that local memory cues will not be sufficient to call up these memories. Rather, the relational structure of the current situation will likely have to lead the process with someone finding common ground. Retrieving memories of this type requires a memory search that isolates remote associations. A remote association is one that does not come to mind easily. Remote associations will often require a deep search of memory. Such searches have been associated with insight.

Insight problems are those that require a very deep and thorough search of memory in order to solve. The classic experience of insight occurs when the appropriate association "pops into mind" without a feeling of deliberate effort, or systematic progress toward finding the association. Karl Duncker (1945) described an example of an insight problem when we posed the challenge that a candle would need to be fixed to a wall and lit. Also the candle could not be permitted to drip wax onto the ground. The participant in this experiment was offered only a box of matches, some tacks, and a candle. The challenge of this and other similar insight problems is that there does not initially appear to be any obvious way to stop the wax from dripping if one simply tacks the candle up on the wall. The solution is to use the match box to catch the wax. If you were to dump the matches out of the box and tack it to the wall, the box could serve as a platform for the candle that would also catch the dripping wax once it is lit (Fig. 10.11). The task

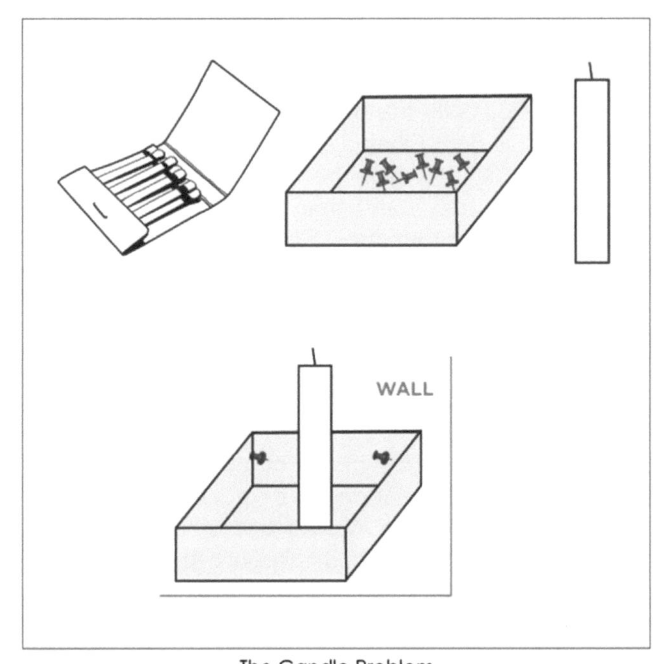

FIGURE 10.11 The solution to Karl Dunker's (1945) "matchbox" problem involves overcoming functional fixedness in order to view the matchbox as a potential surface that can be attached to a wall as a platform for the candle to rest upon. *From https://commons.wikimedia.org.*

requires insight and what some would call creativity. To solve this task you must overcome functional fixedness, a term we discussed in regard to animal reasoning in Chapter 4. Functional fixedness refers to the tendency for people to view an object as having only one function. In Duncker's insight problem, the matchbox does not outwardly appear to have the platform capacity. Platforms and matchboxes are not well associated for most people and therefore they may miss this solution. Like many insight problems, the solution seems to arrive with a surprising feeling. What some might call an "aha" moment. This is different from the feeling one gets from routine explorations of semantic memories where one can simply think through the most associated properties of a given item.

Another insight task that is commonly used in psychology labs is called the remote associates test, or RAT (Mednick & Mednick, 1967). In this task people are provided with three words and their task is to find a common association word that links the three other words together. For instance, consider an example with the three words being "head," "street," and "dark." What word comes to mind when you think about these? Take a moment to think this over. The answer to this problem is the word "light." You can have a headlight, a street light, and light and dark are opposites. Notice that street and light, as well as head and light, are relatively unrelated. If you tried this example, you may have gotten the answer by examining the dark-light association, which would be the strongest of the three for most people. In such a case, you can hypothesis test these possible candidate matches to see if they work with the other words. Classically, one cannot find a great deal of success trying that iterative strategy. Rather, you typically need to just sit with the three words for a little while and focus on other things. At some point the common associate will pop into mind. It is this "aha moment" that likens the RAT to an insight problem-solving task.

The RAT has been studied using neuroscience methods revealing a neural basis for the feeling of insight in the right temporal lobe. Kounios and Beeman (2009) conducted fMRI and EEG studies investigating what they called a compound-remote-associates task. This task followed the RAT problem format in which three words were presented. Kunious and Beeman described an example in which participants were given the words "crab," "pine," and "sauce." Participants were asked to generate a single word that can result in either a compound or a familiar two-word phrase when it is combined with each of these three words. The answer to this problem is "apple," which combines with the three test words to make up the compound words "crabapple," "pineapple," and "applesauce." After a solution period, people were asked to judge if they had arrived at the solution via an insight leap, in which the answer just occurred to them, or whether they had been able to use a trial-and-error analytical method to arrive at a correct solution. When participants reported a feeling of insight, the right superior temporal gyrus was activated in the fMRI study, which you will recall from Chapter 3 has very good spatial resolution. Likewise, when measured with EEG, which has a stronger temporal resolution than fMRI, people's solutions that had been accompanied by a feeling of insight were associated with high-frequency (gamma-band) activity starting around 300 ms prior to reporting the arrival of a solution. This EEG activity occurred at electrodes positioned over the right anterior temporal lobe, nicely complementing the right temporal lobe finding in fMRI.

The temporal lobes have been associated with semantic memory as we discussed in Chapter 3. Individuals who experience semantic dementia from atrophy of the temporal lobes will gradually experience a progressive decrease in the ability to recall semantic memories and name pictures. Notably, the right temporal cortex appears to be preferentially active in support of these insight solutions for RAT problems. Some researchers have associated the right temporal cortex with storing semantic information in a loosely associated way. This has been termed coarse coding (Jung-Beeman, 2005), capturing the idea that rougher or more distant semantic information is placed into correspondence in the right hemisphere.

The Coarse Coding Hypothesis

The coarse coding hypothesis refers to a situation in which more distant associations occur within the right temporal lobes when people consider the relatedness of different words or concepts (Beeman and Chiarello, 1998). The effect of the coarse code is to provide a more distant association, or one that broadly fits with the goals of a task. This has been described in relation to insight problem solving and in reasoning more broadly.

Beeman and Chiarello (1998) described coarse coding as a process characteristic of the right hemisphere, which is nonspecialized for language in most people. Right-handed individuals in particular have a language production dominance in the left hemisphere. Regions such as Broca's area, involved in the motor production of speech, and Wernicke's area, involved in comprehension of language, reside in the brain's left hemisphere in most people. The language focus extends to enhanced language skill when verbal information is presented to the left hemisphere compared to when it is presented to the right hemisphere. The left hemisphere is considered to govern fine-grained semantic searches when language cues are presented. In other words when you read the word "dog," the left hemisphere of your brain is quick to produce the strongest few associates to that word. In this case "cat," "bone," and "bark" probably come to mind quickly. These are strongly associated words that occupy a semantic space very close to the word "dog." By contrast the distant and coarse semantic search process engaged by the right hemisphere is thought to be broader in the sense that the neural networks representing knowledge concepts in this hemisphere activate in a weaker and more widely distributed manner. For instance, with the presentation of the word "dog" a right hemisphere spread of activation in semantic memory might trigger words and concepts that are more remote and weekly associated such as "canine," "mammal," and "wolf." These types of associations may complement the targeted and focused semantic activation characteristic of the left hemisphere. When we need to answer a question about literal meaning, we rely upon retrieval of a concept that is both strongly associated with the target word and also one that is close by within the network of active associations. What would be most useful for retrieval of source analogs from semantic memory?

Isolating a strong associate may be important for many situations in daily life, but reasoning by analogy and solving insight problems may rely more heavily on these coarse coding associations. When we find a literal match, we can compare it to the concept at hand. For example, when we consider the Affordable Care Act we can compare it literally to the healthcare policies of other nations, but are these really analogies, or rather more appropriately considered to be literal comparisons. Certainly for analogies that are creative, we instead must draw upon relational correspondences that come from distant fields of knowledge. Such associations are likely to be weakly activated in much the manner that Beeman and Chiarello (1998) has proposed as a right hemisphere function. President Obama's analogy mentioned earlier in this chapter, in which he likened rising healthcare costs in the United States to a flood with waters rising is one such distant analogy. The flood situation is only weakly related to the cost of healthcare and that association is made through the correspondence between a rising dollar figure and the physical rise of water in the flood. Thus, the analogy appears creative and provides a new way of thinking about the crisis of rising costs in a vivid and memorable way.

Additional evidence for right hemisphere involvement in analogical relationships comes from studies of metaphorical meanings. When we recall a strongly associated word, such as "shoe" in response to "foot," we may rely on basic word retrieval. If however, we make a long-distance connection as "the girl is a peach" we engage supervisory areas within the prefrontal cortex (PFC) in association with a broader search that involves the right temporal cortex (Yang, Edens, Simpson, & Krawczyk, 2009). My colleagues and I studied metaphor comprehension using fMRI and found that novel metaphorical statements such as "he is a kiwi fruit" activated the left ventrolateral PFC and areas within the right temporal cortex to a greater degree than literal statements such as "he is a man" or commonly used metaphors, such as "the job is a piece of cake." These patterns of activation in response to processing new metaphors are consistent with a broader search within semantic memory that allows us to understand a possible symbolic and analogical comparison between the man and the fruit. Such activation is necessary when we compare the two domains in a way that we are not accustomed to. Such processes are likely to be involved in creative problem solving and making distant analogical mappings.

ANALOGICAL REMINDING

Do People Notice Analogies Between Situations?

In order to fully benefit from our past experiences, we must be able to remember a source analog at the critical time at which it is needed. This will ensure that we are able to draw valid inferences that follow from appropriate analogs. Sometimes analogical reasoning will go astray if we are considering a target analog and several semantically similar source analogs come to mind, but these source analogs share little relational similarity with the new situation that we would like to understand.

An excellent example of human problem solving involves the "radiation problem" (or "ray problem"), which is a problem introduced by the Gestalt psychologist Karl Duncker (1945). The setup of this problem proceeds as follows:

> Suppose you are a doctor faced with a patient who has a malignant tumor in his stomach. It is impossible to operate on the patient, but unless the tumor is destroyed the patient will die. There is a kind of ray that can be used to destroy the tumor. If the rays reach the tumor all at once at a sufficiently high intensity, the tumor will be destroyed. Unfortunately, at this intensity the healthy tissue that the rays pass through on the way to the tumor will also be destroyed. At lower intensities the rays are harmless to healthy tissue, but they will not affect the tumor either. What type of procedure might be used to destroy the tumor with the rays, and at the same time avoid destroying the healthy tissue? *Gick and Holyoak (1980, p. 280)*

The ray problem is thought to require insight in order to solve it. The answer is unintuitive and people often take considerable time trying out ineffective solutions before they eventually arrive at an appropriate answer. This is a classic example of an insight problem, one that is difficult to piece together and may require coarse semantic associations in order to solve.

The answer to the ray problem is to use numerous rays at a reduced intensity level and arrange the multiple rays in a circular pattern so that their point of convergence hits the tumor only. The combined strength of the numerous rays will then successfully destroy the tumor only, while otherwise harmlessly passing through the surrounding healthy tissue (Fig. 10.12). This is an excellent solution, but one that very few people will generate. Can analogical reasoning help people to improve their abilities on difficult insight problems?

Mary Gick and Keith Holyoak investigated the role of analogical reasoning in insight problem solving. Gick and Holyoak (1980) studied Duncker's radiation problem finding that under 10% of participants were able to generate the circular convergence solution using multiple weak rays. This result is in line with the original version of the radiation problem in which only about 5% of participants solved the ray problem by employing weakened rays to converge on the tumor. Notably, these successful participants had needed a hint in order to arrive at this solution (Duncker, 1945).

Gick and Holyoak (1980) offered an interesting analogical solution for their participants. They described a source analog that they might later make use of when confronted with the ray problem. The key story began with a description of the "Attack-Dispersion" situation in which a general is attempting to invade and occupy a fortress. The fortress has numerous roads that lead outward from it and these roads contain landmines that will detonate destroying the army if they all cross at one point. Critically the army needs all to arrive at the same time in order to conquer the fortress. The general decides to divide up his forces into smaller units and send these small numbers toward the fortress each using a different road. The smaller units will be able to pass over the landmines without triggering them, and the army will be able to arrive in full force at the fortress and overtake it. In

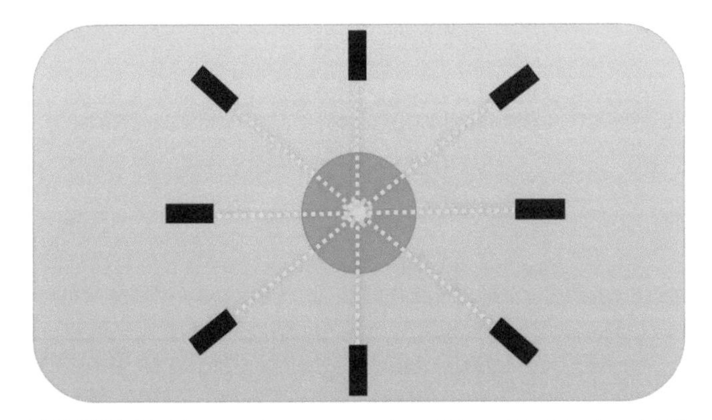

FIGURE 10.12 The "radiation problem" from Karl Dunker (1945) proves to be a challenging thought problem. The answer to this problem is to weaken the strength of the incoming rays so that they converge at a point of maximal strength aimed at the tumor. The weak rays will not harm the surrounding tissue when they are not operating at maximal strength outside the convergence point.

one of the key experiments participants were instructed in memory for text. Three stories were provided and participants were asked to recall the story to the best of their abilities. Among the three stories offered was the Attack-Dispersion story. Participants attempted to solve Duncker's radiation problem after a short delay period. When a hint was provided in the form of a statement that "one of the stories you read will give you a hint for a solution to this problem." The hint was highly effective allowing 92% of the participants to arrive at the convergence solution to the radiation problem. This demonstrates that people can understand the analogy and use with great effectiveness. The critical trials in this experiment were ones that offered people the three stories (one of which was the Attack-Dispersion story) and then attempted the radiation problem without a hint. Performance on this case was far worse, with only 3 out of 15 people achieving the correct answer (20%). Furthermore, only one of these three individuals provided a complete solution to the radiation problem. This study provides evidence that it is rare for people to spontaneously recall of an analogical solution even when one had been provided just minutes earlier. More optimistically the results obtained after the simple hint had been provided indicate that people comprehend analogies adequately. Note that there is very little semantic overlap between the Attack-Dispersion story and the radiation problem. The analogical correspondences in this case are purely relational with groups of soldiers mapping to amounts of radiation and a fortress manned by people being aligned with a cancerous tumor. This may represent an uncommon situation, in which there are very few memory cues to remind people that the story about the fortress could inform them about a radiation situation in medicine.

For a source analog to be very useful, it will typically need to be accessible from long-term memory. Memories tend to become less accessible over time, so it is discouraging that so few people spontaneously found the Attack-Dispersion analog in the study by Gick and Holyoak (1980) when it had been presented just minutes earlier. A more likely scenario is that weeks, months, or years pass between a source analog being encoded and understood before it is recalled in order to assist in one's reasoning about a new situation. Gentner and Landers (1985) performed an initial investigation into access from memory of a potential source analog. Such experiments are difficult to perform, as they require multiple sessions and leave open the possibility that individual differences in memory abilities will compromise the results. In this experiment, participants were initially presented with 32 very short stories consisting of a couple of paragraphs. Examples of these are shown in Fig. 10.13 consisting of a sample source analog along with a true analogy (with only higher-order relational similarity to the source), a false analogy (with only first-order, nonrelational similarity to the source), and a mere appearance match (consisting of only superficial attribute similarity to the source). Eighteen of those stories were potential source analogs for a second experimental session held about 1 week later. The remaining stories were irrelevant to the target stories that would later be presented in the second session and were offered for distraction purposes only. Participants in the second

SOURCE ANALOG

Karla, an old hawk, lived at the top of a tall oak tree . One afternoon, she saw a hunter on the ground with a bow and some crude arrows that had no feathers . The hunter took aim and shot at the hawk but missed . Karla knew that hunter wanted her feathers so she glided down to the hunter and offered to give him a few . The hunter was so grateful that he pledged never to shoot at a hawk again . He went off and shot deer instead.

ANALOGICAL TARGET STORY

Once there was a small country called Zerdia that learned to make the world's smartest computer. One day Zerdia was attacked by its warlike neighbor, Gagrach . But the missiles were badly aimed and the attack failed . The Zerdian government realized that Gagrach wanted Zerdian computers so it offered to sell some of its computers to the country. The government of Gagrach was very pleased . It promised never to attack Zerdia again.

MERE APPEARANCE TARGET STORY

Once there was an eagle named Zerdia who donated a few of her tailfeathers to a sportsman so he would promise never to attack eagles. One day Zerdia was nesting high on a rocky cliff when she saw the sportsman coming with a crossbow . Zerdia flew down to meet the man, but he attacked and felled her with a single bolt . As she fluttered to the ground Zerdia realized that the bolt had her own tailfeathers on it.

FIGURE 10.13 Gentner and Landers (1985) presented a source analog along with multiple potential target analogs. One provided a true analogy, another provided a false analogy, and there was also a match based only on appearance.

session read stories and performed a reminding task in which they were asked to describe whether there was a prior story that came to mind when they read about the target analog story. In this task, the mere-appearance matches were called to mind at the highest levels. These matches were reported at a significantly higher rate than either the true analogy stories or the false analogy stories. Notably, the true analogies were recalled to a greater degree than the false ones. These results indicate that mere appearance, or attribute similarity, is the most effective memory cue resulting in these uninformative examples being most available at the period of considering a source analog.

In a second judgment task, participants were asked to rate the soundness of the similarity between the target analog story and the potential matches. Soundness was defined as "providing a sound match between stories," or in other words that the stories provided a convincing argument about interactions when they were considered together. In these cases, the results reversed and now people claimed that the true analogies had the most informative, or sound, basis for similarity. This tells us that people can see the value in making higher-order relational matches; regrettably, these types of matches are not nearly as available as simple mere-appearance matches. These results are less inspiring when one considers the fact that the stories were all presented in a lab context and only 1 week apart. In daily life conditions, matters may be even worse for analogical reminding. One can imagine that months to years may pass between potential relational matches being observed. Furthermore, higher-order relational similarity does not require common attributes. Therefore, very distant analogies from remote domains are likely to be quite improbable compared to more mundane analogies that share considerable relational overlap and attribute similarity.

A common assumption about reasoning by analogy is that it can allow someone to experience a leap of insight about a new situation based on a very dissimilar situation experienced at some earlier point in time. Such a situation has been referred to as *remote analogical reminding*. Charles Wharton, Holyoak, and Lange (1996) studied this type of analogy situation. In these studies, Wharton and colleagues delivered analogies to participants through short-story materials. The stories included similarity at the level of themes and situations. A theme, such as "sour grapes," was used in which a desirable experience was determined to be not worth having after a failed attempt to obtain it. An example of a *close analogy* described a pair of situations involving the same theme among people who were trying to achieve competitive goals. In one situation, a student who is set up on a meeting with a college recruiter is later denied admission to the recruiter's college; he then decides that the college was not worth getting into anyway. The close analog to this situation involved a woman who is set up on a date; the potential partner fails to appear at the restaurant, and then the woman decides that he was not worth having a relationship with anyway due to his job. The remote analog to this situation involved a unicorn who unsuccessfully tried to cross a river to explore the land beyond, only to decide that there was nothing of particular value across the river anyway. Critically, these short stories did not include attribute similarities, so any use of analogical reasoning would have to be achieved through analogical reminding. The experiment included a delay factor that varied the length of time between the presentation of the first set of stories (the source analogs) and the target analog. Delays varied across three conditions: A 5-min delay, a 11-day delay, and a 1-week delay. Participants were asked to judge how easily each story could be imagined. Next, when assessing target analog stories, the participants indicated which of these stories made them think of other situations and to list those source stories that they could recall. The closeness of the stories had a large effect. The details of theme and situation mattered with more similar themes and situations being retrieved at higher rates than the more remote analogs (which were those sharing only relational similarity). In the close analogies, true relational analogs were retrieved at a rate of approximately 65% considered across the three delay conditions. Meanwhile non-analogous stories that were relationally dissimilar were retrieved at a rate of only around 25%. Distant analogs (such as the student-to-unicorn comparison) were retrieved at only a rate of around 30%. The structure of the stories clearly mattered, but the similarity between the actors in the stories also determined reminding success to a large degree. Delay time had a much less dramatic effect on the results providing a small reduction in accurate analogical reminding over time and a small increase in retrieval strength of non-analogous stories over time.

Overall the Wharton et al. (1996) results indicate that relational similarity is an important source of reminding. Meanwhile, the overall surface features of the stories also matter with more similar stories leading to greater levels of reminding than ones that are only relationally similar, but share few similarities in actors and setting. This may be an adaptive feature of our memory systems. If two situations are overly diverse, there may be little reason to connect them even if the overall structure is the same. For example, students may serve as poor indicators about the likely behavior of unicorns. Therefore, we may be more successful in using analogies for reasoning if we stick to mappings among elements that are at least relatively similar overall.

ANALOGICAL REASONING IN EVERYDAY LIFE AND THE LABORATORY

Analogies as Cognitive Tools

In this section we will consider a final set of studies addressing how analogical reasoning operates outside the confines of the research laboratory. There are several experimental approaches to the use of analogy. Thus far, we have reviewed studies regarding how people assess similarity among situations, notice relational correspondences, and retrieve potential source analogs to apply to a new situation. Now let's address how people actually make use of analogies in real life.

Analogies can be used for a variety of purposes, as we will see in this next set of studies. One of the clear purposes for drawing an analogy is for making new inferences about a new situation based on what is already known about a prior situation. A classic case of this situation can be observed in many scientific studies, in which new groundbreaking discoveries are made and related back to other situations. Such was the case with the Rutherford model of the atom, as we have already discussed in this chapter. Additionally, analogies can be powerful teaching tools. If someone is unfamiliar with a new domain of knowledge, an effective teacher can draw comparisons to relate the new material to older information that is already known to the audience. This use-case of analogy alleviates the need for the thinker to retrieve an appropriate source analog, as one has already been considered and presented to the person. Analogies can also be used to convince people of a particular viewpoint. This is common in political discourse, as candidates point out analogical situations in order to inform, convince, or even to mislead people in their opinions about a particular situation. This type of analogizing was discussed in relation to the US Affordable Care Act earlier in this chapter, with the president and his detractors each crafting colorful analogies to offer to the public. Lastly, humor can be another purpose of analogies. Analogical comparisons can be clever and witty, and they can sometimes be used simply to make speech more interesting to an audience. Many great orators use metaphorical or analogical stories in order to make a point more memorable to their audience.

Daily Life Studies of Analogical Reasoning

The term *in vivo* literally translates to "in the living" and applies to situations that take place in the real world. *In vivo* studies of analogical reasoning can be contrasted with *in vitro* studies meaning "in glass," or within a test tube in the context of scientific studies. Kevin Dunbar and Isabelle Blanchette applied the *in vivo* and *in vitro* labels to studies of analogical reasoning. The *in vitro* term applies to the types of studies we have reviewed thus far; studies in which people are asked to read information and form analogies, select relational matches, or make use of prior analogical information. The *in vivo* term applies to situations in day-to-day life when we make use of analogies. These can include many of the situations we have discussed involving wartime comparisons, communicating political viewpoints, or making scientific discoveries.

In his pioneering work on analogical reasoning in the field, Dunbar (2000) investigated analogical reasoning as it occurred within lab meetings during which scientists discussed experiments and their work more broadly. In these cases, Dunbar attended meetings of several different laboratory groups in either molecular biology or in the field of immunology. He observed and recorded instances of analogy over the course of these meetings. Dunbar categorized a variety of reasoning approaches used by molecular biologists including induction, deduction, and causal reasoning. He also noted several instances in which scientists tried to understand unexpected results. Analogical reasoning was noted, especially in cases in which there was not a straightforward answer to a particular question. The use of analogies applied to many aspects of the work including formulating theories, designing experiments, and interpreting results. Four labs were observed over the course of 16 sessions at which the scientists made use of analogy 99 times that Dunbar and his colleagues recorded. These analogies were then coded based on four different categories: (1) formulating a hypothesis, (2) designing an experiment, (3) fixing an experiment, and (4) explaining a result.

The goals of the individual making the analogy determined which category it fits within. When people formulated hypotheses they tended to make mappings between similar situations. For example, relating the human immunodeficiency virus (HIV) to the Ebola virus. In these cases, the analogy would most often be considered to be close semantically. A virus-to-virus correspondence appears to be close semantically, but this may not appear the same way to an expert on viruses who is sensitive to many differences among various virus types and their properties than a nonexpert. When one considers the purpose of hypothesis formulation, the close comparison makes sense. Viruses do act differently and have different structures. In predicting the properties of a particular virus, another virus makes sense as a match, as it will have plausible features that could also be applied to a different virus that is less well-understood. When microbiologists designed experiments they were more likely to make very near matches, sometimes drawing analogies between the same category, such as using knowledge of the HIV virus as a direct comparison to predicting and designing an experiment to test something else about the HIV virus. In this case, the domain of the two analogs is so similar that more nuanced descriptions are likely needed. If for instance, the virus replicates in one particular way, it may be important to test if this occurs under different types of conditions. The conditions themselves may vary enough that they would be analogical. This same type of domain comparison also applied toward fixing experiments. Again, the domain appears to be very close, but to an expert could be quite different. Lastly, when explaining a result to others the scientists used quite distant analogies. For example, Gallo (1993) made an analogy of the HIV virus to a pearl necklace. In the analogy, the pearls represent nucleotides in a multiplying string as the viral RNA transcribes into DNA adding nucleotides repeatedly as pearls could be added to a necklace in a sequence. This analogy represents two domains that are quite distant. It has that same creative feel to it that we observed with the Higgs field-to-cocktail party analogy. As with the Higgs-cocktail analogy, the HIV-pearls analogy is for illustrative purposes and for explaining the broad structure of something very new in relation to a similar well-known structure that is already likely known to the audience. Dunbar's work suggests that the analogical purpose is essential to determining the distance of an analogy that a reasoner will make. Fig. 10.14 displays the proportion of analogies falling within different categories in the Dunbar (2000) study. As we have observed previously in this chapter, expertise factors heavily into the choice of analogy as well. If one has a wide array of knowledge about molecular biology, then there is likely to be a greater depth for comparisons and such a person may draw analogical correspondences between seemingly close domains, that actually turn out to be more remote in the mind of the expert.

In vivo studies of analogical reasoning in politics were reported by Blanchette and Dunbar (2001). In this study, the source of analogies was news articles describing a referendum that took place in Canada in 1995 (both investigators were based at McGill University in Montreal, Quebec, Canada, at the time of this referendum). Canadian votes were presented with two options, keep Canada as one country, or allow the Province of Quebec to secede and become a separate country. In this *in vivo* study, Blanchette and Dunbar searched 400 articles from major Quebec newspapers that reported on the referendum and found over 200 analogies that were described. The findings indicated that over three-quarters of these analogies came from a domain outside of politics. Domains included magic, religion, sporting events, and farming. Additional probing of analogies made at live political rallies also indicated that a large majority of analogies came from domains outside of politics.

If we compare the study of scientific analogies by Dunbar (2000) to the political analogy study reported by Blanchette and Dunbar (2001), we see some interesting differences. Based on the scientific study, experts in molecular biology tended to draw analogical comparisons between rather close domains for the purpose of generating hypotheses and designing experiments. Meanwhile, the distance of the source and target domain increased substantially when scientific analogies were made in order to inform others about a particular situation. In those instances, a nonexpert was

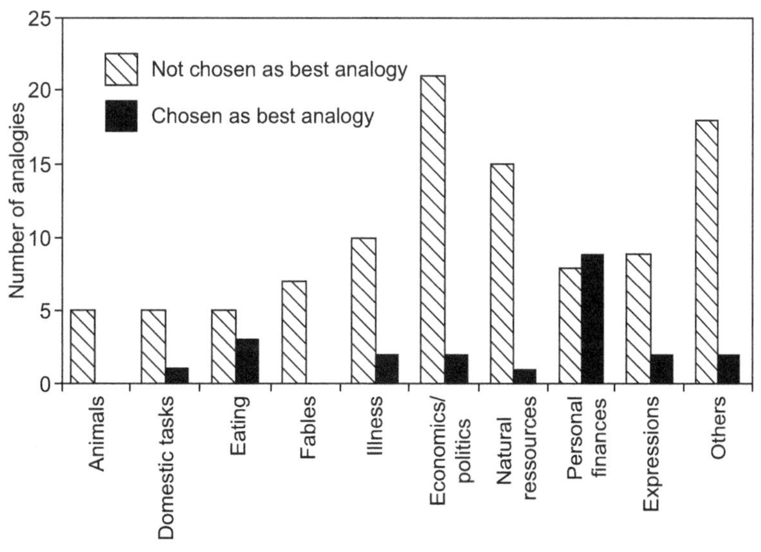

FIGURE 10.14 The goals and types of analogies produced by scientists at molecular biology lab meetings. *From Dunbar, K. & Blanchette, I. (2001). The in vivo/in vitro approach to cognition: The case of analogy.* Trends in Cognitive Science, 5, *334–339.*

presented with a more remote domain, one that they would be able to identify with and relate the new knowledge to. Similarly, in the domain of politics, remote analogies dominated the comparisons made in news stories and at live political rallies. Notice that journalism and political speeches often involve telling other people about a particular situation in terms that they will understand. As with the scientist attempting to explain a finding or mechanism to a naïve audience, the politician or journalist makes use of a previously understood domain and relates new and potentially complex information to that more familiar domain. In this way, the audience will gain some new appreciation of the novel situation. We should consider the fact that the sole purpose of analogical comparisons is not always to inform. Rather, in political situations, especially, the purpose of drawing an analogy can be to win someone over toward a particular point of view (refer to Fig. 10.15). Such is the case when there are two sides to a story and the better outcome is not clearly known or even able to be predicted accurately. In these cases, which include much of what occurs in politics, an analogy can be drawn in order to suggest that one particular alternative is better than the other in the mind of the analogical thinker.

My colleague, Don Kretz, and I decided to embark on an *in vivo* investigation of our own. We had been doing research that was relevant to behavioral economics. This is a discipline that seeks to understand how and why people make certain decisions. We were struck by the differences between Dunbar's (2000) findings, in which molecular biologists seemed to rely on close analogical domains when hypothesizing, designing, and fixing experiments, and the distant analogies reported in Canadian politics by Blanchette and Dunbar (2001). Kretz and I felt that the domain of behavioral

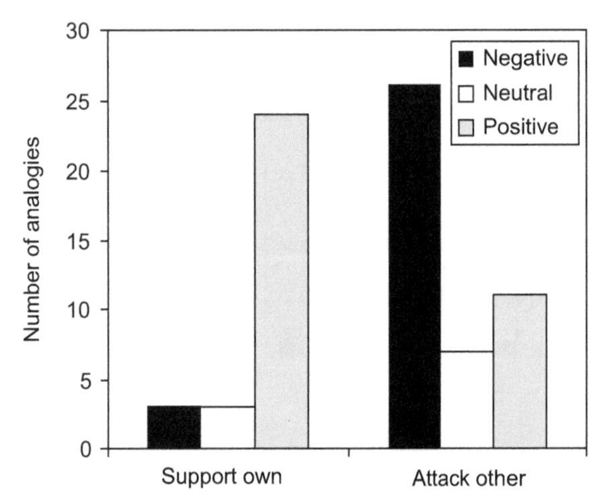

FIGURE 10.15 The emotional connotation of within- and other-domain analogies made in political domains. *From Blanchette, I., & Dunbar, K. (2001). Analogy use in naturalistic settings: The influence of audience, emotion, and goals.* Memory and Cognition, 29, *730–735.*

economics could be another interesting discipline in which we could break further ground on how analogies are used outside of the experimental psychology lab environment.

Economics struck us as a highly analogical discipline. One of the main occupations of economists is to explain and predict the monetary behavior of individuals in society. This involves almost constant reference to different situations that have occurred in the past and relating elements of past behavior to what is occurring in the present. Further, economists make heavy use of quantitative models to represent monetary situations. They regularly modify and update these models on the basis of incoming new information. All of these features fundamentally involve considering aspects of current situations in relation to similar situations that have occurred in the past. Behavioral economics focuses heavily on individual choice behavior. This subdiscipline of economics focuses on how people perceive value under different conditions, how those values can be represented overall, and on modeling values to enable predictions for new situations. Again, this work relies strongly on comparisons that operate at both superficial and structural levels. The reliance on equations and quantitative models in behavioral economics emphasizes the relational qualities of monetary and value-based behavior. We were also excited by the fact that behavioral economists consider social conditions, emotion, and sentiment when they model behavior. Lastly, behavioral economists are scientists who frequently conduct experiments with human subjects, but they are also highly attuned to political and public policy trends that impact and explain aspects of human behavior.

Our work was carried out in the tradition of Dunbar and Blanchette's (2001) previous work investigating *in vivo* analogical reasoning. Colleagues in the School of Economics and Public Policy at the University of Texas at Dallas were generous enough to allow us to observe and record the audio from five weekly reading group sessions that centered on discussing behavioral economics topics. The group consisted of approximately 20 individuals ranging in expertise from graduate level students (75% of the group) to faculty (approximately 25% of the group). Sessions lasted roughly 90 min yielding approximately 7 h of audio recording—a great deal of data to analyze. In order to maximize the chances of isolating as many analogical comparisons as possible, we had two experimenters, who were not trained in economics, find probable analogical comparisons with an attitude of "when in doubt, include it." These potential analogical comparisons were then rated by the experimenters to first determine if it was indeed an identifiable analogy. Coders then identified the source analog, the target analog, and determined the domains that each of these came from. Each analogy was then

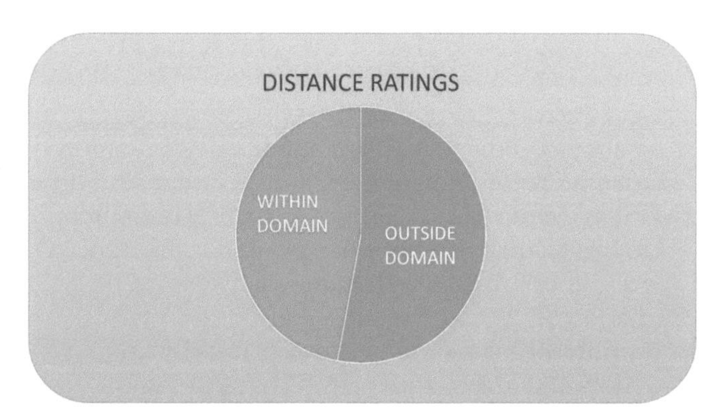

FIGURE 10.16 The ratings of analogical distance observed in the analogies of economists at meetings were roughly balanced with 53% of the source analogs occurring within the same domain as the target and 47% of the analogical sources occurring outside the target domain. *Figure based on results reported in Kretz, D. R., & Krawczyk, D. C. (2014). Expert analogy use in a naturalistic setting.* Frontiers in Psychology, *5, 1333.*

coded along the three key dimensions: distance, depth, and purpose.

The results of Kretz and Krawczyk (2014) indicated a rich array of analogical comparisons that occurred within the discussions of the behavioral economists. Ninety-seven analogies were discovered with the discourse data from the sessions. A majority of these analogies were determined to be deep showing 78% structural similarity over superficial similarity. The distance ratings were roughly balanced with 53% of the source analogs occurring within the same domain as the target, while the remaining 47% of the analogical sources occurred outside the target domain (Fig. 10.16). The purpose of the analogies varied considerably with some purposes being more common than others. The most common analogy observed (37%) involved the speaker providing an example to the audience. The second most common use of analogy was to allow the audience to visualize a concept (25% of analogies fell into this category). Twelve percent of the analogies were exaggerations, and an additional 12% were cases of inclusion. Differentiation, abstraction, and critiques rounded out the analogies gathered and were less common. Lastly, we found a relationship between purpose and distance. Twenty percent of the analogies that served to provide examples had sources and targets that came from the same domain. By contrast, when a visual analogical correspondence was drawn, 27% of the time the source came from a different domain as the target. This finding suggests that the distance of the analogies may not be a pure measure of creativity, but rather varies based on the purpose of the analogy. This finding is consistent with those reported by Dunbar (2000) and Blanchette and Dunbar (2001) with all three of these studies reporting that deeper analogies occur in instances in which a speaker makes an analogy in order to inform, provide an example, or to explain an unfamiliar concept to an audience.

To summarize, the domain of behavioral economics proved to be a rich source of diverse analogies. Most of these came as a speaker attempted to relate a current, novel concept to a concept that was more familiar and better understood. Speakers also regularly used analogies to aid other participants in understanding a concept by presenting a vivid mental image of that idea. As in the Dunbar (2000) study, our participants in the economics study were experts in their field. As experts, there were numerous analogies that would be considered within domain, as an economist referred a particular situation to a prior well-established model or economic game that was known to the group. When speaking to a more naïve audience, we would anticipate that the results would more closely resemble the distant, out-of-domain analogies reported in political speeches and articles by Blanchette and Dunbar (2001).

When we began that study we had hoped that we might get a sense of the audiences reaction to analogical comparisons based on further discussion, but this was not the case. Rather, most analogies occurred within the flow of conversation. The speaker often appeared to generate a source analog on the fly, mention it in conversation, and then continue speaking. Some analogies were merely for entertainment value or to make the discussion more memorable and interesting. This captures the essence of many analogies. The source domain appears to pop into memory without obvious buildup. The analogy may be articulated and then quickly dropped as the reasoner goes on to other aspects of a topic. People frequently mention only partial analogies without articulating a clear source, or a clear target. Sometimes these are just assumed to be known to the audience. Lastly, people did not mention highly fleshed out or elaborate system mappings. Even in the most elegant and distant analogies, we observed only partial reports of some of the corresponding features with many details left out for the audience to infer. Many of these tendencies call to mind the earlier section of this chapter on insight problem solving and long-distant analogies. Source analogs can be retrieved quickly on the fly, and they appear to be retrieved from no one specific identifiable memory cue. Analogies are indeed highly creative and are shaped dramatically on purpose by the reasoner, for their intended audience (Fig. 10.17).

SUMMARY

Analogical reasoning is common in everyday life. Analogies can serve a variety of functions. They may provide us with an opportunity to form new inductive inferences about a novel situation. Such is the case in scientific studies. Analogies can also be used to help, convince, clarify, or to inform us. We frequently use

FIGURE 10.17 Analogies can shape the way we think, solve problems, and see the world around us.

analogical correspondences to make speech more memorable, colorful, and interesting. Furthermore, analogies come in many forms and regularly rise to the level of making national news or being included in political speeches.

Analogies represent a category of similar items that is abstract, as they are based on relations among items that need not be visible perceptually. When we see objects as similar, this can be based on lower-order features, such as having similar visual properties, colors, or shapes. Situations that occur over time may be seen as similar on the basis of the relationships among items, objects etc. that share common relational properties. This makes analogies a highly flexible form of thinking and a basis for some of our strongest inferences and complex thoughts. The more remote the distance becomes between source and target analogs, the more difficulty we tend to have retrieving an appropriate analog.

Lastly, the purpose of an analogy will often be a determining factor about its properties. Structural correspondences come in many forms, but we need a reason to draw an analogy. Sometimes this will lead to a calculated thought experiment, such as when we consider the possible similarities between international wars that occurred over different time periods. Other times analogies will simply arrive in our minds, perhaps due to remote analogical reminding, which is highly similar to insight in which we experience a revelation with an accompanying "aha moment." Analogies have many purposes, and these may be linked to the distance and depth characteristics of these comparisons.

END-OF-CHAPTER THOUGHT QUESTIONS

1. Analogies allow us to learn about a new situation based on prior knowledge about a similar situation. This is classified as a form of inductive reasoning. How is analogical reasoning similar or different from other forms of induction we discussed in Chapter 9?
2. Analogical reasoning is based primarily on structural correspondences. What are some comparisons in news stories that correspond on structural grounds?
3. For most analogical comparisons to be successful there are constraints that apply. What are the main constraints on analogical reasoning that we discussed in this chapter?
4. Superficial similarity is based on perceptual features or properties that are shared between two situations that may be placed into correspondence in an analogy. Can you think of some cases in which superficial similarity is more informative than structural similarity?
5. Analogies have alignable properties that can be placed into correspondence. They also have non-alignable properties that do not allow correspondences. Think of some examples of alignable and non-alignable properties.
6. Source analogs can be difficult to retrieve. What factors make it more difficult for someone to retrieve an appropriate source analog?
7. Analogies in daily life come in many forms. Do these tend to be long-distance creative analogies or simpler ones?
8. Insight problem solving may yield creative analogies through coarsely coded information. How does the brain appear to support this type of reasoning?
9. How might expertise influence the types of analogies that someone forms in his or her specialty area?

References

Artificial best explanation of the Higgs boson?. (July 4, 2012). Retrieved from http://www.bbc.com/news/science-environment-18707698.

Beeman, M., & Chiarello, C. (Eds.). (1998). *Right hemisphere language comprehension: Perspectives from cognitive neuroscience*. Psychology Press.

Blanchette, I., & Dunbar, K. (2001). Analogy use in naturalistic settings: The influence of audience, emotion, and goals. *Memory & Cognition*, *29*, 730–735.

Datta, A., Bansal, V., Diaz, J., Patel, J., Reato, D., & Bikson, M. (2009). Gyri-precise head model of transcranial direct current stimulation: Improved spatial focality using a ring electrode versus conventional rectangular pad. *Brain Stimulation*, *2*, 201–207.

Dunbar, K. (2000). How scientists think in the real world: Implications for science education. *Journal of Applied Developmental Psychology*, *21*, 49–58.

Dunbar, K., & Blanchette, I. (2001). The in vivo/in vitro approach to cognition: The case of analogy. *Trends in Cognitive Science, 5*, 334–339.

Duncker, K. (1945). On problem solving. *Psychological monographs* (Vol. 58). American Psychological Association.

Fregni, A., & Pascual-Leone, A. (2007). Brain stimulation in poststroke rehabilitation. *Cerebrovascular Disease, 24*, 157–166.

Gallo, R. C. (1993). *Virus hunting: AIDS, cancer, and the human retrovirus: A story of scientific discovery*. Basic Books.

Gentner, D. (1983). Structure-mapping: A theoretical framework for analogy. *Cognitive Science, 7*, 155–170.

Gentner, D. (1988). Metaphor as structure mapping: The relational shift. *Child Development, 59*, 47–59.

Gentner, D., & Landers, R. (1985). Analogical reminding: A good match is hard to find. In *Proceedings of the international Conference on Cybernetics and Society* (pp. 607–613). New York, NY: Institute of Electrical and Electronics Engineers.

Gentner, D., & Markman, A. B. (1997). Structure mapping in analogy and similarity. *American Psychologist, 52*, 45–56.

Gentner, D., & Rattermann, M. J. (1998). Deep thinking in children: The case for knowledge change in analogical development. *Behavioral and Brain Sciences, 21*, 837–838.

Gick, M. L., & Holyoak, K. J. (1980). Analogical problem solving. *Cognitive Psychology, 12*, 306–355.

Goldstone, R. L., Medin, D. L., & Gentner, D. (1991). Relational similarity and the non-independence of features in similarity judgments. *Cognitive Psychology, 23*, 222–264.

Green, A. E. (2017). Creativity, within reason: Semantic distance and dynamic state creativity in relational thinking and reasoning. *Current Directions in Psychological Science, 25*, 28–35.

Green, A. E., Fugelsang, J. A., Kraemer, D. J., Shamosh, N. A., & Dunbar, K. N., (2006). Frontopolar cortex mediates abstract integration in analogy. *Brain Research, 1096*, 125–137.

Green, A. E., Speigel, K. A., Giangrande, E. J., Weinberger, A. B., Gallagher, N. M., & Turkeltaub, P. E. (2017). Thinking cap plus thinking zap: tDCS of frontopolar cortex improves creative analogical reasoning and facilitates conscious augmentation of state creativity in verb generation. *Cerebral Cortex, 27*(4), 2628–2639. http://dx.doi.org/10.1093/cercor/bhw080.

Jung-Beeman, M. (2005). Bilateral brain processes for comprehending natural language. *Trends in Cognitive Sciences, 9*(11), 512–518.

Kounios, J., & Beeman, M. (2009). The Aha! moment: The cognitive neuroscience of insight. *Current Directions in Psychological Science, 18*(4), 210–216.

Krawczyk, D. C., Holyoak, K. J., & Hummel, J. E. (2004). Structural constraints and object similarity in analogical mapping and inference. *Thinking & Reasoning, 10*, 85–104.

Krawczyk, D. C., Holyoak, K. J., & Hummel, J. E. (2005). The one-to-one constraint in analogical mapping and inference. *Cognitive Science, 29*, 29–38.

Kretz, D. R., & Krawczyk, D. C. (2014). Expert analogy use in a naturalistic setting. *Frontiers in Psychology, 5*, 1333.

Mednick, S. A., & Mednick, M. T. (1967). *Examiner's manual: Remote associates test*. Boston: Houghton Mifflin.

Philip, N. S., Nelson, B. G., Frohlich, F., Lim, K. O., Widge, A. S., & Carpenter, L. L. (2017). Low-intensity transcranial current stimulation in psychiatry. *American Journal of Psychiatry*. http://dx.doi.org/10.1176/appi.ajp.2017.16090996.

Richland, L. E., Morrison, R. G., & Holyoak, K. J. (2006). Children's development of analogical reasoning: Insights from scene analogy problems. *Journal of Experimental Child Psychology, 94*, 249–271.

Slovic, P., & MacPhillamy, D. (1974). Dimensional commensurability and cue utilization in comparative judgment. *Organizational Behavior and Human Performance, 11*, 172–194.

Smirnova, A., Zorina, Z., Obozova, T., & Wasserman, E. (2015). Crows spontaneously exhibit analogical reasoning. *Current Biology, 25*, 256–260.

Spellman, B. A., & Holyoak, K. J. (1992). If Saddam is Hitler then who is George Bush?: Analogical mapping between systems of social roles. *Journal of Personality and Social Psychology, 62*, 913–933.

Wharton, C. M., Holyoak, K. J., & Lange, T. E. (1996). Remote analogical reminding. *Memory & Cognition, 24*, 629–643.

Yang, F. G., Edens, J., Simpson, C., & Krawczyk, D. C. (2009). Differences in task demands influence the hemispheric lateralization and neural correlates of metaphor. *Brain and Language, 111*(2), 114–124.

Further Reading

Brown, T. E., LeMay, H. E., & Bursten, B. E. (1994). *Chemistry: The Central Science* (6th ed.). Englewood Cliffs, NJ: Prentice-Hall.

CHAPTER

11

Decision Making and Abductive Reasoning

KEY THEMES

- Decision making was originally a specialty area of philosophy. The study of decision making has evolved to include fields such as psychology, economics, and medicine.
- Decision making is related to reasoning and together these are often called "higher cognitive processes." Many reasoning tasks are judged on the basis of an end decision and many decision-making tasks require the use of reasoning skills.
- Decision making can involve simple or complex decisions. Multi-attribute decisions are those that involve multiple aspects that are present in the options available to someone.
- Many simple decisions occur with limited conscious thought. Electrophysiology research indicates that we are not always consciously aware of a decision we make prior to the motor planning activity that precedes the decision. Such studies call into question our capacity for free will.
- Rational models of decision making maintain that someone should always choose according to his or her own best interests. Such models also indicate that people should choose consistently over time and under similar circumstances.
- Heuristics are rules of thumb that enable us to quickly and efficiently make decisions that work most of the time. Such heuristics include deciding in favor of representative outcomes, deciding based on readily available information, and adjusting from an initial fixed value point.

Reasoning
http://dx.doi.org/10.1016/B978-0-12-809285-9.00011-9

- Dual-systems have been proposed to support human cognition. According to this viewpoint, fast decisions arise from an automatic system, while slower, deliberative decisions involve a controlled processing system.
- Abduction occurs when we reason to the best possible explanation. We typically do not have access to all of the facts when we reason abductively. Abductive inferences are not guaranteed to be valid.
- The ventromedial prefrontal cortex is activated in neuroimaging experiments when we process value. Damage to this area impairs people's decision-making ability often leading them to make risky decisions.

OUTLINE

HOW DO WE DECIDE?

Introduction

Decision making is one of the fascinating intellectual problems that people wish to solve. Core cognitive questions involve decision making. How do we make decisions? Why did we choose the one option over another? Why do people choose differently than one another? Our decisions shape us, determine what we will become, determine what opportunities we take advantage of, contribute to whom we befriend, and ultimately influence how we spend our lives (Fig. 11.1).

There has been much interest from the scholarly community in understanding decision making and this topic has a long intellectual history. Ancient Greek philosophers puzzled over questions of free will and determinism. Early psychologists sought to understand the basis for how and why we make the decisions we do. Much of the psychological research focuses on understanding choices that are made by people and other organisms. Frequently the experimenters who conduct psychological studies seek to carefully control the content or inputs to a decision and then examine the outputs. Neuroscientists have also increasingly taken to analyzing decisions and their neural correlates. If we can understand how decisions are made within the brain, we could theoretically predict decisions that people will make in the future. Relevant questions in decision neuroscience include: how do neural systems compute and predict the value of favored options? And how does the brain represent those options that are discarded? Economists and researchers in business schools are increasingly interested in modeling and predicting how people make financial choices and consumer decisions. A major focus of decision research has centered on evaluating how people represent value, as values are central to determining the decisions that we make.

FIGURE 11.1 Our daily life decisions determine many important aspects of our lives. This process is central to both cognition and our social lives. *From Shutterstock.com.*

Decisions vary in complexity. Decisions can be simple. These decisions include what type of dressing we would like on our salad today? Should we wear grey pants or black pants today? Should I turn off the car, or let it idle while I check my phone? Decision making also includes sorting among some of the most complex situations that we face. Where should I invest our family's money? What type of person should I marry? What is the appropriate course of treatment for a rare and poorly understood illness? Complicated decisions of this sort are called *multi-attribute decisions*, as they involve numerous dimensions, or attributes, that likely vary in value among individuals. When those options are considered among other multi-attribute alternatives, the choice can hinge upon which specific attributes are most salient at the time the decision is made.

Decisions also have varying consequences. At first blush, simple decisions may initially appear to make less of a difference in our daily lives. Later on, some of these simple decisions may trigger chains of events that result in much larger consequences. If you choose to attend a breakfast event in college because you happened to be in the area and were hungry at that moment, you may chance upon meeting a future employer who then proves to be a key contact in furthering your career development. When doctors make decisions about prescribing a drug that impacts someone's health, they can change the course of that person's life for many months or years if unforeseen side effects occur. Supreme Court Justices and politicians make decisions that affect millions of people's lives and influence the future of our nation. We engage in decision making constantly, and it can be difficult to think of a cognitive mechanism, or part of the brain, that does not take part in shaping our decisions.

In this chapter we will review the state of research on human decision making. It is a multidisciplinary endeavor involving psychologists, neuroscientists, mathematicians, engineers, and economists. The science of decision making continues to grow and expand into many other diverse fields as well. We will begin by discussing how decision making is studied in the laboratory. We will next move on to a discussion of abduction, which is reasoning to the best possible option that we have available. The study of abduction is strongly linked to decision making, as abduction influences how we decide under very complex conditions, often when the stakes are very high. Lastly, we will focus on how the brain specifically contributes to decisions that are made in daily life.

THE SCIENCE OF DECISION MAKING

How Is Decision Making Related to Reasoning?

The past several chapters have been devoted to distinguishing different types of reasoning from one another and reviewing the evidence about where and when people engage in these particular types of reasoning. As a research lab exercise, or a course topic, decision making is often grouped together with reasoning under the general heading "higher cognition." This reflects the fact that decision making is complex. It can have important consequences, and complicated decision making is consciously accessible to people, as are many of the mental processes that govern reasoning. In many ways the distinction between what we choose to call "reasoning" and what we call "decision making" is an artificial one. We can appreciate this observation by considering how these two topics converge.

Many decisions are the products that are output from the process of reasoning. When a financial analyst concludes that the rising price of oil is being caused by a shortage at some of the offshore drilling locations, she will then make decisions about whether to buy or sell stock on the basis of that causal reasoning process. The actual decision to buy or sell can appear to be a secondary task that occurs after that line of reasoning has been established. The cognitive workload of financial decisions is reasoning about the perceived value and the actual value of a particular commodity. The decision to sell stock may merely represents that end-state of our thinking after causal reasoning and induction have been performed. The point of commitment to a particular decision may not really be the interesting feature of the mental processing in this situation. Alternatively, one could also consider the process of reaching the decision to buy or sell as being part of the act of decision making. Accordingly, decision making and reasoning are intimately intertwined such that decision making involves reasoning and reasoning is undertaken for the purpose of outputting a decision (Fig. 11.2). In terms of our descriptions either scenario can be considered to be correct much of the time.

Reasoning may make up much of the cognitive processing that needs to occur when we make a decision that takes time. In laboratory tasks that we have discussed throughout the earlier chapters, an individual's performance is typically measured by the decision that they reach and how long it takes them to reach it. Should we select a particular shaded square because it shares relational similarity with a sample shape? Should we select a conclusion stating that all birds have spleens because we have heard that certain other birds do? In Chapter 9 we discussed the numerous deductive reasoning possibilities that occur with the Wason card selection task (Wason, 1966) and it's real world versions. In that task people are asked to evaluate the statement "If a card has a vowel on one side then it has an even number on the other side." Participants are asked to select among four cards (displaying an "E," a "6," a "Z," and a "7") to determine which need to be turned over to invalidate the rule (refer to the section on the Wason card selection task in Ch. 9). The behavior in this deductive reasoning task is actually measured by the decisions made. Which cards to flip over? Which cards can be left alone, as they are not relevant to testing the validity of the rule provided? In those instances, the process of making the decision can be a deductive process, but applying the label "decision making" captures the fact that a majority of people are not effective at deductive reasoning to

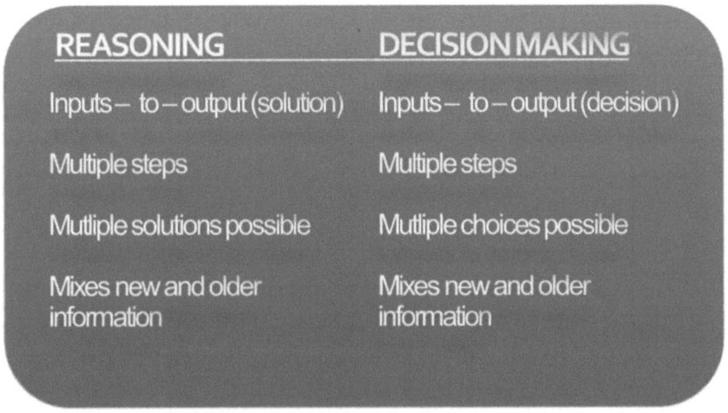

REASONING	DECISION MAKING
Inputs – to – output (solution)	Inputs – to – output (decision)
Multiple steps	Multiple steps
Mutliple solutions possible	Mutliple choices possible
Mixes new and older information	Mixes new and older information

FIGURE 11.2 The mental processes involved in reasoning and decision making are intimately intertwined. It is difficult to pinpoint where the process reasoning comes into play when decisions are complex.

FIGURE 11.3 The Wason card selection task, a well-known deductive reasoning test, is measured by the decisions that people reach about which cards to flip over in order to test a rule. Is this a reasoning task or a test of decision making? The task combines both processes.

find the exception to the rule, which is the card displaying the "7" in the standard Wason card selection task. Without the decision, discussion of these card selections cannot occur (Fig. 11.3).

You may recall that in Chapter 9 we also discussed the drinking age problem variation on the Wason card selection task (Griggs & Cox, 1982). In this variation, participants are asked to make sure that people are conforming to the rule: "If a person is drinking a beer, then the person must be over 19 years old." Under these conditions, participants are much more accurate at spotting the rule violation and frequently decide to turn over the card displaying someone "16 years of age." This card is equivalent to the much more difficult card displaying the "7" in the standard version of this task. Indeed, people completing these two variations of the Wason card selection task may simply decide that they need to check the 16 year old, but the card displaying the "7" just feels intuitively wrong, so they do not choose it. These participants may actually be carrying out decision making, not deductive reasoning. The fact that people are so much better at the drinking age problem variation demonstrates that they can reason deductively and will do so provided that the appropriate schema about the situation is retrieved. With the abstract materials, people apparently fail to engage in a type of deductive reasoning that they are clearly capable of under the more familiar conditions. Is the Wason card selection task a deduction task or a decision task? Most would probably argue that it is a deduction task, but a strong case can be made that task performance is defined by the type of cognitive processing that someone engages in. If someone engages in deductive reasoning they should perform the task correctly noting the two correct contingencies (affirming the antecedent and denying the consequent). If that same individual simply makes an intuitive decision, based on his best estimation at the time, that person often fails to notice denying the consequent card and therefore misses the correct card pair.

Applied situations in life also require decision making even when those situations appear to fall into the category of what we would consider to be reasoning problems. Should we treat these particular symptoms with dopaminergic drugs because the likely cause of the tremor is Parkinson disease? Should we decide to invade Iraq based on analogical reasoning that aligns Iraq with World War II-era Germany? In these last examples we can take note that medical personnel, judicial systems, and governments must regularly commit to decisions that impact people's lives in dramatic ways. They may make those decisions after a careful reasoning process that is informed by causal reasoning, analogical reasoning, or other inductive reasoning or deductive reasoning. In practice, people may use whatever cognitive tools are needed in order to arrive at the best possible choice under the circumstances. In this way, we may ultimately call the whole set of cognitive processes in the situations above to be acts of "decision making." Judges decide in favor of a particular side in a dispute. Governments commit to one course of action after a process of reasoning. Ultimately, these two concepts are linked much of the time and the term that the research community applies has more to do with the background of the study than an exclusive focus on a single cognitive process.

Simple decision making may represent one exception to the fusion of reasoning and decision making. When a fan at a baseball game ducks out of the way of a rapidly approaching foul ball, she has "decided" to move. This is an example of a simple decision. It lacked planning and the fan likely cannot account for exactly when or how she decided to move. She will likely report that "It just happened…I ducked." Such a case resembles a reflex. Indeed some scholars have applied the term "reflexive" thinking to those decisions that occur without much prior deliberation or reasoning (Satpute & Lieberman, 2006). Typically, a reflexive decision is one that is arrived at within a rapid period of time. Reflexive decisions are often simple, as they lack complexity in the form of numerous attributes, or numerous alternatives. Reflexive decisions do not involve heavy amounts of conscious thought, as the label implies. Rather, reflexive decisions may appear to come to mind effortlessly. Sometimes a complex decision can be made reflexively, but such decisions may prove to be illogical or inappropriate for the circumstances.

Reflexive decisions can apply to many of the behaviors exhibited by other species. When we discussed the behavior of an octopus in Chapter 4, we noted that it could open jars, change color, and move in response to the presence of a person. Are these choices? Did the octopus *decide* to carry out these behaviors? These instances may also be called reflexive. Bret Grasse, an octopus expert at the Marine Biological Laboratory in Woods Hole, characterizes many of the apparently complicated behaviors of the octopus as being stereotypical behaviors that are predictable in their occurrence. Grasse proposed that human handlers ascribe higher thinking and reasoning processes to these behaviors based upon an analogy to human behavior. We might term these actions to be automatic, predictable, or reflexive. Perhaps the best mark of a reflexive decision may be one that is made rapidly with the involvement of fewer executive functions including planning, working memory, and attention.

Academic Fields That Study Decision Making

Several disciplines have become involved in decision research because decision making is relevant in so many areas. The roots of studying decision making can be traced back to the discipline of philosophy. Indeed philosophy is still a highly active field of study for decision research (Kagan, 1998). Over the course of philosophical inquiry the topics of thinking, reasoning, and decision making extended outward to include numerous studies of judgment and decision making as they are carried out by people either in laboratories or in the real world.

Experimental psychologists became interested in how people make decisions and decision inquiry became a vibrant and important area of the field by the late 20th century. Again, decision making is difficult to define and shows up repeatedly under different research area labels. It can be accurately stated that researchers studying reward seeking behavior in animals are studying decision making. Frequently animal studies in this area have involved training a rat or monkey to seek out a reward in a maze or make a button response. In the case of a rat in a maze, the dependent variable is often which route the animal chooses. Go right or go left. Because researchers have carefully controlled the animal's prior history with the maze, the probabilities that they have experienced do not change the fact that they are fundamentally asking the animal to make a decision. We can also consider the animal behavior studies of relational matching discussed in Chapter 4 to be decision-making tasks. In those studies, a monkey or pigeon can readily decide to select a red triangle target rather than a blue square target, in response to a red triangle sample. Likewise, when the hooded crow is asked to decide between food wells covered by either a picture of two circles or a picture of a square and a rhombus in response to a sample depicting two rectangles, the animal will decide in accordance with the relational properties of the task (Smirnova, Zorina, Obozova, & Wasserman, 2015).

Social psychology emerged as a vibrant discipline for the study of decision making in the middle of the 1900s. Robert Zajoc is now famous for his research on the *mere-exposure effect*, though the basic phenomenon had been reported in the late 1800s (Fechner, 1876; James, 1890). In these studies, Zajonc (1968) would expose people to a particular stimulus, such as a tone. He would then observe that people would later prefer that same tone to a novel one. Similar shifts of preference toward a particular stimulus have been observed toward stimuli based on nothing more than a very brief prior exposure. The basic phenomenon is that prior exposure to stimuli results in more positive evaluations of those stimuli at a later presentation, compared with less frequently presented stimuli. Several studies indicate that the mere-exposure effect may induce changes that involve little cognitive effort, as the effect can be obtained with nonhuman animals (Taylor & Sluckin, 1964), and can even be obtained with subliminal exposure times to the sample stimulus, which are not able to be detected by most people (Kunst-Wilson & Zajonc, 1980). Such studies suggest that preferences can be altered even without accurate recognition memory of the stimuli suggesting that emotional or other reflexive processes are at least partially responsible for the effect.

Another famous line of research in the experimental psychology of decision making also comes from the social psychology domain and involves a phenomenon that is known as *cognitive dissonance*. Leon Festinger's (1957) cognitive dissonance theory included the proposal that preferences toward a particular idea occur as we set about reconciling dissonant elements of cognition. More formally, Festinger claimed that when two cognitive elements are determined to be related, they can either be consonant or dissonant to one another. If two ideas are dissonant, or at odds with one another, then the importance of each opinion will determine the amount of dissonance felt by the person, as a sort of cognitive discomfort level. Thus, important beliefs that are dissonant to one another cause a need to alter either behavior, or the belief itself. The alteration of the environment is one way of reducing cognitive dissonance; however, under conditions when behavior or the environment cannot be changed, the belief itself is likely to be changed to fit with the discrepant environment, behavior, or context.*cognitive dissonance*

Within the context of decision making, Festinger (1957) described multiple ways that dissonance can be reduced when a difficult choice is made. First, a person can change the decision, altering the overall attractiveness or importance of the attributes of different options by increasing them for a chosen option and also decreasing attractiveness or importance assessments for a nonchosen option. A second route to dissonance reduction is to increase the overlap of the two alternatives. This can be achieved by reclassifying them such that they are more similar to one another. This process also has the effect of allowing people to sit well with having made a particular choice, as this rationalization leads to the conclusion that either option will result in a similar outcome.

Cognitive dissonance theory accounts for several attitude adjustment results in decision making. For example, let's consider Brehm's (1956) classic *free choice paradigm*. Participants in Brehm's task provide ratings of desirability toward possible gift items and then are allowed to choose one of the gifts that is theirs to keep. When Brehm gave participants a choice among highly similar items, the choice of one option would result in a high dissonance level. If all of the gifts are similar, it is difficult to make a clear decision. Those people selecting among the equivalent options showed a greater degree of preference shift in favor of their chosen option compared to people set up with a low dissonance choice in which some items were clearly superior to the others. A selection of the high quality gift in that case yields low dissonance. Similarly, Leon Festinger (1962) found that people gave moderate preference ratings to sets of music albums, but those same albums were rated more highly after participants had chosen them. This same increased value after one has made a choice applies to bigger items such as car purchases. Ehrlich, Guttman, Schonbach, and Mills (1957) demonstrated that people who had recently bought a car were more likely to notice car ads about their new model over other alternatives. These findings demonstrate an upward shift of importance of the product that one has obtained after commitment to the choice.

Dissonance theory has been criticized for moving away from a study of decision making and instead focusing on how people rationalize situations in which they have been manipulated to perform unusual actions. Some investigators have claimed that there became an overemphasis of behavior in dissonance studies and a lack of attention toward cognition (Abelson, 1983; Rosenberg, 1968). Others have noted that the approach changed over time to become a theory of inconsistency (Simon & Holyoak, 2002). There had also been relatively little work addressing the specific processes by which people reduce dissonance. Daryl Bem (1967) proposed a self-perception alternative to cognitive dissonance, in which once a decision or behavior has taken place, people infer that their attitude or belief had supported it, rather than work toward revising their assessments.

The movement toward cognitive dissonance as an explanation for decision making has reduced over the years. Festinger's theory has been tested by studies of arousal accompanying cognitive dissonance in the post-decision or post-behavior phase of task. Zanna and Cooper's (1974) counterattitudinal essay study asked participants to write an essay in favor of tuition increases, a position that would have hurt these students financially. Writing an essay countering one's genuine attitudes should create dissonance and frequently people adjust their preferences after such an action, claiming that they do support tuition increases, thereby maintaining consistent behavior in line with beliefs. In this particular study a placebo drug was included to test for the influence of arousal that would accompany dissonance related to the misalignment between preferences and behavior. In this study participants were instructed that they would receive a drug, but that would have no physiological effect. This group showed an adjustment of beliefs related to their behavior as cognitive dissonance theory predicts. By contrast, a separate participant group was told to expect a tense reaction from the drug. These participants failed to show differing reports of belief. The authors suggested that the dissonance generated by their writing of the essay against the beliefs that was felt by this second group could be explained away as being the result of the drug; thus, they would not be motivated to alter opinion. Other cognitive dissonance studies that had employed physiological recording methods, used after a preference rating, supported the idea that increased arousal levels occur during the dissonance reduction period (Elliott & Divine, 1994).

Cognitive psychologists have also taken a strong interest in decision making. Many of the studies emerging from this discipline have evaluated and identified conditions in which people make irrational decisions. Cognitive psychologists have also focused on cases in which people make decisions that are inconsistent with prior behavior. This work is similar to the studies by Leon Festinger and other social psychologists. Decision-making studies emerging from the cognitive psychology area have also focused on the comparison between economic utility theory and the behavior of people in controlled situations. Economic utility theory emphasizes situations when people's behavior is predictable and consistent over time. At times the behavior of the individual differs from one person to the next and from one situation to another. Daniel Kahneman and Amos Tversky, two psychologists who became very well-known across multiple disciplines for their work on heuristics and biases, studied this type of inconsistent behavior extensively. Another well-known figure in the decision sciences from cognitive psychology is Herbert Simon. Simon famously noted that people experienced a state of bounded rationality, meaning that the limits of human perception, memory, and processing ability limits the extent to which they can operate rationally. In other words the processing limitations of the human brain place limits on our ability to make decisions consistent with a rational model across all circumstances.

Neuroscience has made increasing strides toward understanding how decisions are carried out within the brain. The brain basis of decision making is a complex endeavor, as the factors governing how and when we decide are numerous and variable, while the brain is highly complex as well. Neuroscientists have focused

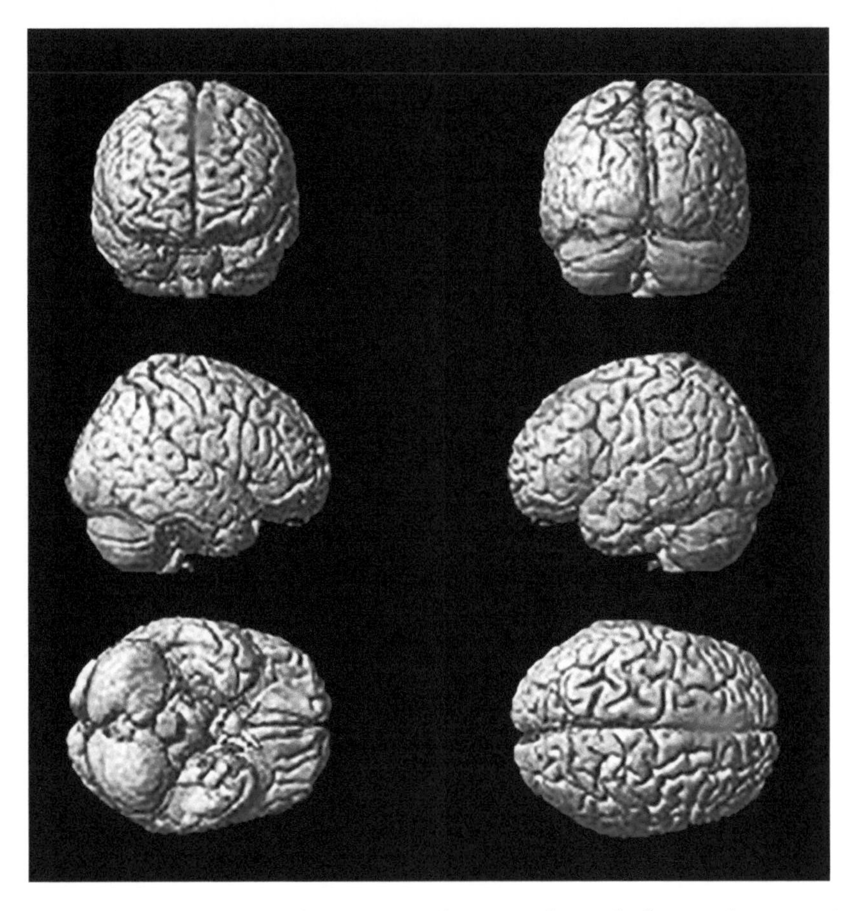

FIGURE 11.4 Functional MRI has become an increasingly important technique used to study the mental processes involved in decision making.

on multiple levels of decision making as a result of this complexity. Some neuroscientists have studied behavior in animals, noting how a particular species makes decisions to navigate their environments (Kepecs, Uchida, Zariwala, & Mainen, 2008). At the other end of the spectrum, functional magnetic resonance imaging (fMRI) has enabled the research community to investigate humans as they make multi-attribute decisions (Zysset et al., 2006) and make complicated social decisions regarding when to trust others to achieve financial goals (Li, Xiao, Houser, & Montague, 2009). Increasingly, the field of decision making uses neuroscience tools for investigations with fMRI having a large effect on the field (Fig. 11.4). Many researchers also use computational modeling to simulate decision processes and then test these predictions using neuroimaging methods.

The medical field may be one of the most critical applied areas for decision making. When doctors decide that they have evidence for a particular condition, the symptoms of the condition and the prior history of treating people with that condition will guide their treatment decisions. When doctors decide on a treatment, they must consult with the patient, as well as with their colleagues, and possibly with family members of the patient in order to make the best decision for that individual. As with many aspects of medicine, the stakes are enormous for the patient. A misdiagnosis can be catastrophic for someone's health. Deciding on the cause for a particular patient remains a critical feature of the job. In some specialties, such a radiology, deciding on a diagnosis makes up the core of the occupation. In accordance with the weight placed on making medical decisions, the field has been actively engaged in trying to determine the best practices for medical decision making over many decades (Gremy, 1980; Weiss, Berman, Howe, & Fleming, 2012).

SIMPLE AND COMPLEX DECISIONS

Free Will

One of the fascinating questions about decision making is the degree to which we are the authors of our own actions. This is similar to some of the causal and correlational studies we discussed in Chapter 8. Our consciousness tells us that the answer to this question is that "yes, of course we have freedom to decide our own fate." When we decide something, it feels as if we have made our conclusion and that we have a sound basis for why we made that choice. There are several challenges with this possibility however. We have probably all experienced situations in which we acted impulsively and wish we could take back what we had just said or did. When we are asked why we did some action or said something that offended someone else, we may really have no good idea why we did what we did. It just happened.

One of the famous challenges to the position of free will comes from work by Libet, Gleason, Wright, and Pearl (1983). These researchers investigated how the timing of the conscious intention to act relates to cerebral activity linked to the decision to act. The idea traces back to the fact that movement is preceded by motor cortical activity that innervates the limbs. Those motor cortical signals in turn have received input from other regions within the cortex that generate planning signals for movement. Those brain regions that are involved in integrating information from the sensory cortex appear to bias the action planning regions toward sending messages related to executing particular movements. At some point we become conscious that we planned to act, or that we made the decision to select one option over another.

Libet and colleagues recorded an EEG negative potential that has been called the "readiness potential," a neural marker associated with voluntary movements. The readiness potential was reported years earlier by Kornuber and Deecke (1965). Libet and colleagues allowed participants to make voluntary motor movements of their own free volition. They presented participants with a revolving dot, which served as a sort of clock allowing people to estimate their subjective experience of a time when they felt that they wanted to initiate a movement. The fascinating result was that the readiness potential was recorded between 150 and 800 ms before the subjective report of the participant's desire to move. This finding was replicated with five individuals across six experimental sessions. Furthermore, participants were asked about the position of the revolving dot when they felt the desire to initiate a movement. This question also confirmed that the readiness potential had occurred before the participants could report a feeling of wanting to move. In turn, the feeling had preceded the subjective feeling of action. Participants were not instructed to move in response to any particular prompt, so the results cannot be explained by a form of anticipation that an instruction was about to occur. Rather, participants reported that the movements were self-initiated, spontaneous, and capricious. The person just simply felt he wanted to move, but his movement was associated with an EEG readiness potential that came first.

The timing differences between the cortical motor readiness signal and the consciously subjective feeling of wanting to move were considerable in Libet's study. In EEG analyses differences above 400ms are considered to be quite large. Neuronal conduction speeds are in the order of milliseconds and signals can be relayed across cortical territory on the order of tens of milliseconds. Libet and colleagues' findings suggest that our subjective will that moves us to action can be predicted by unconscious signals within our brain that occur before our subjective sense.

Extending this "neural pre-destiny" to more complex decisions that we make has rather startling implications. If we feel we are ready to move and therefore do so, it may be that our brains instructed our conscious mind that it was time to perform this action. When we decide what we'd like to order off of a menu, we feel as if we are contemplating the options, considering new information, weighing the price, and mulling over our particular knowledge of the options. But do we really have this ability? Might it alternatively be possible that we are pre-wired to order a particular dish based on our tastes and knowledge at a given time? Could we possibly derive a formula that would predict what we will order? There is not a simple answer to this question. The real issue is not whether *we* make decisions, as we clearly do. The relevant question is *when does conscious processing enter into the decision*? This question forces us to consider the possibility that much of decision making can be automatic, consuming little of our attention, or working memory resources. This probably sounds much less controversial already. We have limited access to much of our neural processing. This is clear when we think about perception, a domain in which we effortlessly allow vast amounts of information in the environment pass by us without paying it much notice. The same applies to actions. We can act without a lot of prior thought. Similarly, we can decide without devoting much conscious processing to the choice. We will revisit this topic in the upcoming section covering dual-systems thinking in decision making.

Multi-Attribute Decision Making

One of the striking features of human decision making is that we can imagine the implications of our choices into the future and can evaluate complicated questions. Deciding among options that have numerous properties or attributes is called *multi-attribute decision making*. Deciding among complicated alternatives requires that we value some dimensions of the decision over others. For example, when you decided what college to attend you might have considered several factors to be most relevant to your decision. Such factors include the size of the campus, the climate of its location, the proximity to a city, the look of the buildings, and the reputation of the degree programs. You could also consider the number of students, the dining facilities, and the athletics programs. These kinds of decisions become difficult, as the focus on each dimension is likely to result in complex combinations mixing advantages with disadvantages. These decisions appear to be quite a long distance away from deductive reasoning, in which we can assure ourselves of a valid answer. Multi-attribute decisions also reflect reality in that there frequently are not clear right answers, rather we seek to find an acceptable solution that gets the job done, acknowledging that we might have decided differently and that would have been possibly better or worse than the way things actually turned out.

The key to evaluating multi-attribute decisions lies in how we quantify the options. All decision options can be said to have a value, which refers to the quality of the option. In the case of multi-attribute options, the overall value of the option is a complicated combination of all of the attribute values combined. In the case of the college decision, you may value the small size of one particular campus, so the college has a high value on that dimension. You might dislike the climate of that particular college, meaning that the value is low for the climate dimension of the option. Thus, overall we can be enthusiastic about some attributes, lukewarm on others, and emphatically against other attributes (Fig. 11.5). How then should we decide?

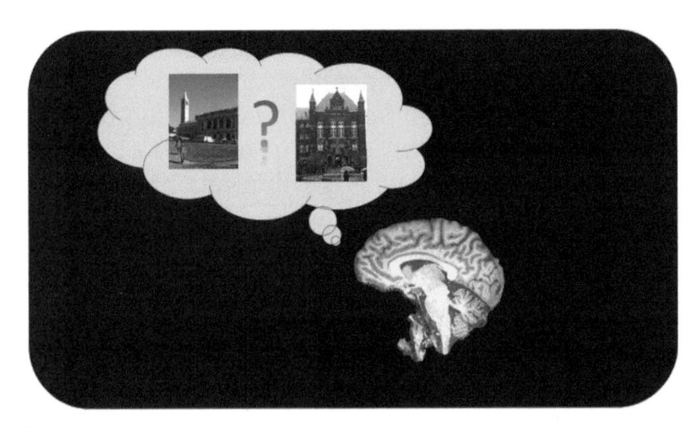

FIGURE 11.5 When we make complex decisions that involve choices with many attributes, we must balance our value on any one of these attributes against that of the others associated with the overall option. It can be difficult to arrive at objectively correct decisions in such cases.

Another important aspect of quantifying an option involves setting the *weights* on particular attribute categories. In addition to values, weights help to determine which values rise to the level of importance necessary to swing a decision one way or another. The weight represents how important a particular dimension is in our decision. For example, I may be very concerned about facing snow and blustery winds as I walk to classes. I have always preferred the idea of sunny weather and strolling around a campus, since I enjoy taking walks and seeing bright natural light from the windows. For me the climate dimension takes on a high level of importance. My lower value setting for a school in a cold climate is coupled with a very high weight on that dimension. Therefore the value of the overall option is dragged down. Meanwhile, I like a smaller campus due to its smaller class sizes, greater frequency of meeting people I know, and for its ease of transportation, since I can walk or bike across the campus quickly. I weight this dimension high as well. This high value/high weight attribute brings up the overall value on the small school option in the cold place. Then there is the issue of buildings. The modern building style at this college is not my preference and therefore holds a low value. I am not especially bothered by architecture and I am not too bothered about how the buildings look. Therefore despite my low value, architecture happens to have a low weight so this attribute does not influence my overall evaluation of this option very much.

The combinations of values and weights help researchers to design experiments to test multi-attribute decision scenarios, and these combinations also allow us to see if people decide consistently across time and how often they decide in favor of the option that stacks up as the higher value alternative (Fig. 11.6).

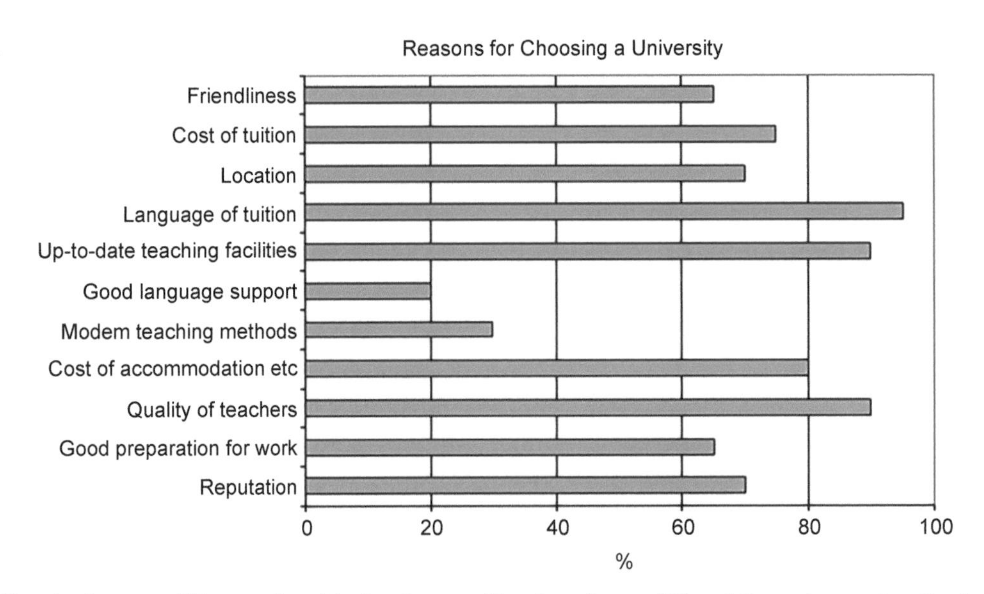

FIGURE 11.6 Choosing between different universities involves consideration of many different dimensions such as the size of the campus, the climate, and the proximity to a city. Decisions that require consideration of several dimensions are called multi-attribute decisions. Such decisions are made on the basis of the values that a given option has on those dimensions and how important each of these dimensions are. Importance can be quantified by the term "weight" that we assign to these dimensions.

People tend to be relatively happy with their decisions despite the complexity and logical inconsistencies that occur when alternatives have multiple attributes. This is somewhat surprising when you consider the fact that many multi-attribute decisions are close calls. We go through life pursuing some options and at the same time forgoing others. People do not tend to have buyers' remorse on a regular basis. Indeed we often make confident decisions despite the fact that the competing alternatives were just about as good on balance. This leads to the question of how rational we actually are in our decision making.

RATIONAL MODELS

Quantifying and Predicting Rational Behavior

The essence of rational decision making is that we go about our lives making the most optimal choices for any situation. The chosen option should always outweigh the alternatives, even if the margin of one choice over another is very small. Furthermore, rational decision makers should also be consistent over time and circumstances in their decisions. If you chose white bread over wheat in one circumstance, you should do so again given that other factors remain the same. Rational decision makers should always maximize their profits, act similarly across situations, and seek maximal gains. The difficulty with this perspective is that psychologically we have different needs at different times and our perception of a situation can be incomplete and even erroneous. As the psychologist Herbert Simon (1957) noted, we experience bounded rationality, in which our ability to be fully rational is bounded by the limitations of our perceptual and cognitive abilities.

Factors That Make Rational Behavior Difficult to Predict

There are several factors that lead people to make nonrational decisions. Such choices include instances in which we decide in favor of a suboptimal decision, or one that should be worse on balance relative to a better alternative. Nonrational decisions can also take the form of choosing one alternative under some circumstances, but another alternative under different conditions. This situation is reminiscent of the logical inconsistencies we discussed in Chapter 9. Contextual factors are one of the primary reasons that we decide irrationally. As we have discussed in previous chapters sometimes cognitive processing is biased by the way information has been presented to us. This may include what decisions we have recently faced, or even how a particular option is worded.

Decision framing is one of the most prominent factors that affects the choices we make. Amos Tversky and Daniel Kahneman demonstrated the framing effect in a very memorable experiment centered on participants solving the "Asian Disease Problem" (Tversky & Kahneman, 1981). Before we move on, take a few minutes to look at Box 11.1 and try to solve this problem in its two different forms.

The "Asian Disease Problem" strongly demonstrates that how a decision is framed can actually affect the options chosen. Remarkably most people when presented with these two alternatives will reverse their position on which program they would like to see in place to deal with the "Asian Disease." Notice that the first frame offering a choice between Program A and Program B presents the information in terms of the number of lives that are going to be saved. In this case, saving lives represents a gain. By contrast, the second frame, which offers the choice between Programs C and D, includes numbers on how many lives will be lost if each program is adopted. This represents framing in terms of losses. Losses are often viewed as a threat and people tend to become more emotional about wanting to avoid the possibility of a loss.

Most people (approximately 75%) will choose to be conservative in their decision, choosing Program A to deal with the "Asian Disease" when information is presented in the gain frame. Program B contains a two-thirds probability that no people will be saved. This is riskier and the risk appears unacceptable in this example. Strikingly, around 75% of people will choose that same riskier option (presented at Program D) when information is presented in the loss-framed version (described as Program D and Program C). When people are presented with the loss-frame, they will take that chance that nobody will live. They do so because Program C has been framed in terms of unavoidable losses. There is just no way to avoid 400 people dying if you choose Program C. In light of the attention to the certain deaths of all of those people, the riskier Program D now looks like the better decision, even though is carries that same disturbing possibility that two-thirds of the time all 600 people will die.

Tversky and Kahneman also reported similar effects when money was on the line framed as a gain or as a loss (see Box 11.1). In the case of possible losses, people are much more likely to gamble and be riskier in order to avoid those financial losses, than when they would stand to make equivalent gains in money. This has led to the observation that "losses loom larger than gains" (Shafir & Tversky, 1995), an indication that there is greater psychological impact for a loss than there is for an equivalent gain. The framing effect is a powerful example of a case in which the context affects people's decisions. This is also a dramatic example of people's decision making violating rational models. If people always decided rationally, there would never be cases in which preference would reverse on the basis of alternatives being worded differently.

A neuroscience study was reported in 2006 replicating the financial framing effects using fMRI. In this study De Martino, Kumaran, Seymour, and Dolan (2006) at

BOX 11.1

DECISIONS ABOUT RISK AND REWARD

An experimental task widely known as the "Asian Disease Problem" was published by Amos Tversky and Daniel Kahneman in 1981. In this task, people were asked to pretend that they were in the role of deciding about public policy actions to be taken to manage a new disease. Take a moment to read through the scenario below and make a hypothetical decision about which program you would choose.

Imagine that the United States is preparing for the outbreak of an unusual Asian disease, which is expected to kill 600 people. Two alternative programs to combat the disease have been proposed. Assume that the exact scientific estimate of the consequences of the programs is as follows (Fig. 11.7):

If Program A is adopted, 200 people will be saved.

If Program B is adopted, there is 1/3 probability that 600 people will be saved and 2/3 probability that no people will be saved.

Which of the two programs would you favor?

Now let's consider another situation that offers two different programs to manage the disease. Take a few moments to consider Programs C and D described below and make a decision, which you would prefer.

Imagine that the United States is preparing for the outbreak of an unusual Asian disease, which is expected to kill 600 people. Two alternative programs to combat the disease have been proposed. Assume that the exact scientific estimate of the consequences of the programs are as follows:

If Program C is adopted, 400 people will die.

If Program D is adopted, there is 1/3 probability that nobody will die and 2/3 probability that 600 people will die.

Which of the two programs would you favor?

Now let's switch gears and consider some financial scenarios that are offered in a financial study:

Imagine that you face the following pair of concurrent decisions. First examine both decisions and then indicate the options you prefer.

Decision 1) Choose between:

A. a sure gain of $250

B. 25% chance to gain $1000 and 75% chance to gain nothing

Decision 2. Choose between:

C. a sure loss of $750 113 %

D. 75% chance to lose $1000, and 25% chance to lose nothing

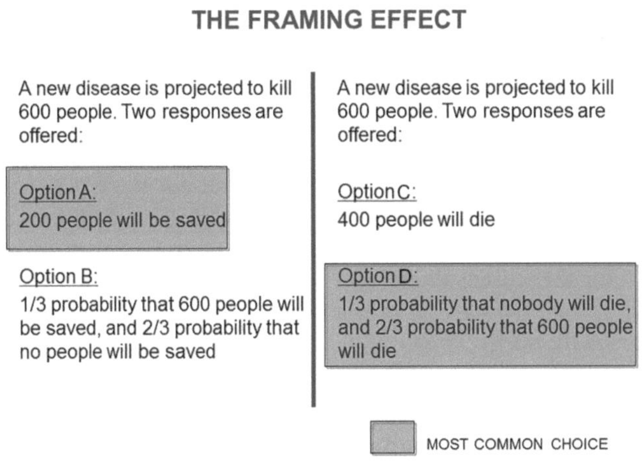

THE FRAMING EFFECT

A new disease is projected to kill 600 people. Two responses are offered:

Option A:
200 people will be saved

Option B:
1/3 probability that 600 people will be saved, and 2/3 probability that no people will be saved

A new disease is projected to kill 600 people. Two responses are offered:

Option C:
400 people will die

Option D:
1/3 probability that nobody will die, and 2/3 probability that 600 people will die

MOST COMMON CHOICE

FIGURE 11.7 The framing effect as demonstrated by Tversky and Kahneman (1981).

University College, London, demonstrated that the financial framing effect could be obtained and studied in the confines of an MRI experiment. People took greater risks when faced with financial consequences framed in terms of losses of money relative to identical cases in which the monetary consequences were framed in terms of money to be gained. Remarkably, people also continued to take risks to avoid financial losses, yet became conservative and chased equivalent gains after dozens of trials. Activation of the amygdala was reported for conservative choices in the financial gain scenarios and for risky choices in the financial loss situations. This result suggests that the amygdala may be a critical region for assigning value to a particular option in a decision. Activation of the orbitofrontal cortex was associated with more rational activity in this study (when people were consistent and not guided by the framing of the problem).

Over the years a variety of framing effects have been reported. One of the most compelling framing effects occurs in the realm of attributes. An attribute framing effect occurs when someone describes himself or herself differently when descriptions are framed positively or negatively (Levin, Schneider, & Gaeth, 1998). My colleague Kevin Murch and I studied the attribute framing effect while recording brain activation using fMRI. In this study we presented participants with a variety of personal attributes and asked them to respond whether the attribute applied to them or not. For example, we asked participants to rate the following statement indicating if it described them or not: "I am honest at least 75% of the time." The majority of people will endorse this description, claiming that it applies to themselves. After all, it is a positive trait and the percentage of time it applies is high at 75%. We also presented the same information in a negative counter-frame as follows: "I am not honest at most 25% of the time." When people evaluate this statement, they are much more likely to claim that it does not describe them. Note that this second statement is precisely equivalent to the first statement about honesty, but it is now framed negatively. Also notice the inclusion of the phrases "at most" and "at least" in these statements. The much less often preferred frame about "not being honest," actually could be interpreted to describe someone who is honest to a fault, though this is a less intuitive way to describe an honest person. A rational decision maker should never be swayed by the context of the framing, but in the actual experiment we found that people were often highly led by framing. Fig. 11.8 depicts the results showing people's agreement with each description. Participants claimed that the positively framed personal attributes (wise, friendly, happy, etc.) were descriptive of themselves over 90% of the time. Meanwhile, the negative counter-frames of these same traits (not wise, not friendly, not happy, etc.) were endorsed less than 60% of the time! There was almost universal agreement with the positively framed items, even when the percentage

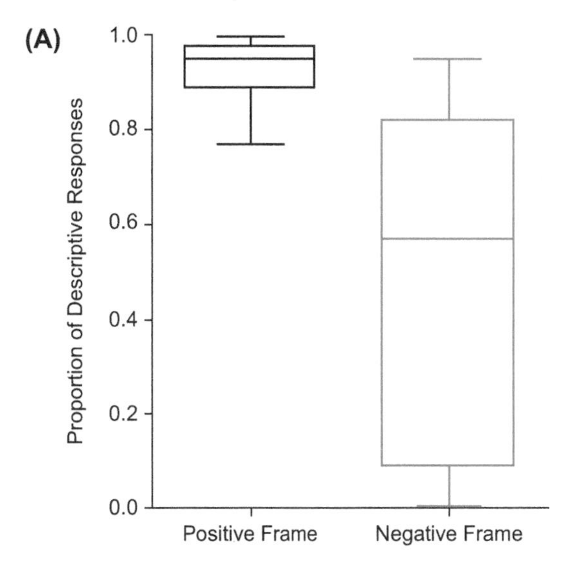

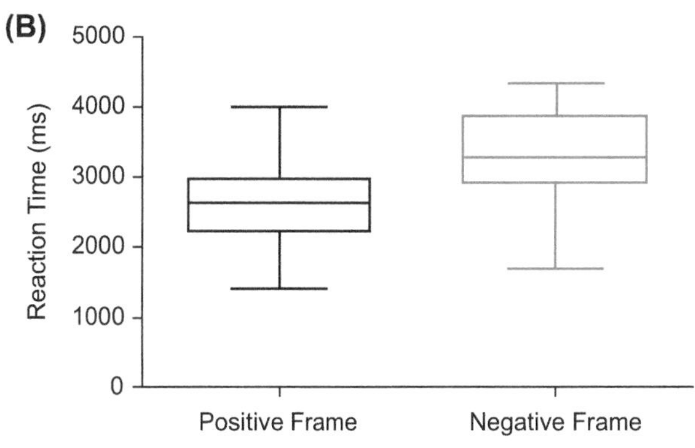

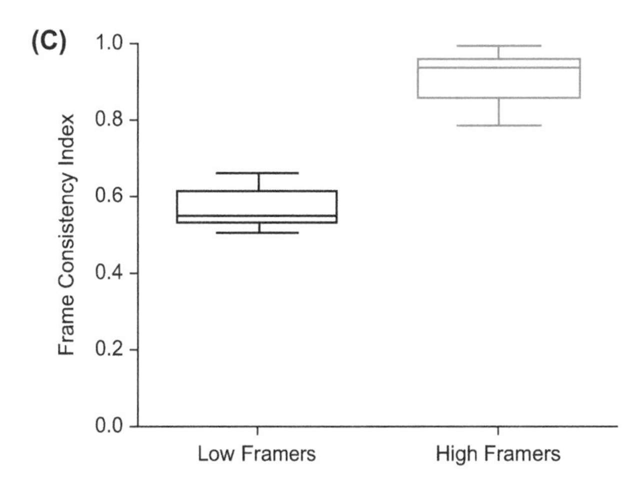

FIGURE 11.8 Results of an attribute framing task. *From Murch, K. B., & Krawczyk, D. C. (2014). A neuroimaging investigation of attribute framing and individual differences. Social Cognitive and Affective Neuroscience, 9(10), 1464–1471.*

listed was relatively low (e.g., "I am wise at least 51% of the time"). There was strong variability about how often people also endorsed the equivalent negative frames as being descriptive of themselves (note the large error bars in the graph of these results, Fig. 11.8).

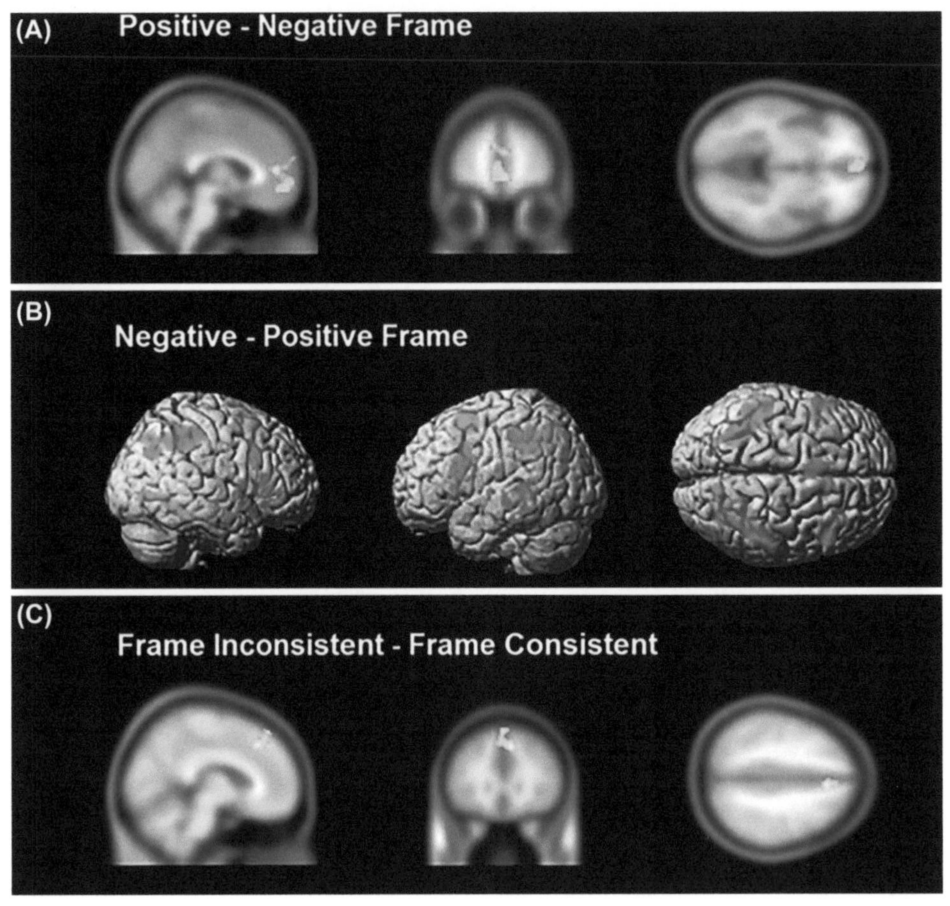

FIGURE 11.9 Functional MRI results showing the activation associated with (A). Evaluations of positively framed items, (B). Evaluations of negatively framed items, and (C). Frame inconsistent responding. *From Murch, K. B., & Krawczyk, D. C. (2014). A neuroimaging investigation of attribute framing and individual differences. Social Cognitive and Affective Neuroscience, 9(10), 1464–1471.*

Kevin Murch and I also recorded brain activation using fMRI during the attribute framing task. When people made judgments about the positively framed items over negative we observed activation within the orbitomedial prefrontal cortex (OMPFC). This same region has been previously associated with reward-based financial decision making (Bechara, Tranel, Damasio, & Damasio, 1996), as well as other positive values (Kable & Levy, 2015). Refer to Fig. 11.9. In a reverse comparison, activation toward negatively framed items over positively framed items yielded considerably more regions of activation including the insula. This region has been commonly reported in association with negative emotion (Harris & Fiske, 2006) and particularly in response to disgust (Wicker et al., 2003). It is possible the negatively framed items generated unpleasant or aversive emotional reactions, as endorsing these negatively framed statements would place the participant in a negative light. We also observed significant activation within areas of the prefrontal cortex (PFC) including the dorsolateral prefrontal cortex (DLPFC), the ventrolateral prefrontal cortex (VLPFC), and the dorsomedial prefrontal cortex (DMPFC). Additionally active were the bilateral parietal lobes and the cerebellum. Each of these frontal and posterior regions have been associated with cognitive control (Lieberman, 2007). Exerting greater control over responses might support people's enhanced tendency to disown attribute statements about themselves when those statements appear in a negative frame.

Activation of cognitive control-related regions was also found for trials that involved the participants choosing against the framing of the problem. Such cases would include instances in which people reject the statement "I am wise at least 87% of the time," though these responses would be rare, as few such positive statements were rejected in this study (Fig. 11.8). More commonly, people accepted statements such as "I am not wise at most 17% of the time." The acceptance of a negative frame was associated with the bilateral DMPFC, the anterior cingulate cortex, and the VLPFC. On these trials, people appropriately avoided being led to reject the statements based on the negative frames when they had already accepted those same statements when framed positively.

Framing effects represent classic cases of the context of the decision situation leading a person to make inconsistent decisions. Some would even call these decisions illogical, as they represent cases in which the individual reverses course simply on the basis of the way in which the choice was presented to them. It is important to note that not everyone falls victim to the framing effect at the same rate. Indeed some people are more able to resist the influence of frames and such people maintain more consistent responding through the course of a series of decisions.

There is a strong role of individual differences in how people approach framing. The study by De Martino et al. (2006), in which the financial framing task was conducted within an fMRI environment, presents an instance of that case. In addition to finding amygdala activity associated with the preferred option (risky options in response to loss framing and conservative options in response to gain framing) there was also activation within the orbitofrontal cortex toward situations in which the individual chose against frame (choosing rationally as one might say). These cases would involve instances in which the gain frame elicits a risky choice and where the loss frames are followed by conservative choices. The ability of people to avoid frames and choose more rationally was positively correlated with activity within this prefrontal region.

Similarly in the attribute framing study conducted by Murch and Krawczyk (2014), an individual difference also existed which enabled us to make comparisons between a high-framing group and a low-framing group (Fig. 11.8C). The high-framing group showed strong susceptibility to being manipulated by frames. The individuals in this group chose with frame over 85% of the time (rejecting negatively framed items and endorsing positively framed ones). Meanwhile a low-framing group showed an ability to resist the framing effect and responded consistently much more of the time by accepting negatively framed items along with their positive counter-framed items. The group that was able to avoid framing effects also showed higher scores on an IQ measure. This would indicate that heightened mental resources might allow people to be less biased in their decision making and avoid frames. The neuroimaging findings also yielded an interesting individual difference. For frame consistent responding (endorsing positively framed items and rejecting negatively framed items), people engaged the right orbital PFC. Notably, the responses were higher in this area for the high-framing group, or the group that was prone to be led by frame. Meanwhile regions of the left PFC were activated when people responded against frame (Fig. 11.10). This is consistent with greater activation associated with avoiding illogical, or frame-based, responding by exerting greater control. Interestingly, the high-framing individuals, who had a greater tendency to be led by frame, activated these regions more than the low-framing group when they responded against the frame of the statement. In other words, it appeared that more activation was required in these frontal control regions for the individuals who tended to be led by frames to overcome the framing effect.

BIASES AND HEURISTICS

The Psychological Impact of Gains and Losses

The psychological impact of losses is generally higher than that of equivalent gains. This is related to the discussion of framing effects from the prior section, wherein people seek to avoid losses and are more risky in pursuit of avoiding losses. This tendency led Tversky and Kahneman (1981) to propose their *prospect theory*. This theory is illustrated by a classic graph charting the relative perception of gains and losses (see Fig. 11.11). The x-axis of the figure illustrates losses to the left and gains to the right, while the y-axis consists of psychological value. Notice that the gain side of the graph moves up steeply in value as the gains increase in magnitude moving to the right. As the gains become larger, their perceived value to the person begins to decrease. This would be equivalent to a situation in which a college student receives an award of $500 relative to a gain of some loose change at a vending machine. The perceived value of the $500 is extremely high. Meanwhile, when you move out toward higher value gains, the difference between $20,000 and $20,500 does not seem nearly as large. The magnitude of the gain has begun to dilute the $500 difference in this second case.

Now let's turn our attention to the loss side of the prospect theory graph. Notice that the curve is similar with equivalent differences in loss magnitudes having a larger change in perceived value at lower levels of loss. Again, if you lose $10 it feels much larger than losing $2. If, however, you lose $510, it likely feels not so much worse than losing $500. Notice also that the loss curve is much steeper than the gain curve. This is meant to capture the idea that a gain feels like it has less value than the equivalent loss. At higher levels of loss the impact on psychological value is enormous compared to an equivalent gain.

Prospect theory captures the tendency that people have to try to protect their existing resources. This appears to be more meaningful than attempting to gain new resources. Another classic example of this comes from Knetsch (1989), who gave participants either a coffee mug or a bar of chocolate as a gift. These gifts were arbitrarily assigned and of roughly equal value (Fig. 11.12). The experimenter offered the participant the opportunity to exchange their gift for the alternative. We should predict that roughly half of the participants would be willing to make this exchange from a purely rational economic perspective based on their value. The data from the experiment indicated that only about 10% of people were willing to make the exchange (regardless of which gift they had been given initially). This experiment

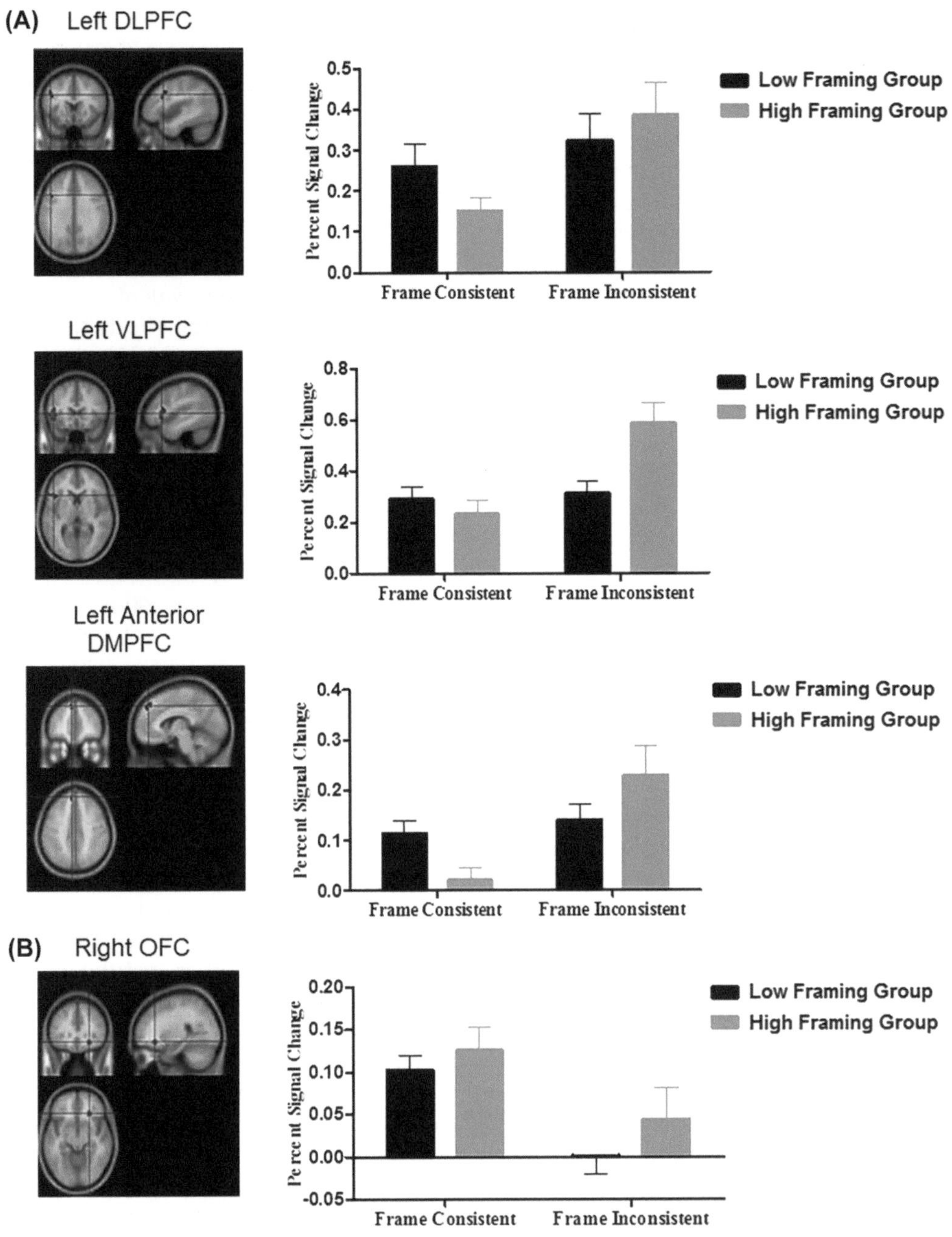

FIGURE 11.10 Individual differences in attribute framing were evident in several brain areas. *From Murch, K. B., & Krawczyk, D. C. (2014). A neuroimaging investigation of attribute framing and individual differences. Social Cognitive and Affective Neuroscience, 9(10), 1464–1471.Murch & Krawczyk, 2014*

illustrates the concept of loss aversion. We find the loss of something we already have to have a higher impact than gaining something we do not currently have.

A related study by Kahneman, Knetsch, and Thaler (1990) demonstrated the phenomenon of loss aversion. They asked people to explicitly set prices for things that they owned or could potentially own. In this experiment the participants were divided into two groups. One group was given a drinking mug and asked to indicate the lowest price that they would sell the mug for. The other group was asked to generate a price that they would be willing to pay for the mug. In both cases, after the participant stated his or her price, the market price would be revealed for the mug and participants could keep the mug if the market price was lower than what they had asked for it. Alternatively, participants could obtain the mug if they had named a price higher than the market value. Note that the value of the mug should be roughly equivalent for people. The only difference was whether the mug was given to you in advance (and you then have a chance to lose it) or whether the mug is for sale and you can attempt to buy it (and gain the mug). Contrary to rational decision making in which both groups ought to value the mug roughly equivalently, the participants that already owned

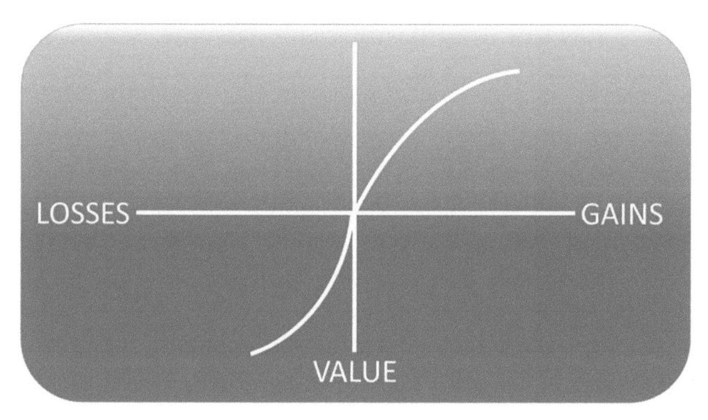

FIGURE 11.11 The graph of Tversky and Kahneman's (1981) prospect theory. The chart shows the steeper curve of perceived loss relative to the magnitude of the situation. This illustrates the concept that a gain feels like it has less value than an equivalent loss. At higher levels of loss the impact on psychological value has a larger associated magnitude compared to the equivalent gain. *Adapted from Tversky, T., & Kahneman, D. (October, 1986). Part 2: The behavioral foundations of economic theory. The Journal of Business, 59(4), S251–S278.*

FIGURE 11.12 The value of items that are roughly equivalent can vary depending on whether someone owns the item. The value of the items we already have tends to be higher than the value of similar things that we do not have.

the mug and could stand to lose it provided a value roughly twice that of the other group that stood to buy the mug. This study represents another instance in which loss aversion drives up the price of an option relative to the situation in which the same item could be gained.

Cognitive Biases in Decision Research

The framing effect and its associated loss aversion tendency represent cases in which people deviate from rational thinking. Such phenomena are irrational, but can be explained in terms of the advantages that they convey for the thinker. We cannot avoid biases as we take in new information and draw conclusions. Bias comes about whenever we are faced with a situation that is similar to one we have experienced in the past. Bias

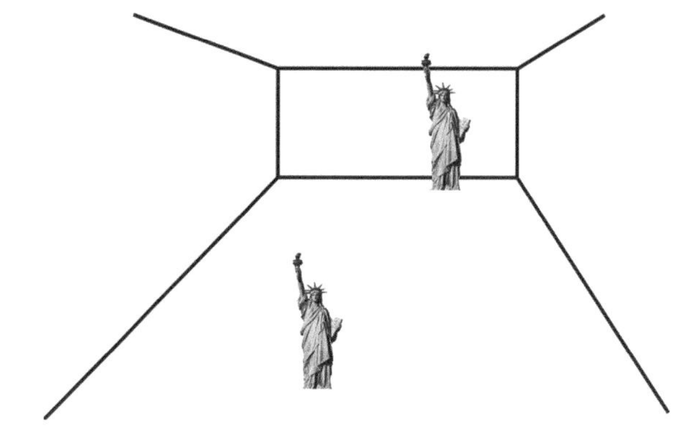

FIGURE 11.13 An illusion in which two identically sized objects that vary in relative position within the image appear to differ in size. We intuitively perceive the image that is lower in the frame as being closer than the other image. In other words we attribute depth to this two-dimensional image. This is a simple case of bias. Similar biases operate on the abstract information that is used in making decisions.

is part of the information processing architecture of our brains. We are highly attuned to recognizing similarities and differences that exist in our environments. We can observe the operation of bias in the case of perceptual illusions. Consider the situation in which individuals are presented with two statues that are identical in absolute size (refer to Fig. 11.13), but vary in perceived size. At a glance you probably quickly ascertain that the figure in the image that his higher up in the frame is further away. This is of course false, as this is a two-dimensional picture and the images on our retinas are actually the same. This illusion of depth or distance comes about because our visual systems are highly experienced in perceiving differences in retinal image size in our environment. This apparent size in the statues is attributable to objects appearing closer or farther away. In other words, a strong set of experiences helps to tune our perceptual systems so that we effortlessly perceive depth and relative distances between objects. Our perceptual systems are biased toward making inferences about relative distance when we are confronted with such examples. Notice that the bias operates effortlessly and mostly without our awareness. This bias is hard to avoid and can lead to errors in perception in rare cases in which someone has created a deliberate illusion to fool our visual systems. Similar biases apply to cognition when we make decisions.

Decision options call to mind related information. This occurs automatically whenever we process information. Consider the example of *priming*, in which a person is provided with a cue word and then asked to rapidly make a simple judgment. Lexical decision tasks demonstrate priming. Participants in a lexical decision task are presented with either words such as "dog" or "cow," along with non-words that can be pronounced such as "dax," or "vab." When we are presented with a relevant prime word just prior to making a lexical decision, we are reliably faster at making that decision. For example, if you see the word "cat" as a prime prior to seeing the target word "dog," you will be faster to verify that indeed "dog" is a word compared to a situation in which the target word "dog" is presented alone. Priming suggests that your cognitive system has been made ready to decide about the upcoming target. This state of readiness is analogous to a situation in which you prime an engine by preloading the gasoline so that it will be faster to start. Priming has been explained in terms of spreading activation within semantic memory. The idea is that the prime word "cat" activates the concept "cat" in your semantic memory. Along with the word, the perceptual image of a cat is activated, the properties of a cat are activated, and close semantic associates of cats, including dogs, are also activated. Word association studies have found that "dog" is one of the strongest associates of the word "cat" (Nelson, McEvoy, & Schreiber, 1998). The activation from a prime does not need to reach our conscious threshold in order to influence performance. We rarely experience a flood of images and words at the mere mention of "cat." Priming may cause weak activation of the concept and its associations. Primes clearly bias our thinking by driving us to respond more quickly.

Biases can operate on more complex information and the properties of the bias are similar to those observed in the case of perceptual judgments and lexical decisions. We can carry strong influences into our decisions by the numerous prejudgments or biases that are evoked by incoming information. This information will activate our prior knowledge.

If we return to our example of choosing a college to attend, which is a highly complex, multi-attribute decision, we can imagine how biases might play a role in this case. Consider a school such as Hendrix College, a small liberal arts school in Conway, Arkansas. If you are relatively unfamiliar with Hendrix and have a limited amount of knowledge about Arkansas, you may be tempted to discount this school as being rural, unsophisticated, and offering a lower quality of education than could be obtained at a more urban school in a more populated state. In this case you would be wrong, as Hendrix College is one of the excellent smaller liberal arts schools in the United States, and it is located in a cultured university town offering an excellent education. Similar biases exist whenever we lack sufficient information. The University of Wisconsin is in Madison, which has a famously cold climate. You might be tempted to assume that it would be an undesirable school to attend if you have not visited it or read about it. In fact, the University of Wisconsin is an excellent school, and it is also widely regarded being located in a highly desirable destination, as Madison is a rich cultured university town with a highly educated population.

Bias is frequently viewed as a negative phenomenon, but biases often help us to make efficient decisions. Sometimes biases serve us well and provide a means toward a rapid and clear outcome. If you passionately despise cold climates, then the bias against the University of Wisconsin will probably help you to discount attending that university. Despite its numerous desirable qualities, Madison does get quite cold in the winter. This situation is again related to decision weights. If we have a bias against something that violates a core value that we hold dearly, we may be alright exercising the bias and getting on with our day, avoiding options that violate that value. Biases are all around us, but we need not fall prey to their influence. The remedy is to further investigate our options, using a more controlled and slower strategy of research and acquisition of new information. Armed with more information, we may reach the conclusion that our bias was correct, or that we will need to adjust and decide differently in light of the additional knowledge.

Dual-Systems and Cognition

A well-established line of thinking in psychology holds that we have two modes of cognitive processing, a fast mode and a slower mode. In his book entitled *Thinking Fast and Slow* (2011), psychologist Daniel Kahneman describes numerous situations in which we engage in rapid, low effort, snap judgments and other situations in which we use a slower and more effortful mode of processing in order to consider information more thoroughly before deciding. The fast and slow modes of thinking have been described in the context of dual-systems or dual-process models. The difference between the terms is that dual-process models emphasize a mode of cognitive function, or in other words a cognitive process. By contrast, dual-system models emphasize that the two functions are represented by distinct systems within the cognitive architecture of the brain (Box 11.2).

These models maintain that we have two modes of information processing. The faster and less-effortful type of processing can be employed when we consider menu decisions in a familiar restaurant. Armed with a sound knowledge of the menu and having tried many of the dishes before, you are likely to rapidly opt for your favorite dish without taking the time or effort to consider all of the alternatives. Such a decision is said to be automatic. Alternatively, if you are in a new restaurant and have no experience with the food on offer, you are likely to take some time to carefully read about the options, ask the server what they would recommend, and devote time and effort to making a choice you are likely to enjoy. These fast and slow modes of processing go by many names. Fig. 11.14 summarizes some of the major labels that have been applied to dual-processing theories. One of the simplest distinctions is between "automatic" and "controlled" processing (Schneider & Shiffrin, 1977). These labels

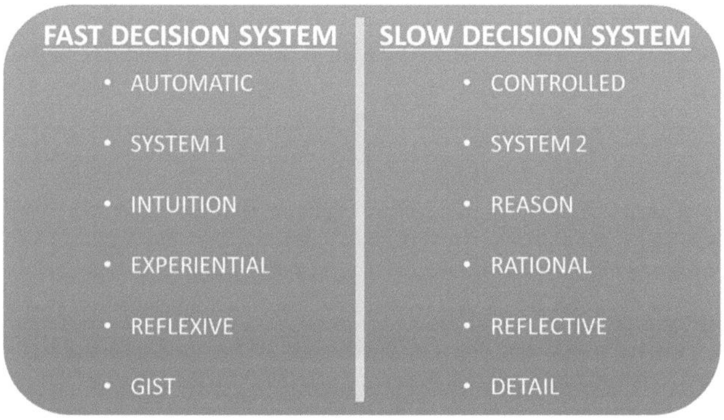

FAST DECISION SYSTEM	SLOW DECISION SYSTEM
• AUTOMATIC	• CONTROLLED
• SYSTEM 1	• SYSTEM 2
• INTUITION	• REASON
• EXPERIENTIAL	• RATIONAL
• REFLEXIVE	• REFLECTIVE
• GIST	• DETAIL

FIGURE 11.14 Categorizing the dual-systems, or dual-process models, of cognition that are relevant to decision making. Over the years, numerous terms have been applied to describe the fast and automatic mode of processing and the slower more thoughtful mode.

capture the rigidity and speed of the fast decision and the flexibility and effort of the slower decision. Kahneman (2011) simply refers to fast decisions as arising from System 1, or the first system to weigh in on the decision, while he uses the term System 2 to describe the slower, more effortful alternative mode of deciding, the second system, to activate if we do not commit to a quick System 1 decision. There is physiological evidence to support the distinction between these two modes of thinking. Jackson Beatty conducted studies in the 1980s using a technique called pupilometry to measure the dilation of the pupils of the eyes while people made decisions. Beatty (1982) reported that people's pupils dilated when they engaged in longer and more effortful processing. When they went through rapid processing associated with System 1, their pupils showed no such dilation.

BOX 11.2

DUAL-PROCESSES IN DECISION MAKING: THE COGNITIVE REFLECTION TEST

An experimental task known as the "Cognitive Reflection Test" (Fig. 11.15) was published by psychologist Shane Frederick in 2005. This task asks people to answer three seemingly simple questions about problem-solving situations. Take a moment to read through the scenarios below and answer these questions as quickly and accurately as you can.

A bat and a ball cost $1.10 in total. The bat costs $1.00 more than the ball. How much does the ball cost? _____ cents.

If it takes 5 machines 5 minutes to make 5 widgets, how long would it take 100 machines to make 100 widgets? _____ minutes.

In a lake, there is a patch of lily pads. Every day, the patch doubles in size. If it takes 48 days for the patch to cover the entire lake, how long would it take for the patch to cover half of the lake? _____ days.

Now let's consider what the answers are:

The answers are: 5 cents, 5 minutes, and 47 days.

How many of these did you get correct? You may have noticed that there is a quick intuitive answer for all three and that answer happens to be incorrect in all cases. If you answer these questions using your System 1, or automatic cognitive processing, you will be wrong in all cases. Success on the Cognitive Reflection Test requires you to think longer and deliberate about the deeper aspects of the problems. In the bat and ball case you have to inhibit the reflexive answer of 10 cents in order to realize that that would only yield a 90-cent advantage for the bat. This is tough to do and few people perform well on the Cognitive Reflection Test.

THE COGNITIVE REFLECTION TEST

- (1) A bat and a ball cost $1.10 in total. The bat costs $1.00 more than the ball. How much does the ball cost? ____cents
- (2) If it takes 5 machines 5 minutes to make 5 widgets, how long would it take 100 machines to make 100 widgets? ____minutes
- (3) In a lake, there is a patch of lily pads. Every day, the patch doubles in size. If it takes 48 days for the patch to cover the entire lake, how long would it take for the patch to cover half of the lake? ____days

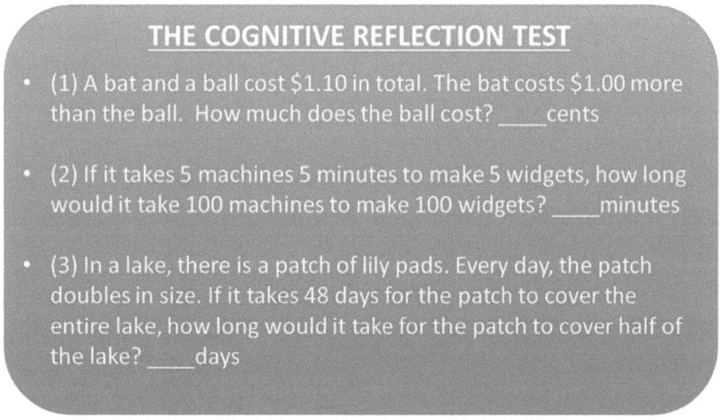

FIGURE 11.15 Example problems from the Cognitive Reflection Task (Frederick, 2005). Intuitive incorrect answers are: 1. 10 cents, 2. 100 minutes, 3. 24 days. Correct answers are: 1. 5 cents, 2. 5 minutes, 3. 47 days. *Adapted from Toplak, M.E., West, R.F., & Stanovich, K.E. (2011). Memory & Cognition, 39, 1275.*

Dual-system models have described the relationship between these two modes of thinking and the circumstances during which they operate. Walter Schneider and Richard Shiffrin (1977) advocated for the terms automatic and controlled modes of processing. They noted that automatic processing is associated with easy tasks that are familiar to us and are well learned. By contrast, controlled processing dominates when a task is complex, difficult, or novel. In such cases, we need to devote greater attention to carrying out the task and in doing so build our knowledge and familiarity with a novel domain. In the case of deciding about colleges, you may already know enough to make a rapid automatic decision about the schools closest to your home, while you may have to engage in controlled processing in order to properly evaluate out-of-state schools that are in climates and cities that you are not familiar with. Schneider and Shiffrin noted that people could move from controlled processing on a task to an automatic mode. Consider the example of driving a car with a manual transmission. Anyone who has attempted this can no doubt recall the stress and awkwardness of attempting to engage the clutch and move the car into the proper gear while also attending to the road and other traffic. I found this to be an act of extreme controlled processing, and I recall thinking through each step explicitly and with great effort. When trying to drive a manual car, I slowly and awkwardly maneuvered the vehicle along neighborhood streets with few cars on them. Stopping at an intersection was an act of extreme challenge for me. Over time driving becomes a highly automatic activity. An experienced driver can control the transmission easily and with little conscious thinking. Decisions can operate the same way. Returning to our restaurant example, when you become familiar with the new restaurant's menu options, tried many of the dishes, and you understand the options, you will be able to make much more rapid and effortless choices about your meal.

Some dual-process models have tended to characterize the fast and automatic system as being irrational, or in other words less accurate than the slower and more deliberative decisions. Epstein, Pacini, Denes-Raj, and Heier (1996) favor the term "experiential" for the rapid and automatic decisions (see Fig. 11.14). This term captures the idea that our experiences guide us when we make rapid and effortless choices. Conversely they use the term "rational" to describe the slower and more deliberative decisions that people make. These terms imply that the experiential decision may be inferior and may be irrational. This is not always the case. Fast decisions arise from a combination of simpler decision contexts and also familiarity with the situations. If we have often experienced a situation before, it is likely that our prior decisions can be a good guide as to what to do under similar circumstances. It may be possible in such cases to make our best decisions using a rapid and effortless strategy characteristic of System 1, or automatic processing. For example, if we have already memorized the menu at a familiar restaurant, it may be insensible to ponder and deliberate for a long time about what choice to make. We would at the very least be wasting valuable time and effort. Even worse, overreliance on System 2, controlled processing, may lead to overthinking a decision and convincing oneself that a new idea would be better than simply employing a solution that is easy to carry out and has worked before. In this way the automatic system may be preferable in the case of certain decisions. There are alternative cases in which we make decision errors by being too quick to decide and relying too much on our gut feeling or System 1, automatic processing. This is precisely the case in the framing effect studies we discussed earlier. The context of potential losses rapidly engages our emotions and a tendency to decide in a riskier manner than we would considering the counter-frame "gain" context.

The Use of Heuristics

A heuristic is a rule-of-thumb, or a guide toward what behavior is appropriate for a certain situation. Heuristics are also known as "mental shortcuts" (Kahneman, 2011). Such shortcuts can aid us when we face time pressure to decide, or when conditions are complex and our attention is divided. A heuristic is a well-learned adaptation that allows us to decide quickly and with low effort. Employing heuristics will yield an outcome that is positive or at least satisfactory much of the time. When Daniel Kahneman and Amos Tversky set up experiments to test whether people followed rational models of decision making, they frequently discovered that people deviated from the rational model, but they did so for good reasons. Participants in these cases appeared to be using heuristics in order to make a rapid decision with less effort that could be relied upon to yield a positive outcome much of the time. Kahneman and Tversky, along with numerous other behavioral scientists, have characterized a wide array of heuristics that occur when people make decisions. In this section, we will review some of the most well-established and interesting heuristics that people use and how these occur in our everyday lives.

Availability

The availability heuristic enables people to make rapid decisions on the basis of the most available information. When we use this heuristic, our decisions are influenced by recent situations that we have experienced. In other words, when you rely upon the availability heuristic you sacrifice accuracy for speed and overweigh the most strongly associated information that you have. Recently processed information tends to be most accessible. For example, if you have been thinking a lot about buying a new car, then recent model, year information will likely be more available to you and influence your choices, than if you have not purchased a car in several years. A classic demonstration of the availability heuristic was reported by Tversky and Kahneman (1973) . They investigated a situation in which people had to make an estimate. Participants were told the following: "suppose one samples a word (of three letters or more) at random from an English text. Is it more likely that the word starts with "r" or that "r" is the third letter?" When people evaluate this statement, they are likely to call to mind all of the cases that they can immediately recall in which the first or the third letter in a word is "r." Doing so would likely generate more instances of words that begin with "r," as these tend to be more available. This greater availability of information makes it more likely that someone will judge these types of words to be more probable. The actual answer is that many more words have an "r" in the third position, so using the heuristic steers us wrong.

The availability heuristic operates whenever information comes to mind more readily for one alternative over another. Estimates increase when famous, or familiar, examples are included in a group. Tversky and Kahneman (1973) reported a word list study in which they presented lists to participants that included both male and female names. There were equal numbers of names for both genders on the lists, but some of the names were those of famous people. Those lists with more famous individuals swayed the judgments of the participants so that these lists were rated as containing more females if famous female names had been included. The reverse was true when more male names had been included. Other characteristics of memory can influence availability as well. These include the word imageability, in which words or concepts that are more easily pictured in one's mind appear to be more probable based on their clearer mental representation.

The availability heuristic has very real consequences in life. Our brains are excellent at calculating statistical probability and often times the information that is called to mind will be highly effective in making a decision. The availability heuristic may help to guide us in our voting behavior when we call to mind times when our governor has made wise decisions. This probably also influences us with regard to political scandals. When a person comes under fire for irresponsible behavior, opinion poll numbers typically drop. If the scandal persists and more evidence of wrongdoing occurs, then the poll numbers drop permanently. If however, the evidence of wrongdoing is thin, then polls are likely to recover to levels similar to those driven by overall job performance.

Availability does have clear negative characteristics as well. Poses and Anthony (1991) asked medical doctors to estimate the probability that patients had bacteremia, a condition in which there are bacteria in the bloodstream. These doctors judged the probability as significantly higher when they had recently cared for several patients who had bacteremia. We would not want doctors to rely purely upon only the most available information, but rather, they should also balance information availability with how probable that information is. This leads us to consider another notable heuristic.

Representativeness

The representativeness heuristic describes a bias that influences our judgments about the probability of a particular instance and whether it is a common or uncommon example of a category. This bias is driven by the way that our minds compute probability based on our stored category knowledge. This heuristic applies in cases in which we are asked to judge the probability of a particular situation. In making probability estimations we can perform quite accurately if we apply some factors toward our judgments. These factors include the number of instances from which a sample is drawn, the prior probabilities of a particular case, and the overall likelihood of a particular event or outcome. In other words we have a variety of statistical tools available to us that would help to improve our judgments. If we do not use these tools and instead rely upon intuition, we are likely to inadvertently fall into using the representativeness heuristic and potentially make a less-optimal decision.

Tversky and Kahneman (1973) provided a classic example of the representativeness heuristic in action. They presented individuals with a question about the likely profession of particular people based on short descriptions. In one of their examples, a man named Steve was described as follows: "He is very shy and withdrawn, invariably helpful, but with little interest in people, or in the world of reality. A meek and tidy soul, he has a need for order and structure, and a passion for detail." Having read this description about Steve, people were asked to estimate the probability that he is a farmer, a salesman, an airline pilot, a librarian, or a physician. According to the description you may notice that the librarian job appears intuitively appealing. It fits nicely with Steve's shy, helpful, and tidy characteristics. It feels right matching Steve to many of the stereotypical qualities that a librarian would have. Claiming that Steve would be most likely to be a librarian in this lineup represents a logical fallacy that follows from the representativeness heuristic. The challenge here is that each of the jobs presented was not equally probable at the outset. There are actually many more farmers and physicians in the population than there are librarians.

Occupations and other characteristics have what are known as base-rates within the population and those need to be accounted for when estimating probability. Base-rate neglect describes a situation in which the base-rates have been ignored in order to make an intuitive judgment of probability that simply feels right based on the quality of match between the characteristics of the person described and those commonly associated in our semantic knowledge for a particular category. Tversky and Kahneman demonstrated this tendency further in an elegant experiment on base-rate neglect and its contribution to the representativeness heuristic.

In their base-rate neglect experiment, Kahneman and Tversky (1973) asked participants to judge probabilities about occupations with questions about lawyers and engineers. Participants saw a set of attributes about people and were told that the individuals described had been randomly sampled from a group of 100 people. The 100 individuals consisted of some proportion of engineers and some lawyers. Participants estimated whether each hypothetical individual that was described was an engineer or a lawyer. Kahneman and Tversky cleverly manipulated the base-rates for these two occupational groups by telling one group of participants that the 100 individuals consisted of 30 engineers and 70 lawyers and another group the reverse odds that there were 70 engineers and 30 lawyers. People should treat these two conditions differently if they are sensitive to base-rates. There will be a much greater chance that an individual in the first group is a lawyer and a much greater chance that an individual in the second group is an engineer. Contrary to this result, Kahneman and Tversky (1973) reported that both groups produced similar judgments about each individual that was described. For example, an individual was described as follows: "a 30-year-old man. He is married with no children. A man of high ability and high motivation, he promises to be quite successful in his field. He is well liked by his colleagues." These attributes are strongly consistent with a stereotypical engineer and participants in both experimental conditions endorsed that stereotype. They overwhelmingly endorsed "engineer" as their answer regardless of whether he was sampled from a group comprised of either 70%, or 30% engineers. The representativeness heuristic is a shortcut that enables us to make rapid intuitive judgments, but one that can lead us to be worse decision makers when we do not appropriately make use of all of the information we have available.

Anchoring and Adjustment

Another major factor in decision making is where we initially set our baseline point. This is known as anchoring. If the price of gas in Dallas is $1.99, people will associate this as roughly the appropriate price of gas. This number becomes the anchor value. If that individual then drives to Los Angeles and is faced with $3.19 gas prices, he will likely find the situation to be unfair and ridiculous to have to pay so much. By contrast, an individual who lives in Los Angeles finding herself in Dallas may be delighted to pay $1.99 for gas by comparison to the higher anchor point that is experienced in California. This difference in the psychological perception of price is an example of anchoring. Once we anchor on a particular number this will set the tone for our future decisions about appropriate pricing.

Anchoring effects can significantly bias people's judgments. Even more problematic is the finding that people fail to adjust their estimates sufficiently when they anchor on a particular value. Tversky and Kahneman (1973) described this phenomenon in an experiment in which people were made clearly aware that the starting point of their estimates was arbitrary. In this case the participants were asked questions such as: "What percentage of African countries are members of the United Nations"? Participants watched as a wheel of numbers one through 100 was spun in their presence. Whatever number the wheel landed on served as a randomly generated numerical anchor point and participants were asked to judge whether the African national percentage was above or below that number. Next the participant would try to generate an accurate number for the number of nations. These participant-generated values were badly skewed by the initial estimate that had been displayed by the spin of the wheel. When the number 10 was offered as a percentage of African countries in the UN, participants increased their estimates to around 25%. Alternatively, when given 60 as a number from the wheel, participants estimated around 45% of countries were UN members! These estimates reflect people's tendency to insufficiently adjust up or down from the initial anchor value. Tversky and Kahneman noted that even when participants were paid for accurate estimates, they were unable to adjust enough from the arbitrary anchor value.

The Benefits of Heuristics

The discussion of heuristics often focuses on how people are irrational and faulty in their decision making. While this is technically true from a rational viewpoint, in which people should make the most logical and consistent decisions possible under all circumstances, this perspective underestimates the advantages that heuristics give us. The major advantage of using heuristics is speed of processing. Heuristics make our mental lives easier and allow us to get through our daily lives without necessitating long deliberation over all of our decisions. We may trade accuracy for speed in carrying out our lives. An experienced driver may allow his mind to wander when driving on a deserted country road, rather than fastidiously devote all of his mental effort to steering and accelerating. We might choose to use cruise control under these driving conditions for this same reason. Heuristics in a sense are like a mental form of cruise control enabling us to quickly and efficiently make decisions that are likely to be good enough most of the time.

The challenge of heuristics may not be that we use them and are sometimes inaccurate, but rather that we need to know when to use them and when not to. We need to appropriately determine when to settle for our heuristic quick estimate, and when we should spend greater time and effort to ensure that we are indeed making our best possible decision. This brings us to the tricky case of expertise, which we will focus on much more in Chapter 13.

ABDUCTIVE REASONING

Relationships Between Induction, Deduction, and Abduction

Abductive reasoning occurs when we have to hypothesize in order to best explain some set of data or evidence. This is sometimes referred to as reasoning to the best possible explanation (Walton, 2005). Like inductive reasoning, there is not a guarantee that abductive reasoning will yield a valid answer, or one that is true in the world.

I used abductive reasoning to try to make sense of something that happened in my car the other night. I typically drive my sixth grade son to school in the morning and had parked the car in the driveway at the back of the house, which feeds into an alley. I got into the driver's seat to find quite a mess in the passenger seat and on the floor. Items from the glove compartment and center storage console had been strewn about. My initial hypothesis was that one of my family members had carelessly looked for something in that car the previous day. I began to wonder which of my children had rummaged through the car looking for something and then failed to replace the items that had been removed. It was early in the morning and I could not place when such search would have occurred. I also considered the fact that nobody in the family was generally careless enough to have left the front of the car in this state. I then began to consider the possibility that the car had been tampered with by an intruder. Had I really locked the car last night? While it was possible I had not and that someone had entered the car looking for objects to steal, there were some aspects of the situation that did not fit with this possibility. I could not detect anything clearly missing from the vehicle, and there was no damage to the windows or doors. Further, our phone chargers and cables remained plugged into the dashboard power outlets in plain view. At a minimum, it seemed to me that an intruder would have taken these. The "intruder hypothesis" was nonetheless my leading guess as to why the car was in this state. I assumed I had failed to lock the car (which made sense, as we had unloaded groceries the previous day), someone looking for a quick theft had found the car unlocked and decided to see if there was anything of value to take from the glove compartment or center consul. When I explained the situation to my wife, she said that one of our dogs had barked at around 4:30 a.m. A couple of days later she read a neighborhood association blog post indicating that someone had been entering cars parked behind houses during that week. I had to assume that I had indeed failed to lock the car, and someone arrived in the wee hours of the morning and rummaged about in my car. Finding nothing of particular value, they appeared to have moved on and left me with a minor mess to clean up (Fig. 11.16).

My interpretations of the small mess in my car involved considerable guesswork, hypothesis testing, and reasoning to find the most likely possible conclusion. Notice a few important features of the situation. It was unlikely that anyone had been in the car earlier than me that day. I almost always lock the car. Nothing was technically stolen, as far as I could tell. There have been other times when our family car has been left in a messy state by myself or other members of family. I eventually arrived at my preferred hypothesis about the would-be thief who merely rummaged through the

FIGURE 11.16 A minor mess in the front of my car. What happened here? We engage in abductive reasoning in cases in which we have to piece together what may have occurred based on a limited amount of information. Abductive inferences are widely used in cases involving detective work to investigate crime scenes.

compartments in my car only through piecing together the limited set of facts and observations that I had. Note that this was not the first theory I developed. I initially assumed it was probably a family member who had rummaged through the car. I came around to the intruder hypothesis after discounting other possibilities based on my memory of the prior day and what appeared likely. Note also that I have no factual hard evidence that it was an intruder. There is always the possibility that I got it wrong. Perhaps a neighbor desperately needed to check his tire pressure early in the morning, opened my car, borrowed the pressure gauge, and then carelessly tossed it back into the car. Putting aside my rather frivolous example, we undergo abductive reasoning to the best possible explanation in much higher stakes situations. When detectives investigate crimes and when intelligence and national security personnel attempt to understand a terrorist act, they often reason to the best possible cause based on limited information. Abductive reasoning is applied in some of the most complex and important cases in people's lives.

Abductive reasoning stands beside its more well-known cousins, deductive and inductive reasoning, which were the focus of Chapter 9. Abductive reasoning has also been referred to as presumptive reasoning, or plausibilistic reasoning (Walton, 2005). We discussed deduction as evaluating statements in the form of "if *p*, then *q*" statements. Notably, the reverse inference "if *q*, then *p*" does not apply in deductive reasoning, as it leads to invalid inferences. Abductive reasoning involves inferring *p as an explanation* for *q*. As a result of this inference, abduction allows the statement *p* to be abducted from the resulting consequent *q*. This highlights a critical difference between deductive reasoning and abductive reasoning. While inferring *p* on the basis of *q*, one is committing a logical fallacy known as affirming the consequent (see Chapter 9). Thus, the directionality of the inference differs in abductive reasoning compared to deductive reasoning. Abductive reasoning, like deduction and induction, has been studied from multiple perspectives including those from philosophy, cognitive science, and artificial intelligence. Scholars have tended to place abduction below deduction and abduction in terms of soundness or validity of the inferences drawn, as abduction is assumed to not only involve a hypothesis, or even a guess, but also it is assumed that the decision maker possesses only a portion of the available information that actually exists in the world. Thus abductive inferences are widely acknowledged to be potentially wrong, or invalid, depending on the complexities and specifics of a given situation.

The philosopher Charles Sanders Pierce is widely credited with introducing abductive reasoning. He noted that "Deduction proves that something must be; Induction shows that something actually is operative; Abduction merely suggest that something may be." (Pierce, 1997). He also observed that abductive reasoning is "the process of forming an explanatory hypothesis." Below is an example of an abductive inference in the formal and generalized forms that we used to discuss induction and deduction in Chapter 9.

The surprising fact, *q*, is observed.
But if *p* were true, *q* would be a matter of course.
Hence, there is reason to suspect that *p* is true.

Notice again that there is no guarantee of validity or correctness when we engage in abductive reasoning. It is typically used when we simply have no other means to employ when seeking an explanation.

Applications of Abductive Reasoning

An example of an experimental approach to investigating abductive reasoning was reported by Green and McCloy (2003). These investigators asked the question, how do people reach a verdict in a courtroom case? Such a situation represents a classic context for the use of abduction. Court cases are complex; they involve uncertainty about information, and they must be decided on the basis of incomplete information. Green and McCloy presented participants with fictional court case scenarios and asked participants to decide in favor of one side or the other. In one of their experimental cases, participants read about the following scenario about a construction company dispute:

Odell is a construction company that has failed to meet its deadline for completing a new factory by two months and is now subject to heavy financial penalties. It is suing Primon a project management company on the grounds that its advice was incorrect and led to the project overrun. The key arguments of the case are summarized below. Your task is to reach a verdict on the case given these arguments.

1. Odell argued that they were broadly on target to complete the project to deadline before hiring Primon.
2. Primon argued that an independent assessment of Odell's projections indicated that they were 4 months over the deadline when they hired Primon.
3. Odell argued that Primon lost staff with the expertise to handle projects of this complexity.
4. Primon argued that they had retained and recruited many staff with well-documented experience of handling projects of greater scale and complexity with highly successful outcomes.

Green and McCloy determined that the arguments are the most critical features that sway people's decisions in courtroom cases. They also noted that people often use two methods to make legal decisions. First, they evaluate the quality of the arguments that each side presents based on the quality of data cited. Second, they assess the situations that are referred to by the arguments. These authors noted that people can imagine different possible situations occurring in the world. When multiple possibilities are entertained, decision confidence is shaped by the quality of arguments that support these different hypothetical situations. This study captures some of the important features of abductive reasoning, notably that people do evaluate multiple possible states of the world in these types of complex, real-world examples. When multiple hypotheses are evaluated, the strength of the particular details or arguments becomes a critical determining factor in how a decision will be made.

DECISION MAKING AND THE BRAIN

Value and Decision Making

We have already discussed the neural basis of decision making when we discussed fMRI studies of the framing effect earlier in this chapter. We also discussed some of the effects of brain injury on decision making in Chapter 7. Over the past decade the field of decision neuroscience has grown by leaps and bounds. The tool of choice for many of these decision making studies has been fMRI. Neuroimaging work has also benefitted from comparisons to cell recording studies that have frequently been carried out in animals.

One of the major findings in decision neuroscience in the 1990s was that reward-related neuronal activity appeared to code the timing and delivery of incentives for behavior. This neural activity was observed within the striatum, comprised of the caudate nucleus and putamen, which are parts of the basal ganglia. The striatal neurons were initially more active over their baseline firing rate when a reward was provided. These same neurons then began to increase their firing rate in response to a reward predictor, such as a light or tone in a conditioning task. Next, the neurons were observed to reduce their firing rate if an expected reward that should follow from a cue was omitted. In other words, these neurons signaled a reward prediction error. This very signal would be critical in decision making, as it is a feedback cue that the value of a particular cue has changed. In turn, behavior should change as result of the decision strategy being no longer effective (Schultz, Dayan, & Montague, 1997). Several later studies conducted using fMRI were able to replicate these findings at the scale of brain regional responses in humans (McClure, Berns, & Montague, 2003). The reward prediction error finding remains one of the cornerstones in decision neuroscience research.

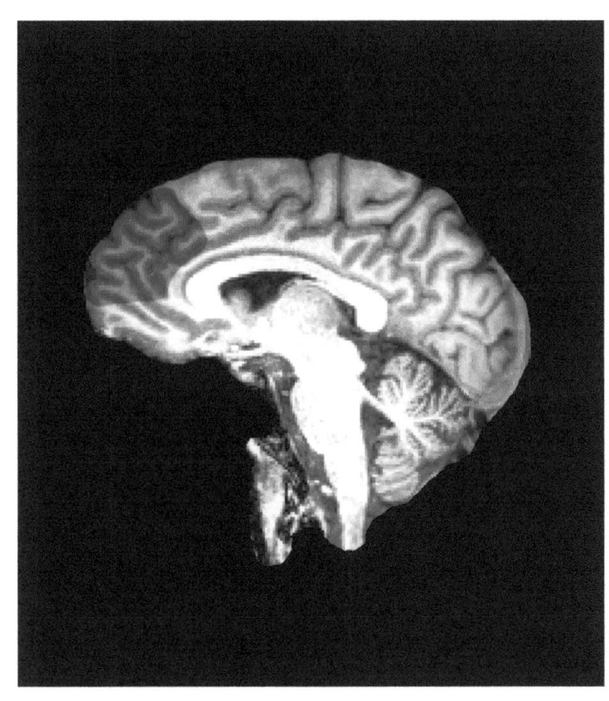

FIGURE 11.17 The medial prefrontal cortex has been identified as a region commonly correlated with people's estimation of value. The region may integrate signals from other brain regions in order to provide a neural coding of the utility of a particular option.

The study of value and how it is represented in the brain may be one of the most fruitful ways to link decision-making behavior to brain systems. In addition to these striatal dopamine system neurons studied by Schultz et al. (1997), other areas have also been found to be sensitive to reward value. Perhaps most notably, the ventromedial PFC has been associated with the expected value of a rewarding option in a decision (Kable & Glimcher, 2007). This is a region that overlaps with the site of damage reported for the patients who struggle to make decisions in the Iowa gambling task (Bechara, Damasio, Damasio, & Anderson, 1994; Bechara, Tranel, Damasio, & Damasio, 1996; Damasio, 2000). This PFC region may be a critical region for integrating value signals with behavior to guide us in our daily life decision making (Fig. 11.17).

SUMMARY

Decision making has been widely studied dating back to the times of ancient philosophy. Over the years, the study of decision making has expanded to include many different disciplines including economics, psychology, medicine, and public policy. Decisions range in complexity from simple decisions that can be made rapidly with little deliberation to complex decisions that require slower controlled processing. Dual-system models of decision making maintain that there are separable systems or processes that lead to fast or slow decision making. Multi-attribute decisions are often complex, as they involve numerous aspects to the alternatives, and we engage in a complex reasoning process when we must make multi-attribute decisions. Context influences our choices considerably and leads us to change our evaluations based on how information is framed or presented.

Our decision making is guided by heuristics, which are employed automatically and with little effort. Such heuristics include representativeness and availability, in which we tend to quickly estimate options as probable if they appear representative of the population from which they are drawn, or if they derive from readily available information. Heuristics are valuable in saving us time and energy, thereby allowing us to output a decision that is likely to work much of the time. There are situations in which our decision making goes astray due to an overreliance upon heuristics. Thus, it is important for us to judge when we need to engage in controlled processing that facilitates deliberative rational decisions, over those that are fast and intuitive.

The field of abductive reasoning focuses on reasoning to the best possible explanation. Abduction tends to yield inferences that are less likely to be valid than deductive reasoning. We tend to use abductive reasoning when we are faced with complicated situations and do not possess all of the relevant facts. Abductive reasoning is critical in complicated decision domains such as national security, intelligence analytics, or in medicine when complicated medical conditions present in the form of partial or incomplete sets of symptoms.

The growing field of decision neuroscience seeks to understand how decision making operates in the brain. Major findings indicate that neurons in the striatum of the basal ganglia are sensitive to rewards. These neurons are responsive to stimuli that predict rewarding outcomes, and they also signal failures of reward delivery. Such neurons may be responsible for guiding our decision making. Another key region in decision neuroscience is the ventromedial PFC that is sensitive to decision values and damage to this region results in overly risky decision making.

Decision making is a fascinating and diverse field. It is intimately linked to the study of cognition more broadly as many of our mental processes lead up to decisions. The continued study of the brain and decision making, along with cognitive modeling of decision behavior, may lead to further advances in understanding how we decide and how we can improve the decisions that people make.

END-OF-CHAPTER THOUGHT QUESTIONS

1. Research in decision making now includes fields such as neuroscience, psychology, philosophy, economics, and medicine. Why do you think the interest in this area has grown so much in relation to the study of reasoning, which is often confined to neuroscience, psychology, and philosophy?
2. How are decision making and reasoning related?
3. Decision making can involve simple or complex decisions. What do simple movement decisions have in common with complex multi-attribute decisions? How are these decisions different?
4. Some electrophysiology studies call into question our capacity for free will. Does the finding that motor physiological signaling precedes conscious awareness indicate that we really do not control our own decisions? Why or why not?
5. Rational models of decision making maintain that someone should always choose according to his or her own best interests. Which circumstances appear to show that this perspective is true? Under what cases do we violate rational models?
6. Heuristics are rules of thumb that enable us to quickly and efficiently make decisions. When are heuristics beneficial to us? What are some circumstances that should lead us away from using heuristics?
7. Dual-systems have been proposed to support human cognition. Given what we know about the complexity of brain responses involved in decisions, does a two-system theory still make sense?
8. Is abductive reasoning the same as decision making or are these different?
9. The ventromedial prefrontal cortex is activated in neuroimaging experiments when we process value. Does the representation of value hold the key to predicting what we will decide? What are some possible exceptions where we decide against our own apparent best interests?

References

Abelson, R. P. (1983). Whatever became of consistency theory? *Personality & Social Psychology Bulletin, 9*, 37–64.

Beatty, J. (1982). Task-evoked pupillary responses, processing load, and the structure of processing resources. *Psychological Bulletin, 91*, 276–292.

Bechara, A., Damasio, A. R., Damasio, H., & Anderson, S. W. (1994). Insensitivity to future consequences following damage to human prefrontal cortex. *Cognition, 50*, 7–15.

Bechara, A., Tranel, D., Damasio, H., & Damasio, A. R. (1996). Failure to respond autonomically to anticipated future outcomes following damage to prefrontal cortex. *Cerebral Cortex, 6*, 215–225.

Bem, D. J. (1967). Self-perception: An alternative interpretation of cognitive dissonance phenomena. *Psychological Review, 74*, 183–200.

Brehm, J. W. (1956). Post-decision changes in the desirability of choice alternatives. *Journal of Abnormal and Social Psychology, 52*, 384–389.

Damasio, A. R. (2000). *Descartes' error: Emotion, reason, and the human brain*. New York: Quill.

De Martino, B., Kumaran, D., Seymour, B., & Dolan, R. J. (2006). Frames, biases, and rational decision-making in the human brain. *Science, 313*, 684–687.

Ehrlich, D., Guttman, I., Schonbach, P., & Mills, J. (1957). Postdecision exposure to relevant information. *Journal of Abnormal and Social Psychology, 54*, 98–102.

Elliott, A. J., & Divine, P. G. (1994). On the motivational nature of cognitive dissonance: Dissonance as psychological discomfort. *Journal of Personality and Social Psychology, 67*, 382–394.

Epstein, S., Pacini, R., Denes-Raj, V., & Heier, H. (1996). Individual differences in intuitive-experiential and analytical-rational thinking styles. *Journal of Personality and Social Psychology, 71*, 390–405.

Fechner, G. T. (1876). *Vorscule der Aesthetik*. Leipzig. Germany: Breitkopf and Hartel.

Festinger, L. (1957). *A theory of cognitive dissonance*. Evanston, IL: Row, Peterson.

Festinger, L. (1962). Cognitive dissonance. *Scientific American, 107*.

Frederick, S. (2005). Cognitive reflection and decision making. *Journal of Economic Perspectives, 19*, 25–42.

Green, D. W., & McCloy, R. (2003). Reaching a verdict. *Thinking & Reasoning, 9*, 307–333.

Gremy, F. (1980). The future of information processing in medicine and public health. *Computer Programs in Biomedicine, 11*, 71–80.

Griggs, R. A., & Cox, J. R. (1982). The elusive thematic-materials effect in Wason's selection task. *British Journal of Psychology, 73* , 407–420.

Harris, L. T., & Fiske, L. T. (2006). Dehumanizing the lowest of the low: Neuroimaging responses to extreme outgroups. *Psychological Science, 17*, 847–853.

James, W. (1890). The principles of psychology. (Vol. 2). New York: Holt.

Kable, J. W., & Glimcher, P. W. (2007). The neural correlates of subjective value during intertemporal choice. *Nature Neuroscience, 10*, 1625–1633.

Kable, J. W., & Levy, I. (2015). Neural markers of individual differences in decision-making. *Current Opinion at Behavioral Sciences, 5*, 100–107.

Kagan, S. (1998). *Normative ethics*. Boulder, CO: Westview Press.

Kahneman, D. (2011). *Thinking, fast and slow*. New York: Farrar, Straus and Giroux.

Kahneman, D., Knetsch, J. L., & Thaler, R. (1990). Experimental tests of the endowment effect and the Coase theorem. *Journal of Political Economy, 98*(6), 1325–1348.

Kahneman, D., & Tversky, A. (1973). On the psychology of prediction. *Psychological Review, 80*, 237–251.

Kepecs, A., Uchida, N., Zariwala, H. A., & Mainen, Z. F. (2008). Neural correlates, computation and behavioural impact of decision confidence. *Nature, 455*, 227–231.

Knetsch, J. L. (1989). The endowment effect and evidence of nonreversible indifference curves. *American Economic Review, 79*, 1277–1284.

Kornuber, H. H., & Deecke, L. (1965). Hirnpotentialanderungen bei Willkiirbewegungen und passive Bewegungen des Menschen: Bereitschaftspotential und reafferente Potentiale. *Pfliigers Archivfur Gesamte Physiologie, 284*, 1–17.

Kunst-Wilson, W. R., & Zajonc, R. B. (1980). Affective discrimination of stimuli that cannot be recognized. *Science, 207*, 557–558.

Levin, I. P., Schneider, S. L., & Gaeth, G. J. (1998). All frames are not created equal: A typology and critical analysis of framing effects. *Organizational Behavior and Human Decision Processes, 76*, 149–188.

Li, J., Xiao, E., Houser, D., & Montague, P.R. (2009). Neural responses to sanction threats in two-party economic exchange. *Proceedings of the National Academy of Sciences, 106*(39), 16835–16840.

Libet, B., Gleason, C. A., Wright, E. W., & Pearl, D. K. (1983). Time of conscious intention to act in relation to onset of cerebral activity (readiness-potential) – the unconscious initiation of a freely voluntary act. *Brain: A Journal of Neurology, 106*, 623–642.

Lieberman, M. D. (2007). Social cognitive neuroscience: A review of core processes. *Annual Review of Psychology, 58*, 259–289.

McClure, S. M., Berns, G. S., & Montague, P. R. (2003). Temporal prediction errors in a passive learning task activate human striatum. *Neuron, 38*, 339–346.

Murch, K. B., & Krawczyk, D. C. (2014). A neuroimaging investigation of attribute framing and individual differences. *Social Cognitive and Affective Neuroscience, 9*, 1464–1471.

Nelson, D. L., McEvoy, C. L., & Schreiber, T. A. (1998). *The University of South Florida word association, rhyme, and word fragment norms*. http://www.usf.edu/FreeAssociation/.

Pierce, C. S. (1997). *Collected papers of Charles Sanders Peirce: 1931–58 edition*. In C. Hartshorne, & P. Weiss (Eds.). London: Burns & Oates.

Poses, R. M., & Anthony, M. (1991). Availability, wishful thinking, and physicians' diagnostic judgments for patients with suspected bacteremia. *Medical Decision Making, 11*, 159–168.

Rosenberg, M. J. (1968). Hedonism, inauthenticity, and other goals toward expansion of a consistency theory. In R. P. Abelson, E. Aronson, W. J. McGuire, T. M. Newcomb, M. J. Rosenberg, & P. H. Tannenbaum (Eds.), *Theories of consistency: A sourcebook* (pp. 73–111). Chicago: Rand McNally.

Satpute, A. B., & Lieberman, M. D. (2006). Integrating automatic and controlled processes into neurocognitive models of social cognition. *Brain Research, 1079*, 86–97.

Schneider, W., & Shiffrin, R. M. (1977). Controlled and automatic human information processing: I. Detection, search, and attention. *Psychological Review, 84*, 1–66.

Schultz, W., Dayan, P., & Montague, P. R. (1997). A neural substrate of prediction and reward. *Science, 275*(5306), 1593–1599.

Shafir, E., & Tversky, A. (1995). Decision making. In E. E. Smith, & D. N. Osherson (Eds.), *An Invitation to cognitive science* (2nd ed.) *Thinking: Vol. 3*. (pp. 77–100). MA: MIT Press.

Simon, H. A. (1957). *Models of man: Social and rational. Mathematical essays on rational human behavior in a social setting*. New York: Wiley.

Simon, D., & Holyoak, K. J. (2002). Structural dynamics of cognition: From consistency theories to constraint satisfaction. *Personality and Social Psychology Review, 6*, 283–294.

Smirnova, A., Zorina, Z. L., Obozova, T., & Wasserman, E. (2015). Crows spontaneously exhibit analogical reasoning. *Current Biology, 25*(2), 256–260.

Taylor, K. F., & Sluckin, W. (1964). Flocking in domestic chicks. *Nature, 201*, 108–109.

Tversky, A., & Kahneman, D. (1973). Availability: A heuristic for judging frequency and probability. *Cognitive Psychology, 5*, 677–695.

Tversky, A., & Kahneman, D. (1981). The framing of decisions and the psychology of choice. *Science, 21*, 453–458.

Walton, D. (2005). *Abductive reasoning*. Tuscaloosa: University of Alabama Press.

Wason, P. C. (1966). Reasoning. In B. Foss (Ed.), *New horizons in psychology* (pp. 135–151). Harmondsworth: Penguin Books.

Weiss, B. D., Berman, E. A., Howe, C. L., & Fleming, R. B. (2012). Medical decision-making for older adults without family. *Journal of the American Geriatrics Society, 60*, 2144–2150.

Wicker, B., Keysers, C., Plailly, J., Royet, J. P., Gallese, V., & Rizzolatti, G. (2003). Both of us disgusted in *My* insula: The common neural basis of seeing and feeling disgust. *Neuron, 40*, 655–664.

Zajonc, R. B. (1968). Attitudinal effects of mere exposure. *Journal of Personality and Social Psychology Monographs, 9*, 1–27.

Zanna, M. P., & Cooper, J. (1974). Dissonance and the pill: An attribution approach to studying the arousal properties of dissonance. *Journal of Personality and Social Psychology, 29*, 703–709.

Zysset, S., Wendt, C. S., Volz, K. G., Neumann, J., Huber, O., & von Cramon, D. Y. (2006). The neural implementation of multi-attribute decision making: A parametric fMRI study with human subjects. *Neuroimage, 31*, 1380–1388.

Further Reading

Tversky, A., & Kahneman, D. (1974). Judgments and uncertainty: Heuristics and biases. *Science, 185*, 1124–1131.

CHAPTER

12

Social Cognition: Reasoning With Others

KEY THEMES

- Theory of mind refers to the ability to understand and appreciate that other beings have thoughts, feelings, and intentions.
- Theory of mind has been observed to some degree in other species including primates and dolphins. Species that show theory of mind abilities tend to have large brains and live in close-knit family groups.
- Theory of mind develops as we age. Most people have a well-developed theory of mind by the time they enter school.
- Social hierarchies are observed across several complex species. Many species also have more complex social dynamics in which different individuals occupy different roles depending on the situation.
- Cetaceans (whales and dolphins) show evidence of social learning in their feeding techniques. The presence of social hierarchies, as well as large complex brains, likely enables these animals to discover new techniques and pass them along to other members of their species.
- Face processing is critical to social interactions. To successfully reason about others, we must be able to recognize the emotions that they display. Automatic processing of facial emotions helps to shape our social interactions.
- Well-developed arguments by single individuals can prevail in group situations provided that the individuals within the group keep an open mind and discuss the reasons for selecting a particular argument.

Reasoning
http://dx.doi.org/10.1016/B978-0-12-809285-9.00012-0

- Complex social and cognitive factors can limit the reasoning of groups relative to individuals, if such factors are not carefully controlled.
- Brainstorming occurs when groups generate ideas. It is most productive when people write ideas down and share them rather than freely discussing them.
- Hormones including oxytocin influence our reasoning when we process emotions and cooperate or compete with others.
- Cultural norms can influence our tendencies to reject or accept economic exchange offers that would be considered to be unfair by individuals with different cultural backgrounds.

OUTLINE

SOCIAL COGNITION: REASONING WITH OTHERS

Introduction

To this point in the book we have discussed the characteristics of different types of individuals and their higher cognitive abilities as they are applied toward reasoning and decision making. We have also reviewed the different types of reasoning that people perform in the last several chapters. For the most part we have reviewed reasoning from the perspective of a single person problem solving, thinking, or deciding using only that person's own personal mental abilities. Things are more complex and dynamic in the societies in which we live, work, and learn. We now turn toward the topic of reasoning in societies.

Social reasoning includes fascinating topics such as how we reason about other people, how people make decisions about others, and how we reason and decide in group situations. This chapter revisits some of the same topics that we have discussed previously, but examines them in the context of how we reason in groups with both the benefits and the limitations offered by input and interactions with other people.

We will begin with a discussion of developmental psychology reviewing the basic cognitive processes needed for social perception. The ability to perceive the social cues around us enables the more elaborate capacity to understand others. Social perception and understanding become guiding forces that influence much of our reasoning in daily life.

We will then examine the evidence for social abilities in several of the nonhuman species that we covered in Chapter 4. We can learn a great deal by considering human behavior through the lens of comparison to other species and what they are capable of. When we examine the social reasoning skills of certain species we are able to see some of the same abilities that people possess.

We next consider face processing and its contributions to social reasoning. The face can be thought of as the gateway to our emotions and feelings. We form first impressions based on such information, and these impressions shape our interactions with others. Information from the face has a wide impact on how we think and reason in our daily lives. These sections of the chapter offer insight into the tools that we have available to facilitate reasoning about other people and consider their input toward making our own decisions.

The next sections of the chapter focus upon the interactions that people have with one another and how our reasoning and decision making occur when we transmit our ideas to others. We are sometimes stronger when we put our heads together and think in groups. Wise counsel and input from trusted colleagues can strongly influence how we approach projects, draw conclusions, and carry out plans. On the flip side, negative influences from untrustworthy individuals can derail our thinking and break down well-laid plans. There can be "too many cooks in the kitchen" in some situations leaving us with group performance that undercuts the valuable contributions of each individual. We will review research that highlights the benefits and the drawbacks associated with reasoning in groups compared to situations in which we think or decide by ourselves. There are increasingly many situations in daily life in which we must trust others and work with them to grow our investments, accomplish complex projects, develop fair standards for the exchange of good and services, and build bonds within our communities (Fig. 12.1).

In the final sections of the chapter we consider the way that individuals and groups reason under dynamic conditions. When we engage our social abilities to gather and integrate information from diverse group members, we can arrive at our best possible outcomes. Some individuals have more influence than others within groups. Such people rise to positions of leadership and have a high impact on others. Certain people gain more influence than others in group situations through their abilities to perceive group dynamics, notice emotions, and translate these into their social behavior. Personality differences dramatically affect the dynamics of a group as individuals collectively solve problems or make decisions. Throughout the chapter we will discuss how group influences impact reasoning outcomes.

FIGURE 12.1 Reasoning in groups can bring out the best in us as we combine our collective wisdom to build a more productive society. *From Depositphotos.com.*

FIGURE 12.2 Increasing globalization has led people to have more opportunities to visit foreign cultures and interact with an increasingly diverse set of individuals. *From Wikicommons.com.*

We will also consider the differences observed among different people who have different levels of social ability and in some cases genetically linked disorders such as autism spectrum disorder (ASD). Nowadays, society has increasingly come to appreciate the diversity of people and social styles that exist within groups. An ongoing project in modern times is to determine how to best educate individuals with varying social abilities to maximize their contributions to society.

We conclude this chapter by considering the effects of culture on reasoning tendencies. This last topic may be particularly timely given the rapid globalization that is occurring across the world. Today more people have jobs and travel opportunities that put them into contact with individuals from other cultures. Such interactions are shaped by the commonalities and differences that exist between people with different backgrounds. Perhaps more than ever, the ability to successfully reason and decide alongside people of other cultures is a prized skill that offers rich opportunities for learning and influence in business and world affairs (Fig. 12.2).

REASONING ABOUT THE MINDS OF OTHERS

Theory of Mind

The term *theory of mind* refers to the ability to understand and appreciate that other beings have thoughts, feelings, and intentions. This ability is also known as *mentalizing*. The more that we can understand the characteristics of another person's thinking, the more we will be able to predict what that individual is likely to do next. Theory of mind underlies our ability to influence others by appealing to their needs and evoking their emotions. If we can "tug at the heart strings" of another person, we may be better able to enlist them to help us solve problems. If we can read and understand what someone wants, we will have opportunities to engage their interest and gain friendships. If we can read why someone is reacting coldly toward us, then we are likely to be more successful at diffusing conflict. Theory of mind strongly determines our opportunities to successfully navigate our social world.

We begin this section by reviewing evidence for theory of mind in other primates. Species with large brains, expanded cortical networks, and integrative, heteromodal cortical areas, such as the frontal lobes, tend to be social species. These species frequently travel in social groups, bands, or tribes. Such species tend to invest heavily in parental behavior and maintain close connections with other individuals. As we discussed in Chapter 4, large-brained species exhibit flexibility in their behavior and an adaptability that enables them to overcome what biologists refer to as stereotyped behavior, or instinctual responding. This same flexibility applies in social situations enabling groups to outperform individuals in a variety of tasks. These may include tasks important for survival such as hunting, gathering, and problem solving. We can observe the growth of theory of mind in children as they age. Age-related differences in theory of mind abilities explain some of the challenges faced by parents trying to reason with their children across the elementary school years.

Evidence for Theory of Mind in Other Species

The term theory of mind is currently applied widely across human development. Some of the original investigations of this ability included attempts to understand the behavior of nonhuman primates. David Premack and Guy Woodruff (1978) questioned whether chimpanzees have a theory of mind. We reviewed this study in Chapter 4 when we considered the capacity of chimpanzees to pass the mirror test (Gallup, 1970). In the mirror test a marking is placed on a body part of an individual or organism. A mirror is made available and an individual "passes" the mirror test when he or she demonstrates the ability to use the reflection to view the marked body part. Passing the mirror test is an indication that there exists an understanding that the image in the mirror is oneself rather than another individual. Notably, the elephant, chimpanzee, dolphin, and magpie all pass the mirror test (Fig. 12.3). The self-concept

FIGURE 12.3 Species that can pass the mirror test demonstrate a self-concept. A range of species can pass this test including elephants, chimpanzees, dolphins, and magpies. *From Shutterstock.com*

is a benchmark capacity needed to enable higher-order social knowledge. Passing the mirror test is not a trivial capability. Numerous studies have been conducted with fish that show aggression toward their mirror image and behave as if it is either a competitor or ally depending upon how the mirror is positioned.

As we discussed in Chapter 4, the chimpanzee is capable of demonstrating some aspects of theory of mind. Chimpanzees are able to demonstrate knowledge of others' capabilities when they beg for food from a human. When presented with a person facing forward and another person facing backward, chimpanzees will request food from the person facing them. This finding indicates that chimpanzees have an understanding that the person must be able to view their movements in order to respond to a food request (Povinelli & Vonk, 2003). In other words chimpanzees can appreciate that people have a field of vision and that blocking their vision negates the ability to meet the chimpanzee's needs. Chimpanzees can also follow a person's gaze in order to discover hidden food (Povinelli, Bierschwale, & Cech, 1999). While these are impressive abilities, there appear to be some limits on theory of mind capabilities in chimpanzees. In a study showing the limitations of chimpanzee theory of mind, Povinelli and Eddy (1996) offered chimpanzees the opportunity to beg for food from either a person who wore a blindfold over her eyes or one with a cloth over her mouth. The chimpanzees demonstrated no particular preference toward either of the people. This behavior suggests that chimpanzees run into a limitation (either in ability or in willingness) that prevents them from simulating the viewpoint of the person wearing the blindfold. It is possible that the blindfold location is a detail that is too subtle for the chimpanzee to notice, but a human would likely notice the blindfold placement easily, especially if it related to another person perceiving a personal need. People are capable of following the eye gaze of another individual quite automatically and gaze is a strong indicator of other's interest and intentionality.

The Development of Theory of Mind

The Sally-Anne test is one of the most famous examples of a developmental task that examines a child's theory of mind ability. Developed by Simon Baron-Cohen, Alan Leslie, and Uta Frith (1985), the Sally-Anne task evaluates the ability of an individual to understand a false belief held by another person. The task requires people to notice the conditions that lead another individual to lack awareness of a critical feature present within a situation. It is a strong test for perspective taking when someone is able to understand that another individual must believe something that is false about the world. Fig. 12.4 illustrates the Sally-Anne false belief task. In this example, Sally has a marble and places it inside her box before leaving the room. Anne observes the marble being placed into the box and then transfers the marble into a different box after Sally has left the room. The individual performing the test is then asked where Sally will look for the marble when she returns to the room. The observer is aware that Anne has moved the marble, but he or she is also able to notice that Sally was not present to observe the marble being moved. Therefore, Sally holds a false belief that the marble is still in the original location where she

FIGURE 12.4 The Sally-Anne test is used to evaluate theory of mind in children. In this task a child is asked to predict where Sally will think the marble is located after she returns to the room. The child clearly knows where the marble really is, but theory of mind ability is needed to realize that Sally does not know the correct location of the marble, as she was not present when Anne moved it into a new box. *From Wimmer, H., & Perner, J. (1983). Beliefs about beliefs: representation and constraining function of wrong beliefs in young children's understanding of deception. Cognition, 13, 103–128. Copyright © 2013 Byom and Mutlu. (CC BY).*

left it. Young children often fail the Sally-Anne test by claiming that Sally will know that the marble is in the new box, as that is its true current location. These children fail to consider the fact that Sally is not privy to all of the knowledge that they possess, as she was not present to see the marble being moved. By age five or six children will typically pass the Sally-Anne test demonstrating a full capability to notice and appreciate the perspective of another person whose knowledge differs from their own (Frith, 1989, pp. 33–52).

Mentalizing, or taking the perspective of another person, involves knowing what is on someone's mind at a given moment or in a particular situation. This ability is complemented by specific knowledge about a given person. Such knowledge may include the personality traits, background, cultural experiences, and temperament of other people. In addition to these knowledge-based clues about behavior, we can perceive emotional expressions, eye gaze, tone of voice, and nonverbal social cues. This all adds up to a wealth of information that we are able to apply toward predicting how someone will behave and what is on their mind at a given moment. Many disagreements arise because people fail to take the perspective of the others present. As an example, many heartfelt disagreements about politics and religion come about because of someone's insensitivity to the background knowledge of another person. Without being able to understand what someone is aware of, or has been exposed to, it will be difficult to predict their behavior in conversation and appreciate their viewpoint. The ability to perceive these features of others' knowledge and a degree of openness are both needed to effectively diffuse conflicts among people.

SOCIAL REASONING IN OTHER SPECIES

Social Dominance Hierarchies

Social dominance hierarchies are important when we consider the behavior of many different species. Social hierarchies apply to many mammal species that live in groups and have to coordinate group hunting behavior. Such behavior is highly organized in species that regularly live in groups such as wolf packs and lion prides. These animals hunt in groups and are keenly aware of their own status in relation to the other individuals present within the group. Dominant wolves are referred to as alphas. There are commonly both alpha males and females. The alpha status assures a wolf the highest priority in getting to feed first and also establishes which individuals will have the opportunity to breed with one another. These factors contribute to determining the survival of a particular individual and their ability to pass on their genes. Similarly in lion prides, an alpha male or sometimes multiple alpha males (that are closely genetically related) hold dominant positions in the pride. An animal in the alpha position spends much of his time in scent marking the territory and patrolling its boundaries in order to prevent rival males from infringing upon the territory. As in wolf packs, the alpha male will be afforded his choice of food and will do the bulk of the mating, thereby enhancing his overall fitness. Female lions also vary in dominance status. Dominance can be determined in part by age and hunting prowess (Fig. 12.5). The alpha status regularly changes over when an animal becomes sick, older, or simply when a stronger animal moves in and successfully challenges the leader. At this point the vanquished alpha will often be forced out of the group and his chances of survival will reduce considerably.

Social Behavior in Nonhuman Primates

Social status is an important factor in nonhuman primate groups. Stress researcher and neurobiologist Robert Sapolsky described the social hierarchies of baboon troops in his book entitled *A Primates Memoir* (2001). Sapolsky described the daily lives of the individuals within the baboon troops that he followed. These primates demonstrated strong social reasoning abilities. The individual occupying the alpha position changed over many times during Sapolsky's observations in Kenya. Sometimes the strongest and most ill-tempered males took over the alpha status of the troop by brute force. At other times, illnesses claimed the lives of some of the strongest animals leaving less capable males in the alpha role. Still other times, circumstances allowed a more benevolent baboon to take over the alpha role. In all cases, the rest of the troop had to fend for themselves, their offspring, and one another while ranking lower in the hierarchy. Sometimes coalitions of baboons formed alliances in response to inappropriate aggression from one of the high ranking troop members. At other times, higher ranking individuals were overthrown by cooperative groups of younger and weaker animals. In all cases, the members of the troop knew their own status and how it compared to the other group members. To survive in groups, it is necessary for the individuals to show deference to the higher ranking members, allowing them priority in feeding and mating. Beyond this, there are many benevolent tendencies that can be of service to the collective welfare of a primate group. Grooming behavior is a bonding tool that baboons use to build cohesiveness, reduce stress, and form social attachments. Often the most successful individuals over the long term were those who were capable of building alliances, reigning in their aggressive tendencies, and benefitting from the social support of the others in times

FIGURE 12.5 Lion prides are highly socially organized. Female lions specialize in hunting for the pride. There is often a hierarchy within the group that is led by a dominant alpha female. Age and hunting ability are important factors in determining the dominance hierarchy in many species. *From Shutterstock.com.*

of illness or injury. Despite the common view that the largest and most aggressive individual will always dominate, it appears that building bridges, bonding with others, and exercising restraint in aggression may lead to the most successful lives in the troop.

Social Behavior in Cetacean Pods

Whales and dolphins (collectively known as cetaceans) are highly social mammals. Like primates, whales have very large brains with high levels of cortical mass (Fig. 12.6). In terms of biological classification whales are part of a larger infraorder called cetacean, which includes the baleen whales (Mysticeti) and the toothed whales (Odontoceti). The baleen whales include many of the largest species, among them the blue whale and the fin whale, which can both exceed 80 ft in length making them the largest animals on Earth. Other baleen whales include the humpback whale, which can reach 50 ft in length and 50 tons. Humpbacks are well-known for their elaborate vocalizations, often referred to as songs. The toothed whales include the dolphins, of which there are many species including the largest member, the killer whale. Another well-known toothed whale is the sperm whale, which can reach enormous sizes of 50 to 60 ft in length similar to those of the large baleen whales.

Despite their enormous size, whales are not particularly well studied scientifically. Whales are difficult to study because they are capable of incredible geographic dispersion and cannot be easily kept in captivity. Those dolphins

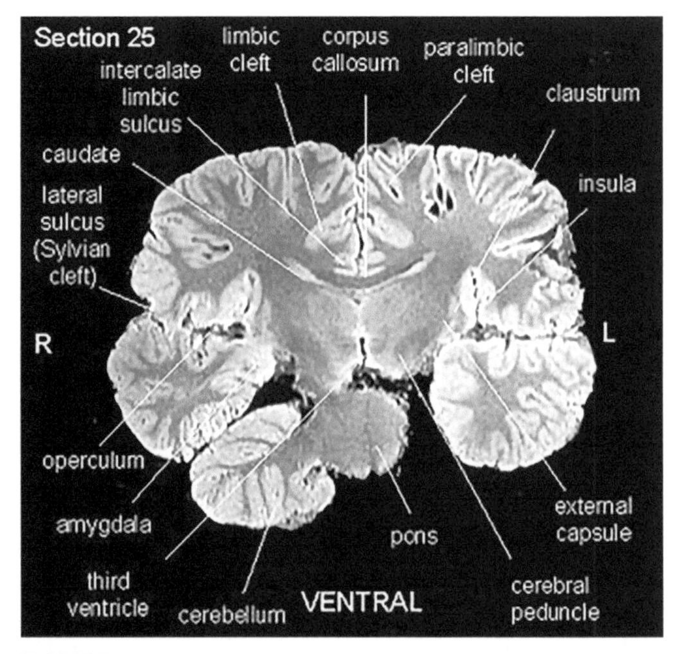

FIGURE 12.6 The brains of whales are large in size and display an extended degree of cortical mass. These complex brains enable whales to communicate acoustically and organize themselves into social groups. *From Marino, L., Sherwood, C. C., Delman, B. N., Tang, C. Y., Naidich, T. P., & Hof, P. R. (2004). Neuroanatomy of the killer whale (Orcinus orca) from magnetic resonance images. The Anatomical Record. Part A, Discoveries in Molecular, Cellular, and Evolutionary Biology, 281(2), 1256–1263. The authors thank Ilya I. Glezer and Peter J. Morgane for their generous donation of the specimen, John I. Johnson for advice and assistance with neuroanatomical identifications, and John C. Gentile for MRI technical assistance.*

and killer whales that are kept in captive environments do not afford researchers the opportunity to investigate the social lives of the species as they occur naturally in the ocean. Cetacean experts Luke Rendell and Hal Whitehead (2001) reviewed evidence of social behavior in their intriguing review paper entitled "Culture in Whales and Dolphins." They focused their review on four primary species due to an overall lack of scientific reporting in other whale species. The four species most studied include the bottlenose dolphin (*Tursiops truncatus*), the killer whale (*Orcinus orca*), the sperm whale (*Physeter macrocephalus*), and the humpback whale (*Megaptera novaeangliae*). These species vary widely in terms of which oceans they inhabit, their range, and their diet, yet all of them possess large brains with densely interconnected cortex (Fig. 12.7).

Rendell and Whitehead considered the social lives of these four cetacean species in relation to three key factors: 1) *The rapid spread of behavior across populations* – this refers to situations in which a complex and novel behavior moves through the population indicating a horizontal transmission of the behavior across a generation of individuals, 2) *Parent-to-offspring transmission of reasoning strategies* – representing cases in which a behavior is passed from a parent to their offspring from one generation to the next vertically, and 3) *Behavioral varieties among distinct groups* – situations in which different groups of individuals show new sophisticated behaviors that are different from one group of individuals to the next. Next we will consider each of these three methods of social learning as they occur within the cetacean species.

FIGURE 12.7 The four major cetacean species that were reviewed by Rendell and Whitehead (2001) in their review paper, making the case that cultural factors operate in the lives of whales and dolphins. (A) the bottlenose dolphin (B) the killer whale (C) the sperm whale (D) the humpback whale. *From Shutterstock.com.*

The Rapid Spread of Behavior Across Populations

The humpback whale may be best known for its complex and elaborate vocalizations. Some of the first of these vocalizations to be recorded were actually released by biologist Roger Payne in the form of an album in 1970. Humpback whales live in groups termed *fission-fusion societies* comprised of 18 individuals or fewer (Clapham, 1993). Such groups are often comprised of a set of core individuals, but the group is loosely organized and likely to split up into subgroups over time due to environmental changes or the availability of food.

Humpback whales are capable of some interesting achievements in complex group problem solving. The humpback feeds upon small fish and krill, a type of small crustacean. Humpbacks feed by gulping large quantities of these organisms into their huge mouths along with a substantial amount of water. The whales then expel the water through their baleen plates while keeping the food inside. The challenge of this feeding method is to make each gulp worth the effort in terms of calories gained. Humpback whales use "bubble nets" to facilitate effective gulp feeding. A bubble net is formed when several humpbacks swim in a circular formation expelling columns of air. The prey will not pass through the bubble columns. The fish or krill become centralized within the middle of the area and this enables effective gulp feeding when enough whales participate in creating a bubble net. A novel adaptation adjusting this behavior occurs when the whales "lobtail" in combination with using the bubble net strategy. Lobtailing occurs when a whale raises its tail flukes above the water and slams them back down again. Lobtailing disorients the prey organisms close to the striking tail flukes. The combination of lobtailing and bubble net feeding was reported to have spread rapidly during the 1980s in the waters off southern Maine (Weinrich, Schilling, & Belt, 1992). The spread of this behavior likely occurred by observational learning in which a large number of individuals watched one another and began to participate in this effective group feeding behavior.

Parent-to-Offspring Transmission of Reasoning Strategies

Humpback whales engage in social learning through their selection of migration patterns. The whales migrate between rich feeding grounds that are often located in colder northern waters and mating grounds in warmer waters closer to the equator. Young whales tend to repeat the same migration patterns that their mothers carried out. This migration behavior appears to be transmitted from one generation to the next, as other feeding and mating grounds are available to the young whales, but these are not utilized (Katona & Beard, 1990).

An interesting adaptation on feeding behavior has been observed in the bottlenose dolphin population that lives off Shark Bay, Australia. These dolphins have learned that humans will feed them if they visit a local beach. Many dolphins within the native population of Shark Bay take advantage of this food source, but the actual learning appears to have been directed by dolphin mothers who show their young this opportunity. The majority of the dolphins who take advantage of the food provided by humans had mothers who had also received this food in the prior generation (Smolker, Richards, Connor, Mann, & Berggren, 1997).

Killer whales that hunt seal pups make use of an innovative hunting strategy. The young seals are particularly vulnerable to predation and remain on the shore much of the time. Certain killer whales in Patagonia on the coast of Argentina have been observed hunting these seal pups by swimming up toward the sand and actually crashing themselves up onto the beach in order to snatch an unsuspecting seal. The huge killer whales then wriggle their way back into the deeper water. This is both an innovative strategy and a risky one. These whales cannot survive if they become stranded on the beach. This kind of risk could deter many young animals from attempting such a move. Fortunately, for the young killer whales the benefits of parental influence appear to facilitate this type of learning (Baird, 2000). Adult killer whales have been observed helping the young to successfully make their way back into the deeper water after a beaching attempt.

Studies of parents and offspring among the cetaceans are difficult to carry out as the genetic similarity must be established to verify that there is actual parent-to-offspring teaching occurring. Without clearly establishing the relatedness of the individuals the behaviors may instead represent instances of social learning by groups of whales that happen to swim together.

Behavioral Varieties Among Distinct Groups

Further evidence of social learning in cetaceans occurs when there are differences in behavior between two or more distinct groups of a particular species. This is evident in killer whale populations. Some of the most widely researched members of this species live in the waters off Vancouver Island in matrilineal groups headed by a female killer whale. These whale pods remain highly stable and involve as many as 18 whales, none of which leave the group for any sustained period of time or migrate to other groups. The resident killer whales off Vancouver Island use a variety of calls to communicate with one another and these calls are distinctive to each group, or pod. Notably, the Vancouver Island resident killer whale pods specialize in eating salmon. Both the vocalization behavior and the specialization of prey type are distinctive within this population.

One markedly different killer whale group is known as transient killer whales. Unlike the resident pods, transients tend to travel a much wider range of the ocean in small groups of around three individuals. The groups of transient whales sometimes reconfigure with an individual member leaving the group and joining another for a period. The transient killer whale groups do not tend to vocalize nearly as often as the resident whales and their vocalizations are qualitatively different than those of the resident killer whales. Lastly, transient whales pursue different prey than the resident killer whales. These transients tend to hunt seals, porpoises, and other large whales. This difference in food selection relative to the resident whales means that the hunting methods and tactics of the groups differ dramatically between these groups.

The differences between resident and transient killer whales offer a glimpse into the effects of social behavior on reasoning abilities in these animals. As we discussed in Chapter 4, obtaining prey is one of the clearest instances of behavior that we might classify as reasoning, provided that the animals are using novel and complex behavior. These hunting techniques frequently appear to have been generated by trial and error, or by social learning. Social learning is a situation in which individuals observe other animals succeed in hunting or foraging and repeat those for themselves. In the case of killer whales, there is good reason to suspect that some of the clever specializations for hunting are communicated by the whales to the other individuals, as the groups differ so dramatically in terms of both prey types and the solutions that different groups of animals have arrived at for obtaining prey. The resident killer whales may be classified as being more social overall, as indicated by the stability of group membership within their pods and the more common vocalizations among pod members. The resident killer whales diet of salmon does not require the elaborate methods and intensive cooperation necessitated by the hunts of the transient killer whales. The transient killer whales exhibit some of the most impressive social hunting methods in part because they specialize in hunting large intelligent marine mammals. These hunting methods include cases in which a pod of killer whales will attack and kill large baleen whales, such as the right whale, gray whale, and even the huge blue whale. Each of these species can reach lengths of over 40 ft and weights ranging from 20 to 50 tons. Each of these prey species dwarfs even the largest individual killer whales. Attacking prey of this size requires intensive communication and cooperation among the killer whales. To take down these ocean giants, the transient killer whales must approach their targeted victims silently. Once their presence is detected, they must then take on different roles in the hunt with some whales chasing and herding their victim, while others jump on top of the whale to disrupt its breathing and eventually drown the victim. Other killer whales will focus on biting and raking the flukes, fins, and flanks of the larger whale to further weaken it. All of these roles are highly taxing to the killer whales as well with an attack spanning multiple hours. The killer whales likely communicate regularly as they have been observed to take turns carrying out the different roles in these hunts (Russell, 2001). The killer whale's brain is both large and highly convoluted with a massively expanded cortex. Such a brain is capable of both generating and receiving detailed acoustic information and using that information to facilitate hunting behaviors.

A fascinating example of group problem solving has been observed in the bottlenose dolphins off the coast of Brazil (Pryor et al., 1990). There are numerous fishing boats that frequent the waters that are home to these dolphins and the boats fish with large nets. The dolphins began working cooperatively with the fishermen, and the earliest examples date back to 1847. In these cases, the dolphins spot the fish underwater. They then signal the fishermen use a rolling motion to indicate the best times to cast the nets. Additionally, the dolphins communicate the number of fish to the people by how much of the dolphins body comes out of the water (Fig. 12.8). It is unclear whether this body signal is explicit and intentional, or whether it is generated by an implicit type of process. The benefit to the people is that they have an excellent spotter in the dolphin that can help them to catch the most fish. The dolphins benefit by getting to catch an easy meal by scooping up the stunned fish that fall out of the nets. This interesting behavior appears to be transmitted from one dolphin to another and probably down at least three generations (Pryor et al., 1990). There are other dolphins in the area that do not participate in this particular fishing technique. In fact, some dolphins have been observed to actually interfere with the technique. This again suggests that only some

FIGURE 12.8 The bottlenose dolphins off the coast of Brazil signal to fishing boats when fish are near. The dolphins then benefit by catching the leftover fish after the boats submerge their nets. *Credit Fábio Daura-Jorge.*

isolated groups of dolphins generated the behavior and passed it on within their group, while others of the same species do not appear to understand the behavior and do not benefit from it.

Implications of Social Abilities for Problem Solving

In the preceding sections we have discussed some advanced problem-solving abilities that are exhibited by primates and cetaceans. These species have large brains and are capable of a complex repertoire of behaviors. The examples we have discussed involve the behavior of groups in which the individuals may not have discovered the behavior on their own. Rather, it appears that many of these sophisticated behaviors are transmitted from one individual to another. Important factors that facilitate this transmission of novel behaviors include the level of parental investment of the species. Those species that spend more of their lives in tight social groups with their relatives tend to be capable of behavior that has been transmitted by parent to their offspring. Other factors include large brains with extended cerebral networks. The brain organization of these complex mammals likely enables successful new behaviors to be discovered, repeated, and modeled for others to acquire. In the remainder of the chapter we will return to human reasoning and examine how the influence of other people facilitate or interfere with our reasoning abilities.

SOCIAL ABILITY AND FACE PERCEPTION

The Face of Communication

Face perception is one of the most important aspects of social cognition. We use information transmitted from the face when we reason about other people and try to understand what may be on their minds. Face recognition is a critical skill that develops early and supports our social abilities. Emotion recognition is perhaps second to face recognition in enabling social reasoning. People's facial expressions give us important clues regarding how they are feeling and reacting to ongoing events. These clues can be subtle, but people who can perceive emotions from faces and from listening to the tone of voice cues are in a privileged position to better understand others and anticipate actions and statements that will help to guide, build consensus, and lead others.

The Basic Emotions and Cross-Cultural Similarities

Early studies of cultural differences indicated that the basic emotional expressions are perceived similarly across cultures. The five basic emotions are happiness, sadness, anger, disgust, and surprise. Paul Ekman and Friesen (1976) conducted pioneering research indicating that these emotional expressions are universally identified across cultures. In Ekman's studies, he presented photos of individuals modeling the five basic emotional expressions. Across cultures people were able to accurately identify these expressions and associate them with the emotions that they represent. Neuroimaging studies have indicated that the amygdala, a primary region supporting emotional processing, is active during the processing of the basic emotional faces. The amygdala appears to be particularly active when we perceive faces that display negative emotional states such as fear, sadness, and anger (Glascher, Tuscher, Weiller, & Buchel, 2004; Wang et al., 2004). The association of the amygdala with facial processing indicates that processing these facial emotions is a highly automatic activity placing these facial emotions in a primary role for influencing social reasoning.

The Role of First Impressions

According to an old saying "You never get a second chance to make a first impression." Is this statement mere folklore, since we form impressions over time based more on people's words and actions? Do we actually place an exaggerated level of importance on a rapid initial impression and the success or failure of that impression changes all interactions from that point onward? Neuroimaging studies are beginning to provide surprising insights into the way that our brains process first impressions. Two fascinating studies by Nicholas Rule and his colleagues have indicated a strong role for the amygdala in processing first impressions of faces.

Rule et al. (2010) asked the intriguing question: do people vote on the basis of first impressions? The researchers showed individuals photos of political candidates. The study was run cross-culturally to avoid having a sample of participants who are overly familiar with the candidates. The researchers asked Japanese and American participants to evaluate candidates from each of these countries respectively. Across both participant groups, amygdala activity was elevated for those candidates who had won elections. This is especially interesting as the participants were not aware of which candidates had won these elections in advance of performing their evaluations. Additional activity within the amygdala was associated with viewing out-group candidates. In other words, American participants exhibited elevated amygdala activation when viewing all Japanese candidates and vice versa. This study suggests that people are highly sensitive to the features of first impressions. The tendency of dominant looking people to appear successful may have a significant bearing on our elections around the globe.

FIGURE 12.9 The amygdala is highly active toward the emotional and dominance features of faces. *From goodfreephotos.com.*

A second study by Rule et al. (2011) investigated the brain imaging effects of first impressions on business leaders. Participants in this study viewed corporate photos of many companies' chief executive officers (CEOs). Participants passively viewed these images while undergoing functional MRI scanning without making explicit judgments about competence or leadership ability. After the scanning session was completed, participants rated the likely leadership quality that each individual would likely provide. Results indicated that activation within the left amygdala was elevated for those executives who would later be rated as better leaders (Fig. 12.9). Perhaps even more fascinatingly, the profitability of each company was associated with the level of left amygdala activity as well. This second result is particularly surprising since corporations are large entities that are full of complexity. The result linking the dominance or leadership appearance of the executives' faces to the company's profitability is downright remarkable under those circumstances. This may speak to the nature of modern business, however. Leaders of major corporations are responsible for quite literally being the "face of the organization." This person is charged with representing the company's image to members of the board of directors and to the public in general. Therefore, it may make sense that CEO's' faces sometimes mirror the profitability and success of a company. Perhaps those companies that are doing best financially gravitate toward hiring CEOs who make excellent first impressions that are associated in a highly automatic way with perceptions of strong leadership.

These studies highlight the importance of face processing in social interactions and the subsequent decisions that we make regarding leadership. Choosing a leader in the domains of politics and business ought to depend upon many performance factors. How has the person operated in prior situations? What are their major promises to the country or organization? How do they interact with people in order to accomplish objectives? These are probably the types of questions people should be asking when they choose leaders, and many of us do care deeply about the answers to these task-related questions. Still, it is difficult to ignore the role of appearance in the leadership domain. The studies by Rule and colleagues add support to a growing literature indicating that we place great importance upon rapid and automatic impressions. These impressions are supported by the amygdala contributing strongly toward supporting our basic emotional processes (Domes et al., 2007). When we perceive others, their faces and the emotions that they project become intimately interwoven into the way in which we perceive their actions. This is critical to the reasoning process. If someone appears competent and trustworthy, we are likely to implicitly assume the best about their behavior. We may be more likely to believe such people and trust that they are informed and capable when they make statements that could appear controversial. By contrast, we are often less likely to give people these same benefits if they appear untrustworthy or less competent. We may instead tend to doubt and distrust their actions and words. The amygdala provides a rapid and highly automatic contribution to our initial impressions of others. This speed and primacy of processing make it difficult for people to disentangle the appearance of a face from the actions of a person. This leads to a bias of sorts in which we sense a type of intuition about others and are likely to trust that feeling. This can become troublesome in cases in which our impressions do not match with the reality of someone's actions. Under those conditions we are disturbingly likely to trust the wrong political leaders or support the wrong people as business leaders.

SOCIAL ASPECTS OF REASONING

Group Problem Solving

Sometimes we value the output of a group more than that of a single individual. In Chapter 11, we discussed biases and heuristics that people commonly exhibit. Such biases make individuals judgments somewhat suspect. What if the person was wrong? Should we strongly trust any one person's opinion? People attempt to avoid taking unnecessary risks by forming committees comprised of several people in order to make recommendations. Common wisdom suggests that "two heads are better than one," and in psychology labs we tend to frame our own opinions as being an "N-of-one," suggesting that any one person's opinion is only as good as a case study, which may not generalize to other people. Alternatively, some people will roll their eyes when it is suggested that a committee should be formed to get something done and that we are always worse off when we try to "do it by committee" (Fig. 12.10).

In this section we consider whether the reasoning of groups is preferable to that of one individual. More specifically, we will discuss situations in which a group can outperform an individual in reasoning and other times and conditions when we should probably trust the judgment of one person rather than a crowd. These are complex questions, as the people involved are key to answering the question. A single chess grandmaster will always give you better advice on your chess game than a group of novices. Likewise few of us would ever trust the single far-fetched opinion of one unknown mechanic over a group of mechanics with similar opinions when getting our car repaired.

Group Problem Solving on the Cognitive Reflection Task

The cognitive reflection test is used to evaluate people's engagement of automatic or controlled processing when they solve problems. The task was discussed in Box 11.2 of the previous chapter and you may recall that it consists of three questions that have an intuitive, but incorrect answer that comes to mind along with a more difficult correct answer that requires controlled processing and the ability to realize that the first answer is wrong. Psychologist Shane Frederick (2005) introduced the cognitive reflection test. Take a moment to refresh your memory about this test by reading through the questions below.

A bat and a ball cost $1.10 in total. The bat costs $1.00 more than the ball. How much does the ball cost? _____ cents.

If it takes five machines 5 minutes to make five widgets, how long would it take 100 machines to make 100 widgets? _____ min.

In a lake, there is a patch of lily pads. Every day, the patch doubles in size. If it takes 48 days for the patch to cover the entire lake, how long would it take for the patch to cover half of the lake? _____ days.

FIGURE 12.10 Groups and committees can get a bad name. There are times when group decision making slows things down and times when groups muddle the vision of an endeavor making things difficult to finish. At other times, group reasoning clearly outperforms any one individual's thinking. *From depositphotos.com.*

The intuitive answers to these three questions are: 10 cents, 100 min, and 24 days. Of course the real answers are: 5 cents, 5 min, and 47 days (see Box 11.2). This is a very unintuitive test that discriminates people who can more readily engage controlled processing over those who cannot do so as effectively. In a compilation of several participant samples, Frederick (2005) reported that only 17% of participants answer all three questions correctly, while 33% solve none of the questions correctly. In those data, 61% of participants solve either one problem or no problems correctly with data coming from many elite engineering university samples! Those who do solve the problems correctly also frequently reported having considered the incorrect answers before generating the correct ones. Thus, the cognitive reflection test proves to be a very difficult measure for most people. How would groups of people working together fare at facing this challenging test?

A study by Trouche, Sander, and Mercier (2014) asked individuals to solve the cognitive reflection test items and then to revisit their answers in groups. The initial phase of the experiment required people to solve the problems, rate their confidence in their answers, and provide a brief justification for their individual answers. Next, the participant was given a written argument either in favor of the correct answer or supporting an incorrect answer. At this point, an opportunity was provided for a change of answer. Most people who had provided the intuitive, but incorrect answer and read about the argument in favor of the correct answer opted to change their response. There was not a strong relationship between confidence and willingness to change one's answer. While people were far from perfect at the cognitive reflection test on their own, they did appear to be willing to listen to reason when presented with a sound argument.

In the second experiment presented by Trouche et al. (2014), participants went through the same exercise trying to solve the cognitive reflection test items on their own providing a confidence rating and justification. Next, rather than be presented with an argument, they were asked to discuss the answers within a group of three to five individuals. If even one of the group members had arrived at the correct answer then the group members were generally willing to change their answers to the correct solution. This finding held even when the individual in the group with the correct answer was not highly confident in that answer and when the correct answer was coupled with a confidence rating that was lower than other possible answers that had been provided. This study suggests that groups can outperform individuals on this highly challenging cognitive measure. It also presents a case in which our biases toward a rapid and intuitive answer can be mitigated by having to discuss the possible answers within a group of people. Extending these findings into daily life, the possibility exists that we will listen to other people and the most sound solutions can win out, even when they are not advocated for by the loudest or most confident people. When a minority presents a more well-thought out answer, that solution can still win over a group of individuals who did not arrive at the answer themselves. This provides a reassuring case in which group-based reasoning can outperform individual thinking.

Group Problem Solving on the Wason Card Selection Task

The Wason card selection task is one of the most well-studied examples of deductive reasoning. This task was discussed extensively in Chapter 9 when we reviewed the literature on behavioral studies that ask people to solve the classical four classes of deductive arguments. The two cases, *affirming the antecedent* and *denying the consequent,* provide valid conclusions. The other two cases, *denying the antecedent* and *affirming the consequent,* yield invalid conclusions. You may recall that the standard version of the Wason card selection task asked people to evaluate an arbitrary rule such as "If a card has a vowel on one side then it has an even number on the other side". An example of this task is depicted in Chapter 9 in the sections on deduction. The goal of the Wason card selection task is for the participant to turn over the *fewest number of cards* to test the "vowel-even number" rule, and the cards will read with a combination that includes a vowel (E) a consonant (Z), an even number (6), and an odd number (7). Most people will readily choose the card that shows the "E" face up as this card must be evaluated in order to assess the forward direction of the rule (known as *affirming the antecedent,* a valid reasoning strategy). Many people will also select the "6" card to turn over. The "6" card appears to be relevant at first glance. It is mentioned explicitly in the rule as fitting the requirement "an even number on the other side."

The problem is that the "6" represents *affirming the consequent,* an invalid reasoning strategy. Indeed we must go against our intuition and select the "7" card in order to check for a rule violation. To select the "7" the reasoner has to try to find a violation and try to disprove the hypothesis going against the common tendency to avoid looking for disconfirming evidence. Indeed the "7" card represents the correct case of *denying the consequent,* another valid reasoning strategy. This task has proved to be immensely difficult for people with the vast majority of individuals ignoring the unintuitive, but correct odd number card that is necessary in order to fully check that the rule holds. How do groups perform when attempting to solve this deductive dilemma?

Moshman and Geil (1998) asked people to solve the Wason card selection task alone and then in groups. People initially had an opportunity to solve the standard vowel-consonant version of the task. Next, they were asked to provide a written justification of their answers. At this point participants were placed into groups of five to six individuals and each person made a verbal justification of their card selections. The groups had to reach a full consensus regarding which cards should be selected. At this point; all group members were asked to provide a written summary of the group's favored answers, which ought to be the best of the five to six people's judgments. Participants were then offered an additional chance to revisit the group discussion after this point. The findings from this study were highly encouraging. As with most Wason card selection task studies, individuals were initially most likely to generate a combination of the intuitive and correct strategy affirming the antecedent (the "E" card) and the intuitive, but incorrect strategy of affirming the consequent (the "6" card). About half of the participants generated this common incorrect solution. Meanwhile, when placed into groups, people arrived at a much higher success rate, between 70% and 80% correct, in choosing the correct combination of affirming the antecedent (the "E" card) and denying the consequent (the "7" card). This is highly encouraging for group performance, as people are often reluctant to choose strategies that disconfirm the hypothesis by looking for negative evidence. The tendency for the groups to succeed in obtaining the unintuitive correct answer suggests that we are likely to put aside our incorrect personal opinions in favor of stronger answers from a group. Much like the cognitive reflection test performance data from Trouche et al. (2014), Moshman and Geil's results indicate that small groups can overcome some of the most significant reasoning challenges that individuals face in lab environments. Hopefully these results translate to real-world environments when decisions are made about safety protocols, financial markets, and managing large organizations.

Factors That Influence Group Creativity

Group dynamics can have large effects on our thinking and reasoning. Groups can change the process that we use to arrive at a decision based on the relationships that exist among the people involved. Some well-documented social psychological phenomena operate under group reasoning conditions. Some of these effects are helpful to group performance, while other group-related effects are detrimental. While groups vary in their composition, expertise, and familiarity, there are some commonly observed trends that are likely to bear out in many group situations.

Brainstorming represents one of the interesting examples of an activity that feels like it ought to work well in groups, but the research on the topic suggests otherwise. Brainstorming is a situation in which people generate as many new ideas or solutions to a problem as they can. Alex F. Osborn, an advertising executive, described brainstorming in 1953 in his influential book entitled *Applied Imagination*. The goal of brainstorming is to generate ideas, so it stands to reason that brainstorming sessions ought to increase in effectiveness as more individuals are added to a brainstorming session. The need to generate ideas is important in a variety of areas of work life. Government groups, companies, and organizations need to reach consensus, and they need to agree on their best ideas. Brainstorming offers one approach to develop the best ideas that are put forth by members of the group requiring a full exchange of ideas by all group members, which is critical to capturing the most effective idea generation (Janis & Mann, 1977).

The realities of brainstorming appear to be quite different, often indicating that groups can severely underperform relative to individuals. Some research suggests that if a group of five or six people were to brainstorm together, they will actually be less productive as a group than if those same individuals were to brainstorm independently (Fig. 12.11).

Osborn's original conception of brainstorming was as an act of creative problem solving that was centered upon group discussion. He presented the method as capable of vastly outperforming the idea generation of individuals acting alone. Group brainstorming was recommended as an additional method that should be used in combination with individual idea generation. The purpose of group brainstorming was to remove the roadblocks and inertia that can characterize some group or committee meetings. He described the common challenges of group idea generation as "driving with the brakes on." Osborn claimed that ideas generated using group brainstorming could yield gains of over 40% more useful ideas relative to individual idea generation sessions. The key ingredient was proposed to be the added benefit of discussion with other group members (Isaksen, 1988).

The major features of Osborn's approach to brainstorming included several guidelines that were critical to the process. These guidelines are intended to remove some of the undesirable social pressure, judgmental attitudes, and feelings of insecurity that can cripple group decision making. Osborn's four major principles are described as follows:

1. Judgments about the ideas generated are not allowed. The judgment should enter the process at a later stage.
2. Idea generation should include wild and outlandish ideas. The ideas are meant to be formative at this early stage, and they can be fine-tuned and scaled back later on after the initial idea generation session.

FIGURE 12.11 Sometimes a group can take the good ideas of individuals and dilute them or ignore them entirely resulting in a worse decision overall. *From depositphotos.com.*

3. Large numbers of ideas are encouraged. The more ideas the better, with the notion that more ideas generated initially will enable more opportunities later on in the process when the ideas are being refined.
4. The group should facilitate the ideas generated by each of the individuals. This step emphasizes that the group should not only help with the idea generation, but they should help to embellish and amplify the ideas that the individuals generate. Any idea can be improved upon and it is also acceptable to merge multiple ideas to improve upon them further.

In addition to the four steps outlined above, Osborn also presented other guidelines for successful brainstorming. These included conducting sessions of 30–45 min in duration by 5 to 10 individuals. The individuals in the group should have had previous experience with the domain of problem solving to which they are tasked. An orientation session was also recommended prior to the commencement of the session in order to place the group on the same page and to clarify what the goal of the task was to be. Additionally, a clear presentation of the problem to be addressed was seen as a critical feature of the session. This presentation should be made prior to beginning the act of brainstorming. These steps were recommended in order to eliminate the potential for confusion within the group members. The group was to have a leader, who would be carefully trained in the brainstorming methodology, but there would otherwise be equity among the group members to avoid any dominance by an expert over novices in the area. These final recommendations were designed to limit undue influence of group factors and social dynamics that could become undesirable if left unregulated.

The topic of brainstorming has been subject to large amounts of research over the past several decades. While the initial plan for brainstorming appears sound, in practice the output has been questionable. The level of skepticism surrounding brainstorming may stem from the numerous social dynamics that can limit the usefulness of this exercise.

Taylor, Berry, and Block (1958) conducted a landmark study indicating that individuals brainstorming independently could generate more quality ideas alone than individuals acting as a group. A challenge with this study is that the group brainstorming sessions included individuals who were randomly assigned. This procedure makes sound experimental sense from a sampling perspective, but it neglects some of the guidelines set forth by Osborn originally. Randomly sampled individuals would not include a leader, and this procedure does not allow for the equating the levels of expertise across group members. Lastly, the groups of participants in this study did not receive prior preparation on each of the problems that they were asked to solve. These key features of the Osborn technique for group brainstorming may have contributed to the limited success of the brainstorming groups in the study by Taylor and colleagues.

A later study by Mullen, Johnson, and Salas (1991) also indicated the limited effectiveness of brainstorming. In this study, the participants generated more quality ideas when brainstorming alone than when they interacted as a group. Thus, there appears to be a reduction in productivity of ideas by a group relative to when individuals generate ideas alone. Studies such as that by Mullen and colleagues point to a broader issue that has proven challenging in this area of the literature: researchers have not always applied the term brainstorming to the same set of processes. Additionally, there are clear social aspects of idea generation tasks that limit the effectiveness of the technique relative to cases in which individuals reason alone. We will now consider these challenges.

1. *Evaluation apprehension*: Individuals are inhibited from generating ideas due to group pressure and the fear of evaluation from the group (Isaksen, 1988). Note that this challenge violates two of the core brainstorming guidelines set forth by Osborn (1957). He recommended that evaluation should take place only after ideas have been generated and that ideas can be wild and outlandish. In real social interactions, people sometimes do feel anxious about being evaluated, and this may serve to undermine the intended methods for brainstorming. There is also a tendency for individuals to conform to the dominant ideas that appear to have gained group consensus (Paulus & Yang, 2000). To avoid this trap, group selection must be carried out in a very rigorous manner with individuals who do not feel pressured to conform to dominant individuals or ideas that have already gained some consensus.
2. *Social loafing*: Some individuals may not fully participate in the group, but instead will allow others to step forward and put forth the majority of the effort (Karau & Williams, 1993). This challenge can arise due to individual laziness, instances in which someone does not feel invested in the outcome, or in groups in which dominant individuals step forward and overly contribute to the process.
3. *Production blocking*: The productivity of an individual may at times be superior when he or she acts alone relative to when that same person is in a group setting, if the group setting introduces conditions that limit one from fully contributing (Diehl & Stroebe, 1987). Blocking can occur when an individual is unable to voice his ideas due to interference from another person talking, or delays created by someone recording another person's idea, thereby derailing the contribution of a group member.
4. *Attentional limits*: Limitations of attention are a cognitive factor that may limit the utility of group brainstorming relative to individual brainstorming. Individuals who are placed into group situations have to divide their attention between the act of generating new ideas and monitoring or evaluating the ideas of others. Dividing our attentional resources in this manner can negatively affect the overall idea productivity of an individual when they are placed into a group situation (Mulligan & Hartman, 1996). This challenge may be particularly damaging due to the fact that it does not rely upon group dynamics, but rather upon individual cognitive capacity. Even the most effectively prepared and assembled group may still suffer from this challenge as a result.

These challenges cast doubt upon the value of brainstorming in a group format and could suggest that individual brainstorming may be a superior way for organizations to generate ideas.

There are conditions that can enhance the effectiveness of group idea generation using methods designed to limit some of the negative effects of group thinking and maximize some of the positive effects. From a cognitive perspective, associations can be made to cue memories. Theoretically, the diverse perspectives of a group should be capable of being harnessed in a way that increases the overall generation of ideas through the collective efforts of the group members. The sharing of ideas within a group may be one route by which to cue the superior collection of associations that a group should be capable of (Brown, Tumeo, Larey, & Paulus, 1998). To access this superior memory and association capability possessed by a group over an individual, it is necessary to develop methods to overcome the challenges that group social dynamics present. Chief among those challenges are the blocking of ideas that comes about due to the divided attention state people are placed into when they have to both generate ideas and listen to and evaluate the ideas of others. A second critical idea blocking factor is that social inhibition can prevent people from fully participating and offering their ideas for fear of negative evaluations being made by others.

Overcoming the Limitations of Generating Ideas in Groups

Methods emphasizing writing down the ideas generated and then sharing them with a group have been termed "brainwriting." Brainwriting methods offer a means to avoid, or to minimize, these negative factors (Greene, 1987). Van de Ven and Delbecq (1974) presented one effective brainwriting technique called the *nominal group technique*. Using this method, a group was tasked with idea generation, but in a highly structured way that involved both individual brainstorming and group evaluation. In the nominal group technique, individuals write down their ideas independently. After the ideas have been generated, they are next shared within a group environment. Ideas are written down one-by-one for the purpose of a targeted discussion focused on clarifying and improving each idea. Lastly, the ideas are then rank ordered by the group in terms of their quality, by means of a vote polling all of the group members. This method was found to generate more ideas and better quality ideas than an unstructured discussion group when conducted with groups of seven individuals. Note that the process of independently writing down ideas limits the attention demands for the people involved. This method also offers people an uninterrupted opportunity to contribute their ideas without the possibility of interference by other individuals talking or interrupting their thinking.

Madsen and Finger (1978) presented another brainwriting technique that combines the benefits of individual and group thinking. In this case, groups evaluated two types of problems. One problem was abstract and fictional, asking people to predict the possible uses of having an extra thumb. The other was a more practical and realistic problem, in which people generated possible brand names for a new type of toothpaste. Madsen and Finger placed participants into cohorts of four individuals. In an independent idea generation condition, the individuals simply brainstormed independently and wrote down their ideas. In a pure group condition, the four individuals freely discussed possible ideas for the two problems without structured techniques or independent writing sessions. The third cohort engaged in a mixture of written idea generation carried out on their own and structured group feedback. This group initially brainstormed independently and wrote down their ideas. Half way through this session the individuals exchanged their ideas with another group member and received feedback. This study met with somewhat mixed results that depended upon the problem type. When generating brand names for toothpaste, both the independent writing condition and the group that exchanged written ideas for feedback outperformed the group discussion condition. This is similar to the finding presented by Van de Ven and Delbecq (1974). Unfortunately, there was no added benefit of the group influence in the ideas exchange condition over the independent brainstorming condition. Further, all groups performed similarly on the "extra thumbs" problem. The Marsden and Finger (1878) study points to some of the complexities that remain when individuals engage in group-based reasoning exercises. The hypothetical "extra thumbs" problem may not be realistic to people and may lead people down a path of unrealistic thinking that is not grounded by real possibilities. Additionally, the group size may enhance or inhibit the production of effective ideas.

As noted by Paulus and Yang (2000) the mere exposure to other individuals does not guarantee a maximization of cognitive resources by group members. They pointed out some simple challenges that may limit the ability of a group to harness their full potential for generating additional associations. One reason has to do with the possible divided attention states when people simultaneously attempt to both generate and evaluate ideas. This division of attention resources can result in less attention being focused on each of the ideas generated by other individuals. Straus (1996) also highlighted that it takes additional attention to appropriately contribute to a group discussion, as timing is critical in contributing an idea. An effective group member with a good idea will need to balance the creative spark of when the idea forms with revealing the idea at an appropriate moment. If the idea is voiced too early, then it may be dismissed by the group before it can be discussed and reach its potential. Additionally, a good idea can be dismissed, or lost in the shuffle if it is presented to a group at a time when people are engaged in intensive discussion about a different idea, or simply not paying sufficient attention when an idea is voiced. Another challenging related issue is what creativity researchers have termed incubation. Some ideas require additional time to fully form and associate with other possible ideas. Evidence collected over many decades dating back to the early period of Gestalt psychology indicates that an incubation period can dramatically benefit creative problem solving. Modern evidence from neuroscience continues to point to the benefits of time and sleep for the consolidation of memories (McClelland, McNaughton, & O'Reilly, 1995). The practical implication for group idea generation is that many ideas that are generated are simply not revealed at a point when their full potential can be realized.

Paulus and Yang (2000) discussed a method for group idea generation designed to simultaneously address the challenges of idea blocking by a group and the negative impacts of limited attention resources. Participants were grouped into cohorts consisting of four individuals to ponder the possible uses of paper clips. In an individual idea generation condition, the participants were asked to generate ideas independently and write those ideas down on paper. In two group conditions, the participants were seated around a table each having their own sheets of paper. Both groups were tasked with generating as many quality ideas as possible, but one of these group conditions additionally emphasized memory through an instruction indicating that there would be a memory test on the ideas generated after the experiment was over. The group members generated an idea and then passed it on to the person to their right. At this point the group members added their own ideas to those generated by the person to their left. Critically, the group members all wrote with a different ink color. This ensured that every group member was accountable for the production of quality ideas, as their contributions could be directly traced back to them. This feature of the brainwriting exercise prevents social loafing, in which a general diffusion of responsibility limits people's willingness to contribute to a group (Fig. 12.12). The experiment was divided into two sessions. In the second session, all participants had an individual opportunity to continue generating ideas for the paper clip use. Results indicated that both of the idea generation

FIGURE 12.12 Brainwriting can be an effective exercise to enhance group idea generation, while also reducing the negative effects that can occur when people reason in groups. By asking people to write down ideas in a unique ink color, Paulus and Yang (2000) were able to reduce the possible negative effects of social loafing.

groups outperformed the independent individuals in terms of productivity. This is in stark contrast to prior studies that showed a reduction in overall idea productivity by a group relative to individuals (Mullen et al., 1991). Additional findings included a questionnaire response in which the group conditions reported feeling more positive about verbal problem solving after the task and perhaps most importantly both group conditions reported greater enjoyment of the task overall relative to the independently conducted condition.

In summary, group problem solving appears to enhance the productivity and enjoyment of the individuals involved relative to independent thinking. Critically, this advantage of groups over individuals does not necessarily happen spontaneously. The presence of complex social and cognitive factors can limit groups relative to individuals if they are not carefully controlled. Optimal brainstorming appears most likely when people write ideas down and share them rather than hold a free form discussion. This writing exercise may allow greater idea incubation and the structure of the method appears to allow people to better time their contributions to overcome negative group dynamics limiting the ability to appreciate an idea. The innovation by Paulus and Yang (2000) by highlighting individual contributions through pen color appears important as well, as this emphasizes individual accountability and requires full participation by each person. A group can also lose its way through diffusion of responsibility if it becomes too large. Adherence to some of these methods takes time and effort, but in organizations needing to get the most out of their employees the effort appears to be worthwhile in terms of maximizing both the productivity and enjoyment of the act of brainstorming.

SOCIAL ASPECTS OF DECISION MAKING

Trust Games

Decision making is another area of higher cognition that is strongly influenced by groups and group dynamics. When we make a decision on our own it is likely to involve many of the biases and heuristics that we discussed in Chapter 11. Personal decisions made in the moment will focus on the information available to us and how we happen to feel about it at the time of the decision. This will likely involve our perception of the context and any memories that are cued by that context. Other people can influence these personal decisions, especially when we are indecisive. Sometimes bouncing ideas off of another person will reveal to us how we really feel about the decision. This is still a relatively indirect social influence on a decision.

There are other situations in which we must make decisions that explicitly involve the influence of others. Such situations have been studied in lab environments under the category of economic exchange games, or trust games. These games set up situations that force an individual to decide in favor of her own interests over those of another person, or alternatively such situations make evoke benevolent feelings and allow someone to put the interests of another person above her own. The degree to which we trust the other person is at the core of many of these types of social decision-making tasks. Trust is established by two major factors. A primary factor in trust is the first impression. First impressions matter a great deal and how someone looks can affect how trustworthy others perceive them to be. An additional secondary factor that influences trust is a personal history that we share with another person. If they have been helpful in the past, then we are likely to believe they will be helpful now and in the future. If they have cheated us in the past, we are much less likely to trust that person in the future. Once someone has revealed themselves to be self-focused, we find it very difficult to fully trust them. Statements such as "Once bitten, twice shy" and "We may forgive, but we do not forget" apply in such cases. Let's now consider two famous decision-making games that take place between two individuals. Trust will influence the decisions considerably, as they must decide whether to cooperate or not.

The Ultimatum Game

The Ultimatum game is a behavioral economics exchange game that is played over numerous trials.

The situation places the monetary interests of two people into close association (Güth, Schmittberger, & Schwarze, 1982). In a standard Ultimatum game, there is an amount of money that can be split between two players, a proposer and a responder. Often a sum of 10 dollars is used. The proposer is placed in control of the money and has to make an offer to the responder. If that offer is accepted, the proposer and the responder each receive his or her agreed upon amounts. If they do not agree and the responder rejects the proposer's offer, then nobody receives any money. For example, let's imagine that a round of the Ultimatum game is going to be played by Joshua and Dominic. Joshua is the proposer and will have a chance to split 10 dollars with Dominic. Dominic knows that there are 10 dollars available and so he will be fully aware of how much Joshua would get to keep and how much he is being offered. Let's imagine Joshua is extremely fair and offers to split the 10 dollars evenly and offers Dominic 5 dollars. Almost everyone will accept this offer and Dominic goes ahead and says, "Yes, Joshua, that is very fair, I accept the 5 dollars." A fair even split in an economic game will emerge as a favored strategy for many people, as it is an equilibrium point where both people receive the most possible without anyone being shortchanged (Haselhuhn & Mellers, 2005). From a purely rational economic perspective, Dominic should theoretically always accept any offer made by Joshua, even if it is one cent, as the alternative option of rejecting the offer results in no money.

Let's imagine these two play the Ultimatum game again. It is in Joshua's best interest from a purely monetary standpoint to see if he can get away with a more selfish and lopsided offer, as long as he thinks Dominic will still accept it. On this round Joshua offers $3.50. Dominic knows that Joshua will get to keep $6.50 for himself. He feels this is a bit shady, but $3.50 is not bad and certainly it is better than nothing. He accepts saying, "Joshua, I'll take the offer, but I'm not thrilled that you're keeping more for yourself." Indeed over 60% of the time people will still accept an offer between 3 and 4 dollars. Now a third round occurs and Joshua really decides to push his luck. Dominic has accepted his past two offers and he's on a roll. He now offers Dominic a sum of 75 cents. Dominic has to think about this one a bit. He could probably pay for very little with that sum of money, and he is just plain irritated with Joshua at this point. He decides to reject the offer saying, "Joshua, you've gone too far. You've got to be kidding me if you think I'm going to accept only 75 cents when I know you're getting to keep 9.25 dollars." In this case, Dominic is violating expected utility theory from a purely rational economic model. Seventy-five cents should always outweigh nothing. Most people will reject this kind of unfair offer that they know is stacked so heavily toward benefitting the proposer. Over 90% of unfair offers below one dollar are rejected (Haselhuhn & Mellers, 2005). Economic exchange behavior in the Ultimatum game represents a case of deciding in favor of punishing the unfair by rejecting an offer that is simply too heavily weighted in that person's favor.

In some versions of the Ultimatum game there are variations on this premise. In some instances many rounds of the game are played. In these cases it is probably a sensible strategy to occasionally reject offers when they are too unfairly weighted in favor of the proposer. Rejecting offers is likely to drive up the next offer or future offers overall. If the responder appears to be too much of a push-over who will accept any offer, then the only incentive for the proposer to make even split offers is to avoid the guilty feeling of constantly taking advantage of the responder. Indeed future proposer offers are likely to be guided by the history that the two individuals experience in the game. In other variations, people play against a computer. In this case, people will typically behave more rationally by accepting much lower offers. Even when the proposer is known to be computerized, people do still choose to reject the most extreme unfair offers in the range of one dollar or less. The impulse to correct unfair offers is strong, even when we know our opponent is not a real person (Blount, 1995)!

The Prisoner's Dilemma

Another well-established reasoning task that involves trusting a partner is called the Prisoners Dilemma. It was originated as a problem posed by Merrill Flood and Melvin Dresher at the RAND Corporation in Santa Monica, California, in 1950. In Flood and Dresher's original version of the task, there was no reference to prisoners and the task was called "A Non-Cooperative Pair" (Flood, 1958). RAND specializes in providing research to the US Government on a wide range of important public policy and defense topics. Famous economists such as John van Neumann and John Nash did considerable work at RAND in this same era. Prior to the more famous "Prisoner's Dilemma" formulation of the task, Flood and Dresher asked two colleagues to play their game in which the two people could either cooperate or defect from one another. Refer to Box 12.1.

Albert W. Tucker developed the most notable versions of the Prisoner's Dilemma task that is framed around two hypothetical prisoners who are attempting to act in their own best interest under police questioning. Tucker described the conditions of the game as follows (Poundstone, 1992):

Two men, charged with a joint violation of law, are held separately by the police. Each is told that

BOX 12.1

COOPERATION OR COMPETITION?

When Merrill Flood and Melvin Dresher originated their two-person reasoning game called "A Non-Cooperative Pair," they had initially framed the scenario within a financial context (Poundstone, 1992).

The task was inspired by a car deal. Flood was looking to buy a Buick car from another RAND employee at the time. They were friends and coworkers and therefore wanted to find a fair price that would not come between them or cause resentment on either side. To solve this little pricing dilemma they went to a used-car dealer and asked what the price would be for trading the car in to the dealer in its current condition. Let's imagine that hypothetically (according to modern day pricing) the car would be worth $3000 to the dealer, who would then go on to sell the car for $4000. The dealer is marking up the car by an additional $1000 for his efforts. So, one fair way to go about the transaction would be to consider the car's lowest value to be $3000 and highest value to be $4000. This leaves Flood and his co-worker with a difference of $1000 and a decision to make about how to set the price between themselves. Flood and his partner in the deal decided that a fair solution was for Flood to buy the car for the hypothetical $3500, thereby allowing Flood to pay less for the car than he would at a dealership and the seller to get more for the car than he would have at the dealership. Everyone seems happy according to that arrangement.

There are other arrangements that could be made regarding that "floating" dollar figure of $1000. Either Flood or the seller could have demanded that a different split be made in order to further their own self-interests. Flood could insist that he pay only what the car is really worth to the dealer, $3000. This would be equal to what the seller would receive for the car at a dealership. Flood could also insist that he will only pay $3100 for the car. In this case the seller probably should sell to Flood at that price, because he gains an additional $100 that he would not be able to obtain from the dealer. Even if Flood drives a very hard bargain and offers a mere $1 more than the dealer, the seller should still take this price from a purely rational economic standpoint. Meanwhile, the seller could decide to be equally stubborn about pricing and offer the car at $3800 or $3900. After all, the best Flood could have done with the dealership was to pay the full $4000. In some sense both parties to the deal have the other "over a barrel" if they choose to insist on a deal more fully weighted toward their own benefit.

In their "Non-Cooperative Pair" game, Flood and Dresher conducted an experiment in 1950. Their two work colleagues who acted as the players were identified in the research as A. A. and J. W. The pair played 100 rounds of a strategy game. Their payoffs are shown in Fig. 12.13. A. A. would select a row (his Strategy One or Strategy Two according to Fig. 12.13). J. W. would select a strategy column representing his Strategy One or Strategy Two in a similar manner. The outcome would then be revealed and each would get his share of the money (in pennies). Notable economist John Nash, who would win the Nobel Prize in Economics for his "Nash Equilibrium" work would have suggested that A. A. should always play his Strategy Two, while J. W. should always play his Strategy One. This solution would yield no net gain for A. A., but he would also not lose any money in the long run. Note that A. A. stands to sustain net losses if he plays his Strategy One, because it is in J. W.'s interest to play his own Strategy One maximizing his own gains and penalize A. A. a full penny on each play. The Nash Equilibrium would yield a net gain of $0.50 for J. W., who has a more favorable set of outcomes overall. The Nash solution assumes that both will be risk averse and assume the other player will not cooperate with them. Therefore they will have to seek to avoid losses rather than seek gains.

In the actual data something very different happened. A. A. and J. W. frequently chose to cooperate with one another. They often allowed the other to make his most maximal gain at least some of the time. The actual data from the way the game was played allowed A. A. to make $0.40 and J. W. to make $0.65. Both are nicely above the expected outcomes from a competitive Nash Equilibrium perspective. This mutually beneficial arrangement resulted from a wide array of behaviors that varied across the 100-trial experiment. In the end, the two people did not choose to always try to maximize their own self-interest. This suggests that when people cooperate they can both come out ahead.

ROUND OF PLAY	STATEGIES		PAYOFFS	
	Player 1	Player 2	Player 1	Player 2
1	2	2	1	-1
2	2	2	1	-1
3	2	1	0	+
4	2	1	0	+
5	1	1	-1	2
6	2	2	1	-1
7	2	2	1	-1
8	2	1	0	+
9	2	1	0	+
10	2	1	0	+

FIGURE 12.13 The payout matrix from "A Non-Cooperative Pair." In this game described by Flood, Dresher, Tucker, and Device (1950), two people have an opportunity to play one of their two possible strategies as described in Box 12.1.

- *If one confesses and the other does not, the former will be given a reward...and the latter will be fined.*
- *If both confess, each will be fined.*
- *At the same time, each has good reason to believe that if neither confesses, both will go clear.*

In experiments on the Prisoner's Dilemma we can imagine the following setup between Prisoner A and Prisoner B in which each prisoner can betray the other by claiming that their partner committed the crime:

- *If A and B each betray the other, each of them will serve 2 years in prison*
- *If A betrays B but B remains silent, A will be set free and B will serve 3 years in prison (and vice versa)*
- *If A and B both remain silent, both of them will only serve 1 year in prison (on the lesser charge)*

Let's imagine two people, John and Craig, have been arrested and are being detained at the police station. John and Craig are both guilty of theft. They are being held independently as the game specifies. John is faced with the decision offered, either betray Craig or stay silent. The stakes of the game are high. If John betrays Craig and rats him out, he can achieve his best possible personal deal in which he is set free. The risk of ratting out Craig is that if Craig betrays as well, then John will get a 2-year sentence. The alternative option, to stay silent, comes with greater costs. If John does not betray Craig his best possible deal is a 1-year sentence, provided Craig stays silent as well. This is a good deal all things considered. A sentence of just 1 year is the second lightest of the possible outcomes and this is a case where John does not have to betray his friend, which is another relevant factor to consider. The risk of silence is very high though, since Craig can choose to betray John and thereby leave John to serve out the largest sentence available of 3 years.

Because the risk of silence is so high, John strongly considers betraying Craig. It seems like a sensible move, but it also comes with the added cost of "throwing his partner under the bus" and risking their future friendship. Also associated with this choice is John's calculation of how Craig is likely to act knowing that he is being faced with the same choice elsewhere in the police station. John begins to question how much he really trusts Craig. Have they been on good terms over the past few months? In retrospect, perhaps repeatedly pointing out Craig's challenges in his dating life wasn't such a good idea. John also thinks back to a time in which Craig attempted to scam free sandals at a beach club when John believed the sandals really should have been his. These facts make Craig appear less dependable under the current circumstances. John starts to lean toward betraying Craig and telling the police that Craig acted alone. As John ponders this action, he remains on the fence. He cannot help but consider all of the fun times he had with Craig in college when they were on very good terms. He likes Craig overall and would feel guilt over betraying his friend. This situation indeed represents a dilemma for John.

Meanwhile, Craig is faced with this same dilemma in another room at the station house. He is leaning toward silence. Despite his criminal past and taking money that he did not earn, he considers himself to be fundamentally a good person. He's a stand-up guy, not a rat. It would be wrong to take the easy way out and betray John. He repeats a mantra to himself that "a man can do the time." Craig leans toward staying silent. On the other hand Craig loves his freedom. He feels that as a good person deep down, he deserves the freedom. He probably deserves to be free more than John whom he also feels has been lazier than he is. He begins to question his relationship with John. Craig begins to convince himself that John will almost surely betray him. That would mean 3 years in the slammer and being in jail would offer far too few entertainment options for him. Craig decides that John is a negative person, and John may assume that Craig will betray him and then the two of them will get stuck with 2-year sentences! What to do?

We observe two strongly competing desires in our hypothetical example examining the thought processes of John and Craig. There is a tendency to want the best deal for oneself, which can only be attained by betraying the other person. On the other hand, silence can lead to a relatively light sentence and has the added benefit of maintaining the friendship. This social cohesion factor of choosing to stand by your accomplice adds complexity. It is critical to know what the other will do, since friendship comes with a price. This requires the ability to take the other person's perspective. If one person is silent and the other betrays, then the loyal friend will receive the worst possible deal of 3 years. That person will also have to sit with the thought that their supposed friend betrayed them. This is a terrible deal all things considered. The challenge for the person who tattles in that scenario is that their partner will have to face the music knowing that they were betrayed. This will likely weigh on the conscience of the person who betrays. There may be an additional challenge that the former accomplice could seek revenge after the 3-year sentence has been served. Additional complexities come with considering one's reputation in the future. The calculation can get quite complex under these circumstances as one considers the short-term and the long-term consequences of these possible actions. There is a very serious long-term price to be paid for the person who rats out his accomplice. The person who tattles will likely feel guilt, but also that person will have destroyed his reputation in the criminal community and there may be retribution by violence associated with that choice.

Findings in the Prisoner's Dilemma game tend to show that people cooperate with one another and choose to remain silent (Poundstone, 1992). This may make sense to us from a human well-being perspective, as we don't want to appear greedy and this option appeals to our kinder instincts. It also may make more sense in the long term to try to preserve our reputation and friendship. Notably this strategy violates the purely rational economic strategy, which is to betray, as betraying without reprisal represents the best possible outcome for the individual in terms of pure self-interest. Even if the other prisoner betrays as well, the sentence is still lighter than the alternative possibility in which you stay silent if your partner betrays you.

In an interesting twist on the Prisoner's Dilemma, the game is sometimes played multiple times in the same way that the original Flood and Dresher monetary game "A Non-Cooperative Pair" was played. This multiple-round version of the game is known as the iterative Prisoner's Dilemma game and is similar to the Ultimatum game (described in the previous section) when it is played for multiple rounds. The multiple-round, iterative Prisoner's Dilemma game offers the possibility to punish the other person for a previous betrayal. In such a game, the optimal rational economic strategy becomes difficult to pin down, as the history of the choices made complicates the matter considerably.

In a fascinating version of Prisoner's Dilemma, a contest was held in the early days of personal computing in 1980. In this contest people built computer programs that were set up to compete in a multiple round Prisoner's Dilemma game. Fourteen programs participated in the tournament and many had very complex strategies that sought to maximize gain based on a series of contingencies. The winning program was developed by Anatol Rapoport, who described the economics of cooperative and competitive behavior in Robert Axelrod's book entitled *The Evolution of Cooperation* (1984). Axelrod reported that the economically optimal self-interested computer algorithms tended to do worse over the course of many rounds, while staying silent yielded better results for the individual, as well as his or her opponent. Rapoport's winning computer program performed a mix of the two strategies. This program implemented a strategy called "tit-for-tat." In the "tit-for-tat" play, the computer initially stays silent on the first round establishing a level of trustworthiness. From that point on whatever the partner does on the prior round, the computer program matches that move on the next round. In doing so, the program remains in a defensive mode never being the pure transgressor, but only acting in retaliation. The Prisoner's Dilemma has no mathematically correct strategy, but the "tit-for-tat" strategy has become famous in behavioral economics circles for being the best possible solution to this complex game. William Poundstone (1992) described the "tit-for-tat" strategy as carrying out a policy of "Do unto others as you would have them do unto you – or else!"

THE NEUROSCEINCE OF TRUST

Hormones and Trust

There has been increasing interest in the role of hormones in how we evaluate the trustworthiness of others. Oxytocin is a hormone that appears to play a particularly strong role in determining behavior based on trust. Studies evaluating the experience of trustworthiness within the brain have implicated the amygdala and ventromedial prefrontal cortex as being particularly important (Winston, Strange, O'Doherty, & Dolan, 2002). Oxytocin receptors can be found within the amygdala. Oxytocin was originally studied in the prairie vole, a small mammal that lives in highly communal groups. Oxytocin levels can be measured in these mammals under a variety of conditions relevant to the social status of the animal. Similar effects of oxytocin occur in people in social conditions, such a pair bonding and parenting (Insel & Young, 2001).

Hormone levels can be manipulated by introducing synthetic hormones, such as oxytocin through a nasal spray procedure (See Box 12.2). The nasal spray introduces additional oxytocin into the brain and artificially increases the overall level of the hormone available to the receptors. When the oxytocin levels are elevated in this manner, researchers can conduct experiments to examine the effects of the hormone on the individual. This emerging literature suggests that providing additional oxytocin increases people's ability on several different aspects of social behavior including directing attention toward people's eyes, recognizing emotional expressions, trusting others, and experiencing social rewards (Guastella & MacLeod, 2012).

Perceiving others faces and facial emotions are core aspects of social judgment. We experience a rise in amygdala activity when viewing negative social stimuli such as pictures of individuals displaying fear or anger. Studies of individuals who have been provided with an artificial boost in oxytocin indicate that the hormone acts to reduce amygdala activation toward aversive stimuli, including faces displaying negative emotions (Domes et al., 2007). This finding could indicate that people with elevated oxytocin levels are in a better position to experience empathy and act positively toward others who are under stress or displaying aggression. The social behavior of someone with this lower amygdala reactivity will influence their interactions, possibly improving their diplomatic tendencies. Interestingly, people have been observed to show increases in amygdala activation toward happy faces displaying positive emotions when oxytocin is artificially delivered. This suggests that the amygdala response can both reduce reactivity toward negative emotions and enhance the experience of positive emotions. Another region, the ventral tegmental

BOX 12.2

NASAL SPRAYS THAT CAN UP YOUR HORMONE LEVELS

What if you could influence someone's judgments and thinking by releasing a chemical into the room? Such a scenario sounds like something that would happen when a comic book villain tries to influence city leaders, or something that would have been attempted by spies during the Cold War between the United States and the Soviet Union. Nowadays, just such a scenario has played out in psychology, neuroscience, and behavioral economics labs when researchers have temporary elevated hormone levels in people using inhaled chemical versions of hormones.

Pharmaceutical companies have developed synthetic versions of the hormones oxytocin and vasopressin. These two hormones occur naturally and their levels change naturally within people. These hormones have effects on the brain and may influence behavior when coupled with particular circumstances. Oxytocin is often considered to influence social behavior. There are oxytocin receptors within the amygdala, a region that influences our emotions. Oxytocin levels change in pregnant women. Oxytocin levels naturally increase when we view images of babies or engage with others in social situations. There are also behavioral effects of the hormone vasopressin. Vasopressin is associated with perceiving dominance and aggression.

Researchers can temporarily enhance the available amount of oxytocin and vasopressin using synthetic versions of these hormones delivered through intranasal atomizers. Synthetic oxytocin has been considered as a potential treatment for social cognitive disorders such as autism. In a typical hormone manipulation study, a relatively low dosage of synthetic oxytocin is mixed with saline solution and packaged within a spray bottle with a dispensing tube attached to it (Fig. 12.14). The same procedure can be performed with synthetic vasopressin. Prior to the study, the participant is asked to inhale the solution through the nose. This procedure is similar to the inhalation of solution to reduce nasal congestion. The procedure allows the synthetic hormones to diffuse within the brain. Interestingly, the hormones do not pass through the blood-brain barrier. This barrier is critical for isolating the brain from the rest of the body to prevent infections from attacking the nervous system. In this case, it performs the opposite role protecting the body from receiving the diffusion of synthetic hormone. After approximately 45 min, the hormone levels have been elevated and the participant can now take part in an experiment while under the influence of the elevated hormone levels. These experiments may be face emotional judgments, trust judgments, or reasoning games played between two or more people. Over the course of the next hour, the hormone levels will return to a baseline level and the participant is discharged from the experiment.

These studies have relatively few risks, but there are some complicating factors. There are differences in oxytocin receptor levels that depend upon genetics. Some people have greater oxytocin receptivity than others. People who have the genetic allele for lower oxytocin receptor activity will be less sensitive to the effects of the synthetic hormone. Another challenge for this type of work is that if someone is experiencing any nasal congestion, then the hormone solution may not be adequately delivered into their system at a sufficient dosage. Additionally, the typical dosages of these hormones delivered in cognitive and social reasoning studies are quite low (24–40 International Units (IU)). Some have questioned whether this dosage is sufficient to adequately modulate our hormone systems in a way that approximates our natural levels. As with all pharmaceutical products, the dosage needed for a given person is likely different dependent upon body size and receptor density levels within the brain. The majority of the studies use a standard dosage for all participants due to limited time and ability to determine an optimal individualized dosage for each participant.

Despite these challenges, there is an increase in interest in these studies, and they continue to reveal complex and interesting effects on people's thinking, trust levels, and risk-based decision making. Intranasal hormones offer a relatively safe and noninvasive way to further probe the social brain.

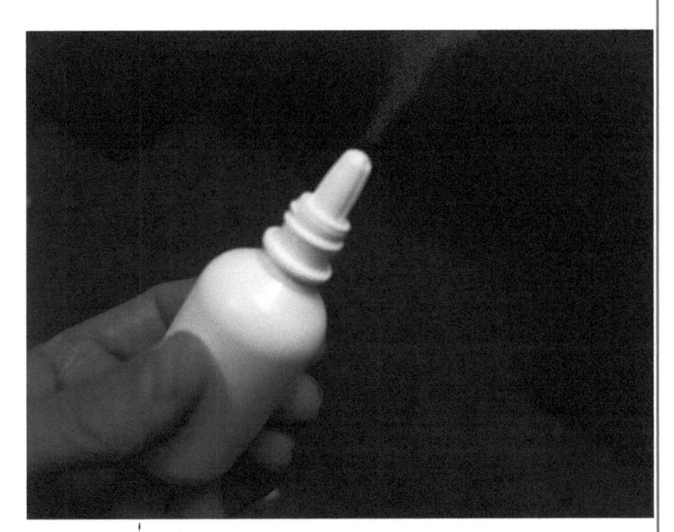

FIGURE 12.14 Synthetic hormones can be delivered through a spray bottle mixed with saline solution. The participant inhales the hormone mixture through the nose prior to participating in a research study.

area, has been shown to activate when people are asked to make a rapid button press response to a cue and receive social feedback in the form of pictures of happy or angry faces in response to their speed of performance. When provided with additional oxytocin, the activation of the ventral tegmental area was elevated indicating that this region may enhance the social motivation of the individual and that additional oxytocin may amplify this response in people (Groppe et al., 2013).

Several studies involving the administration of oxytocin indicate that hormone levels can influence people's rapid and automatic information processing about others. The way that we react to facial expressions, emotions, and their salience will influence our later decisions about those individuals. More research is needed to better understand the differences that occur with these hormones among men and women and among people with different genetic phenotypes for oxytocin receptivity. There is not yet a clear use case for intranasal oxytocin in everyday life, but researchers are interested in whether enhanced oxytocin levels through intranasal sprays may show promise for enhancing social abilities for people who have autism spectrum disorders (ASD) (Domes et al., 2013) (see Box 12.2).

Oxytocin levels have an effect on our actions toward others in social exchanges. The Ultimatum game has been played by two individuals for money, while the players undergo neuroimaging. Alan Sanfey and his colleagues carried out an fMRI version of the Ultimatum game in 2003. In this study, the decision to reject an unfair offer was associated with activation in the anterior insula, a region commonly active in response to disgust. Sanfey and colleagues concluded that this brain activation indicated a negative emotional reaction to the unfair offer. We may experience a degree of satisfaction in punishing someone for his or her unfair behavior. There can be value in having the opportunity to teach an unfair proposer a lesson. Not all of our expected utility comes in the form of monetary gain. We may also feel we are benefitting society by not allowing a transgressor to get away with bad behavior. Reacting to an unfair individual may encourage that person to act differently toward others in the future.

Additional neuroimaging studies have revealed other patterns of brain activation that are associated with economic trust decision making occurring between two people. Xiang, Lohrenz, and Montague (2013) prepared participants for particular types of offers from economic exchange partners in a repeated trials version of the Ultimatum game. People played 60 rounds of the game in the role of the responder, having the opportunity to accept or reject an offer. Participants were told that each offer was coming from a new partner during each trial. In some cases people were set up to expect somewhat unfair offers of around $4 out of $20. Other participants were set up to expect a history of fair offers averaging $12 out of $20. When undergoing neuroimaging, the participants who were set up to expect unfair offers were more likely to accept low ball offers from a proposer and felt better about these offers than those who had a history of partners who had made more generous offers. When offers violated people's expectations, activation was observed in regions including the ventral striatum and ventromedial prefrontal cortex that are often associated with processing violations of expected rewards. Areas within the medial prefrontal cortex and posterior cingulate cortex were active when people were asked about their subjective feelings toward the offers. This activation pattern indicates that the context of the decision engages top-down cognitive control regions that impact the subjective feeling about the offered reward. This study indicates that people are adaptive toward different partners and our expectations are molded by the history of offers that we experience. Brain activation supporting economic exchange decisions appears to involve our basic reward expectation circuitry in the striatum and ventral prefrontal cortex, as well as areas associated with subjective feelings that have been tuned by the prior social exchange history that we have experienced.

On the other side of the economic exchange, oxytocin, which modulates brain activation in areas associated with the Ultimatum game, increases people's tendency to make fair offers in the Ultimatum game. Barraza and Zak (2009) conducted a study in which they enhanced people's empathy levels by presenting participants with video clips displaying emotional scenes. Viewing these clips temporarily enhanced people's oxytocin levels and they were then assigned to the proposer role of an Ultimatum game. The raised oxytocin levels associated with increased empathy yielded increases in subsequent generosity toward strangers. This finding demonstrates that empathy can be altered through social emotional processing. The empathy level is linked to increases in oxytocin levels occurring naturally. Lastly the empathy-driven oxytocin increase improves people's tendency toward acting generously toward strangers.

THE EFFECTS OF CULTURE ON SOCIAL REASONING

Attributions About Others Are Influenced by Culture

To this point in the chapter we have discussed the effects of other individuals on one's reasoning and decision-making processes. Most of the research that we have reviewed in this chapter has been conducted at educational institutions in Western industrialized nations. Clearly, this is not a representative sample of all people, and there are strong reasons to believe that cultural factors play a critical role in how we reason and decide. We discussed the dramatic cultural differences that are observed when people consider the Wason card selection task in contexts that are either culturally unfamiliar or familiar. Similar situations can be considered in the context of economic exchange games. While we have focused heavily on the biological side of social decisions to this point, there is clearly an environmental nurture side to this story in which our biological tendencies can be shaped by our experiences.

Trust Games Across Cultures

Economic exchange is shaped by the context and cultural rules that we are familiar with. Joseph Heinrich (2000) described a dramatic example of this situation. Heinrich studied a nonindustrial society called the Machiguenga, who live in the rain forests of Peru. This group is highly isolated from other people and their culture has largely existed in small groups consisting of immediate family members. These family groups have been organized into hamlets of extended family members and their lifestyle features hunting, fishing, and subsistence farming. There are few political hierarchies, and cooperation and competition are rarely conducted outside of the immediate family. These conditions are dramatically different from those of Westernized industrial nations. The Machiguenga people live in groups of around 300 individuals, and their exchanges tend to be made in material goods such as tools, food, and clothing. Heinrich asked the fascinating question: how do these isolated subsistence farmers cooperate or compete in economic exchanges? Would their small, isolated, and family-centered lifestyle alter how they approach monetary exchanges among individuals? Heinrich evaluated the behavior of Machiguenga people when playing both the proposer and the responder roles. He compared their performance to that of university students from several parts of the world. While the Machiguenga tend to trade for goods, they do have some familiarity with Peruvian money, so they were not entirely unfamiliar with making monetary exchanges. Results indicated that the Machiguenga treat the monetary exchange very differently than people in industrialized societies. University students offered on average 46% of the money when they were in proposer role. Strikingly, the Machiguenga proposed very low offers of only 26% of the money, the types that would be very likely to be rejected by industrialized responders. The Machiguenga also differed in that they tended to accept these reduced offers when in the role of the responder. This study illustrates that the norms of a given population can strongly influence economic exchange. As we discussed with regard to the study by Xiang et al. (2013), brain regions such as the dorsomedial prefrontal cortex and the posterior cingulate are active in economic exchange games when processing the subjective feeling of the situation (Fig. 12.15).

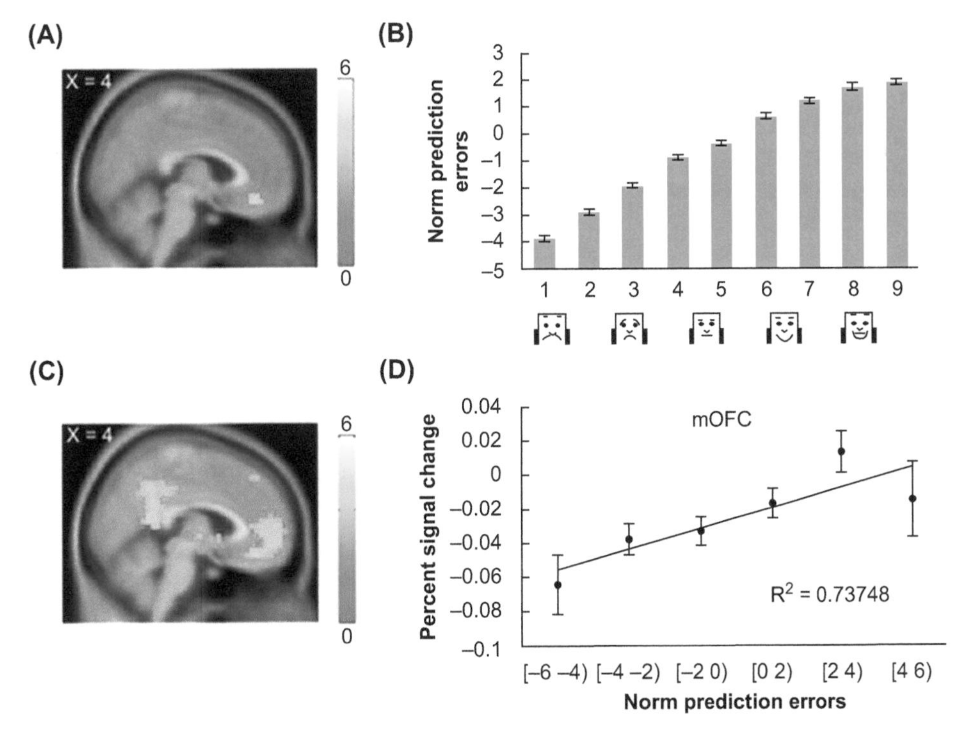

FIGURE 12.15 Results from Xiang et al. (2013). Subjective feelings of participants playing an Ultimatum game correlated with norm prediction errors, both involving medial orbitofrontal cortex and ventromedial prefrontal cortex activation. (A) Areas correlated with norm prediction errors. (B) Emoticon ratings displayed a linear relationship with norm prediction errors. (C) Areas correlated with emoticon ratings of the offers received. (D) Region of interest analysis of the medial orbitofrontal cortex and ventromedial prefrontal cortex. The averaged regional response displayed a linear relationship with norm prediction errors. *From Xiang, T., Lohrenz, T., & Montague, P. R. (2013). Computational substrates of norms and their violations during social exchange. Journal of Neuroscience, 33(3), 1099–1108.*

When combined, the results of these two studies suggest that our expectations are shaped by the context in which we have learned about monetary exchange. The cortical regions supporting our evaluations about exchange contexts are modified to support the expectations that we have been conditioned to experience. This pair of studies represents another example in which social experiences can influence biological responses. The reverse of this can also be found in which our biology helps to shape our expectations.

SUMMARY

We live within a social context and this influences our reasoning considerably. When we consider the roots of social reasoning abilities, we must first consider our ability to perceive and react to the emotional expressions of others. These inform us about the mental states of others and our knowledge of other people's perspectives is known as theory of mind. Some other species including chimpanzees, elephants, and cetaceans also exhibit some degree of theory of mind. Other species that have large brains and live in social groups show the ability to learn from one another. The learning can take place across species, or from parent to offspring within family groups. Some of these learning abilities resemble human reasoning, as they can be innovative, unique to one population that discovered the technique, and able to be passed from one individual to others.

When people engage in group-based reasoning tasks, they can be highly successful. As we have discussed throughout the book, individuals can struggle with many unintuitive cognitive tasks. When groups take on these tasks, they can outperform the individuals, especially when each person has an opportunity to attempt the problem prior to the group discussing the outcomes. Less optimistic outcomes are sometimes found for idea generation in groups. This is frequently due to the complex social dynamics that occur in association with generating ideas. There can be dominant individuals that inhibit the contributions of others. There can also be timing challenges and limitations of attention that undermine group-based idea generation.

When people reason in pairs they are often faced with the decision to compete and act in their own best interests, or cooperate and attempt to arrive at a solution that fits everyone. Economic theories predict that people should usually compete to obtain the best outcome for oneself. In many instances, research shows that people will choose to cooperate and allow both members of a pair to come out ahead. Factors including empathy and considering possible future opportunities for mutual benefits tend to drive us to cooperate. The hormone oxytocin modulates group reasoning and can stimulate cooperation in bargaining or negotiation tasks. Cultural norms and experiences can also influence the outcomes of cooperative or competitive game situations.

END-OF-CHAPTER THOUGHT QUESTIONS

1. Theory of mind has been observed to some degree in other species. What are some of the characteristics and lifestyles of the species that exhibit some aspects of theory of mind?
2. Theory of mind is mostly developed by the end of the preschool ages. Consider what other cognitive characteristics children at this age have. It may help to reconsider the work by Piaget that we discussed in Chapter 5.
3. Social hierarchies are important for determining dominance in many animal species. Do humans follow similar rules in Westernized societies? In tribal societies?
4. Does the social learning of cetaceans and other primates qualify as "reasoning"? What examples suggest that it does or does not?
5. How does interpreting basic emotional expressions contribute to reasoning in our social interactions?
6. In exchange or negotiation games, economists have specified that we should look to preserve our own self-interest. Why might we see results that differ from economic predictions in which people act more altruistically?
7. Social behavior represents an excellent domain to observe the interactions between biology and our environment. Which do you think has a greater influence over our social reasoning?
8. Brainstorming occurs when groups generate ideas. Why does brainstorming so frequently produce results that indicate problems for the reasoning of groups?

References

Axelrod, R. (1984). *The evolution of cooperation*.

Baird, R. W. (2000). The killer whale: Foraging specializations and group hunting. In J. Mann, R. C. Connor, P. L. Tyack, & H. Whitehead (Eds.), *Cetacean societies: Field studies of dolphins and whales*. University of Chicago Press.

Baron-Cohen, S., Leslie, A. M., & Frith, U. (1985). Does the autistic child have a "theory of mind"? *Cognition, 21*, 37–46.

Barraza, J. A., & Zak, P. J. (2009). Empathy toward strangers triggers oxytocin release and subsequent generosity. *Annals of the New York Academy of Sciences, 1167*, 182–189. http://dx.doi.org/10.1111/j.1749-6632.2009.04504.x.

Blount, S. (1995). When social outcomes aren't fair: The effect of causal attributions on preferences. *Organizational Behavior and Human Decision Processes, 63*, 131–144.

Brown, V., Tumeo, M., Larey, T. S., & Paulus, P. B. (1998). Modeling cognitive interactions during group brainstorming. *Small Group Research, 29*, 495–526.

Clapham, P. J. (1993). Social organization of humpback whales on a North Atlantic feeding ground. *Symposia of the Zoological Society (London), 66*, ,131–45.

Diehl, M., & Stroebe, W. (1987). Productivity loss in brainstorming groups: Toward the solution of a riddle. *Journal of Personality and Social Psychology, 53*, 497–509.

Domes, G., Heinrichs, M., Glascher, J., Buchel, C., Braus, D. F., & Herpertz, S. C. (2007). Oxytocin attenuates amygdala responses to emotional faces regardless of valence. *Biological Psychiatry, 62*, 1187–1190.

Domes, G., Heinrichs, M., Kumbier, E., Grossmann, A., Hauenstein, K., & Herpertz, S. C. (2013). Effects of intranasal oxytocin on the neural basis of face processing in autism spectrum disorder. *Biological Psychiatry, 74*, 164–171.

Ekman, P., & Friesen, W. V. (1976). *Pictures of facial affect*. Palo Alto, CA, USA: Consulting Psychologists.

Flood, M. M. (1958). Some experimental games. *Management Science, 5*, 5–26.

Flood, M. M., Dresher, M. Tucker, A. W. & Device, F. (1950). Prisoner's dilemma: game theory. *Experimental Economics*.

Frederick, S. (2005). Cognitive reflection and decision making. *Journal of Economic Perspectives, 19*, 25–42.

Frith, U. (1989). Autism and "theory of mind". In C. Gillberg (Ed.), *Diagnosis and treatment of autism*. New York: Plenum Press.

Gallup, G., Jr. (1970). Chimpanzees: Self recognition. *Science, 167*, 86–87.

Glascher, J., Tuscher, O., Weiller, C., & Buchel, C. (2004). Elevated responses to constant facial emotions in different faces in the human amygdala: An fMRI study of facial identity and expression. *BMC Neuroscience, 5*, 45.

Greene, R. J. (1987). Brainwriting: An effective way to create more ideas. *The Quality Circles Journal, 10*, 33–36.

Groppe, S. E., Gossen, A., Rademacher, L., Hahn, A., Westphal, L., Gründer, G., et al. (2013). Oxytocin influences processing of socially relevant cues in the ventral tegmental area of the human brain. *Biological Psychiatry, 74*, 172–179.

Guastella, A. J., & MacLeod, C. (2012). A critical review of the influence of oxytocin nasal spray on social cognition in humans: Evidence and future directions. *Hormones and Behavior, 61*, 410–418.

Güth, W., Schmittberger, R., & Schwarze, B. (1982). An experimental analysis of ultimatum bargaining. *Journal of Economic Behavior and Organization, 3*, 367–388.

Haselhuhn, M. P., & Mellers, B. A. (2005). Emotions and cooperation in economic games. *Brain Research Cognitive Brain Research, 23*, 24–33.

Heinrich, J. (2000). Does culture matter in economic behavior? Ultimatum game bargaining among the Machiguenga. *American Economic Review, 90*, 973–979.

Insel, T. R., & Young, L. J. (2001). The neurobiology of attachment. *Nature Reviews Neuroscience, 2*, 129–136.

Isaksen, S. (1988). *A review of brainstorming research: Six critical issues for inquiry* (PDF). Creative Problem Solving Group Buffalo.

Janis, I., & Mann, L. (1977). *Decision making: A psychological analysis of conflict, choice, and commitment*. New York: Free Press.

Karau, S. J., & Williams, K. D. (1993). Social loafing: A meta-analytic review and theoretical integration. *Journal of Personality and Social Psychology, 65*, 681–706.

Katona, S. K., & Beard, J. A. (1990). Population size, migrations and feeding aggregations of the humpback whale (*Megaptera novaeangliae*) in the western North Atlantic Ocean. *Reports of the International Whaling Commission, 12*, 295–305.

Madsen, D. B., & Finger, J. R. (1978). Comparison of a written feedback procedure, group brain- storming, and individual brainstorming. *Journal of Applied Psychology, 63*, 120–123.

McClelland, J. L., McNaughton, B. L., & O'Reilly, R. C. (1995). Why there are complementary learning systems in the hippocampus and neocortex: Insights from the successes and failures of connectionist models of learning and memory. *Psychological Review, 102*, 419–457.

Moshman, D., & Geil, M. (1998). *Collaborative reasoning: Evidence for collective rationality*. Educational Psychology Papers and Publications. Paper 52.

Mullen, B., Johnson, C., & Salas, E. (1991). Productivity loss in brainstorming groups: A meta- analytic integration. *Basic and Applied Social Psychology, 12*, 3–23.

Mulligan, N., & Hartman, M. (1996). Divided attention and indirect memory tests. *Memory and Cognition, 24*, 453–465.

Osborn, A. F. (1957). *Applied imagination* (1st ed.). New York: Scribner's.

Paulus, P. B., & Yang, H. C. (2000). Idea generation in groups: A basis for creativity in organizations. *Organizational Behavior and Human Decision Processes, 82*, 76–87.

Payne, R. (1970). *Songs of the humpback whale*. CRM Records.

Poundstone, W. (1992). *Prisoner's dilemma*. Anchor.

Povinelli, D., Bierschwale, D., & Cech, C. (1999). Do juvenile chimpanzees understand attention as a mental state? *British Journal of Developmental Psychology, 17*, 37–60.

Povinelli, D. J., & Eddy, T. J. (1996). Chimpanzees: Joint visual attention. *Psychological Science, 7*, 129–135.

Povinelli, D. J., & Vonk, J. (2003). Chimpanzee minds: Suspiciously human? *Trends in Cognitive Sciences, 7*, 157–160.

Premack, D., & Woodruff, G. (1978). Does the chimpanzee have a theory of mind? *Behavioral Brain Science, 1*, 515–526.

Pryor, K. W., Lindbergh, J., Lindbergh, S., & Milano, R. (1990). A dolphin-human fishing cooperative in Brazil. *Marine Mammal Science, 6*, 77–82.

Rendell, L., & Whitehead, H. (2001). Culture in whales and dolphins. *Behavioral and Brain Sciences, 24*, 309–382.

Rule, N. O., Freeman, J. B., Moran, J. M., Gabrieli, J. D. E., Adams, R. B., Jr., & Ambady, N. (2010). Voting behavior is reflected in amygdala response across cultures. *Social, Cognitive, and Affective Neuroscience, 5*, 349–355.

Rule, N. O., Moran, J. M., Freeman, J. B., Whitfield-Gabrieli, S., Gabrieli, J. D. E., & Ambady, N. (2011). Face value: Amygdala response reflects the validity of first impressions. *NeuroImage, 54*, 734–741.

Russell, D. (2001). *The eye of the whale*.

Sanfey, A. G., Rilling, J. K., Aronson, J. A., Nystrom, L. E., & Cohen, J. D. (2003). The neural basis of economic decision-making in the Ultimatum Game. *Science, 300*, 1755–1758.

Sapolsky, R. M. (2001). *A primates memoir*. New York: Scribner.

Smolker, R. A., Richards, A. F., Connor, R. C., Mann, J., & Berggren, P. (1997). Sponge-carrying by Indian Ocean bottlenose dolphins: Possible tool-use by a delphinid. *Ethology, 103*, 454–465.

Straus, S. G. (1996). Getting a clue: The effects of communication media and information distribution on participation and performance in computer-mediated and face-to-face groups. *Small Group Research, 27*, 115–142.

Taylor, D. W., Berry, P. C., & Block, C. H. (1958). *Administrative Science Quarterly, 3*, 23–47.

Trouche, E., Sander, E., & Mercier, H. (2014). Arguments, more than confidence, explain the good performance of reasoning groups. *Journal of Experimental Psychology: General, 143*, 1958–1971.

Van de Ven, A., & Delbecq, A. L. (1974). The effectiveness of nominal, delphi, and interacting group decision making process. *Academy of Management Journal, 117*, 605–621.

Wang, A. T., Dapretto, M., Hariri, A. R., Sigman, M., & Bookheimer, S. Y. (2004). Neural correlates of facial affect processing in children and adolescents with autism spectrum disorder. *Journal of the American Academy of Child and Adolescent Psychiatry, 43*, 481–490.

Weinrich, M. T., Schilling, M. R., & Belt, C. R. (1992). Evidence for acquisition of a novel feeding behaviour: Lobtail feeding in humpback whales, *Megaptera novaeangliae*. *Animal Behaviour, 44*, 1059–1072.

Winston, J. S., Strange, B., O'Doherty, J., & Dolan, R. (2002). Automatic and intentional brain responses during evaluation of trustworthiness of face. *Nature Neuroscience, 5*, 277–283.

Xiang, T., Lohrenz, T., & Montague, P. R. (2013). Computational substrates of norms and their violations during social exchange. *Journal of Neuroscience, 33*, 1099–1108.

Further Reading

Abelson, R. P. (1983). Whatever became of consistency theory? *Personality and Social Psychology Bulletin, 9*, 37–64.

CHAPTER

13

Future Directions in Reasoning: Emerging Technology and Cognitive Enhancement

KEY THEMES

- Reasoning is influenced by technology as new computerized approaches become more common in everyday life situations.
- The first generation of computers were enormous in size, generated high heat levels, and cumbersome to operate. This limited their accessibility in the 1950s through the 1960s.
- The second generation of computers were powered by transistors, which limited their size and increased the reliability of the machines.
- The third generation of computers are powered by microchips making the machines small, portable, reliable, and highly powered.
- Artificial intelligence (AI) has developed alongside computer hardware. Modern AI is capable of delivering information at key moments with minimal effort by the user.
- Landmark moments in AI have included high profile matchups between a computer program and a top human expert. To date, computers have been victorious in games including checkers, chess, Go, and Jeopardy!
- The IBM Watson uses deep learning approaches that involve the computer, in effect, training itself. These approaches will likely see widespread application complementing human reasoning abilities and augmenting what we can do as information processors.
- The development of virtual reality (VR) technology has enabled researchers to build immersive and realistic training environments and experiments to assess cognitive abilities and reasoning skills in new ways.

Reasoning
http://dx.doi.org/10.1016/B978-0-12-809285-9.00013-2

OUTLINE

FUTURE DIRECTIONS IN REASONING

Introduction

To this point in the book we have covered a wide array of topics and previous approaches that have been applied toward the study of reasoning and higher cognition. During this journey we have frequently emphasized techniques and experiments that have laid the groundwork for our current understanding of the field. In this chapter we turn toward the future. There are numerous exciting technological innovations that will likely change the face of how we reason and decide as we move forward in the 21st century. These advances include the development of enhanced artificial intelligence (AI) made possible by deep learning algorithms. Deep learning and machine assistance are already augmenting human cognition in our daily lives by enabling the rapid access of information and analysis of large datasets. Advancements in AI in the future have the potential to dramatically enhance our capabilities and achievements in reasoning. This may also translate into dramatic improvements in speed and efficiency in our daily lives.

AI began with the birth of the computer in the 20th century and still remains an emerging field that promises additional capabilities. In modern times many forms of AI exist. These include the circuit boards that control our microwave ovens, the search engines that enable the rapid delivery of information to our smartphones, and computerized algorithms that help to filter and monitor the pricing of stocks during trading hours around the world. In this chapter we will review some AI history and examine how technology is changing and augmenting the ways that we reason and decide. We will also discuss the growing set of applications for AI and computerized support that have been changing our capabilities and potential for the past several decades.

We are witnessing startling new developments in technology that are moving at a staggering pace. Computers can now frequently compete with humans on higher cognitive tasks that place high demands on our reasoning capacity. In some areas computerized models and AI can already far exceed our own cognitive abilities. While these competitions can sometimes lead AI to be viewed as a potential threat, as it has in science fiction writing for many decades, a more likely scenario is that AI will continue to enhance human reasoning ability by making information more readily available and relevant due to enhancements in filtering capacity. Advances in technology will also likely enhance the probability of delivering the correct type of information at a time when it is most needed based on greater input of context-relevant information.

One of the most dramatic and rapidly advancing areas of our reasoning ability is being driven by the increases in available information. As we have discussed throughout this book, the information that we have available when solving a problem or making a decision will often have a profound effect on the outcome. The accessibility of information at critical points in the reasoning or the decision-making process is equally important. At this time in human history we have access to orders of magnitude of more information at the click of a mouse or tap of a touchscreen than any other people in the past have ever had. Because of this widespread availability of information, there has never been a time when people could be so well informed. Alongside this vast wealth of information is the means of delivering the necessary information at key moments when it is most needed. Filtering information remains one of the most pressing challenges, as the more information that there is available, the more difficult it is to make that information match the current circumstances or needs of an individual. There are still a wide variety of difficulties and inefficiencies in people's daily reasoning. For instance, why do people make poor and uninformed decisions when they could simply look up the answer? Why do people continue to pay too much for automobiles? Why don't we all fix things using the information and step-by-step guidance available on our devices instead of calling in a repair person? Why do economists in charge of government financial policies still make errors of judgment when so many more facts and figures are theoretically accessible? Why do we still struggle to predict the weather at a given location at a particular time despite the massive quantity of atmospheric data that we now collect around the globe? These challenges can be overcome by augmenting our abilities with information search tools.

The Internet search engine is perhaps the most obvious of the technological tools designed to help with the delivery of information. Search engines have evolved to a point at which they are highly effective at predicting what we are likely to be searching for based on the input of only a few key words supplied. The widespread use of mobile devices with search engines capable of rapid access to information is perhaps equally important for the future of our reasoning ability. In many instances we are able to make more informed decisions and generate higher quality solutions to problems through the use of devices such as smartphones and tablet computers than we would have in the past (Fig. 13.1).

FIGURE 13.1 Devices such as smartphones and tablet computers have enabled people to access information more rapidly and efficiently than at any other time in history. The dramatic increase of available information will continue to reshape how we reason and decide.

Technological devices can deliver information to us in a format that allows us make use of the collective wisdom of the human race, rather than having to reason our way through all new situations using trial and error. We do not have to reinvent the wheel generating our own personal best solution if we can simply look up the top previous solutions that have been collectively developed and posted for others to read about. Tools such as Yelp, TripAdvisor, and Grubhub allow us to quickly learn which restaurants are nearby when we find ourselves hungry and visiting a new city. Search tools such as Yahoo Answers, Med MD, and health.com can be a great way to learn more about what might be causing joint pain or insomnia and what health tips might best stop these conditions.

Other important technological developments include advancements in the presentation of information for people. This can involve visualizing information that was not previously possible. For several years virtual reality (VR) has been promoted as a novel game-changing technology that will alter our lives. On the 1990s television show *Star Trek: The Next Generation*, characters enjoyed the use of the holodeck, a completely immersive three-dimensional simulation that could be used for both training purposes and entertainment. We currently appear to be a long way from development of such a contraption, but VR is advancing at the quickest pace in years. Emerging VR technologies include small and light headsets that are commercially available and are capable of providing striking three-dimensional immersive environments. These headsets demonstrate the incredible capabilities that this technology will enable in the near future (Fig. 13.2).

We are beginning to see new capabilities emerge for measuring our cognition and our biological signals. These capabilities include rapid advances in wearable

FIGURE 13.2 A modern virtual reality (VR) headset capable of rendering impressive three-dimensional graphics to the user through a comfortable interface.

FIGURE 13.3 A modern wearable EEG set capable of rapidly acquiring and processing different frequency bands associated with arousal states and cognitive activity.

technology to measure our vital signs through smart watches, wristbands outfitted with sensors, as well as brainwave measures captured by light and wearable EEG headsets (Fig. 13.3). These methods are likely to enhance our reasoning skills through providing additional access to health-related data and applying those signals toward the development of cognitive training tools. Cognitive training tools are already commonly used for enhancing working memory, goal-directed reasoning skills, speed of processing, and training in life skills, such as interacting with other people (Didehbani, Kandalaft, Allen, Krawczyk, & Chapman, 2016). The research community has been actively seeking to address the best ways to ensure that gains in reasoning abilities transfer toward our capacities used in our daily lives. These tools may help to make us more efficient and leave us with greater free time and energy.

In this chapter we will discuss the development of technology with a particular eye toward areas that are likely to impact our reasoning abilities. We will discuss how emerging technologies impact us in real-world conditions and how they may alter and enhance the things we are able to do in laboratory environments.

ADVANCES IN AUTOMATED COMPUTING

First Generation Computers

The computer revolution began in the middle of the 20th century. The war effort associated with World War II was instrumental in sparking advances in the ability to automate computing. Enhanced calculation speeds also prove to be very helpful as a reasoning aid. As we have discussed in previous chapters, attention and working memory abilities are often limiting factors that negatively influence our processing speed and our reasoning capacity. Outsourcing laborious computations to machines was a great leap forward enabling people to automate once time consuming processes. Some examples of early successes in computing include improved efficiencies in routing mail and breaking enemy communication codes during the war. An effective computer can perform the work that had been done previously by dozens of individuals. It can perform these same functions in a fraction of the time as well. Computers also eliminate the problems associated with human error. For example, errors in mathematical calculation are an incredibly common problem limiting people's abilities. The error rate for a computer carrying out calculations is effectively zero when it operates using a well-written program.

Early (first generation) computers were capable of making work faster and more efficient, but these machines were not widely available. Early computers were large, cumbersome, and available only to select government or university research groups. The Colossus was one of the first computers created in the 1940s. Developed by mathematician Alan Mathison, the Colossus used the vacuum tube as its primary processing component (Fig. 13.4). A vacuum tube is about the size of a small lightbulb and consists of a filament encased within glass. Computer programming is fundamentally binary code with "one" and "zero" states making up the two binary elements. Binary code in the form of zeros and ones was accomplished by heating the filaments to release electrons into the tube producing an "on" state and an "off" state in which to represent the code. The Colossus used over 18,000 vacuum tubes

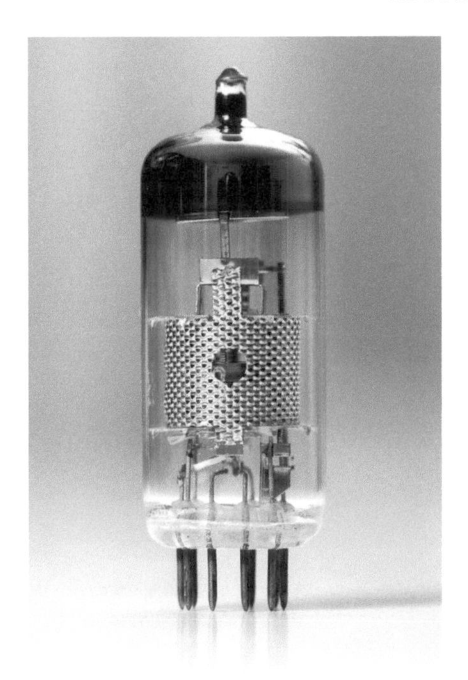

FIGURE 13.4 The vacuum tube was the primary processor in early computers. These tubes measured about the size of a modern day lightbulb consisting of a filament encased within glass. Binary code in the form of zeros and ones could be implemented using these tubes by heating the filaments to release electrons. Electrons moving within the tube produced an "on" output state which would be equivalent to a "one" in binary code, while the "off" output state represented a zero. Each binary state is called a "bit." Computers can represent additional numbers by combining multiple binary code bits. Eight bits make up the representation known as a byte. More complex representations including the kilobyte, megabyte, gigabyte, and terabyte are further expansions of processing capability with each increase being 1024 times the size of the next lowest number. The use of vacuum tube technology led to an expansion of processing power, but this came at the cost of massive increases in size.

to achieve its processing capability. The huge machine was programmable and demonstrated some technical achievements not previously seen before in terms of its flexibility and speed. Its name serves as an indicator of the enormous size of the device and demonstrates how far the computer industry has traveled given the light and tiny computing devices that are available today.

John W. Mauchly and J. Presper Eckert at the University of Pennsylvania developed the Electrical Numerical Integrator and Calculator (ENIAC) computer that could process inputs in the form of punch cards. A punch card was a piece of cardboard that contained coded instructions in the form of holes punched into it. Punch cards allowed computers to read these codes and produce automated outputs. The ENIAC was capable of a variety of useful numerical processing operations that had not been automated before. These operations included determining the sign of a number, carrying out large-scale addition, multiplication, division, and deriving square roots. ENIAC, like the Colossus before it, used vacuum tubes for its processing hardware. Like the Colossus, the ENIAC was remarkably large using approximately 18,000 vacuum tubes for information processing. Storage of these vacuum tubes and the machinery required to cool the ENIAC required over 1800 sq ft, or roughly the equivalent of a mid-sized house! Computers had a long way to go before they would be routinely available to the public, as these large machines consumed vast amount of power and required constant maintenance and cooling.

A few years later in 1951, Eckert and Mauchly created the first commercially available computer known as the Universal Automatic Computer (UNIVAC). The UNIVAC was considered to be the first true multipurpose computer as it could perform all of the numerical information processing operations that had been performed by prior computers, and it could also receive text inputs. Magnetic tapes were used to input data to the machine and to receive its output. While it was commercially available, the size of this machine remained extremely large at 50 ft long by 25 ft wide. Computers remained relatively inaccessible to most people. This machine did reduce the number of vacuum tubes used to under 6000; however, size would not reduce substantially until more advanced hardware was developed.

Second Generation Computers

Landmark events in hardware development enabled the massive computers of the past to give way to smaller and more powerful devices. The development of the transistor was critical in this progression toward smaller machines. The transistor led to the second generation of smaller and more powerful computers. A transistor is an electrical component that replaced the vacuum tube allowing computing power to grow exponentially, while size and complexity shrunk at an equally remarkable rate. Transistors are tiny components typically having a central body about the size of a pencil eraser with three electrical leads attached (Fig. 13.5). The transistor is capable of modifying the voltage or current of an electrical signal. Transistors represent information bits allowing the implementation of binary code. Transistors can operate both efficiently and reliably.

One of the important second generation computers was created in 1955. The Hartwell Transistor Electronic Digital Automatic Computer (CADET) was the first machine to complete its processing entirely transistors. While older devices required thousands of vacuum tubes for processing, transistor-based devices demonstrated greater processing capability with only a fraction of the processing units. The CADET used only 324 transistors. Computers using transistors were

FIGURE 13.5 The transistor consists of a silicon semiconductor with three electrical leads that are soldered to a circuit board. Transistors can perform operations on electricity such as modifying the voltage and current. Transistors replaced vacuum tubes as the central processing units within computers during the period between 1956 through 1963.

FIGURE 13.6 The integrated circuit, or microchip, became prevalent in the middle of the 1960s. These microprocessors could do the work that had previously been accomplished by dozens of transistors. Microchips sped up processing time into the nanosecond range (millionths of a second) and enabled the development of the modern computer.

dramatically smaller and more reliable than the earlier generation of machines. The transistor has a much longer operational lifespan than the vacuum tube and outputs much less heat. This reduced the cooling and mechanical failure rates of computers. The reduction in size, power consumption, and heat generation enabled transistor-based computers to move into much smaller operating spaces making them more versatile and widely available. Memory systems in second generation computers consisted of magnetic cores. Secondary storage devices were developed at this time using magnetic disks, a component that would be implemented for decades to accomplish data storage.

Programming capacity increased in these second generation transistor-based computers. These machines used Assembly language, which used abbreviations rather than numbers in order to implement instructions. Now the abbreviation ADD could be substituted for the operation of addition and SUB could be substituted for the subtraction operation. Assembly language made programming more efficient. Programming languages such as COBOL and FORTRAN were developed for second generation computers. These systems relied on punch cards for input and allowed users to print outputs. Many senior experimental psychologists will reminisce fondly of their days hauling massive quantities of punch cards around campus so that they could process their data at a special centralized university computing facility!

Third Generation Computers

The third generation of computers emerged with the development of the integrated circuit or microchip (Fig. 13.6). A single microchip can range in size from a centimeter up to a few inches in diameter. The microchip can perform the same function accomplished by dozens of transistors. This device enabled two major developments in computing. First, speed was enhanced dramatically and resulted in nanosecond processing (one billionth of a second). Secondly, the size of the computer was shrunken dramatically, this time allowing for the development of the personal computer. Microchip-based computers were now capable of fitting on a desktop making them able to be widely adopted in the business workplace. With the increase in processing speed, decrease in size, and the continued development of programming languages, the computer was now capable of becoming a standard desktop office product. This generation of computer had overcome several of the key challenges related to availability and distribution that had limited the usage of the previous generations of computers. Microchip-based computers led to new user-friendly interfaces including standard keyboards, monitors, and mice. This third generation of computers included operating systems that allowed many different applications to run simultaneously. These machines also allowed users to monitor memory usage and coordinate multiple tasks at the same time (time sharing). These small machines were versatile and from this point forward programming developments and software capabilities would define the utility of a computer more than the hardware of the machine.

By the end of the 1970s personal computers came into widespread existence enabling consumers to purchase and use their own devices. Box 13.1 summarizes several of the key developments in computers beginning in the 1950s through the present.

BOX 13.1

COMPUTERS THROUGH THE YEARS 1950–2017

Early computers were capable of computations that had never before been possible. These machines replaced the work previously done by entire teams of people using prior counting devices. Despite their potential these computers were massive in size, expensive to build, cumbersome to operate, and were only accessible to large organizations. Some of their defining characteristics are listed below:

- Early computers implemented binary code using vacuum tubes (Fig. 13.4)
- Early computers were fast relative to previous human computation methods
- These computers moved computation times into the milliseconds (one thousandth of a second)
- Early computers were massive in size and were not portable
- The use of vacuum tubes led to the computers generating large amounts of heat
- These computers consumed a large amount of electricity
- Early computers experienced frequent hardware failures due to the low reliability of vacuum tube technology
- Early computers were relatively difficult to program and use

Transistor-based computers evolved dramatically from the early vacuum tube-based machines. Some of their defining characteristics are listed below:

- Transistor-based computers were dramatically smaller than the first generation vacuum tube-based computers
- These computers enabled a dramatic decrease in processing time reducing computations from milliseconds (one thousandth of a second) down to microseconds (one millionth of a second)
- Transistor-based computers were more reliable than the machines of the previous generation
- The transistor reduced the amount of heat generated by the machine
- Transistor-based computers could implement Assembly language for programing making them more user-friendly
- Specialists were still needed to assemble transistor-based computers

Third generation, or modern, computers advanced radically compared to the earlier machines (Fig. 13.7). Some of their defining characteristics are listed below:

- Modern computers rely upon integrated circuits or microchips (Fig. 13.6).
- These were able to reduce computational time from microseconds to nanoseconds.
- Modern computers implement binary code using transistors and microchips enabling vast amounts of processing power, rapid speed, and large memory
- Modern computers process information quickly and are capable of performing numerous processes at the same time
- Modern computers are extremely small compared to any other period in history with smartphones representing the current state-of-the-art in high speed portable computing versatility
- Modern computers consume few resources
- Modern computers are easy to use, widely distributed, and continue to advance at a rapid pace
- Cloud computing services have enabled massive storage capability
- The wide availability of user-friendly software applications make today's computers accessible to almost anyone

Continued

BOX 13.1 *(cont'd)*

FIGURE 13.7 The development of computers over a roughly 60-year period. While initial computers were enormous, they were marked by fast processing speeds, were made of thousands of vacuum tubes, and consumed massive quantities of electricity in order to run. The development of the transistor led to a dramatic decrease in size, as well as a reduction in the power consumption of the device. By the year 2000 standard laptops weighed just a few pounds, had long battery life, consumed little power, and were much more versatile in their operation than computers of prior decades. *From The History of Early Computing Machines, from Ancient Times to 1981 (via University of Cambridge, Dr. Maurice Wilkes and NASA).*

ARTIFICIAL INTELLIGENCE AND HUMAN REASONING

The Development of Artificial Intelligence

AI has been an ongoing project in computing since the first period of technological development. From the beginning computers were able to complement the human user's cognitive abilities. The initial computers primarily took on calculations that would have been laborious and prone to human error. These early machines accomplished computations in fraction of the time the same workload would have taken a human. These machines could operate error-free provided that the human working behind the scenes supporting the machine had been accurate. Indeed computers are only as good as the engineers and programmers behind them. Bugs in software that have been caused by humans remain one of the few Achilles' heels of the modern computer.

The quest to achieve AI has gained momentum over the years. This section of the chapter is devoted to describing some of the key achievements that have been made by computerized AI in the quest for engineers and programmers to build thinking machines that are on the same cognitive level with humans. Many of the public achievements by AI have occurred when a machine was able to outcompete a human. This is particularly striking when a machine is able to defeat an expert human at their specific intellectual endeavor. The series of AI achievements that follow describe cases in which machines can outmaneuver human experts at strategy games and lastly compete with a human at a task demanding high storage and accessibility of semantic knowledge.

FIGURE 13.8 The starting position in a game of checkers. Checkers pieces advance forward diagonally moving one square at a time. Each piece can only move to an empty square. Checker pieces can capture the other player's pieces by jumping over one of the opponent's pieces. The checkers can become promoted to "kings" if they are able to proceed to the opponent's back row. Kings are able to advance in any direction on the board.

Man Versus Machine

One of the benchmark tests for success in the development of AI is to place a computer into competition against a human expert in a game that requires reasoning and the application of strategy. People have been able to enjoy engaging in a game of checkers or chess against a computer program since the early days of personal computers. A stronger test of the caliber of AI occurs when an advanced computer system (often involving both innovative software capabilities and advanced hardware) competes against a human player who has been determined to be an expert as measured in competitions against other human expert players. Over the past several decades high profile man versus machine competitions have been held demonstrating the rapidly advancing sophistication of AI. Landmark moments in the development of AI occur when a computerized program evolves to the point that it can defeat a human expert. Such events inform us both about the structural capacities of intelligence and how human thinking and reasoning are carried out in the process. The following sections review some of the intriguing developments in AI that led to computerized opponents rising to meet the challenge of human reasoning experts in progressively more complex intellectual games.

The Chinook Project

Checkers is a board game consisting of simple pieces and relatively simple rules, but out of this simplicity grows a remarkable level of complexity. The complexity increases as a game is played from start to finish. Each player begins with 12 pieces and moves diagonally across a board full of checkered squares attempting to capture the other players' pieces. Checkerboards consist of 64 squares, but only 32 of those squares are used in the game as locations to which the pieces can move. Each side is equipped with 12 pieces of each color (Fig. 13.8). Computerized checkers games have been developed since the early days of computing in the 1950s. It took decades of advancement for computers to match up well against humans. Since the 1990s people have been able to play a game of checkers against a highly capable computerized opponent.

The game of checkers is challenging for a variety of reasons. There are an estimated 500 billion billion (5×10^{20}) possible situations that can arise in a single game. With all of these possibilities, expert checkers players regularly play to a draw. One of the most noteworthy man versus machine competitions in checkers history came about in 1992 when a program called Chinook took on checkers champion Marion Tinsley. Chinook was developed in Canada and named after a series of winds that famously bring about changes in weather. Metaphorically Chinook represented a wind of change in the capabilities of AI.

The Chinook program plays checkers using a combination of some degree of strategy and some degree of pure computation. Computer scientist Jonathan Schaeffer at the University of Alberta developed Chinook in 1989 (Schaeffer, 2009). Chinook's strategy is derived from a stored database that allows it to open with successful moves that had been played by checkers grandmasters in prior tournaments. This gives Chinook a human-like ability to start the game in a strong manner. The program also has a database of endgame possibilities that was developed by starting with endgame outcomes and tracing backward so that Chinook will be able to repeat previously successful endgame situations. Coupled with these libraries of previously successful possibilities Chinook also has the sheer computing power to select among the best probable moves based on its ability to evaluate the likely outcomes of many moves several as they could play out into the future (Sreedhar, 2007). Chinook could look ahead approximately 13 moves from the machine's current game position. This was made possible in part due to the advancement of computing power during the time of Chinook's development. Most human players could not begin to approach such a vantage point in terms of seeing numbers of possible moves ahead, as the numerous possibilities available in each checkers turn would exceed our limited working memory capacity.

One of Schaeffer's inspirations for Chinook was a prior checkers program that had been developed by Arthur Samuel in the 1950s and 1960s. Samuel's program also faced off against a checkers expert many years before Chinook, but this program would tend to fall into traps that had been laid by its human opponent. For example, Samuel's program would often end up playing to a draw when the program had previously held a clear advantage and seemed to be on the verge of winning. A human expert in that same situation would likely have been able to finish off the opponent. Samuel's program could examine the game from a perspective three to six moves forward, so computing power limited the effectiveness of the program at least in the earlier days of AI carried out by second generation computers. In addition to superior hardware that allowed for a processing advantage not seen previously, Schaeffer was in correspondence with a variety of helpful checkers experts who suggested ways that human experts play that could benefit a checkers playing AI program. One such expert was Derek Oldbury, a checkers world champion who recommended to Schaeffer and the Chinook development team that they implement endgame databases in order to improve upon the program's strategies to finish off an opponent late in the game. This interactive process would help Chinook to overcome the endgame difficulties that had been a challenge for Arthur Samuel's program years earlier (Schaeffer, 2009).

Chinook proved highly successful in 1990 when the program competed in the World Checkers championship. Success in this high profile competition allowed for a potential matchup with then checkers champion Marion Tinsley. Tinsley had reigned as a champion of checkers for over 40 years only losing five games between 1950 and 1992! Tinsley faced Chinook first in 1992 in a matchup billed as the "Man versus Machine World Championship." Tinsley emerged as the victor of this public challenge by defeating Chinook four games to two. The remaining games resulted in draws, a very common outcome in championship checkers. The fact that Chinook was able to defeat Tinsley two times is an achievement unto itself, as this loss total for Tinsley was equal to nearly half of the losses that Tinsley had been handed in the past 40 years during which he had played the top human competitors in the game. The matchup was successful enough that a rematch was staged 2 years later in 1994. In this rematch Marion Tinsley withdrew after playing Chinook to a draw in several games. Tinsley's withdrawal led to Chinook being pronounced the overall victor. Sadly, Tinsley was not in good health at this point and died within months of this historic matchup, so there were to be no more rematches between this top human competitor and Chinook. A milestone had been reached in checkers playing AI based on the capabilities that the Chinook team built into this remarkable machine.

Several factors led to the success of the Chinook project. First, advances in hardware and processing power enabled the AI to search many moves ahead of what a human is capable of. This gave the computer a processing advantage that could not be matched by a human. We simply do not have the working memory capacity to store and retrieve so many possible hypothetical moves into the future no matter how much we play or study checkers. Perhaps the most decisive factor that led to Chinook's success was interestingly the human programming behind the AI. Without the tireless determination of programmer Jonathan Schaeffer and his team, the computational abilities of Chinook may not have been enough to stand up to the top human players. Indeed Schaeffer's careful analysis of opening and endgame situations and feeding this information to Chinook's program enabled some of the critical capacity of this remarkable program. Schaeffer continued his scholarly work on checkers carefully tracing successful moves back and forward in time as they occur within the context of a checkers game. He would eventually publish a scholarly paper detailing a proof that the game of checkers, when played perfectly move-by-move, would always result in a draw. In a very real way the draw outcomes that Chinook achieved against Tinsley count very much toward the achievement of the AI programmers and of the advancement of machine intelligence. The checkers success of the Chinook project remains one of the milestone moments in AI history demonstrating that computers and AI had advanced to a striking degree.

Deep Blue

Chess has long been considered to be one of the ultimate games of reasoning and intelligence. A chessboard consists of 64 squares, all of which are used in the game. Each side has a set of 16 pieces. The pieces each have their own particular appearance, name, and manner of moving around the board in order to capture pieces from the other player (Fig. 13.9). Each side has a single king piece and the ultimate goal of the game is to place the opponent's king into *checkmate*, a situation in which the king has no means of escaping capture. At that point the opposing player must resign admitting defeat. Like

FIGURE 13.9 A starting position in a game of chess. The pieces move about the board in a variety of different ways. At the start of the game each player has eight pawns, two rooks, two bishops, two knights, one queen, and one king. Chess pieces capture the opposing player's pieces by landing on the same square as an opponent's piece. The pawns are the weakest pieces, but these can be promoted to additional queens (the strongest piece) if they are able to move all the way to the opponent's back row. While the king is more limited in moves than the queen, its capture marks the end of the game.

checkers, chess can also be played to a draw in which the game does not necessarily end outright, but one of the players has fewer pieces and can only resort to repeatedly moving defensively. The player who is at a disadvantage in this situation typically withdraws conceding defeat.

Computerized chess programs are considered to have an advantage in an area called brute force computing. Brute force calculations essentially analyze every possible move equally without prioritizing the most likely moves or most probable means of victory. Only after a large set of potential moves has been evaluated does the computer hone in on the best probable moves. The brute force approach is strikingly different from that used by human chess champions. Human chess grandmasters are capable of quickly and efficiently analyzing the most likely means to achieving victory. People are so quick that a popular chess pastime is known as playing blitz games. Blitz chess involves each player making his or her moves as quickly as possible. When chess experts play blitz games, they are often able to determine their best possible move in just seconds. It would take a player far too long to use brute force computing during a game of blitz chess.

Chess experts demonstrate remarkably efficient storage of the piece locations, recognition of groupings of pieces, and exploiting strategic opportunities that arise in the game. Chess experts can read these situations at a glance. In classic chess research, Adriaan de Groot (1946, 1965) showed chess players a variety of chessboards depicting various arrangements that might appear in a game. Adriaan de Groot showed these experts the boards for just a few seconds. The chess players were able to accurately remember and reconstruct the appearance of the boards after just a short display. This ability indicates that the players must be able to take in a huge amount of information and process it very rapidly in order to achieve this remarkable memory feat.

Chase and Simon conducted a now famous follow-up to the de Groot study in 1973. These investigators asked players at three skill levels to recall chess board positions from actual games (middle games, in which many of the pieces were still on the board, but had been moved around from their original positions) and scrambled chess boards that involved random arrangements of the pieces in configurations that could not have arisen during a game. As in the earlier work by de Groot, Chase and Simon's participants saw each board for only a few seconds and then were offered a physical board of their own on which they could arrange the pieces to emulate the board they had seen. In this experiment, expert players were compared to intermediate players and to chess novices. The results were striking with experts being able to reconstruct the entire board after approximately three attempts. Meanwhile intermediate chess players took an average of about five viewings in order to reconstruct the complete board of roughly 24 pieces. The novice players took on average about six attempts to complete the reconstruction. This result suggests that chess experts do indeed encode a great deal of visual information present in a chessboard in just a few seconds.

Chase and Simon (1973) also conducted an interesting second part to their study using scrambled chess boards. The chess novices and intermediate players were able to reconstruct about 16 pieces over four attempts. This equates to getting about four pieces in the correct placement on each try. Meanwhile, the experts advantage had been completely neutralized, and they were performing at a level worse than the chess novices in their attempts to re-create these scrambled board arrangements in which the pieces were randomly positioned in ways that did not occur in actual games (Fig. 13.10). The findings from this second portion of the study indicate that the advantage in information storage and processing that chess experts possess is mostly limited to real chess games and does not extend to general purpose visuospatial memory.

In my own chess research with top ranked intercollegiate players (see Box 13.2) I have found that grandmaster level players actually do show an advantage over novice and intermediate players on randomized scrambled chess board memory as well. At a certain level of learning, the chessboard arrangement still enables people to succeed despite the nonsensical arrangement of the pieces in randomly arranged chessboards. I once interviewed former British chess champion and Grandmaster Jonathan Rowson, who stated that chess positions became more like

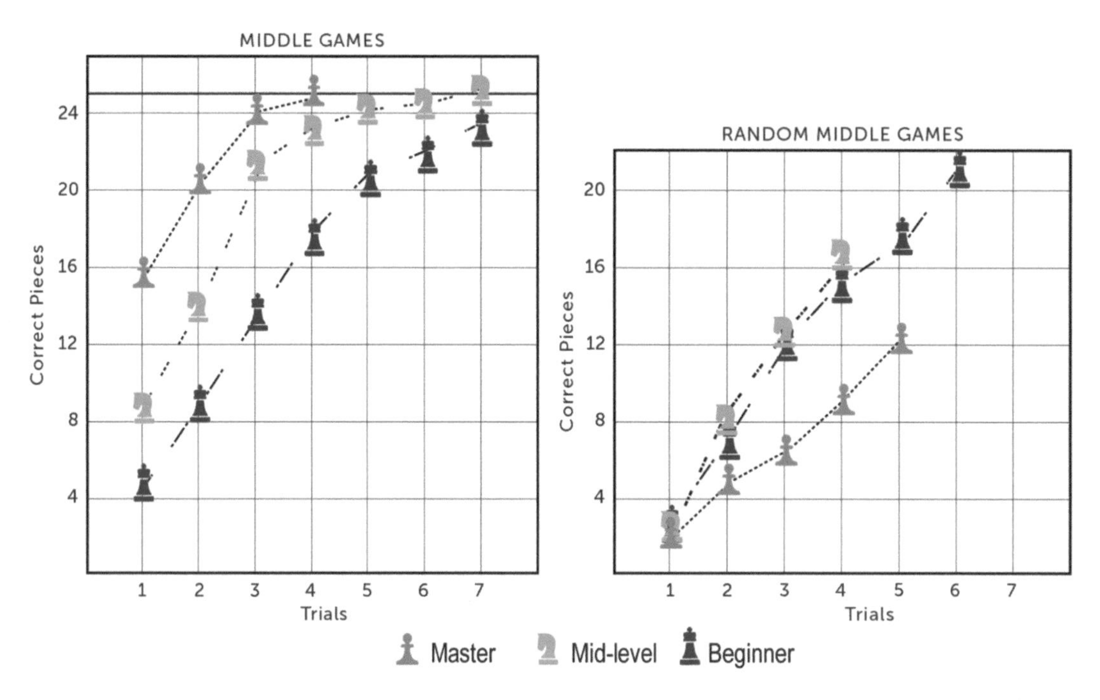

FIGURE 13.10 Results from "Chase and Simon's (1973) Perception in Chess": The "experience recognition" hypothesis. *Reproduced by Ashton Miller.*

FIGURE 13.11 Former British chess champion and author Jonathan Rowson described the jump from international master to grandmaster as being enormously significant. At the peak of his chess career, Rowson suggested that thinking about chess was more like thinking in words than mental images. *By Frank Hoppe.*

language to him than visuospatial memories (Fig. 13.11). This statement highlights the fact that the final jump from international master to grandmaster may involve a qualitative shift in the way that the board is processed in human experts at the highest levels of competitive play.

One of the most noteworthy events in competition between a human expert and an AI chess program occurred in the 1990s when renowned Russian chess Grandmaster Gary Kasparov faced a computer named Deep Blue that had been developed by IBM. The Deep Blue project evolved from the company's interest in pushing the boundaries of deep computing. IBM had been a leader in the computer industry emerging in the 1980s as the top seller in the battle to dominate the personal computer market and hold the major market share in business computing. By the 1990s many other companies had overtaken much of IBM's market share by building computers that were able to run the Microsoft operating system and associated software that had once been available in the earlier generations of the IBM PC. IBM hired Feng-hsiung Hsu, a Carnegie Mellon-educated computer scientist who had been working in the field of computerized chess. Hsu had already built an AI software system called ChipTest to compete against humans in games of chess for his doctoral dissertation. Other members of the Deep Blue programming team included a former classmate of Hsu named Murray Campbell (Hsu, 2002). The Deep Blue team set out to build a program that could defeat the greatest human chess players. This was an important mission in the development of AI.

As we had discussed earlier, human chess experts rely upon their vast store of knowledge about the game, but do not laboriously think through every possible scenario, as this would take too long and possibly exceed working memory capacity. Rather, human experts appear to rapidly notice moves that are most likely to be successful.

FIGURE 13.12 Deep Blue Versus Gary Kasparov. *Credit Mentalfloss.*

Instead of searching exhaustively, the human expert appears to narrow down the search space quickly and then devote attention only to a few key possibilities. This allows the human expert to use his attention and working memory resources efficiently. By contrast, computers long used the brute force calculation method, which required them to analyze all possible moves equally before narrowing down to the best selection.

In the 1990s Russian chess Grandmaster Gary Kasparov was at the peak of his career and widely considered to be the greatest living chess player. Kasparov was in his thirties, but had dominated the competitive chess scene for years holding the title of world champion for over a decade already. Chess expertise is quantified using the Elo rating, which factors in a player's competitive record along with the caliber of opponents that they play. In the 1990s, during the Deep Blue era, Kasparov had achieved the highest rating in chess history. If a computer could compete against Kasparov, this would represent the ultimate in man versus machine AI achievement.

There were two widely publicized matches held between Deep Blue and Gary Kasparov. The first match occurred in 1996 and Kasparov won the overall series convincingly. Despite Kasparov's overall win, Deep Blue had scored an early victory in the first game, which was an impressive AI achievement in and of itself, as Kasparov was considered to be such a dominant human expert.

The rematch between Kasparov and Deep Blue held in May of 1997 proved to be an even bigger moment in AI and the history of computing in general (Fig. 13.12). The Deep Blue team had been encouraged by their early success in Game One of the 1996 matchup. This match had proved that the program could win against a human expert of Kasparov's stature. The question remained, could the human programmers build an AI player that would consistently beat the best human in the world? Deep Blue and Gary Kasparov were set to face off in New York and would play six games over a 6-day period. This matchup received considerable media attention including reports on national news channels, Internet coverage, and would later become the source of a documentary film. The 1996 version of Deep Blue was capable of processing as many as 200 million moves per second! An astounding number given that a human expert could be expected to only evaluate the top two or three most probable moves over that same period of time. The rules for this matchup were that each victory would count as one point and each draw provided the players both with one half of a point.

Gary Kasparov led off with a victory over Deep Blue in Game One, but the matchup was already proving to be intriguing. On the 44th move of the game Deep Blue did something unexpected. The computer moved a rook to a strange place on the board, a move that appeared to provide no particular advantage to the machine. The programming team had only included this strikingly strange move in order to prevent the game from stalling or timing out in cases in which the computer did not have any strong move available. The oddball move was referred to as "random legal move" by the Deep Blue team, and it got Gary Kasparov thinking (Finley, 2012). He puzzled over why the computer had made this strange play. This move did not fit with what Kasparov remembered about the version of Deep Blue that he had defeated the previous year. Kasparov started mulling over just how many moves ahead Deep Blue was capable of considering. He began to psych himself out and the computer began to use his human emotions of fear and anxiety against him. Interestingly, the Deep Blue programmers freely admitted that the strange 44th move was nothing more than a failsafe to ensure that the program kept a dead-end game moving forward.

Kasparov's reaction to the strange move in Game One points to a bigger issue that started to wear on the champion as the multi-day match continued. Game Two was tightly contested with both Kasparov and Deep Blue making classic conservative moves. Gary Kasparov was a characteristically emotional player who regularly made alarming facial expressions conveying emotion as games moved on. Kasparov's emotional expressions coupled with his dominant reputation served him well in intimidating his human rivals, but they had no effect on Deep Blue. The computer was simply unable to process these personal characteristics, thereby eliminating one of the advantages that Kasparov often held over human opponents. Emotionality may have led Kasparov to withdraw and admit defeat a bit prematurely in Game Two. Experts indicate that if he had continued, Kasparov might have successfully battled back in that match. Instead he chose to end the game and expressed frustration after the contest even going so far as to accuse the Deep Blue team of possibly cheating! Deep Blue was merely executing

FIGURE 13.13 Gary Kasparov's frustration was evident after his Game Six loss to IBM's Deep Blue in 1997. This was a landmark moment in artificial intelligence (AI), which gave way to a new era of deep computing. *From Wikimedia commons.*

its program, but Gary Kasparov was showing a range of human emotions that would ultimately prove costly for him in this contest.

The remaining games between Gary Kasparov and Deep Blue served to accomplish an AI milestone with Deep Blue's overall victory over the top human player. Games Three through Five had ended in draws affording each player half of one point. Despite these outcomes, Gary Kasparov did not appear to be himself. He appeared emotionally rattled at times and played uncharacteristically. The stage was set for a thrilling winner take all Game Six with each player tied with two-and-one-half points. In this game Deep Blue emerged victorious having successfully laid a classic trap for Kasparov. The image of a visibly frustrated Gary Kasparov abruptly standing up and brusquely walking away from the board symbolized the downfall of human expertise and the takeover by AI computing power (Fig. 13.13).

Lead computer scientist Feng-hsiung Hsu would go on to write a book on the Deep Blue project entitled *Behind Deep Blue: Building the Computer That Defeated the World Chess Champion* (2002). The book details the story of Deep Blue and some of the unique features that enabled the computer to finally defeat one of the best human experts. For many years it had been noted that people use a heuristic approach to play the game of chess. Heuristics are helpful rules of thumb that prove to be successful much of the time. We use heuristics in much of our decision making. In Chapter 11 we discussed several of these heuristics including relying upon the most available information, the representativeness of information, and anchoring estimates upon a numerical representation. Chess experts also

FIGURE 13.14 Deep Blue was a large computer in the 1990s. The Deep Blue team of programmers consisting of computer scientists who achieved an AI milestone when their program defeated human Grandmaster Gary Kasparov. *From Wikimedia commons.*

use heuristics in order to narrow down the search space of likely successful moves. By contrast computers use an algorithmic approach to choosing moves in chess. Chess playing computers search massive numbers of potential moves in search of the best one through sheer brute force computing power. The Deep Blue project led to later high capacity computing methods. These efforts led to other important AI developments and enabled IBM to become a leader in computing for business solutions. The same software developments that allowed Deep Blue to defeat Kasparov allow the machines of today to identify trends, perform risk analyses, perform large database searches, and massive calculations (*Deep Blue*, 2017).

As a tribute to the magnitude of the achievement of the Deep Blue team, the computer itself was displayed in the Smithsonian Institute in Washington, DC. Today almost any widely sold chess program is more powerful than Deep Blue had been in the 1990s and the notion that a computer would not be capable of defeating an expert chess player is no longer seen as a relevant idea. What should not be lost is the fact that humans were behind the programming of Deep Blue, and they used their own ingenuity to help the program defeat Kasparov (Fig. 13.14). This is quite different from modern conceptions of AI in which the computer itself learns the relevant information in order to overcome its limitations without substantial human input or supervision.

BOX 13.2

CHESS RECOGNITION IN EXPERTS

Working at the University of Texas at Dallas has afforded me a remarkable opportunity that I had not anticipated when I first took the job here. Unlike other large universities, which pride themselves on having excellent basketball, football, or baseball teams, UT Dallas has one of the top intercollegiate chess teams in the world. The school is regularly ranked in the top three nationally. Chess players at UT Dallas are quite similar to division I athletic scholars who attend universities with big powerhouse sports programs. My colleagues and I have had the opportunity to study individuals who are on scholarships for chess and who are leading players on the collegiate tournament scene. These fascinating individuals come from all over the world including Canada, Poland, Zambia, the Philippines, New Zealand, and Serbia. Most players have Elo ratings that place them at the international master or grandmaster level (Fig. 13.15).

The director of the chess program is James Stallings who is a highly accomplished player. Mr. Stallings is involved in all of the chess operations including recruiting some of the most promising young talent to join the UT Dallas chess program, often on scholarship. For many student players, the team offers them the opportunity to live abroad along with the chance to become educated. They also have excellent opportunities to excel at the game they love and to travel around the world to various tournaments.

FIGURE 13.15 The UT Dallas chess program is consistently ranked among the top intercollegiate teams in the nation and regularly features some of the most talented young players in the world. *Courtesy of UT Dallas Chess Program.*

Many of these experts have participated in the research studies that my colleagues and I have carried out. In some of this research, we were able to demonstrate that chess experts can process a chessboard so automatically that it is similar to the way that they recognize faces. When you ask experts to match only the top halves of chessboards on subsequent presentations they experience great difficulty due to their extreme expertise. We asked these experts to try to process only the top half of each board, but they were compelled to look at the bottom half and experienced heavy interference when the top of the board was different, but the bottom half was a repeat from the previous board (Fig. 13.16 depicts this complex task) (Boggan, Bartlett, & Krawczyk, 2012). It is just as difficult for the expert to avoid attending to this repeat bottom half (which should be ignored) as when a face is presented with the bottom half repeating and the top not repeating (a situation that leads everyone to experience high interference). We also asked chess experts to report how much of the chessboard they were able to process while viewing rapidly presented examples of chessboards. Most experts reported that they were able to process the entire board in just seconds! Their recognition was nearly perfect on both real game arrangements of the pieces and on randomly arranged nonsensical board arrangements. This study (Krawczyk, Boggan, McClelland, & Bartlett, 2011) further demonstrated that modern grandmasters are so skilled at chessboard recognition that they do not experience interference from randomly arranged board formations, as had been long reported in some of the previous literature (Chase & Simon, 1973).

I will relay one last interesting story involving the UT Dallas chess program. I had boldly invited Chess Director James Stallings to help me re-create the famous Chase and Simon (1973) procedure in a public talk. I was confident that I could play the role of the chess novice, while Mr. Stallings would be able to serve as the chess expert. For my demonstration we both viewed an arrangement of a real board from a chess game for approximately 3 s. Each of us then attempted to reconstruct the board we had just seen using an actual physical chessboard. I was able to get about three pieces in the correct place and made a couple of errors by putting pieces on the incorrect squares

Continued

BOX 13.2 *(cont'd)*

conveying my overall lack of chess prowess. I was out of ideas after these five pieces. I looked over to see the expert in action. Stallings had indeed placed every one of the 23 pieces in the correct place. While my performance solidly replicated that of the novices in the Chase and Simon study, Stallings' performance outstripped that of the experts in that study accomplishing in one trial what it took those master level players around three trials to achieve. We then tried the same demonstration again, but this time viewing a randomly arranged chessboard for just a few seconds. Again my performance reinforced my novice status achieving about three or four correctly arranged pieces. To my astonishment Jim Stallings managed to reconstruct the entire board with just a single error of placement having one piece off by one square! Again, this indicates that modern experts possess a strong cognitive advantage with chessboard recollection in both re-created games and in scrambled arrangements. During our collaborations, James Stallings has regularly emphasized that his game is nowhere close to that of his top grandmaster scholarship players.

These studies serve as a clear indication of the superior recognition and memory that human experts are capable of. This same skillset enables them to excel at exercises such as blitz chess games in which moves are made in just seconds.

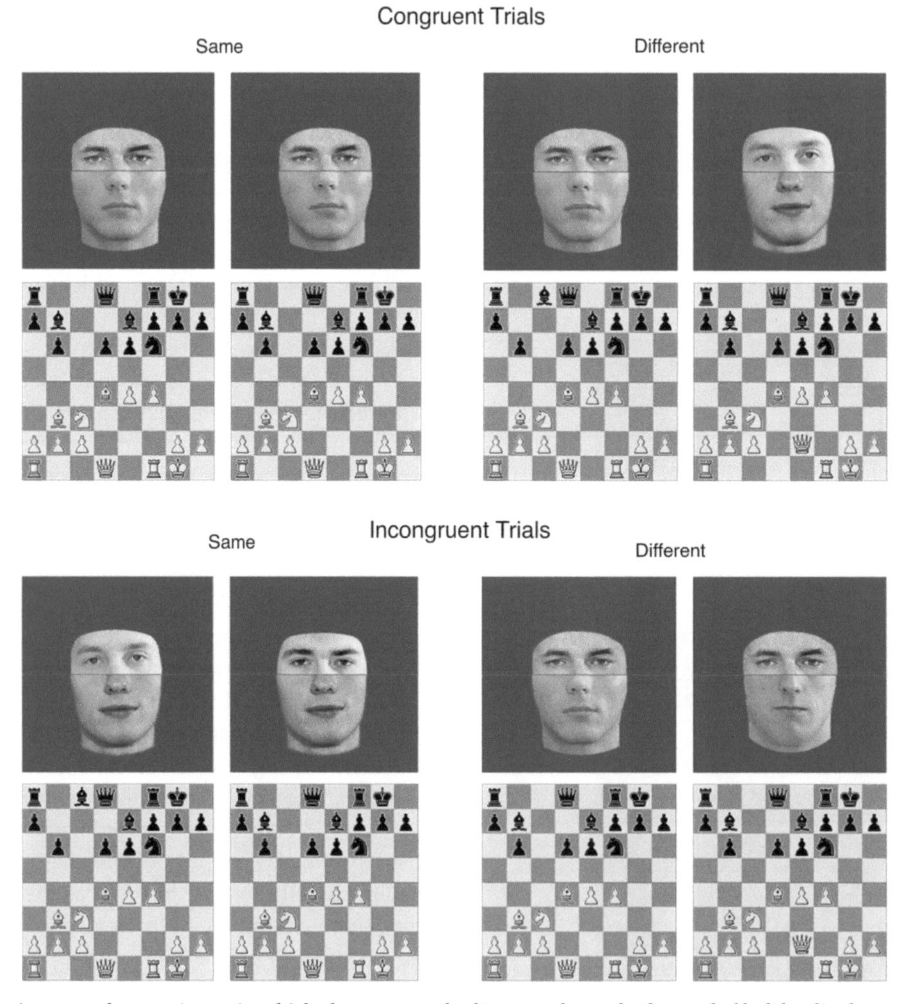

FIGURE 13.16 An example experiment in which chess experts had to attend to only the top half of the display when both chess and faces were presented. Experts experience great difficulty due to their extreme expertise, which can interfere with the task. Experts were compelled to look at the bottom half and experienced heavy interference when the top of the board was different, but the bottom half was a repeat from the previous board (Boggan et al., 2012).

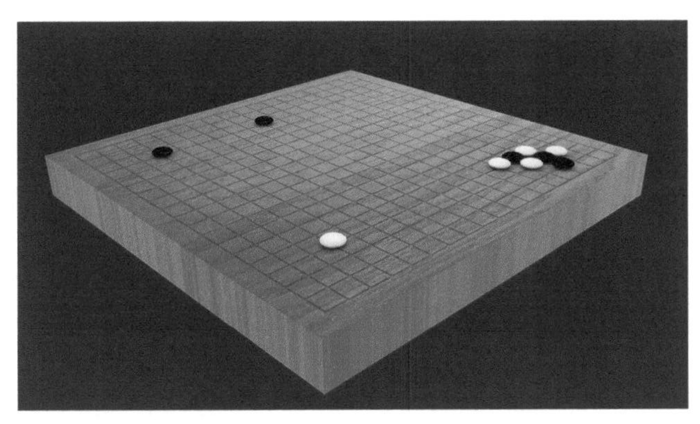

FIGURE 13.17 A board from the game Go which originated in China. Go is widely played throughout east Asia. It is thought by many to be the most challenging game for AI to master and a program that could defeat a human expert would represent a major accomplishment in AI. *From Shutterstock.*

Alpha Go

Once AI programs could successfully defeat the greatest human experts at reasoning games such as chess and checkers, theorists wondered what would be the next major reasoning challenge that might be tackled by AI? Some felt the answer might lie in the Asian board game of Go. Go originated in ancient China and is played on a board measuring 19-by-19 squares divided by grid lines. Black and white pieces called stones are placed on the board at the intersections of the grid lines (Fig. 13.17). Go players attempt to place their stones on the majority of the board, which is analogous to capturing territory in warfare. Players can capture their opponent's pieces by surrounding an opposing player's stone on the grid. The game proceeds until one player decides to resign at which point the player holding the most territory on the board along with having captured the most stones is declared the winner.

Go may be the most difficult board game to master. This becomes evident when it is compared to chess. The move possibilities offered to a Go player are orders of magnitude larger than those offered to a chess player. In chess there are 20 moves available to the player at the start of the game. After the first chess moves there are roughly 400 possible board configurations that are possible. By contrast Go players are faced with a blank board initially and this leads to a much more open-ended set of potential arrangements. At the start of a game of Go the person playing black (black moves first) has 361 possible moves, while the person playing white has 360 possible moves. The depth of Go's complexity can also be measured in terms of the number of moves available to a player on each successive turn. Chess offers a player 35 moves available on each turn, while Go offers players 250 possible moves per turn. To put the complexity difference in to further perspective, a Go board has an incredible 129,960 possible board configurations after the first two moves have been played. Lee Sedol is widely considered to be one of the greatest living Go players. Of Go, Sedol remarked "There is chess in the Western world, but Go is incomparably more subtle and intellectual."

A London-based AI lab called DeepMind developed a remarkable program that was able to successfully compete at the game of Go against top human experts. This was a landmark that many theorists had assumed would not be possible perhaps until the 2020s. The DeepMind Company was bought by Google in 2014, and the developers called their program AlphaGo. AlphaGo was able to initially compete with the 633rd ranked player in the world, Fan Hui, in October of 2015. AlphaGo was simply too much for Hui defeating him by a count of five games to none. Clearly the AI programmers at DeepMind were onto something new in the realm of intelligent computing.

Next up for AlphaGo was a high profile match against Korea's Lee Sedol, the number five ranked Go player in the world with a prize of one million dollars to the winner (http://www.bbc.com/news/technology-35785875). Sedol was widely considered at the time to have been the best overall player of the prior decade (Metz, 2016a). The match took place in Seoul, South Korea, and was widely publicized by the international press and followed online. Sedol and AlphaGo would play a best of five game series. Lee Sedol opened the first game aggressively and AlphaGo responded back with its own aggressive play. AlphaGo made an unusual move in the middle of the game forcing Sedol to pause a full 10 min before he resumed play. AlphaGo would go on to defeat Sedol in the first game. Game Two featured even more theatrics as AlphaGo made a very strange play 37 moves into the game placing a stone on a wide open area of the board. This unusual move led Lee Sedol to excuse himself from the game briefly. The move forced Sedol to contemplate his artificial opponent's strategy for 15 min before resuming. Onlookers who knew Go well were also shocked by this unconventional move (Metz, 2016b). It was reminiscent of the strange "random legal move" that Deep Blue had made in its historical Game One against Gary Kasparov nearly 20 years back. Such moves can throw a human expert off his or her game and force them to question exactly what type of AI their computerized opponent is equipped with. By the end of Game Three, AlphaGo was able to secure the overall victory with a resounding third straight defeat of Lee Sedol. Google went on to donate the million dollar prize money to charity (Fig. 13.18).

Lee Sedol was able to regain some respectability for the human Go players of the world by winning one game in the series. In Game Four Sedol managed to thwart AlphaGo by making some key moves that reduced the machine's opportunities for victory. The series did continue with AlphaGo polishing off Sedol one final time

FIGURE 13.18 Go champion Lee Sedol versus DeepMind's AlphaGo program in 2015. *Photograph Yonhap/Reuters.*

in Game Five. Overall, Go experts and AI theorists alike were astonished at the completeness of AlphaGo's dominance over this top human expert (Metz, 2016c). This was indeed a rare situation in which a milestone achievement in AI was accomplished at a remarkable pace.

Observers within the AI community were impressed with the speed at which a computer was developed that could dominate at Go. A key aspect of DeepMind's efforts comes from the fact that AlphaGo uses multiple approaches to achieve success. First, the program used a machine learning approach. Machine learning has been considered to be analogous to biological cognitive systems, as it uses the same principles. Major features of machine learning include sets of connections that can be trained to respond to different patterns. In the case of reasoning or strategy games, training the machine's algorithm with numerous game-relevant inputs allows the program to categorize the patterns that it is presented with. AlphaGo was originally trained with a database of over 30 million expert Go moves that had been played by experts (*Artificial intelligence: Google's AlphaGo*, 2016). The game was then able to learn additional techniques by playing different versions of itself millions of times! DeepMind honed AlphaGo's database of moves further by providing it with the information gained from the vast array of game-specific experiences that have been accumulated over a long period of time by many people playing the game. Such practice would not be possible for a person unless they devoted themselves to play Go almost exclusively for a period of many years. Remarkably, AlphaGo was able to begin its play armed with the knowledge of human experts and then use its algorithm to further refine its own capabilities through a form of self-driven enhancement.

The man-versus-machine matchup of Lee Sedol against AlphaGo may emerge as an AI landmark moment on par with Gary Kasparov versus Deep Blue. The program demonstrates the rapidly accelerating innovation that AI is currently undergoing. Note that the reasoning ability of an AI program such as AlphaGo derives from an incredibly elaborate database of possible moves and their associated values. In addition, the program has a quick learning capability that allows it to further shape its database by experience. This is quite different from heavily scripted software that is hard-coded one time by a human programmer in order to repeatedly carry out particular actions. Rather, the computer learns on its own in this case and in a sense programs itself based on the regularities that it absorbs over the course of training by experience. These same characteristics of a large knowledge base that can be rapidly updated are often attributed to humans and also explain some of our most sophisticated reasoning abilities. As time goes on the big questions in the AI field are migrating from asking whether computers will be able to outperform humans to instead asking new more specific questions. Now we may wonder how will intelligent computers be able to help people do their work in the future? How long will it be before we can begin to outsource important reasoning problems to machines? How might self-learning AI systems facilitate human understanding of complex systems including predicting weather and world financial markets? Only time will tell how and when people will be able to leverage these intelligent computing systems to accomplish these and other complicated reasoning tasks.

IBM Watson

As computing continues to advance, the importance of knowledge structures has arisen repeatedly as a key to future advances. When we think about human reasoning, it is often our superior knowledge that allows us to be successful. The richness and structure of human knowledge may be the ultimate factor that sets human cognition and reasoning apart from the capabilities of other species. Human knowledge is organized in networks of activation that we call semantic memory. Semantic memory is the hallmark feature of our experience. This leads us to one more of the remarkable stories of AI success when matched against a human.

Jeopardy! is a game show that has been a television success since the 1980s. Jeopardy! consists of three players who are able to choose from a variety of trivia categories. At the start of a game, one of the players selects a category and all three are allowed to attempt to answer an associated fact from that category. To answer a question, the player must be the first one to trigger a button after the question has been asked. The game proceeds with each player getting to select the category after successfully answering a question. As the game progresses the questions become more difficult and are associated with higher dollar values. After two rounds, the players have an opportunity to risk as much of their money as they would like on a single "final jeopardy" question from a particular category. If they succeed in answering this last question they can as much as double their money, but if they answer incorrectly, they lose the dollar amount that they had bet on their final answer. The game ultimately represents a strong test of both knowledge and memory retrieval with successful players needing to know a wide array of facts about diverse areas. In addition, successful Jeopardy! players need to have a fast mind, as the questions must be answered within seconds of pressing an indicator button that one would like to respond. When Jeopardy! initially aired on network television it was thought to be a strong intellectual challenge. It would have been very doubtful that anyone would have considered a computer capable of competing against top players decades ago.

In 2011, IBM's Watson computer played against two expert humans in a game of Jeopardy!. The players were widely considered to be the best that the game had seen. Ken Jennings had set a record for the most Jeopardy! wins in a row at 74. This is exceptionally impressive given that many Jeopardy! champions could only defend their title one or two more games before being displaced by another champion. One source of difficulty in Jeopardy! is that only one error can derail an entire game's worth of work, especially if an error happens in the final Jeopardy! round when much of one's earnings are on the line. Brad Rutter was the other human champion that played Watson. Rutter ranked as the all-time biggest money winner in Jeopardy! history having amassed 3.2 million dollars in winnings. These were undeniably some of the best people who had ever played the game of Jeopardy! An AI system that could compete successfully against Jennings and Rutter would represent an advance in the areas of speech recognition, language recognition, and knowledge retrieval.

One of the strategies needed in standard Jeopardy! matches concerns when a participant can "buzz in" by pressing a button to indicate that they would like to answer a question. The participants are required to wait until each question has been fully read before pressing the button. A light visible to the players cues them that they can register to answer a question. The ability to buzz in at the earliest possible time has been critical to the player's success. In the Jeopardy! game featuring the IBM Watson, the computer "buzzed in" when it had completed a search process and had an answer at the ready. This points out one of the interesting features of the game. It has been understood that anxiety can reduce reasoning ability, probably by disrupting one's focus of attention and blocking retrieval of relevant information (Tohill & Holyoak, 2000). The challenge for a human player is that they will have to be capable of rapid retrieval and determine almost instantly whether they know what the answer is to a given question. Given the fast pace of the game and the social pressure of competing on a nationally televised scale, it is remarkable that some people are such effective players.

Several complex things need to happen for Watson to answer a question on Jeopardy!. First, the computer has to recognize the words involved in a given question. It then needs to be able to determine the overall meaning, or implied meaning of a question, which is challenging given that speech is inexact and can be misleading (Cook, Feuz, & Krishnan, 2013). Lastly, Watson would have to generate the most probable answers, which would appear in a hierarchical order according to Watson's confidence levels about each possible answer. In effect, Watson graded its array of possible responses. Whichever was the top choice became the answer Watson was credited with giving. Sometimes Watson gave the correct answer. For example, when the clue was "four-letter for the iron fitting on the hoof of a horse or a card dealing box in a casino" Watson generated the correct answer "shoe" with 68% confidence. It also generated "boot" at 18% and "stud" at 16%.

There were other times when Watson was incorrect. When the clue read "this trusted friend was the first non-dairy creamer," Watson generated an incorrect answer "milk" with a confidence level of 72%, with Watson's other two answers being "Coffee-Mate" at 49% (the correct answer) and the somewhat bizarre answer "lactase enzyme" at 12%. Another oddity was observed when Watson attempted to answer the item "the first modern crossword puzzle is published and Oreo cookies are introduced." Ken Jennings registered first to answer this question and stated incorrectly "1920s." Watson followed second with its own answers: "1920s" at 57%, "1910s" at 30%, and "1904" at 4% (the correct response was 1910s). This error provided a telling clue about the limitations of Watson's AI, namely that it was not able to track and update in real time. No human would have made some of these errors, and they reveal some of the limitations of the AI that ran Watson. The complexities of language and extracting the meaning of the question were some of the greatest challenges for the computer.

Watson played three exhibition Jeopardy! matches against the two human champions. The first match saw Watson get off to an excellent start. Watson won the match with a prize money total of $35,734. Rutter finished in second place with $10,400 with Jennings in third place with $4800. This is a remarkable result given the large margin of victory by Watson and the fact that Watson had even missed the correct response in the final Jeopardy! round, but managed to avoid trouble by having risked only a small dollar value on the single question. The second match saw Watson win in an even more impressive fashion totaling $77,147 to Jennings $24,000

FIGURE 13.19 IBM Watson wins on the Jeopardy! game show against two top human players. This achievement serves as an indication of how far AI has evolved in the past several decades. Machine learning approaches such as the Watson project are now seeing widespread usage in the business world. *From Carol Kaelson/Jeopardy Productions Inc., via Associated Press.*

and Rutter's score of $21,600. The overall outcome stood as a resounding success for the Watson program and another benchmark of success for AI overall (Fig. 13.19).

Deep Learning and the Future of Artificial Intelligence

Deep Learning algorithms such as those used by the Watson computer give hints about the future of AI. Since 2011, Watson has gone on to become a successful business consulting tool for IBM. Companies can hire the Watson team to consult for them. IBM will provide Watson's computing power to be applied toward your own interests. For example, an oil exploration group might hire IBM to feed Watson information about possible oil exploration sites. In such a case Watson would apply deep learning algorithms toward generating new possible ideas for the customer. A challenge with this type of service is to determine what Watson will need to be informed about. People may wonder what type of questions to ask Watson. Another challenge may be for Watson to generate answers that people can interpret. While the business future of IBM's Watson program will likely continue to advance, the major achievement of Watson was to indicate that AI had conquered one of the daunting challenges. Watson's AI could interpret a language-based statement and generate correct answers in a manner that ranked and settled upon the most probable best answer.

FUTURE DIRECTIONS

Experiments Using Technological Approaches

Throughout much of the book we have discussed reasoning as it occurs either in the real world as encountered in daily life, or in a laboratory-based environment. We have discussed the general problems associated with each approach. Real-world field studies offer little experimental control and therefore offer limited opportunities to draw causal inferences about factors that contribute to reasoning. These uncontrolled factors can be either related to features of the context or the individual. Furthermore, the ability to measure aspects of reasoning behavior in field studies is very limited. By contrast, laboratory-based experiments excel at offering control and an ability to measure behavior precisely, but they offer limited ability to provide realistic situations limiting their generalizability. Technological solutions may offer an excellent compromise to address these challenges.

As an example, imagine a situation in which you need to do some shopping. To accomplish this task you need to visit a store, locate and purchase a set of grocery items that are needed. Further, you likely need to watch your spending and cover the essential items without overspending on your food budget. Another common constraint on shopping trips is that they should not take an excessively long time. You need to balance your time and be efficient in the market in order to get the shopping done and move on with other errands in your day.

Patients who have experienced cortical or white matter damage after disease or brain injury may experience difficulties when shopping. Daily living tasks include a mixture of reasoning skills that all have to be solved effectively in order to get the task completed. Patients with neurological damage may fail at the planning stage of such a task. As Goel, Grafman, Tajik, Gana, and Danto (1997) demonstrated, frontal lobe damage can lead people to experience difficulties when planning a household budget. In this study one patient spent considerable time pondering how to reduce the food and clothing budget.

This patient offered a solution that one could eliminate housing from the budget by buying a tent cheaply! This type of difficulty appears to come about due to a lack of ability to manage the demands of a situation. Damasio (1994) noted that patients with frontal lobe executive function deficits will often still test within the normal range on intelligence and intellectual function measures. Tasks such as the financial planning test developed by Goel et al. or the Iowa gambling task (discussed in Chapter 7) are needed to capture the subtle deficits that occur in such patients. Critically, those deficits are likely to impair daily living functions.

Attempts have been made to bridge the gap between real-world reasoning performance and that measured within a lab environment. The Multiple Errands Test (MET) was developed by Tim Shallice and Paul Burgess (1991) to measure the real-world reasoning and planning deficiencies that occurred after patients experienced frontal lobe damage. This task was developed to address the fact that clinical neuropsychological assessments may miss the cognitive challenges that people actually experience in real-world daily life. The original version of the MET involved a clinician taking the participant outside of the lab or hospital and into a shopping center environment. The participant was given a set of items that they had to purchase at different stores and a set of rules that had to be followed in the course of buying the items. For example the rules stated that a store could not be entered multiple times. Performance on the MET was quantified by a clinician noting a combination of errors, rule violations, and inefficiencies by the participant. The benefit of a measure such as the MET is that it provides a more ecologically valid measurement of a person's reasoning ability and executive function capacity. Sometimes a person can perform within the typical range on a digit span or Tower of London planning task conducted in a quiet room. This same person may exhibit difficulties when faced with the fluid reasoning needed to spend money, stay on budget, complete a task efficiently, and interact with the complexities and multitasking of everyday life.

A downside of the MET is that it involves taking a patient out into the real world, which is likely to be variable and difficult to control. These features limit the replicability of the MET, as weather, people, and time of day may dramatically impact the performance of an individual on the task. Likewise, the task is costly to administrate owing to the time taken to prepare the participant, travel to the shopping center, and observe the person carefully. Lastly, the MET requires some subjective ratings to be carried out by the administrator, who may be distracted or fail to notice something about a person's performance. This limits confidence in the scoring of the task.

VR technology can help to address the challenges experienced in the MET. One such example of this is the Virtual Multiple Errands Test (VMET) that has been developed by Pietro Cipresso et al. at the Instituo Auxologico in Milan, Italy (Cipresso et al., 2014). The investigators noted the difficulties in administering the original MET, while also noting the value of such a task for addressing real-world reasoning functions in patients with executive function challenges. This prompted Cipresso et al. to create a virtual shopping center that could be navigated by a participant within a lab environment while seated at a computer, or situated within a VR cave setup to enable more immersive three-dimensional simulation (Fig. 13.20). When administered with either method, the VMET limits the need for the participant to travel and controls for many of the challenging environmental variables that can occur when completing the MET in an out-of-lab environment. The VMET has a set of rules like the original MET and also provides a straightforward and realistic way to navigate and engage with items in the store. Furthermore, the VMET allows for electronic data capture so that quantitative measures can be obtained. The interface for the clinician was also improved, as they do not have to focus on following the participant or worrying about crowds and other variables interfering with task performance. Tools such as the VMET offer an excellent compromise balancing the complexity of daily life (which is controlled through computerized technology) with the reliability and replicability of a laboratory or neuropsychological task.

FIGURE 13.20 The Virtual Multiple Errands Test (VMET) is conducted within a virtual shopping center environment. The task requires participants to efficiently navigate through the market acquiring a specified set of items, while avoiding distracting items. Features include automated data capture so that the test administrator has an electronic record of timing and performance data for each participant.

Cipresso et al. (2014) conducted an evaluation of the VMET comparing this test to other executive function measures that evaluated planning, language, and memory. Participants in this study had Parkinson disease, which leads to a deficit of dopamine receptors in the basal ganglia. Some participants with Parkinson disease were cognitively impaired, while others showed normal cognitive ability despite having the disease. These groups were compared to a healthy age-matched group of individuals on the VMET along with a clock drawing test, verbal fluency measures, and Tower of London planning task. Among the tests administered, only the VMET was able to discriminate performance between all participants with Parkinson disease relative to unimpaired individuals. These participants showed evidence of less efficient methods of completing the task compared to healthy control individuals. The groups with Parkinson disease also used less effective strategies in going about the navigation of the VMET. An effective strategy would involve planning the particular subtask (such as acquiring fruit) before going about navigating within the virtual market. Overall, this result suggests that lifelike tasks such as the VMET are capable of emulating the actual complex demands of everyday life to a greater degree than other types of measures, including rather complex planning tasks such as the Tower of London, requiring people to place various sized disks on pegs while following an arbitrary set of rules.

In social reasoning tasks VR technology can be advantageous, as it can provide a visually rich environment simulating the immersive qualities of social engagement. In some of my own work we have employed VR technology in order to train people who have social deficits on a variety of lifelike problem-solving tasks. For these studies (Didehbani et al., 2016; Kandalaft, Didehbani, Krawczyk, Allen, & Chapman, 2013) we had developed VR-based simulations of schools, offices, coffee shops, playgrounds, and stores set within a town that can be navigated by avatar. An avatar is a representative figure that can be piloted in virtual space by a participant. Our participants in these studies were adolescents and young adults who have Asperger syndrome, a form of autism that limits people's ability to make social connections. Individuals with Asperger syndrome tend to be less sensitive to other people's emotional expressions, tend not to make eye contact, and struggle with understanding others intentions. Such individuals may exhibit theory of mind or mentalization difficulties. All of these challenges manifest themselves and problems reasoning about other people.

We provided Virtual Reality Social Cognition Training (VR-SCT) for participants in these studies (Didehbani et al., 2016; Kandalaft et al., 2013), training people with Asperger syndrome on social challenge problems. For young adults we asked our participants to attempt age-appropriate daily life tasks such as trying to get a job on a virtual job interview. We also asked them to negotiate a lease agreement with a landlord at a virtual apartment. We asked our younger participants to interact with others on playground environments and join others for lunchroom situations. These are exactly the types of situations that people with Asperger syndrome find challenging in their daily lives (Fig. 13.21). By engaging with other avatars, these individuals were able to practice their skills in negotiation and efforts to make friends. In all cases humans behind the scenes operated the avatars. The other children and adults that the participants engaged with were actually clinical psychologists whom the participants had met previously. Voice altering software was used along with body alterations in order to allow the clinicians to simulate a variety of different types of people having different traits. The VR technology allows people to play through these different social problem-solving challenges multiple times.

FIGURE 13.21 A screen capture of the Virtual Reality—Social Cognition Training (VR-SCT) software. This software provides engaging environments for people to undertake social interaction training. Age-appropriate situations include job interviews and café experiences for adults and playgrounds and schools for children.

Another important feature of the VR environment that we used in this study is that we could easily record all of the social interactions that had taken place in VR space. These scenarios could then be played back and viewed by a participant and an experimenter. Such situations can become potent learning opportunities, as people could see how they had performed previously and the clinical assistant could give guidance and make suggestions about what had been done well or poorly. The effects of the social training within VR appeared on real-life measures of social ability. For example, participants rated their ease of interactions much higher after the training than before and indicated that they felt less anxious about the social situations after the training. Furthermore, we saw improvements in areas such as reading emotional expressions, mentalizing, and even analogical reasoning about scene situations (Didehbani et al., 2016).

In the future these types of VR toolkits can also be used for other situations that involve training. Such situations include job-specific training for mechanical operations, training in international business with individuals from other cultures, as well as social situations such as dating and making friends. The immersive ability of this software provides a compelling and lifelike simulation. The ability to measure and record events and replay these for later learning is another promising feature of this technology. The future is bright for VR applications. These will necessitate strong collaborations between technology developers and psychologists and will likely come about through industry and academic partnerships.

In the future, it is likely that VR-based tasks for reasoning will continue to be developed. Some of the types of reasoning and decision making that we have discussed in the previous chapters would likely make excellent candidates for task development for a new generation of technology-savvy researchers. For example, one can imagine simulated versions of trust games, such as the Prisoner's Dilemma or the Ultimatum Game. VR interfaces could make AI partners in the game exceptionally realistic. You can probably also imagine many of the decision-making scenarios that we discussed being implemented in VR. Choosing a college or job could be simulated with realistic three-dimensional visual input. Basic reasoning abilities such as attributing cause could also be achieved in VR with simulated physical interactions. Analogies can be drawn between scenes that unfold in time using VR technology. For many years the field has relied heavily on text-based descriptions of static situations. These provide less than realistic contexts for certain types of reasoning that change over time. We may find that attention and working memory abilities are more relevant to reasoning skills than previously estimated, as simulations with complex visual information will likely place a greater burden on people's capacities.

Applications of Technology to Improve Reasoning

We stand at a unique point in history with regard to computerized technology. We have seen glimpses of the possibilities for the future. There is already strong realization of the high potential in areas such as information search. The combination of enhanced computing power and high reliability of electronic components has enabled massive storage and speed in modern computing devices. These capabilities have enabled search engines and deep learning algorithms to rapidly deliver information to people at an unprecedented speed. This impacts the way we think and reason.

In previous times in history, people simply had to rely on their own memories for solving problems. Things changed once information could be readily copied down and stored. The challenges associated with the physical storage on paper included a space problem, as large volumes of paper required whole buildings to store, as well as a search problem as people would have to physically sort through increasingly large amounts of paper. With more information stored and physically available, it would take a person much more time to actually sort through it and find an answer. The Internet has demonstrated the incredible importance of generating the most probable desired response quickly. Before the wildly successful Google search engine arrived on the scene in the early 2000s, other search engines such as Excite, Netscape, and Ask Jeeves varied in their ability to generate the key results desired in response to Internet search queries. People are now able to have their computers carefully tuned to learn about their personal preferences and deliver tailored content based on their web search history. For instance, if you regularly search for new automobile models and descriptions of scientific technology, your YouTube sidebar will automatically populate with likely content that you will enjoy. Similarly Netflix has become a widely used service for movie and TV show viewing. Netflix keeps a profile for any individual who can recommend likely content of interest.

Will these advances in search capabilities enhance our reasoning abilities? Only time will tell, but my hunch is that the answer will be a resounding yes. While immediate information delivery may make us lazy reasoners in some cases, most of the time rapid information delivery will enable us to possess the facts faster and more accurately than ever before. Similar to the advantageous social effects on reasoning that we discussed in Chapter 12, in which people can benefit from discussion with others on difficult tests including the Wason card selection task and the Cognitive Reflection Test, we can benefit from computers. They will continue to provide us with answers that save time, money, and effort.

While writing this book, I encountered a daily life example. I had an automatic garage door opener fail. When I pressed the button for the door to open, I could hear a relay click and saw a flashing LED light. A quick Internet search indicated that a consensus view for the problem I was having involved re-soldering a couple of key points on the unit's circuit board. I was able to quickly pop the board out and perform the needed solder work in a relatively rapid fashion. Annoyingly, this solution did not work for my opener. I opted to scrap the old model and find out what a new opener costs, as it was either buy a whole new one or invest in a rather costly board for my old worn out opener. After another web search I efficiently found a new unit locally available for a reasonable price. The reasoning process would have been more difficult without electronic help from the World Wide Web-based community. I might have spent more time trying to diagnose the problem, searched through a library for a manual for my old opener, and possibly wasted a lot of time trying out ineffective solutions.

Perhaps the other major frontier for reasoning in the future will be computerized assistance in the form of visually compelling and realistic three-dimensional environments. Such environments will likely be able to simulate almost real-world perception. Additionally, these tools will likely be able to help us to see spatial and visual features in greater detail. Learning anatomy has already been advanced through the use of the HTC Vive VR product, which incorporates a realistic three-dimensional rendering of the human body in which complex organ systems can be easily seen and understood without the need for physically dissecting an organism or cadaver.

SUMMARY

The rapid advances in technology in the past several decades influence our daily lives in many ways. Reasoning and decision making are now informed by much more accurate and up-to-date information than in previous times. This new availability of information is largely attributable to the widespread availability of powerful computing devices that are small, light, and portable. In addition, the advances in software that have taken place over the past several decades have enabled accurate information to be delivered quickly based on sparse search queries. So, the information is there, we just still need to be able to look for it and act on it at the appropriate time.

The development of computers and the information processing approach in psychology largely coincided. First generation computers were large, cumbersome, and inaccessible to most people. Technological components became refined to a point at which computers were reliable, fast, powerful, and small. We then saw unprecedented levels of automation in society and the widespread use of computers in our lives.

AI has continued to advance alongside developments in computer hardware. This endeavor can ultimately help us in our reasoning processes by providing the correct information when we most need it. In some cases entire jobs can be outsourced to computers with remarkable effects. A series of high profile man versus machine contests have taken place over the past 30 years in which a computer program played a human expert at a strategy game. The results have continued to impress society as computerized opponents became increasingly formidable to a point of defeating even the top humans.

As we look toward the future there are a variety of exciting developments. VR technology promises to help with visualization and immersion. This will no doubt lead to more immersive and realistic computerized reasoning tasks as we move into the future. Another area of progress is the capability of VR to emulate daily life conditions, which will enable researchers to better gain an appreciation of how someone with a neurological impairment is doing in their daily lives. These VR tools will also help us to assess and deliver training methods for cognitive skills.

END-OF-CHAPTER THOUGHT QUESTIONS

1. Reasoning is influenced by technology in everyday life situations. Name some situations that involve technology assisting or changing the way that we solve a problem.
2. The first generation of computers were inaccessible to most people. Name some factors that made these machines difficult to access and use. Note that there are technological reasons for these challenges. Try to think of how reliance on vacuum tubes impacted these machines.
3. What effect did the transistor and integrated circuit (microchip) have on computers?
4. What may have been the most important feature of modern computers that allowed them to become widely distributed?
5. What was a key property of the Chinook computer program that allowed it to defeat human checkers expert Marion Tinsley?
6. IBM's Deep Blue was a computer program that defeated legendary chess expert Gary Kasparov. How did emotion play into that matchup and ultimately influence the result?
7. When the DeepMind program AlphaGo defeated Go expert Lee Sedol, experts were surprised. Many believed it would be many more years before a computer won at Go over an expert. Why was Go considered to be so difficult for AI?
8. The development of VR over the past several years is enabling new experimental designs to be possible. What are some of the areas in which VR may have a large impact in training reasoning skills?

References

Artificial intelligence: Google's AlphaGo beats Go master Lee Se-dol. (March 12, 2016). Retrieved from http://www.bbc.com/news/technology-35785875.

Boggan, A. L., Bartlett, J. C., & Krawczyk, D. C. (2012). Chess masters show a hallmark of face processing for chess. *Journal of Experimental Psychology: General, 141*, 37–42.

Chase, W. G., & Simon, H. A. (1973). Perception in chess. *Cognitive Psychology, 4*, 55–81.

Cipresso, P., Albani, G., Serino, S., Pedroli, E., Pallavicini, F., Mauro, A., et al. (2014). Virtual multiple errands test (VMET): A virtual reality-based tool to detect early executive functions deficit in Parkinson's disease. *Frontiers in Behavioral Neuroscience, 8*, 405.

Cook, D., Feuz, K. D., & Krishnan, N. C. (2013). Transfer learning for activity recognition: A survey. *Knowledge and Information Systems, 36*, 537.

Damasio, A. R. (1994). *Descartes' error: Emotion, reason, and the human brain*. New York: G. P. Putnam. Chicago.

Deep Blue. (2017). Retrieved from http://www-03.ibm.com/ibm/history/ibm100/us/en/icons/deepblue/.

Didehbani, N., Kandalaft, M., Allen, T., Krawczyk, D. C., & Chapman, S. B. (2016). Virtual reality social cognition training for children with high functioning autism. *Computers in Human Behavior, 62*, 703–711.

Finley, K. (2012). *Did a computer bug help Deep Blue beat Kasparov?*. Retrieved from https://www.wired.com/2012/09/deep-blue-computer-bug/.

Goel, V., Grafman, J., Tajik, J., Gana, S., & Danto, D. (1997). A study of the performance of patients with frontal lobe lesions in a financial planning task. *Brain, 120*, 1805–1822.

de Groot, A. D. (1946). *Het denken van den schaker*. Amsterdam: Noord Hollandsche.

de Groot, A. D. (1965). *Thought and choice in chess*. The Hague: Mouton Publishers.

Hsu, F. (2002). *Behind Deep Blue: Building the computer that defeated the world chess champion*. Princeton, New Jersey: Princeton University Press.

Kandalaft, M. R., Didehbani, N., Krawczyk, D. C., Allen, T., & Chapman, S. B. (2013). Virtual reality social skills training for young adults with high-functioning autism. *Journal of Autism and Developmental Disorders, 43*, 34–44.

Krawczyk, D. C., Boggan, A. L., McClelland, M. M., & Bartlett, J. C. (2011). The neural organization of perception in chess experts. *Neuroscience Letters, 499*, 64–69.

Metz, C. (January 27, 2016a). *In a huge breakthrough, Google's AI beats a top player at the game of Go*. Retrieved from https://www.wired.com/2016/01/in-a-huge-breakthrough-googles-ai-beats-a-top-player-at-the-game-of-go/.

Metz, C. (March 11, 2016b). *The sadness and beauty of watching Google's AI play Go*. Retrieved from https://www.wired.com/2016/03/sadness-beauty-watching-googles-ai-play-go/?mbid=social_fb.

Metz, C. (March 9, 2016c). *Google's AI wins first game in historic match with Go champion*. Retrieved from https://www.wired.com/2016/03/googles-ai-wins-first-game-historic-match-go-champion/.

Schaeffer, J. (2009). *One jump ahead computer perfection at checkers*. New York, New York: Springer.

Shallice, T., & Burgess, P. W. (1991). Deficits in strategy application following frontal lobe damage in man. *Brain, 114*, 727–741.

Sreedhar, S. (July 2, 2007). *Checkers solved!*. Retrieved from https://webdocs.cs.ualberta.ca/~jonathan/publications/ai_publications/checksolved.pdf.

Tohill, J. M., & Holyoak, K. J. (2000). The impact of anxiety on analogical reasoning. *Thinking & Reasoning, 6*, 27–40.

Further Reading

ITL Education Solutions Limited. (2011). *Introduction to computer science*. New Delhi, India: Dorling Kindersley.

Kandalaft, M. R., Didehbani, N., Cullum, C. M., Krawczyk, D. C., Allen, T., Tamminga, C. A., et al. (2012). The Wechsler social perception test: A preliminary comparison with other measures of social cognition. *Journal of Psychoeducational Assessment, 30*, 455–465.

Index

'*Note*: Page numbers followed by "f" indicate figures, "t" indicate tables and "b" indicate boxes.'